STATISTICS of the GALAXY DISTRIBUTION

Vicent J. Martínez
Observatori Astronòmic de la Universitat de València
Burjassot, 46100, València, Spain
e-mail: Vicent.Martinez@uv.es

Enn Saar
Tartu Observatoorium
Tõravere, 61602, Estonia
e-mail: saar@aai.ee

Cover: The top and bottom panels show an equal-area Hammer–Aitoff projection of the galaxies in the PSCz catalog within the range of radial velocities 610 km/s $< v <$ 1810 km/s. The central panel shows the reconstructed density field. Dark shading indicates underdense regions; light shading indicates overdense regions. (Courtesy of Will Saunders and collaborators of the PSCz team.)

STATISTICS of the GALAXY DISTRIBUTION

Vicent J. Martínez
Enn Saar

CHAPMAN & HALL/CRC
Boca Raton London New York Washington, D.C.

Library of Congress Cataloging-in-Publication Data

Martínez, Vicent J.
Statistics of the galaxy distribution / Vicent J. Martínez and Enn Saar.
p. cm.
Includes bibliographical references and index.
ISBN 1-58488-084-8 (alk. paper)
1. Galaxies—Clusters. 2. Cosmology—Statistical methods. I. Saar, Enn. II. Title.
QB858.7 .M37 2001
523.1′07′27—dc21 2001028885

Direct all inquiries to CRC Press LLC, 2000 N.W. Corporate Blvd., Boca Raton, Florida 33431.

International Standard Book Number 1-58488-084-8
Library of Congress Card Number 2001028885
Printed in the United States of America 1 2 3 4 5 6 7 8 9 0
Printed on acid-free paper

To Laura, Albert, Clara, and Jordi, and the memory of Nuria

V.J.M.

To my family. To the memory of my son Veikko, hoping that he would have found the book useful

E.S.

Contents

Preface

> You may ask "What can a hard headed statistician offer to a starry eyed astronomer?" The answer is, "Plenty." One normally associates statistics with large numbers, and astronomy is full of large numbers. The number of stars in our galaxy, the so-called Milky Way System, is more than a hundred thousand million. The number of galaxies in the observable universe is upwards of a thousand million. Surely these large numbers justify a *prima-facie* case for the use of statistical techniques! ... I have every reason to believe that increased interaction between statistics and astronomy will be to the benefit of both the subjects.
>
> — *J. V. Narlikar*

The last half of the twentieth century saw cosmology develop into a very active and diverse field of science. This was largely due to the development of observational techniques that allowed astronomers to observe extremely distant regions of space. This motivated a flow of new theories about the evolution of our universe and the formation of the large-scale structure we found in it. The main tools to compare theoretical results with observations in astronomy are statistical, so the new theories and observations also initiated an active use of spatial statistics in cosmology.

Many of the statistical methods used in the analysis of the large-scale distribution of matter in the universe have been developed by cosmologists and are not too rigorous. In many cases, similar methods, sometimes under different names, had been used for years in mainstream spatial statistics. In the late 1950s, when the Berkeley statisticians J. Neyman and E. Scott carried out an intensive program for the analysis of galaxy catalogs, the connection between spatial statisticians and cosmologists was a fruitful one. However, in the following 30 years cosmologists were not, in general, aware of developments in statistics, and vice versa. Fortunately, recent years have brought the resumption of a dialog between astronomers and mathematicians, led by the Penn State conferences. We hope that this dialog will continue and will be useful. Cosmology is a good field for applications of spatial statistics. Its well-defined and growing data sets represent an important challenge for the statistical analysis, and therefore for the mathematical community.

The very influential book by Peebles (1980), a milestone in the field, gave the first complete description of statistical methods for study of the spatial distribution of galaxies and of the essential cosmological dynamics. In the years that followed new data have been collected, new methods have been developed, and new discoveries have been made.

This book describes the presently available observational data on the distribution of galaxies and the application of spatial statistics in cosmology. It provides a detailed derivation of the basic statistical methods used to study the spatial distribution of galaxies and of the cosmological physics needed to formulate the statistical models.

We have delineated the basic ideas and practical algorithms and have cited original articles for a more detailed description: there is always more information in articles than can be collected in a book. We have tried to select articles that present the subject in the clearest possible way (frequently they are review articles); thus, our selection is not meant to give a full history of the development of the methods.

Cosmological statistics is a rapidly developing field. We have tried to give the most up-to-date results (at least up to the year 2001 at the preprint level). After working through this book, the reader should be able to understand new research articles and to set up her/his own research projects.

The book is meant to appeal to two different communities of scholars:

- graduate students in cosmology and practicing cosmologists
- mathematicians interested in methods of spatial statistics and their applications

Because the prevalent statistical approach in cosmological statistics has been frequentist, a large collection of different statistics has been developed and studied. Many have been applied only in specific problems and models. This book describes the most general and the most widely used methods of this collection. Personal bias is unavoidable in such a selection, although we have tried to be as objective as possible. But we also describe Bayesian techniques that have become popular recently in large-scale structure studies (especially for the estimation of the power spectrum).

The outline of the book is as follows: after a short description of galaxies (the main objects of cosmological statistics), magnitude systems, and distance indicators in astronomy, we provide in Chapter 1 an overview of current knowledge of the structure in the universe, based on the catalogs of galaxies observed thus far. In Chapter 2 we describe briefly the standard hot Big Bang model of the universe, with special emphasis on measuring distances in cosmology. In Chapter 3 we explain how the galaxy distribution can be analyzed using techniques developed for point processes. Correlation functions are explained in detail together with the different estimators used in their evaluation. Since second-order intensity functions have been a topic of intensive research in spatial statistics lately, we have tried in this chapter to break the language barrier between both fields.

Fractal methods have become very popular in the analysis of galaxy clustering and Chapter 4 is devoted to this subject. In Chapter 5 we review the history of the field, introducing the Neyman–Scott processes and other related geometrical models. We also give special attention to the Voronoi models and to a physically motivated Saslaw distribution function. In Chapter 6 we discuss the dynamics of structure formation, which is important background for formulating contemporary statistical models of galaxy clustering. The theory of random density and velocity fields is described in Chapter 7, with emphasis on its cosmological applications. We focus our attention on both Gaussian and non-Gaussian random fields, with special attention to the properties and clustering of peaks. We end this chapter by explaining the mass (intensity) functions predicted by the Press–Schechter theory and the recent halo model of galaxy clustering that is close to the ideas of Neyman and Scott.

Chapter 8 describes the statistical measures of clustering in Fourier space, in particular methods of estimation of the power spectrum from observational data. In the last sections, the properties of lower dimensional fields are discussed, studying what can be inferred from the projected catalogs as well as from one- or two-dimensional surveys (pencil-beams or slices). In Chapter 9 we explain the methods of reconstructing the density field in the nearby universe (cosmography), trying to remove velocity distortions and making use of the statistical knowledge of velocities and positions of galaxies. We briefly review the application of gravitational lensing for studying the large-scale structure of the universe. In the last chapter we present the statistical tools that have been developed to highlight morphological features of the galaxy distribution. The topology of the galaxy distribution, used to judge whether we have a cellular or a sponge-like distribution of galaxies, can be found here together with other morphological descriptors, such as the Minkowski functionals, minimal spanning trees, and wavelets. Algorithms for finding clusters, voids, and possible periodicities in the galaxy distribution are explained in detail. Appendix A gives a short introduction to spherical astronomy and to the coordinate systems used in the catalogs of galaxies, with some practical formulae to perform coordinate transforms. Appendix B provides a brief summary of the basic statistical terminology.

Vicent Martínez and Enn Saar

Acknowledgments

This book has benefited from valuable discussions and, in some cases, from fruitful collaborations with our colleagues, some of whom read drafts of the book and made useful suggestions for improving it. In particular, we must mention Rien van de Weygaert, who was one of the originators of the idea for the book and who participated in creating its overall structure, Guillermo Ayala, Adrian Baddeley, Stefano Borgani, Peter Coles, Rosa Domínguez-Tenreiro, Matthew Graham, Bernard Jones, Martin Kerscher, Sabino Matarrese Belén López-Martí, Chris Miller, Rana Moyeed, Jesper Møller, José Antonio Muñoz, María Jesús Pons-Bordería, Silvestre Paredes, John Peacock, Jim Peebles, Nurur Rahman, José Luis Sanz, Jean-Luc Stark, Dietrich Stoyan, Michael Strauss, Istvan Szapudi, Max Tegmark, Luis Tenorio, and Licia Verde.

This book is profusely illustrated. We created some of the figures; others are from our earlier papers. Most of the illustrations, however, are from works by other authors previously published in journals or books. Thanks are due to all those authors and publishers who generously granted permission to us to make use of this material. Those authors who have contributed in this way are A. Baleisis, M. Bartelmann, C. Baugh, A. Bijaoui, S. Borgani, A. Canavezes, P. Coles, M. Colless, L. Christensen, G. Christianson, S. Colombi, R. Croft, L. da Costa, P. de Bernardis, A. Dekel, V. de Lapparent, J. Einasto, G. Efstathiou, H. El-Ad, A. Fairall, E. Falco, H. Feldman, K. Fisher, G. Goldhaber, L. Guzzo, A. Hamilton, S. Hatton, W. Hu, G. Kauffmann, M. Kerscher, J.-P. Kneib, O. Lahav, H. Lin, D. Malin, H. McCracken, S. Maddox, C. Miller, S. Moody, B. Moore, V. Müller, A. Nusser, J. Peacock, J. Peebles, S. Perlmutter, W. Percival, E. Pierpaoli, T. Piran, D. Pogosyan, V. Sahni, W. Saunders, R. Scaramella, W. Schaap, D. Schlegel, I. Schmoldt, R. Scoccimarro, T. Souradeep, U. Seljak, E. Slezak, F. Sylos-Labini, A. Szalay, I. Szapudi, M. Tegmark, M. Turner, H. Valentine, R. van de Weygaert, L. Van Waerbeke, E. Vishniac, M. Vogeley, S. Webb, D. Weinberg, M. White, N. Wright, K. Wu, N. Yoshida, and S. Zaroubi.

We have tried to provide a balanced overview of the statistical techniques currently used to quantify galaxy clustering. It has been necessary to deal with a huge amount of written information. We are grateful to those who created and maintain ADS and ArXiv, the marvelous archives for electronic communication of research results.

A few parts of the book rely upon, or are extended versions of, some of our earlier published papers, cited in the reference list. We are grateful to the editors and publishers of the relevant journals and proceedings for permission to use the material: *The Astrophysical Journal*, the American Astronomical Society, *Monthly Notices of the Royal Astronomical Society*, Blackwell Science Ltd., Astronomical Society of the Pacific, Società Italiana di Fisica, and Springer-Verlag.

We have made every effort to obtain permissions to use copyrighted material. We apologize for any errors or unintentional omissions, and would be grateful if they were communicated to us.

This book would not have been possible without the support of many institutions. V. Martínez acknowledges the support of Spanish Dirección General de Enseñanza Superior Project PB96-0797 and Ministerio de Ciencia y Tecnología Project AYA2000-2045. V. Martínez also thanks the Instituto de Astrofísica de Andalucia and the Observatoire de Genève where parts of the manuscript were written, and the University of Valencia for providing financial support for these visits. E. Saar spent several months at the Department of Astronomy and Astrophysics of the University of Valencia, during which time large portions of the book were written. He is grateful to the Department for its hospitality and for the invited professor positions funded by the Vicerrectorado de Investigación de la Universitat de València and by the Conselleria de Cultura, Educación y Ciencia de la Generalitat Valenciana. He also acknowledges the Estonian Academy of Sciences for a monograph grant and the Estonian Science Foundation for support in the form of grant 2882.

We thank our editor, Kirsty Stroud, and her assistant, Sharon Taylor, for their continued interest in the progress of our book, and our production editor, Christine Andreasen, for her encouragement and, particularly, for her efforts to translate our Spanish and Estonian English into "proper" English.

CHAPTER 1

The clumpy universe

In the first sections of this chapter we introduce the basic astronomical notions that a statistician should know before embarking on the analysis of the galaxy distribution; some technicalities, such as coordinate systems and transformations, are given in the appendices. We present here an overview of galaxies and a description of the magnitude systems, followed by the construction of the distance scale ladder.

The last sections are devoted to a qualitative description of our current knowledge of structure in the universe. After a brief argument about which structures are cosmic fossils that contain information on the formation of structure in the universe, we focus on one of the main fossils: the galaxy distribution.

We then discuss the observational strategies for obtaining statistically useful samples of galaxies. The different selection effects and biases in the compilation of redshift surveys are examined. The advantages and disadvantages of the different kinds of galaxy surveys are also discussed.

We then continue with a description of the "morphology" of the galaxy distribution in light of what the projected galaxy distribution reveals. A general description of groups and clusters of galaxies follows. We then provide an overview of current knowledge of the galaxy and cluster distribution on supercluster scales. We call attention to the often flattened or elongated nature of these structures and present a summary of our observational knowledge of the voids in the galaxy distribution. This overview concludes with a discussion of how all these features fit in as ingredients of the general sponge-like or cellular network that conspicuously shows up in large redshift surveys. Finally, we conclude with the indications for the existence of structures on scales of 100 Mpc or more, which points to the problem of determining the size of the largest deviations from the homogeneous universe.

1.1 Galaxies

1.1.1 The Milky Way Galaxy

Our Galaxy is a rather flat structure in which we can distinguish three different parts: the *nuclear bulge*, in which the *galactic nucleus* lies, the *disc*, and the *halo* (see Fig. 1.1).

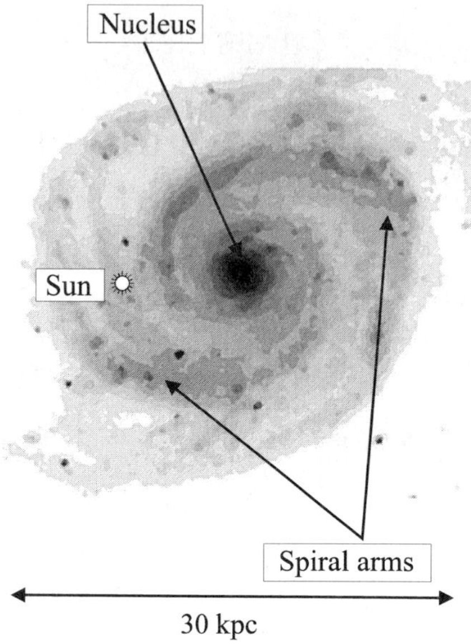

Figure 1.1 *Sketch of the Milky Way Galaxy indicating the position of the sun.*

The natural unit of length for describing galaxies is the kiloparsec (kpc): 1 kpc = 1,000 pc $\simeq 3.086 \times 10^{19}$ m $\simeq 3,261.6$ light years.

The disc has a diameter of around 30 kpc and is about 0.7 kpc wide. The disc, with a spiral structure, is populated by young metal-rich stars, dust clouds, and interstellar gas. The sun lies in the inner part of a spiral arm at 8.5 kpc from the galactic center. The sun rotates around the center of the Galaxy at a velocity of 220 km s^{-1}.

The nuclear bulge at the central regions of the Galaxy is 5 kpc wide. The stars in this region are older than the disc stars. Close to the galactic center lies a strong radio source (Sagittarius $A^{\star}$) that could be related to a supermassive black hole.

A spherical halo with a diameter of around 50 kpc surrounds the nuclear bulge and the disc. The halo is populated with *globular clusters*, concentrations of thousands of old stars forming very dense and nearly spherical structures. A larger dark halo surrounds the visible part of the galaxy. Its mass and extent are not yet completely known.

An Aitoff projection of the Milky Way in galactic coordinates is shown in Fig. 1.2.

Figure 1.2 *A view of our own Galaxy in galactic coordinates (see Appendix A). The center of the Milky Way lies at the center of the diagram. The interstellar dust prevents a clear view of the shape of the galaxy. The two spots on the right part of the southern galactic hemisphere are the Magellanic clouds. (Courtesy of Lund Observatory.)*

1.1.2 Morphological classification and properties

In 1929 Hubble classified the galaxies into several types according to their overall observed shapes. Fig. 1.3 shows a drawing of this classification. This tuning-fork diagram divides galaxies into ellipticals (E's) and spirals (S's). A third group is formed by the irregulars (Irr's). Elliptical galaxies have the shape of an ellipsoid. The surface brightness is rather uniformly distributed. These galaxies are brighter at the central region, like a fuzzy blob with no well-defined edges. Fig. 1.4 shows an image of the giant elliptical galaxy M87 lying in the Virgo cluster of galaxies (also shown in the figure). This galaxy lies roughly 18 Mpc from the Earth and has a diameter of 90 kpc. The classification of the elliptical galaxies corresponds not to their real shapes, which cannot be known completely, but rather to the degree of roundness or elongation of their images from the Earth. With α and β the lengths of the semimajor axis of the projected elliptical image of the galaxy, its type according to the Hubble scheme is En where

$$n = 10\left(\frac{\alpha - \beta}{\alpha}\right).$$

Fig. 1.5 shows how we can classify an elliptical galaxy as one type or another, depending upon the orientation from which the galaxy is observed.

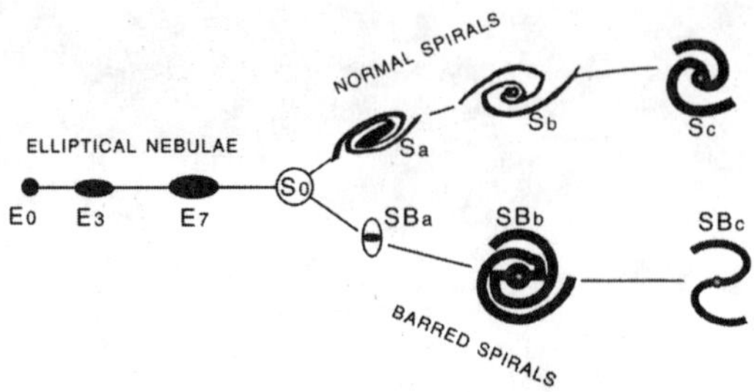

Figure 1.3 *Hubble's morphological classification of galaxies. (©Gale E. Christianson, reproduced with permission.)*

Figure 1.4 *On the left we can observe the giant elliptical galaxy M87, in the Virgo cluster. This is a nearly circular elliptical galaxy classified as an E1 galaxy with a diameter of about 90 kpc. On the right we can see a wide field view of the Virgo cluster (M87 lies at the lower left corner of the photograph). (©Anglo-Australian Observatory. Photographs by David Malin, reproduced with permission.)*

The S0 galaxies or lenticular galaxies are midway between ellipticals and spirals. Like spiral galaxies, they have a nuclear bulge surrounded by a flat disc, but no spiral arms.

There are two kinds of spiral galaxies, normal and barred. In normal spirals the spiral arms originate from the nuclear bulge, while in barred spirals the arms appear at the end of a bar crossing the nucleus itself. Within each group of spirals several types can be distinguished according to their overall shape. They are referred to as Sa, Sb, and Sc for normal spirals and SBa, SBb, and SBc for barred

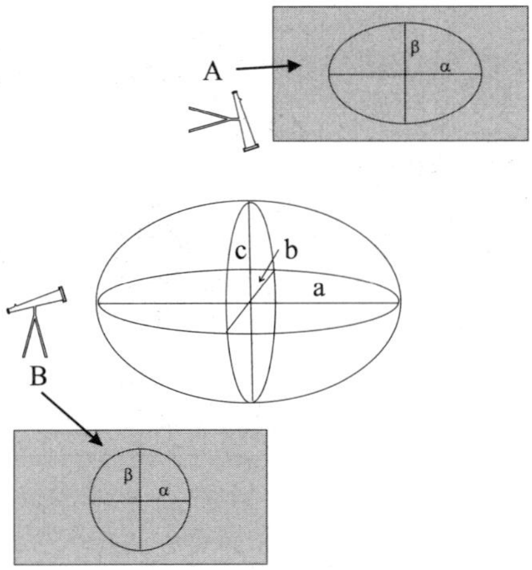

Figure 1.5 *Elliptical galaxies are classified as a function of their projected image. An ellipsoidal-shaped galaxy with semiaxis* $b = c$ *and* $a > b$*, seen from A, is classified as E3, while an observer at B would consider it an E0 galaxy. (Adapted from a figure of B.W. Carroll and D.A. Ostlie, An Introduction to Modern Astrophysics, Addison-Wesley, 1996.)*

spirals, although intermediate types also exist. In both sequences the latter types have a relative smaller nuclear bulge, thinner spiral arms, and a more open overall shape as shown in Fig. 1.3. Fig. 1.6 shows the images of three normal spiral galaxies following the sequence Sa, Sb, and Sc. In addition, those galaxies that cannot fit well into spirals, lenticulars, or ellipticals, having an amorphous shape without any kind of symmetry, are known as irregulars.

The morphology of galaxies in relation to their spatial distribution allows us to consider the distribution of galaxies as a marked point process where the mark is, in this case, a qualitative one. Galaxies of different morphological types present different degrees of clustering. This property is known as morphological segregation (Giovanelli, Haynes, and Chincarini 1986; Binggeli, Tarenghi, and Sandage 1990). In the astronomical literature elliptical galaxies are known as early-type galaxies and spirals as late-type galaxies. This is because Hubble thought that galaxies evolved in his diagram following a sequence from left to right. This is not the case, but the terminology is still in use. Therefore, it is common to find statements such as "early-type galaxies cluster more strongly than late-type ones."

Galaxies have varying luminosities of a wide range of values, from the luminosity of a dwarf elliptical galaxy of 3×10^5 $L_\odot$ to 10^{12} $L_\odot$ corresponding to the supergiant ellipticals ($L_\odot$ is the luminosity of the sun). The luminosity can be regarded as a quantitative mark. Luminosity segregation is an observable effect when analyzing the clustering of galaxies with different luminosity. It seems

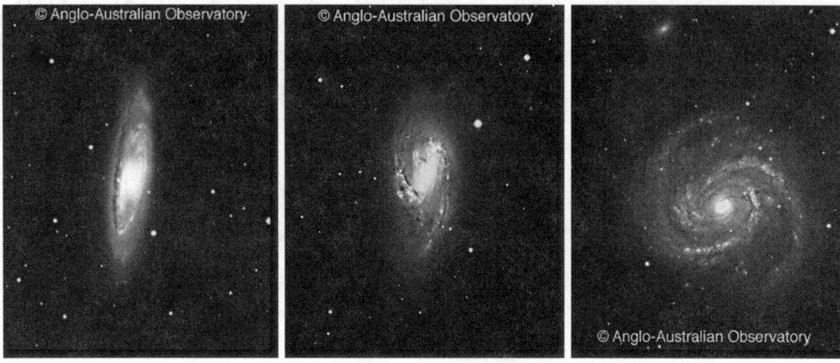

Figure 1.6 *Three spiral galaxies: at the left, M65, an Sa galaxy; at the center, M66, Sb; and at the right, M100, Sc. (©Anglo-Australian Observatory. Photographs by David Malin, reproduced with permission.)*

that brighter galaxies are more clustered than fainter ones, although this matter is still controversial (Alimi, Valls-Gabaud, and Blanchard 1989), and depends on the luminosity cut and on the analyzed scale. It seems, however, that morphological segregation and luminosity segregation are independent characteristics of the galaxy distribution (Domínguez-Tenreiro and Martínez 1989).

1.1.3 Brightness and magnitude systems

An important characteristic of a galaxy for a statistician is its brightness — this is frequently used for selecting subsamples from an observational catalog. Astronomers use a historically defined logarithmic scale for describing the brightness of luminous objects — if the total observed energy flux from a galaxy is S, then the galaxy is assigned an apparent magnitude m by

$$m = -2.5 \log_{10} S + \text{const}, \tag{1.1}$$

where the constant defines the zero point of the magnitude scale. Note that the brighter a galaxy, the smaller its apparent magnitude. The absolute magnitude M of a luminous object is defined as its apparent magnitude if it were at a distance of 10 pc. The radiation produced by a galaxy with bolometric (total) luminosity L emitted isotropically in all directions arrives to an observer at a distance d with a flux density S. Therefore,

$$S = \frac{L}{4\pi d^2}, \tag{1.2}$$

and hence the relation between its apparent magnitude and its absolute magnitude is

$$m - M = 5 \log_{10}(d/1 \text{ pc}) - 5. \tag{1.3}$$

The value of $m - M$ depends only on the distance to the galaxy and is known as the distance modulus. Expressing distances in Mpc, we can write

$$m = M + 5\log_{10}(d/1 \text{ Mpc}) + 25. \tag{1.4}$$

Due to instrumental limitations, brightness of galaxies is usually measured for fixed spectral intervals (filters), and both m and M carry the index of a specific filter. Differences between apparent magnitudes as measured with different filters are known as color indexes, for example, $B - V = m_B - m_V = M_B - M_V$, where B stands for the blue filter and V for the photovisual (yellow) filter. It follows obviously from (1.3) that the color indexes do not depend on distance.

1.1.4 Distance estimators

One of the most challenging problems in astronomy is to determine the distance at which the astronomical objects lie. Our concept of the distances involved in the universe has changed drastically in the last 80 years: a change comparable to, perhaps, the Copernican revolution. This change is rather well illustrated by the debate that took place in Washington in April 1920 between the astronomers H.D. Curtis and H. Shapley. Shapley defended the idea that galaxies, at that time still called nebulae, were part of our own Galaxy, the Milky Way. The Galaxy would have a diameter of about 300,000 light years. Curtis supported the idea that galaxies were *island universes*, a term coined by I. Kant 165 years earlier. Curtis' opinion was that spiral nebulae were objects similar to the Milky Way, but lying much farther away. The size of our own Galaxy would be about 30,000 light years. Only a few years later, E. Hubble showed unambiguously that the Andromeda galaxy was indeed an extragalactic object. Shapley accepted that Hubble's results were "the end of his universe."

How are distances measured in astronomy? Several books are devoted to this matter, for example, the monographs by Rowan-Robinson (1985) and by Webb (1999). In this section, we briefly summarize the most important methods of constructing the cosmological distance ladder. These methods provide the rungs of the ladder, and calibrating each method against the other is the standard way of determining distances in the universe.

Trigonometric parallax

As a consequence of the orbital motion of the Earth around the sun, nearby stars describe a small ellipse on the celestial sphere in 1 year. This is illustrated in Fig. 1.7. The angle P is called annual parallax or trigonometric parallax, a is the semimajor axis of the Earth orbit, i.e., 1 astronomical unit (1 AU $= 1.49597870 \times 10^{11}$ m), and d is the distance from the star to the center of the sun. Since d is always very much larger than 1 AU, P is an extremely small angle; in fact, it is smaller than a second of arc for even the nearest star. Therefore, expressing P in

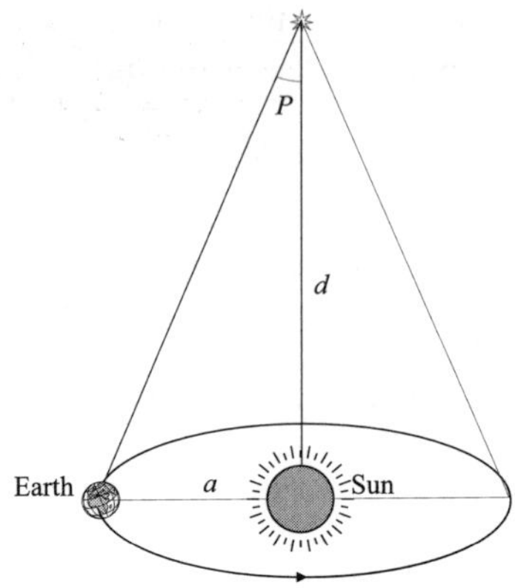

Figure 1.7 *In this diagram, we show the annual parallax of a star, P, at a distance, d, from the sun, where a denotes the semimajor axis of the Earth's orbit.*

radians, the distance $d = a/P$. In Fig. 1.7, we see that P is the angle subtended by 1 AU as seen from the star. This allows us to define the most important scale unit in astronomy and cosmology, the parsec. One parsec (pc) is the distance at which a star would present an annual parallax of one arcsecond and therefore 1 pc $\simeq$ 206,265 AU $\simeq$ 3.26 light years $\simeq 3.086 \times 10^{16}$ m.

The first parallax of a star was measured by F.W. Bessel in 1838. It was for the star known as 61 Cyg and its annual parallax was $0.3''$. The nearest star, after the sun, is Proxima Centauri, whose parallax is $0.76''$, lying therefore at a distance of 1.31 pc from the sun. By means of this method, astronomers have measured distances of about 1,000 stars up to 20 pc from ground-based observatories. Nevertheless, the space mission Hipparcos has already measured the annual parallax of about 100,000 stars within a distance of 200 pc and with an accuracy of $0.002''$.

Moving clusters

Open clusters are groups of stars formed more or less at the same time and linked together by their mutual gravitational force. The stars within the cluster move across the Galaxy with the same speed and pointing in the same direction. These parallel trajectories, due to an effect of perspective similar to parallel railways converging at the horizon, seem to converge on, or diverge from, a single point of the sky. Once we know the position of the convergent point we can determine the distance to the cluster, first measuring the radial velocity (along the line of sight) of a star from its Doppler shift v_r (in km s^{-1}), and then applying the equation

$$d = \frac{v_r \tan\theta}{4.74\eta},$$

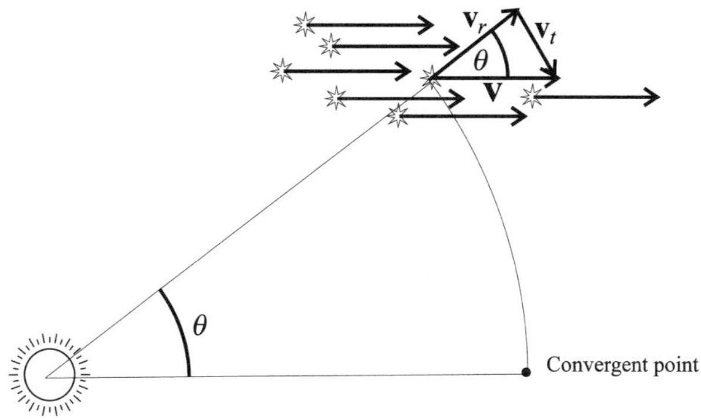

Figure 1.8 *The real motions of an open cluster are shown as parallel arrows. The angle between the velocity of a star and its radial component is just θ, the angle between the star and the convergent point.*

where θ is the angle between the star and the convergent point and η is the proper motion in the standard units of seconds of arc per year. The factor 4.74 in Eq. 1.1.4 matches the different units used for the angular velocities. With this method astronomers have measured distances up to 100 pc, the most important of which is the distance to the Hyades cluster in the constellation of Taurus. The distance to this cluster is 45.75 ± 1.25pc.

Main sequence fitting

In the early twentieth century, the Danish astronomer E. Hertzsprung and the American astronomer H.N. Russell discovered independently a strict correlation between the spectral type of a star, which is an indicator of its effective surface temperature, and its absolute magnitude. This is known as the main sequence of stars in the HR diagram. Knowing the distance to the Hyades cluster, we deduce the absolute magnitudes of each star, and we can plot in an HR diagram the magnitude versus the spectral type. This is a calibrated HR diagram. Looking at a remote cluster of stars, we can now plot on an HR diagram the apparent magnitudes of the observed stars against their spectral types. Since we do not know the distance to the cluster, we therefore ignore the absolute magnitude. Now we shift this diagram vertically until its main sequence matches the calibrated one (see Fig. 1.9). The difference between the absolute and the apparent magnitude of the two vertical axes provides an indicator of the distance to the remote cluster by means of (1.3).

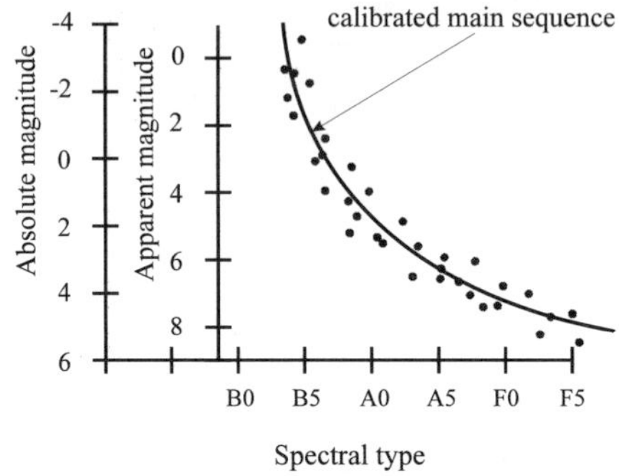

Figure 1.9 *By shifting the HR diagram of apparent magnitudes versus spectral type of a distant cluster to fit to the calibrated main sequence, we can determine the distance modulus of the remote cluster.*

Period–luminosity relation

Stars whose luminosity changes periodically are known as pulsating variable stars. One of the best-known types is the Cepheid variable. The name comes from the star δ-Cephei. This star changes its brightness following a cycle of variation of 5.34 days, as shown in Fig. 1.10. In 1912, H. Leavitt found a strict correlation between the period of variability of the Cepheids, P, and their absolute luminosity, L. This is the so-called period–luminosity relation, which is basically $\log L \propto \log P$. Fig. 1.10 shows a modern version of this relation for different types of variable stars. Therefore, we can deduce the absolute luminosity of a variable star from its period, and hence its distance, by measuring its apparent magnitude and applying (1.3). Hubble used this method to establish the distance to the Andromeda galaxy. Recently, the Hubble Space Telescope has made it possible to measure distances to Cepheid variables in other galaxies up to 20 Mpc.

Other distance indicators

Like Cepheids, there are other astronomical objects whose intrinsic brightness is known. They are generically named *standard candles*. Apart from Cepheids, other good standard candles are globular clusters, the most luminous supergiant stars, HII regions, and type Ia supernovae. For nearby galaxies several independent methods can be applied and astronomers can cross-check their results or take averages. For more remote galaxies, this is not the case, and in most cases only the brightest standard candles, type Ia supernovae, are observable.

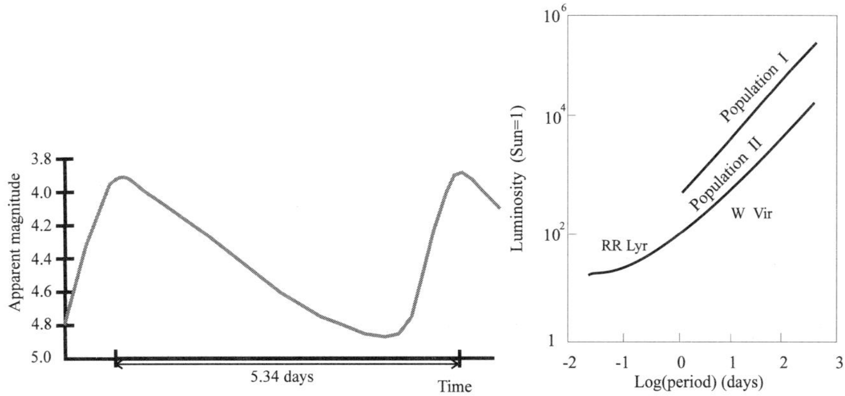

Figure 1.10 *On the left, we show the variability of apparent brightness of δ-Cephei. On the right, the period–luminosity relation for different kinds of variable stars has been plotted.*

The Tully–Fisher relation

There are other methods to estimate distances to remote galaxies that are not based on standard candles. Such methods are very important in the context of building the cosmography of the local universe, as we shall learn in Chapter 9. Spiral galaxies rotate and their rotational velocities can be measured using radio telescopes that look at the Doppler broadening of the 21-cm line. The width of this line is proportional to the speed at which a galaxy is rotating, and, by Kepler's third law, we know that the greater the rotational velocity, the greater the mass the galaxy contains and therefore the more luminous it is. In the late 1970s, R.B. Tully and J.R. Fisher found a relationship between the intrinsic optical luminosity of a spiral galaxy and its rotational velocity and used it to estimate distances (Tully and Fisher 1977). Basically, the relationship is

$$L \propto V_{\rm max}^4.$$

The fundamental plane relation

Elliptical galaxies have no rotation, but it is possible to measure the internal velocity dispersion of the stars in the galaxy. The mass, and hence the luminosity of the galaxy, is a consequence of its velocity dispersion. In 1976, S.M. Faber and R.E. Jackson found a relationship between the luminosity of an elliptical galaxy and its velocity dispersion, σ, $L \propto \sigma^4$. This relation, however, has a large scatter. A refined version was proposed by S.G. Djorgovski and M. Davis in 1987. This revised Faber–Jackson relation connects three parameters: the effective radius of the galaxy, R_e, the average surface brightness within that radius, I_e, and the velocity dispersion, σ. The relationship, also known as the fundamental plane for elliptical galaxies, is

$$R_e \propto \sigma^{1.36} I_e^{-0.85}$$

and can be used to estimate relative distances to elliptical galaxies.

The redshift

For the statistical analysis of the large-scale structure of the universe, the redshift catalogs are the primary data to be analyzed. The redshift z of a galaxy is the relative variation between the wavelength (or frequency) of the observed and the emitted radiation $z = (\lambda_0 - \lambda_e)/\lambda_e$. The emission and absorption lines of the spectrum of a distant galaxy are displaced toward the red in relation with the spectrum of a nearby one. By measuring this displacement astronomers estimate the redshift of the distant galaxy.

Due to the expansion of the universe, galaxies are receding from us, and for small recession velocities, the redshift can be interpreted as a Doppler shift $z = v/c$ (this expression is only approximatively valid for $z < 0.1$) where v is the line-of-sight recession velocity. The Hubble's law (Hubble and Humason 1931) states that the recession velocity is proportional to the distance ($v = H_0 r$). With this simple relation, redshifts can be used to estimate distances (see Fig. 1.11). However, the measured recession velocities are not due to the Hubble flow alone; peculiar velocities contaminate this measurement, as will be explained in the next section. The relationship between the redshift z and the distance r at larger scales depends strongly on the adopted cosmological model, as explained in detail in Chapter 2.

1.2 Mapping the universe

There are two primary observations of the universe with cosmological implications:

- The characteristics of the cosmic microwave radiation (CMB)
- The clustering of the bound virialized objects such as galaxies and clusters of galaxies

The hot Big Bang model predicts a universe very dense and hot in the early stages. In this primordial fireball, matter and radiation rest in thermal equilibrium, and the universe is opaque. As a result of the expansion, the universe cools down. When the temperature drops below 4,000 K, photons produced in early phases from matter–antimatter annihilation can freely escape and the universe becomes transparent. This is the surface of the last scattering that we can observe because when we look toward the edge of the observable universe, we are looking back in time, as a consequence of the finite speed of light. The CMB, therefore, represents our view of the universe when it was about 300,000 years old. This image of the early universe is very important to understand the development of the large-scale

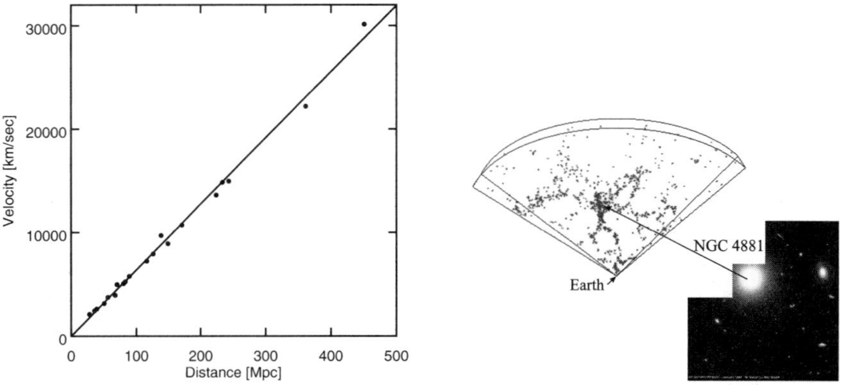

Figure 1.11 *On the right, we see a modern version of Hubble's law using data from Type Ia SNe obtained by Riess, Press, and Kirshner (1996). (©Edward L. Wright (UCLA), reproduced with permission.) Once the recession velocity of a galaxy has been measured, as for example, for the elliptical galaxy, NGC 4881 (HST image by W. Baum, AURA/STScI/NASA), we can build a three-dimensional map of the distribution of galaxies, using velocity as indicator of distance. This is shown in the right panel, where this galaxy, with a radial velocity of 6,691 km* s^{-1}, *is placed into the first published slice of the CfA2 catalog (de Lapparent, Geller, and Huchra 1986).*

structure. This fossil radiation was predicted by a group of physicists (R.H. Dicke, P.J.E. Peebles, P.G. Roll, and D.T. Wilkinson) at Palmer Physical Laboratory in Princeton, New Jersey, more or less simultaneously with its detection by A.A. Penzias and R.W. Wilson at Bell Laboratories.

The main features of the CMB are the black-body radiation spectrum and its extreme isotropy. As a consequence of the latter property the universe, at very large scales, must be smooth. In fact, Ehlers, Geren, and Sachs (1968) showed that the CMB isotropy combined with the assumption that the Earth is not placed in any privileged place in the universe – the so-called Copernican principle – implies homogeneity. The observations of the distribution of galaxies both in projected maps and redshift surveys show, however, that the universe is clumpy. Galaxies and groups of galaxies have a clear tendency to cluster. The theories of structure formation need to account for both observed opposite trends, namely, the large-scale homogeneity and the small-scale clustering of galaxies. The scales probed by the deepest redshift surveys available today are too small to fill the gap unambiguously, but a gradual tendency to homogeneity has been also detected (Wu, Lahav, and Rees 1999; Martínez 1999).

The measurements of the tiny deviations of the isotropy in the temperature of the CMB provided by the Cosmic Background Explorer (COBE) satellite are of the order of 10^{-5} measured on scales of about 1000 Mpc, corresponding to the large angular scales at which the satellite was operating. The measurement of the CMB temperature fluctuations by the COBE satellite in the 1990s provided a fundamental insight into the inhomogeneities of the universe at very early stages. These anisotropies originated from small perturbations in the gravitational potential. They provide information about the inhomogeneities at the linear evolution stage, avoiding most of the complications associated with the study of the nonlinear evolution of structures.

Since the discovery of the anisotropies, several statistical analyses of the data have been proposed. The most widely used ones are the two-point correlation function and the angular power spectrum. These statistics contain a complete description of the anisotropies if they are Gaussian distributed, and thus are the most natural choices to test cosmological models leading to Gaussian primordial fluctuations. However, there are possible physical mechanisms that produce non-Gaussian distributed anisotropies. Motivated by this possibility, other statistical analyses of the anisotropy maps, which are not shaped for Gaussian fields, have been developed. These include the study of higher-order correlation functions, morphological characteristics of hot and cold spots, genus and density of spots, Minkowski functionals, wavelets, and multifractal studies. Some of these techniques are rather similar to the ones this book examines, although in the context of the galaxy clustering. Of course, they have been adapted to the special character of CMB data, a topic that is beyond the scope of this book. Instead, in the following sections, we shall concentrate on the clustering of objects, describing the main galaxy catalogs and the qualitative features observed in them.

1.2.1 Redshift surveys

The best way to construct a three-dimensional map of the universe is to use the redshift of an object to estimate its distance. As we have already seen, other distance indicators exist, but they are not useful enough to build large and deep catalogs of galaxies because their uncertainties are too large when applied to objects lying at more than 50 Mpc.

In the 1980s different groups of astronomers started systematic observational programs to construct a true three-dimensional fair sample of the universe (see a review in Giovanelli and Haynes 1991). The task consisted of measuring the location in space of galaxies lying in the studied region of the sky. In addition to its angular position, the distance to each object was estimated from its redshift.

We denote by $\{\mathbf{x}_i\}_{i=1}^{N}$ the position of the N galaxies in a portion of the universe with volume V. Techniques of point fields statistics may be used to describe the statistical features of the distribution of such galaxies. Some words of caution are in order before we embark on the different techniques of the statistical analysis. Catalogs of galaxies are not simple point samples as are many of the planar point processes usually studied in the literature of spatial statistics (such as the positions

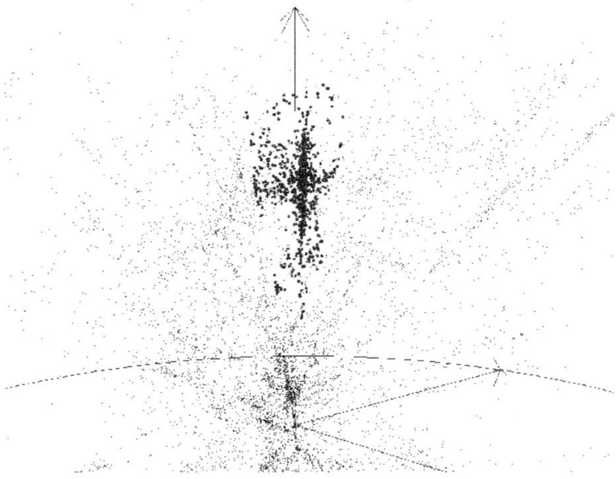

Figure 1.12 *An example of a finger-of-God feature in redshift space around the Coma cluster. (Reproduced, with permission, from Christensen 1986, Compound Redshift Catalogues and Their Applications to Redshift Distortions of the Two-Point Correlation Function, University of Copenhagen.)*

of trees in forests). To handle galaxy samples properly, we need to bear in mind the characteristics of their construction.

1.2.2 Peculiar motions

The observed recession velocity $v_{\rm rec}$ is not only due to the Hubble expansion. Other components have to be considered to obtain $v_{\rm rec}$. Bulk flows or large-scale streaming motions or local velocities within clusters might not be negligible. The peculiar velocity is the velocity of a galaxy with respect to the Hubble flow. Let us indicate its component along the line-of-sight by $v_{\rm pec}$; then the observed recession velocity is

$$v_{\rm rec} = cz = H_0 r + v_{\rm pec},$$

and therefore we have to distinguish between "redshift space" and "real space." The former is artificially produced by setting each galaxy at the distance r obtained by considering $v_{\rm pec} = 0$, and is therefore a distorted representation of the latter. The effect of this radial distortion is clearly illustrated when dense clusters of galaxies, almost spherical in real space, appear as structures elongated along the line of sight in redshift space. These structures, known as "fingers-of-God," are well-known features of redshift surveys. A cone around the Coma cluster (see Section 1.5.1) is highlighted in Fig. 1.12. In the plot, a Cartesian frame has been displayed, with the observer at the origin. The vertical axis points to the north

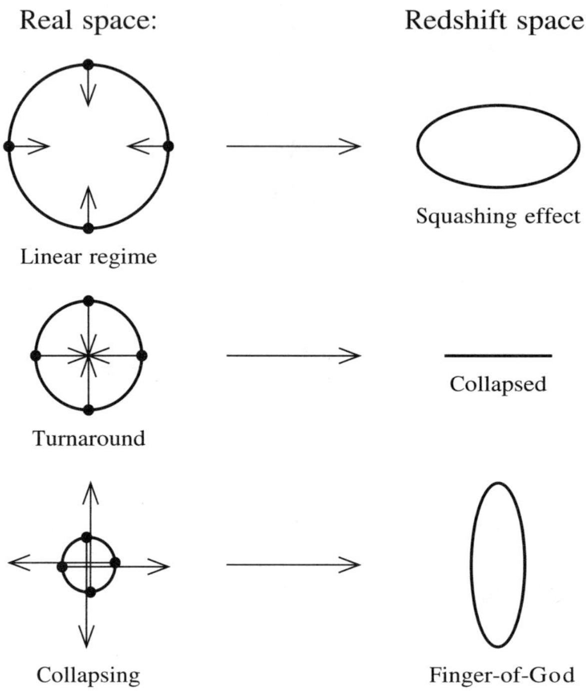

Figure 1.13 *A schematic representation of the distortion caused by the peculiar velocities produced by an overdensity in redshift space. (Reproduced, with permission, from Hamilton 1998, in The Evolving Universe, Kluwer Academic Publishers.)*

galactic pole. Dots represent positions of galaxies in a compilation of redshifts. Those points lying within a truncated cone centered in the direction of the vertical axis and with redshift in the range ($0.01 < z < 0.032$) are highlighted.

How the peculiar motions give rise to the fingers-of-God is nicely explained in Hamilton (1998). We reproduce here Figure 2 of Hamilton's review (Fig. 1.13). This diagram displays a schematic illustration of the process causing this feature. A spherical shell containing an overdensity is considered. At large scales, the peculiar velocities (infall arrows in the diagram) are small compared with the Hubble expansion; therefore, the effect in redshift space is an apparent squashing of the shell. At smaller scales, shells that are collapsing in real space appear turned inside out in redshift space, giving rise to the fingers-of-God. Details of how these redshift distortions have to be considered in the statistical analysis of the galaxy clustering, such as measuring the power spectrum, are given in Section 8.3.

1.3 Selection effects and biases

1.3.1 Galaxy obscuration

The extragalactic optical light does not reach the Earth uniformly from all directions. The plane of the Milky Way, our own Galaxy, is filled with interstellar dust that absorbs most of the light coming from extragalactic sources. Therefore, catalogs are incomplete below galactic latitudes $|b| < 20°$ to $30°$. This fact implies that the band of the sky corresponding to very low galactic latitudes is usually not considered in the analyzed optical samples: it is the so-called zone of avoidance. The geometry of the three-dimensional regions often becomes irregular because of this and other observational constraints. The infrared radiation is not absorbed by dust as much as radiation is absorbed in the optical band. For this reason, observing in the infrared, as the IRAS satellite does, allows us to safely identify galaxies closer to the galactic plane. The brightness of the galaxies is also affected by the galactic absorption. This effect is usually modeled by the cosecant law in latitude $\Delta m = A \operatorname{cosec} b$, although it is also known that the absorption is more inhomogeneous than that and the best solution is to map the galactic extinction from the observations (Burstein and Heiles 1982). The latest dust maps were built by Schlegel, Finkbeiner, and Davis (1998), combining the COBE/DIRBE and IRAS/ISSA data. Fig. 1.14 shows the dust distribution in the galactic sky (the left panel shows the northern galactic hemisphere and the right panel shows the southern galactic hemisphere). The extinction conversion is given as

$$E(B - V) = pD,$$

where $B - V$ is the color index, D is the 100μm flux in MJy/sr, as given in the map, and $p = 0.0184 \pm 0.0014$. The maximum infrared flux in the map gives the extinction $E(B - Y) \approx 0.55$, and the extinction toward galactic poles is $E(B - V) \approx 0.015$, averaged over a $10°$ patch. The magnitude–distance relations tell us that a change of 0.1 magnitude in extinction gives a 10% change in distance; it is clearly necessary to take extinction into account. Dust maps are readily available from a special Web page* (*Dust*).

1.3.2 Flux limit, luminosity function, and selection function

Galaxies in a redshift survey have different intrinsic brightness. Catalogs are usually built by fixing an apparent magnitude limit $m_{\rm lim}$. Therefore, galaxies with $m > m_{\rm lim}$ are not seen by the telescope or are not considered because of observational strategies. An apparent magnitude-limited sample is therefore not uniform in space, as intrinsically faint objects are only seen if they are close enough to the Earth, while higher luminosity galaxies are observed even at larger distances. This

* All Web addresses cited in this book are listed in the Web Site References section. Web citations in the text are given in italics.

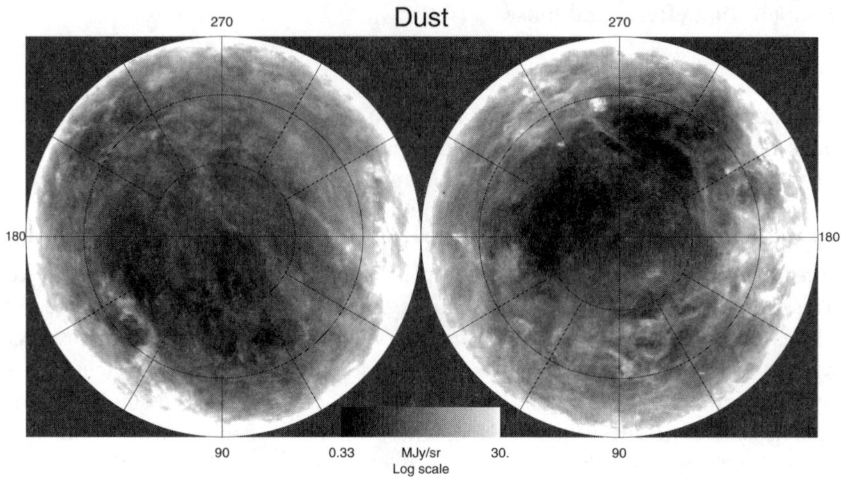

Figure 1.14 *Our windows to the universe; the distribution of dust in galactic coordinates (courtesy of D. Schlegel, D. Finkbeiner, and A. Kriegel). The left panel shows the northern galactic hemisphere and the right panel shows the southern galactic hemisphere (Lambert equal-area projections). (Reproduced, with permission, from Schlegel, Finkbeiner, and Davis 1998, Astrophys. J., 500, 525–553. AAS.)*

nonuniformity of the apparent magnitude-limited samples must be taken into account when performing the statistical analysis of the sample. To analyze this kind of flux-limited sample we can follow one of two strategies:

1. We can extract volume-limited samples by fixing a value of the depth $D_{\rm max}$ (in h^{-1} Mpc, h being the Hubble constant in units of 100 Mpc^{-1} km s^{-1}). Then, from (1.3), we can see that keeping only galaxies brighter than the absolute magnitude limit

$$M_{\rm lim} = m_{\rm lim} - 25 - 5\log_{10}(D_{\rm max}). \tag{1.5}$$

the resulting sample is uniformly selected in the sense that the gradient in number density of galaxies with distance because the flux limit is no longer present (see Figs. 1.15 and 1.16). $D_{\rm max}$ should be a luminosity distance (see Chapter 2). $M_{\rm lim}$ could be affected by other factors such as the galactic extinction and the K-correction (see below). A more general expression for (1.5), incorporating those effects, must be used, specially when $D_{\rm max}$ is large (see Eq. 2.36).

Let us illustrate how volume-limited samples are extracted from magnitude-limited catalogs. For example, if the catalog has an apparent magnitude limit $m_{\rm lim} = 15.5$ and we want to take as maximum depth of the volume to be studied $D_{\rm max} = 101.11\,h^{-1}$ Mpc, only galaxies with absolute magnitude $M \leq -19.70 + 5\log h$ will remain in the volume-limited sample. With this strategy, however, we disregard a huge part of the hard-earned information contained in

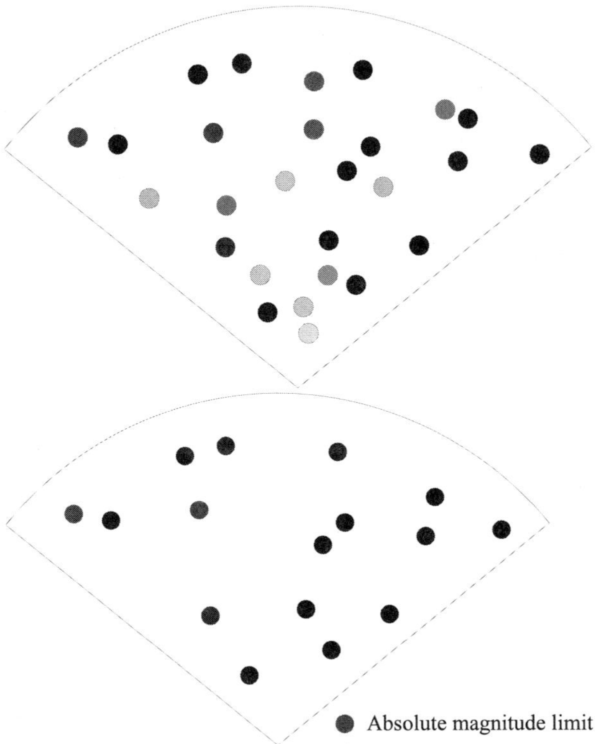

Figure 1.15 *This illustration shows how, from a magnitude-limited sample (top panel), we select a volume-limited sample (bottom panel). The intrinsic brightness of a galaxy is represented by gray scale (darker gray means brighter). In the magnitude-limited sample only intrinsically brighter galaxies are seen at large distances. This creates a gradient of the number density of galaxies with distance. To draw a volume-limited sample, we first have to find the absolute magnitude limit corresponding to the depth of the sample and the apparent-limit cutoff (see Eq. 1.5), and then we have to keep in the sample only those galaxies with* $M \leq M_{\rm lim}$.

the parent magnitude-limited survey (see Figs. 1.16 and 1.17). To avoid this problem we can follow the second strategy.

2. The second procedure is based on the knowledge of the selection function $\varphi(x)$. The function $\varphi(x)$ gives an estimate of the probability that a galaxy more brilliant than a given luminosity cutoff, at a distance x, is included in the sample. If the sample is complete up to a distance R, $\varphi(x) = 1$ for $x \leq R$.

The selection function is derived from the *luminosity function* $\phi(L)$. The luminosity function is defined by requiring that the mean number of galaxies per unit volume with luminosity in the range L to $L + dL$ is $\phi(L)dL$, independent of the

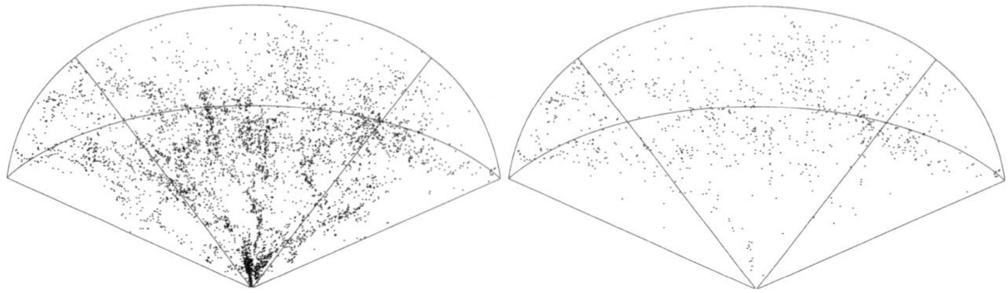

Figure 1.16 *The left panel shows part of the CfA2 redshift survey. It is a magnitude-limited sample. All the galaxies in this panel have apparent-magnitude* $m \leq 15.5$ *and the catalog is complete up to this limit. There are 4,933 galaxies. The right panel shows a volume-limited sample with depth 101.11* h^{-1} *Mpc, with only galaxies brighter than* $-19.70 + 5 \log h$ *remaining in the sample. Now there are only 905 galaxies.*

location of the volume. The empirical luminosity function is often fitted by the analytical expression (Schechter 1976; Felten 1977)

$$\phi(L)dL = \phi_* \left(\frac{L}{L_*}\right)^{\alpha} \exp\left(-\frac{L}{L_*}\right) d\left(\frac{L}{L_*}\right), \tag{1.6}$$

where L_* and α are the fitting parameters, with ϕ_* related to the number density of galaxies. The previous expression in terms of magnitude is

$$\phi(M)dM = A\,\phi_* \left(10^{0.4(M_*-M)}\right)^{\alpha+1} \exp\left(-10^{0.4(M_*-M)}\right) dM\,, \tag{1.7}$$

where $A = \frac{2}{5}\ln(10)$. Therefore, the selection function is just the ratio

$$\varphi(x) = \frac{\int_{-\infty}^{M(x)} \phi(M)dM}{\int_{-\infty}^{M_{\max}} \phi(M)dM}, \tag{1.8}$$

where $M(x) = m_{\lim} - 25 - 5\log(x)$, $M_{\max} = \max(M(x), M_{\rm com})$, and $M_{\rm com}$ is the absolute magnitude for which the catalog is complete. For the Schechter luminosity function and when the parameter $\alpha > -1$, the selection function can be written as

$$\varphi(x) = \frac{\Gamma(\alpha + 1, 10^{0.4(M_*-M(x))})}{\Gamma(\alpha + 1, 10^{0.4(M_*-M_{\max})})},$$

where Γ is the incomplete gamma function. Within this strategy, we can assign to each galaxy a weight $w = 1/\varphi(x)$ depending on its distance x from us.

There are different methods to estimate the luminosity function of a galaxy sample. The most popular method is the maximum likelihood approach, presented in Yahil et al. (1991). We start with writing the conditional probability density that a galaxy at a given distance x_i has the observed absolute magnitude M_i. It is given by the luminosity function $\phi(M_i)$, normalized by the integral over all

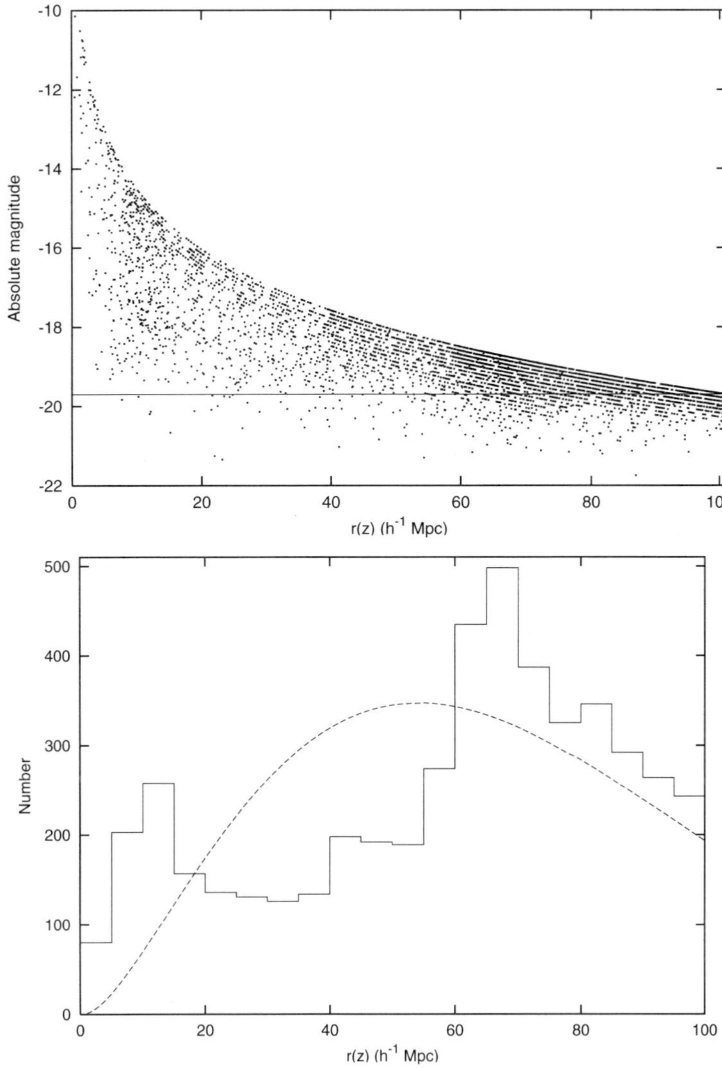

Figure 1.17 *The top diagram shows the absolute magnitude versus distance for all galaxies in the apparent-magnitude limited survey shown in Fig. 1.16. Only galaxies below the horizontal line are bright enough to be members of the volume-limited sample. The bottom panel shows a histogram with the number of galaxies observed as a function of their distance. The huge concentrations around the Virgo and the Coma clusters are clearly shown. The dashed smooth line indicates the expected number of galaxies we would observe if they were distributed uniformly in the apparent-limited magnitude sample, calculated by integration of the luminosity function provided in de Lapparent, Geller, and Huchra (1989).*

absolute magnitudes it could have at that distance, given the magnitude limit of the survey:

$$f(M_i|x_i) = \begin{cases} \dfrac{\phi(M_i)}{\int_{-\infty}^{M(x_i)} \phi(M)\, dM}, & M_i \le M(x_i), \\ 0, & M_i > M(x_i). \end{cases} \tag{1.9}$$

To find the luminosity function we minimize the log-likelihood for all sample galaxies to have the observed absolute magnitudes

$$\mathcal{L} = -\sum_i \ln f(M_i|r_i).$$

with respect to the parameters of the luminosity function.

Because the probability (1.9) does not depend on density, we can find only the shape of the luminosity function. The normalization factor (the mean number density of the sample) has to be found independently.

The parameters of the luminosity function depend on the sample. One of the most accurate measurements of the luminosity function has been performed on the Las Campanas redshift survey† (Lin et al. 1996). The fitting parameters (solid line in Fig. 1.18) are $\alpha = -0.70 \pm 0.05$, $\phi_* = 0.019 \pm 0.01\ h^3$ Mpc $^{-3}$, and $M_* = -20.29 \pm 0.02 + 5 \log h$.

1.3.3 Segregation

The point field formed by the galaxies in the surveyed volume is clearly a *marked point field*, in the sense that qualitative marks, such as morphological type, and quantitative marks, such as intrinsic luminosity, distinguish different objects within the same catalog. The statistical properties of the spatial distributions of different kinds of objects can be different. In fact, it is well established that elliptical galaxies are more frequent in denser regions such as rich clusters, while spirals are more often found in low-density environments (Davis and Geller 1976). Statistical descriptors such as the two-point correlation function or multifractal measures might provide us with different results if they are applied to different categories of galaxies. Although less evident, it seems also established that a certain degree of luminosity segregation exists, at least for galaxies with absolute magnitude $M_B \le -20 + 5 \log h$ (Phillipps and Shanks 1987; Hamilton 1988). Bright galaxies are more strongly correlated than faint ones. It appears that both kinds of segregation exist but are independent effects. The segregation mechanisms must be understood on the basis of a convincing structure formation theory.

† A description of this catalog is provided in Section 1.4.2.

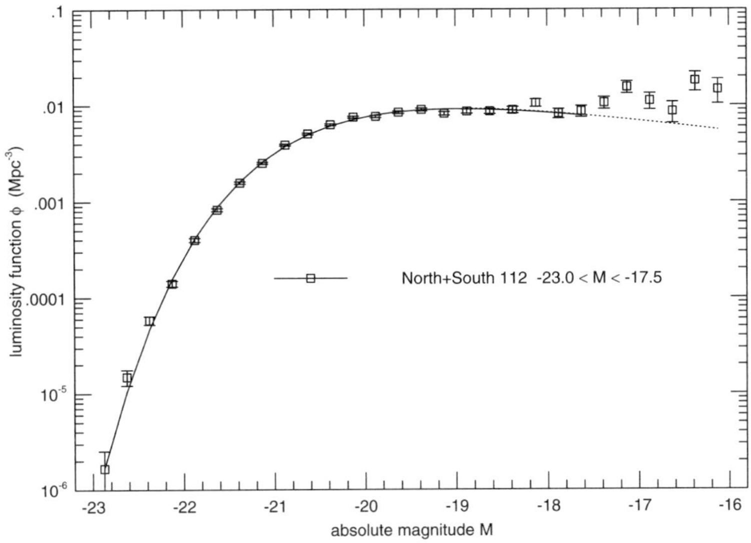

Figure 1.18 *The luminosity function of the Las Campanas redshift survey. (Reproduced, with permission, from Lin et al. 1996, Astrophys. J., 464, 60–78. AAS.)*

1.3.4 Cosmic variance

Obviously, all the statistical measures will be applied to a portion of the universe. Let D be the size of such a portion and let L be the scale to which the measure refers. If L is not much smaller than D, and we apply the same measure to another portion of size D, we expect to find different results. This is sometimes referred to as "cosmic variance." If, on the contrary, $D \gg L$, we shall have many realizations of the probability distribution within our sample, and therefore we expect that the results do not depend too much on the studied region. Statisticians would say that we are assuming *ergodicity*, in the sense that our sample is enough to obtain statistically reliable results, as it contains an adequate number of independent realizations. Strictly speaking, the ergodic property means that averages taken over an infinite spatial domain tend toward averages over a probability distribution or ensemble. In cosmology, however, we will always deal with a finite sample, so the property that we expect to be satisfied is that averages over some finite domain D should be within some acceptable bound of the ensemble averages.

1.3.5 Malmquist bias

We have already explained how standard candles are used to measure distances. However, the issue is more complicated since the luminosity of a standard candle

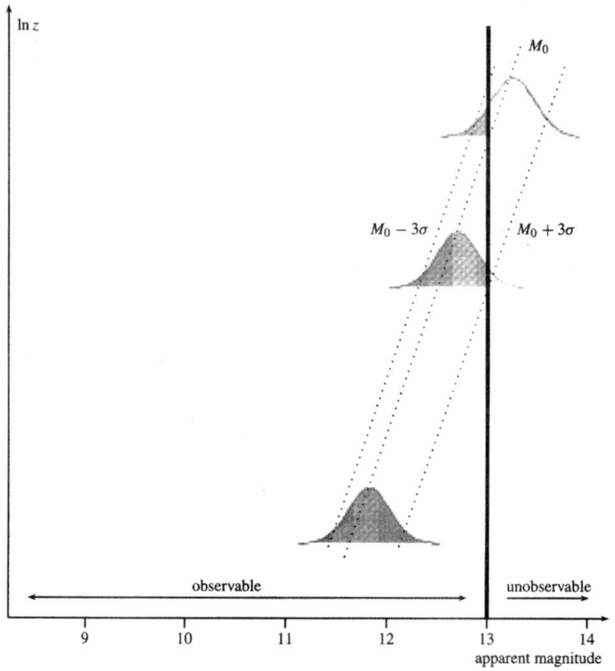

Figure 1.19 *An illustration of the Malmquist bias. All nearby standard candles with a Gaussian luminosity function are included in the sample, but at large distances only the intrinsically brighter objects will be included. (Reproduced, with permission, from Webb 1999, Measuring the Universe. The Cosmological Distance Ladder, Springer-Verlag in association with Praxis Publishing.)*

is not a single number but rather a spread of luminosities usually well parameterized by a Gaussian distribution function with mean absolute magnitude M_0 and dispersion σ. Let us illustrate this bias with an example: Let us suppose that we are looking at standard candles within a magnitude-limited sample of galaxies with apparent magnitude limit $m = 13$, as shown in Fig. 1.19. When we are close to the magnitude limit, only the brightest cases of the standard candles are observed. In other words, galaxies in the distant regions of the sample have to be systematically brighter in order to be included in the sample. The derivation of the magnitude of this effect was anticipated by the Swedish astronomer K.G. Malmquist in 1920. A modern discussion of this derivation can be found in Webb (1999) for a uniform distribution of the candles in the sample, all of which have the same Gaussian luminosity function independent of the distance. Under these conditions the bias is $1.38\sigma^2$ magnitudes.

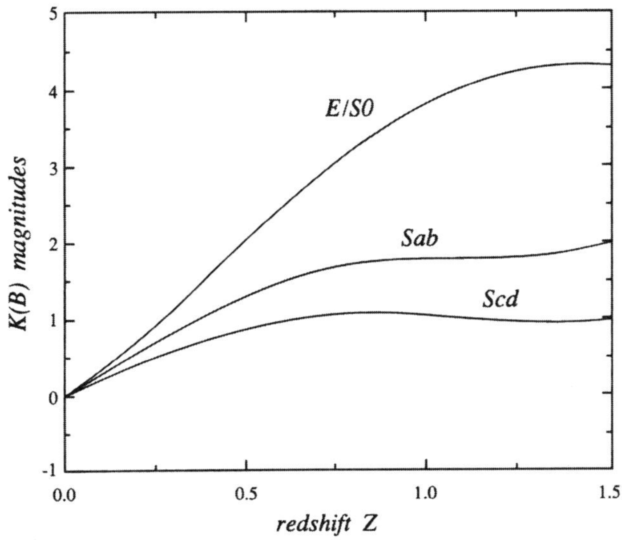

Figure 1.20 *The K-correction as a function of redshift and morphological types. Data adapted from Shanks (1990) and from King and Ellis (1985). (Reproduced, with permission, from Peebles 1993, Principles of Physical Cosmology, Princeton University Press.)*

1.3.6 K-correction

The luminosity of galaxies observed at large redshifts is detected at a longer wavelength than was actually emitted. This shift of the galaxy spectrum toward the red produces a change in the relation between the absolute and apparent magnitudes of a galaxy and (1.4) becomes

$$m = M + 5\log_{10}(d/1\ \mathrm{Mpc}) + 25 + K(z). \tag{1.10}$$

The K-correction depends on the redshift, the galaxy type, and the waveband in which the observation has been performed. In Fig. 1.20 we show the K-correction in the blue B band (Shanks 1990; King and Ellis 1985) for different morphological types, obtained from the analysis of the true spectra of these sources.

1.3.7 Velocity corrections

Redshift catalogs usually list the velocity cz of each galaxy. Several corrections are applied to this quantity to adapt it to a particular reference frame:

1. Heliocentric frame. The Earth moves around the sun in a nearly circular path in 1 year at a speed of 30 km/s. Thus, velocities have to be corrected for this motion to have them referred to a heliocentric frame. This is a rather small

correction and generally catalogs already list the heliocentric velocities of the galaxies.

2. Local group frame. The sun rotates around the galactic center, and the whole Galaxy has a motion within the Local Group (this is a small gravitationally bound group of galaxies in which the Milky Way lies; details are given in Section 1.5.1). A standard correction of the heliocentric velocities cz for the solar motion with respect to the centroid of the Local Group is to add to the heliocentric velocities the quantity $300 \sin l \cos b$, where l and b are the galactic longitude and latitude, respectively, of each galaxy (see Appendix A for definitions of these celestial coordinates).

3. Cosmic microwave background frame. The frame of reference established by the thermal background radiation is considered an inertial frame, and therefore some authors correct the velocities listed in the redshift catalogs for our motion with respect to the rest frame of the CMB. This motion has a velocity of 371 km s^{-1} in the direction of $l = 264.7°$ and $b = 48.2°$ (Fixsen et al. 1996).

1.4 Current and future galaxy catalogs

1.4.1 The galaxy distribution in projection

Before any distance indicators are applied to an observed galaxy in the sky, its accurate angular position is recorded. This information is the basis for the compilation of different catalogs of galaxies known as angular surveys. Typically each entry of this catalog consists of the angular coordinates of a given object together with its visual magnitude. Very early in the history of modern cosmology these catalogs were analyzed using statistical techniques. For example, Hubble (1934) studied the frequency distribution of galaxies in small angular fields and found that a lognormal distribution was a good fit for them.

The Lick catalog

The compilation of the Lick survey (Shane and Wirtanen 1967) was an important challenge for the statistical analysis of projected galaxy maps. The pioneering work of the Berkeley statisticians Neyman and Scott (1952, 1955) was greatly motivated by the Lick catalog. The survey consists of about 1 million galaxies counted over photographic plates. The task was enormous if one considers that digital scanning and high-speed computer resources were not available at that time. The authors counted the number of galaxies brighter than $m \sim 18.9$ in cells of size $10' \times 10'$ at $\delta > -23°$. In the representation of Seldner et al. (1977) reproduced in Fig. 1.21 we can appreciate for the first time structures on scales of a few tens of Mpc; in particular, the filamentary structure of the cosmic web can be glimpsed.

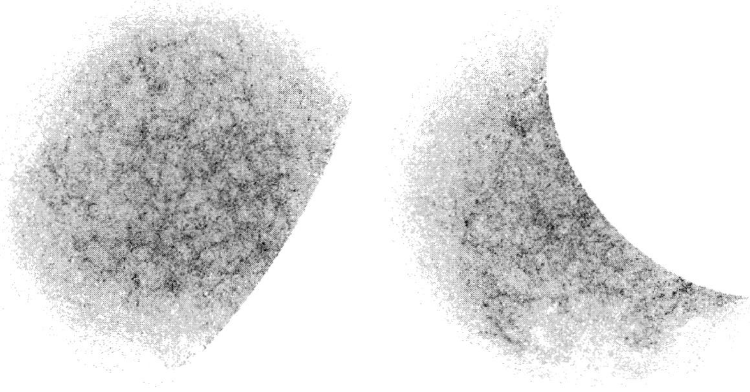

Figure 1.21 *An equal-area projection of the Lick catalog. (Reproduced, with permission, from Seldner et al. 1977, Astronom. J., 82, 249–256. AAS.)*

The Zwicky catalog

The Zwicky catalog (Zwicky et al. 1961–1968) lists the positions on the celestial sphere of about 30,000 galaxies brighter than $m \sim 15.5$ having declination $\delta > -2.5^{\circ}$, thus covering the whole northern sky. Apparent visual magnitudes and in some cases morphological types were listed in addition to the angular coordinates of each object. An updated version of this catalog merged with the Nilson catalog (Nilson 1973) was recently published (Falco et al. 1999).

The importance of this projected catalog is reinforced by the fact that it has been the source of positions for measuring redshifts of galaxies in the northern hemisphere. For example, the Center for Astrophysics redshift surveys (Geller and Huchra 1989) and the Perseus–Pisces redshift catalog used the positions of the galaxies from the Zwicky catalog. Fig. 1.22 shows the sky projection of an updated version of this catalog having 19,367 galaxies with $m \leq 15.5$, the limit for which Zwicky estimated his catalog was complete. In *TDC* the reader can find additional information regarding these catalogs. In most cases, when catalogs have been released into the public domain, the data can be freely downloaded from the Web pages listed in this book.

The APM catalog

Maddox et al. (1990a) compiled a modern version of this kind of projected catalog using the Automatic Plate Measuring (APM) machine in Cambridge. From scans of 185 photographic plates taken with the UK Schmidt telescope in Australia, and after extensive computer processing, positions and apparent magnitudes of

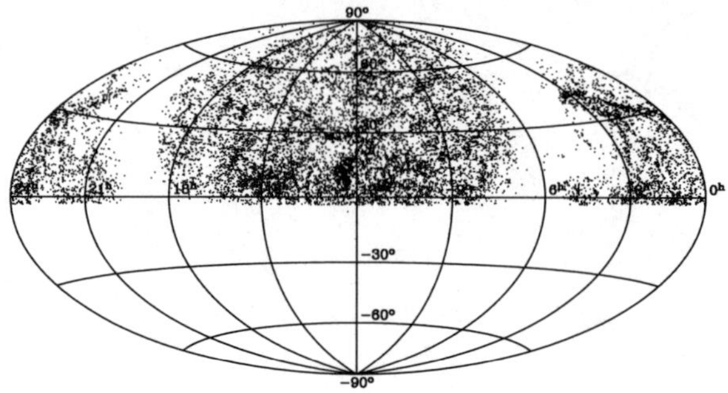

Figure 1.22 *A Hammer–Aitoff projection in equatorial coordinates of the updated Zwicky catalog containing 19,367 galaxies with* $m \leq 15.5$. *The empty region is the zone of avoidance close to the galactic plane (Reproduced, with permission, from Falco et al. 1999, PASP, 111, 438–452. Copyright 1999 by the Astronomical Society of the Pacific.)*

around 2 million galaxies brighter than $m = 20.5$ have been listed. The galaxies lie within a solid angle of 1.3 steradians around the south galactic cap.

Fig. 1.23 shows the APM catalog using pixels with brightness scaled to the number of galaxies in it. A galaxy catalog with the brighter objects was published by Loveday (1996).

1.4.2 The three-dimensional galaxy distribution

During the past two decades systematic collections of redshifts have been compiled following different strategies. Nowadays, use of multifiber spectrographs increases the number of available redshifts very rapidly. In this section we describe briefly some of the wide-angle redshift surveys used to date and some of the projects already under way to map the universe.

Center for Astrophysics (CfA2) redshift survey

The first version of this catalog, the CfA1 (Huchra et al. 1983), sampled only the Zwicky galaxies brighter than $M_B = 14.5$ in a region of the sky with galactic latitude greater than $40°$ or smaller than $-30°$ to avoid the zones of the sky with high galactic extinction. The extension of the CfA1, the CfA2 catalog (see *TDC*), was done by measuring redshifts of Zwicky galaxies brighter than $M_B = 15.5$. The strategy was to complete contiguous slices $6°$ wide in declination and 9 hours

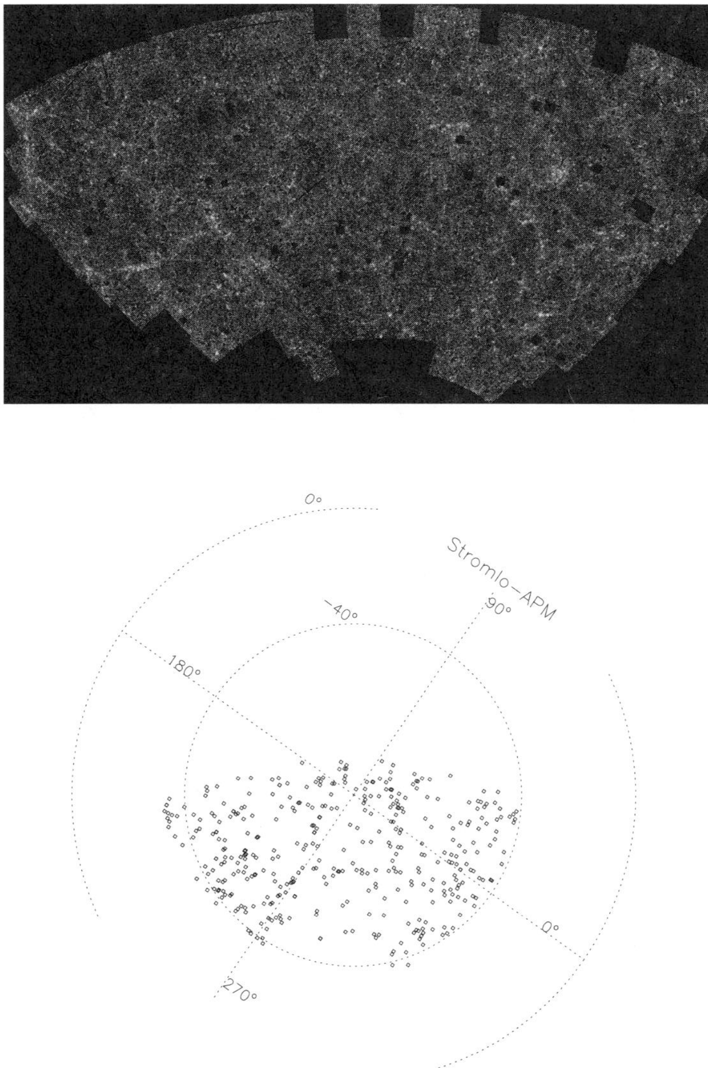

Figure 1.23 *The top panel shows the APM catalog. Brighter areas are more populated. The dark holes are excluded areas around nearby objects. (Courtesy of S. Maddox, W. Sutherland, G. Efstathiou, J. Loveday, G. Dalton, and the Astrophysics Dept., Oxford University.) The bottom panel shows the projection of the Stromlo–APM redshift survey. (Reproduced, with permission, from Martínez et al. 1998, Mon. Not. R. Astr. Soc., 298, 1212–1222. Blackwell Science Ltd.)*

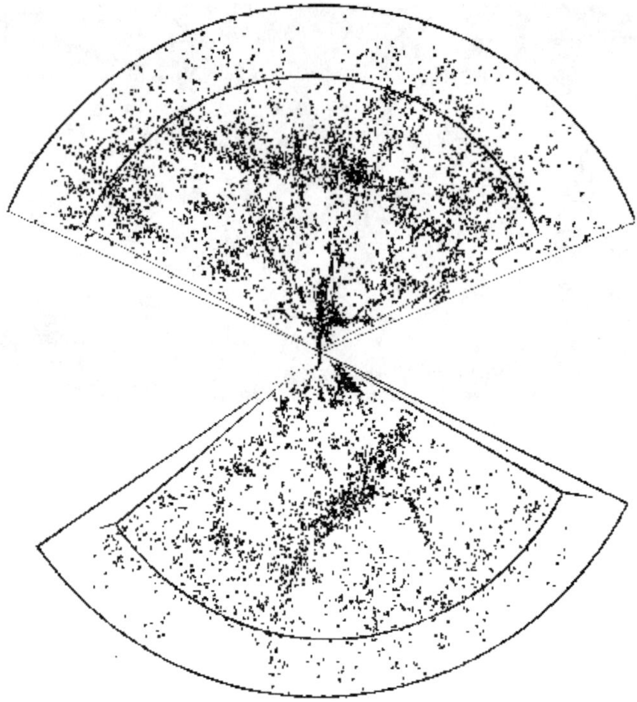

Figure 1.24 *A combination of the CfA2 redshift catalog with the SSRS2 survey. The limit in depth of the slices is* $z = 0.04$ *and the equatorial coordinates of the northern slice (top) are* $8^h < \alpha < 17^h$ *and* $8.5° < \delta < 44.5°$, *while for the southern slice* $20.8^h < \alpha < 4^h$ *and* $-40° < \delta < -2.5°$. *9,325 galaxies are shown. The Great Wall with the Coma cluster in the northern slice and the Southern Wall with Pisces–Perseus supercluster in the southern slice are the most remarkable features. (Reproduced, with permission, from de Costa et al. 1994, Astrophys. J. Lett., 424, L1–L4. AAS.)*

in right ascension (de Lapparent et al. 1986; Geller and Huchra 1989; Huchra et al. 1990). Today it covers a large region of both northern and southern galactic hemispheres, mapping a solid angle of 2.95 sr: $8^h < \alpha < 17^h$, $8.5° < \delta < 44.5°$ in the north with 6,500 galaxies and $20^h < \alpha < 4^h$, $-2.5° < \delta < 48°$ in the south with 4,283 galaxies (see Fig. 1.24).

Southern Sky Redshift Survey (SSRS2)

In the southern hemisphere ($\delta < 0°$), there is no equivalent to the Zwicky catalog. In fact, the first version of the Southern Sky Redshift Survey (SSRS) was diameter limited. The task performed by L. da Costa and collaborators was to use the diameter of the galaxies taken from the ESO/Uppsala Survey as an approximation for its magnitude (da Costa et al. 1991). The extension of this catalog, the SSRS2 (see *TDC*), is already magnitude limited to $m_B \leq 15.5$ (da Costa et al. 1994; da Costa et al. 1998), and therefore it is directly comparable to the CfA2 catalog (see Fig. 1.24). It is complete in the declination range $-40^\circ \leq \delta \leq -2.5^\circ$ and galactic latitude $b \leq -40^\circ$ in the southern galactic cap, and in the region $\delta \leq 0^\circ$ and $b \geq 35^\circ$ in the northern galactic cap. The northern part makes it contiguous with the CfA2. It covers a solid angle of 1.70 sr and contains 5,369 galaxies.

Perseus–Pisces sample

The Perseus–Pisces sample is also based on the Zwicky catalog, and points into the direction of the huge Perseus–Pisces supercluster. Redshifts have been measured at 21 cm using the Arecibo radio telescope. It is complete in the region of the sky with right ascension $22^h < \alpha < 4^h$ and declination $0^\circ < \delta < 45^\circ$, comprising 5,183 galaxies (Giovanelli and Haynes 1991).

The average redshift of these three catalogs is $\langle z \rangle \sim 0.02$. Redshifts generally have been measured one at a time.

IRAS catalogs

Several catalogs are based on the data obtained by the IRAS Infrared Astronomical Satellite (see Rowan-Robinson 1996 and references therein). This satellite has mapped a great portion of the sky detecting the point sources at wavelengths of 12, 25, and 60 μm. In the infrared part of the spectrum the galactic extinction is rather low and therefore the sky coverage can be greater (more than 80%). In this way, regions with low galactic latitude have been mapped. In the infrared part of the spectrum, as well as in radio observations, galaxy fluxes are measured, not in optical magnitudes within a waveband, but in Janskys (1 Jy = 10^{-26} watt m^{-2} Hz^{-1}).

Although the sky coverage is larger, the main problem with the infrared observations is that elliptical galaxies are undersampled. There is a reason for that. The infrared emission comes mainly from the clouds of interstellar dust, which are abundant in the discs of spiral galaxies. Elliptical galaxies, however, are poor in this kind of material, and therefore their infrared emission is lower. Four redshift surveys have been extracted from this database:

1. *QDOT.* This is a sparse sample. It contains one in six (randomly selected) of all IRAS-point sources with 60 μm flux, S_{60}, exceeding 0.6 Jy and galactic latitude $|b| > 10^\circ$. Once the random selection has been performed, the redshifts of the selected galaxies have been measured. There are 2,387 sources (Rowan-Robinson et al. 1990; Martínez and Coles 1994) in the redshift catalog. Some masked areas of the sky are not covered by the satellite. These kinds of sparse samples are worth compiling to obtain preliminary results from the surveys by means of statistical descriptors that are invariant under thinning, i.e., those

statistics that provide reliable results when applied to a subsample randomly selected from the whole sample. Using this strategy, meaningful physical and statistical results are obtained before the whole survey is complete, but at the expense of larger uncertainties.

2. *2 Jy.* This catalog is complete down to the flux limit $S_{60} > 1.936$ Jy and with galactic latitude $|b| > 5°$. It contains 2,685 galaxies (Strauss et al. 1990, 1992).
3. *1.2 Jy.* This is the extension of the previous sample to galaxies brighter than 1.2 Jy. Many statistical analyses of the galaxy distribution have been based on this sample. It contains 5,339 galaxies (Fisher et al. 1995).
4. *Point Source Catalog (PSCz).* This is the extension of the QDOT sample to all galaxies with flux $S_{60} > 0.6$ Jy. It contains 15,411 galaxies in a region covering 84% of the whole sky (see *PSCz*). The catalog is virtually complete and uniform at high galactic latitudes and 90% complete at low latitudes (Saunders et al. 2000). An equal area projection of all galaxies in this catalog is shown in Fig. 2.4.

The Stromlo–APM redshift survey

This is a sparse sample of 1 in 20 of the APM galaxies brighter than $b_J = 17.1$. The survey covers 4,300 square degrees centered at the south galactic pole. The region is approximately defined in equatorial coordinates by $\alpha \in [21^h, 24^h] \cup [0^h, 5^h]$, $\delta \in [-72.5°, -17.5°]$. The redshift survey contains 1,769 galaxies (Loveday et al. 1996).

Las Campanas redshift survey (LCRS)

This is a very deep survey with an average redshift of $\langle z \rangle \sim 0.1$. Redshifts have been measured using multifiber optics. The catalog consists of six slices of $1.5° \times 80°$ each and contains 23,697 galaxies. Fig. 1.25 shows the position of the galaxies in the southern slices of the LCRS together with the first CfA2 slice. As Kirshner (1996) has proposed, this view shows "the beginning of the end," because although we can see in the Las Campanas slice the same structures (walls, filaments, and voids) observed in the CfA2 catalog, the size of these structures is not larger for the deepest slice as should happen in a fractal pattern. In some sense, this diagram indicates that homogeneity is being attained. The whole catalog is now available together with some references and scientific results in *LCRS*.

2 degree field (2dF) survey

This survey is being constructed using the 2dF instrument built by the Anglo-Australian Observatory. This instrument is a two-degree diameter field spectrograph situated at the prime focus of the Anglo-Australian telescope (AAT). Redshifts of about 250,000 galaxies brighter than $m_B = 19.5$ and selected from the APM galaxy catalog will be measured. The survey covers a region of about 1,700

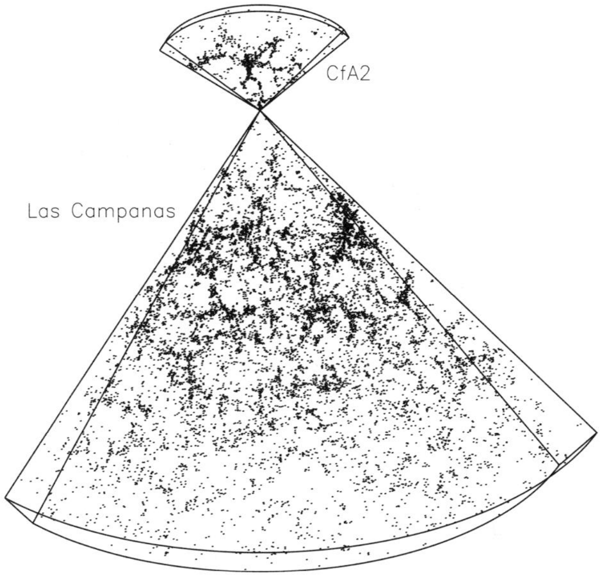

Figure 1.25 *The galaxy distribution for the southern slices of the Las Campanas redshift survey together with the first slice of the CfA2 catalog at the northern hemisphere. Although the depth of the Las Campanas is four times (in redshift) the depth of the CfA2 slice, the size of the structures is the same in both samples, contrary to what is expected for an unbounded fractal. (Reproduced, with permission, from Martínez 1999, Science, 284, 445–446; ©1999, American Association for the Advancement of Science.)*

square degrees of the sky (Folkes et al. 1999). Fig. 1.26 shows the present status of the survey after the first 2 years of the observing period. More than 150,000 galaxies already have had redshifts measured (see *2dFGRS*).

Sloan Digital Sky Survey (SDSS)

This survey will be the largest galaxy redshift catalog ever compiled (Gunn 1995; Margon 1999). The observations will be carried out using a 2.5 m telescope in New Mexico totally dedicated to this task. The telescope carries two double fiber spectrographs with 320 fibers. The solid angle covered by the survey will be around π sr. Redshifts for more than 1 million galaxies will be measured (see *SDSS*).

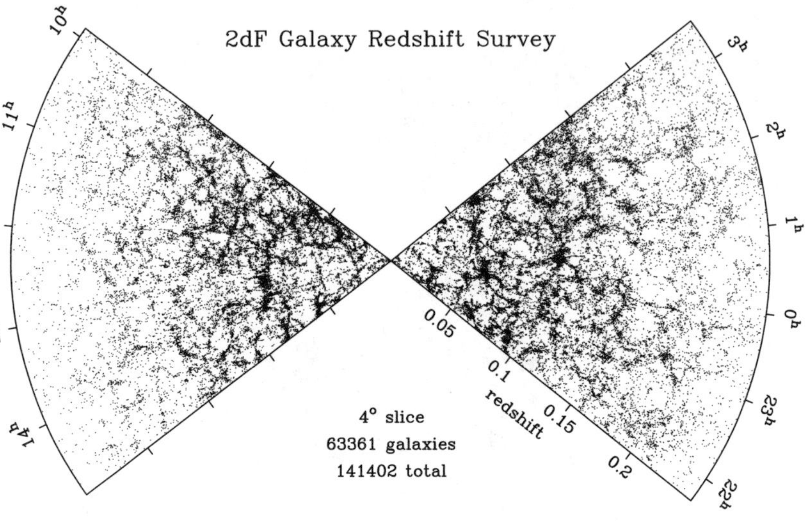

Figure 1.26 *A 4° slice containing 63,381 galaxies with measured redshift, drawn from the 2dF galaxy redshift survey. This is the largest view to date of the distribution of galaxies in the universe. The map shows the very long filamentary structures of galaxies, which could be cross sections of big walls, clusters, and superclusters, and voids up to 60 h^{-1} Mpc in diameter. It substantiates the impression that the size of the cosmic structures already has been attained. (Reproduced, with permission, from Peacock et al. 2001, Nature, 410, 169–173. Macmillan Publishers Ltd.)*

1.5 The observed structures: clusters, filaments, walls, and voids

As we have already seen in the previous sections, galaxies are arranged in interconnected walls and filaments forming a cosmic web encompassing huge, nearly empty regions between the structures. Voids, like the bubbles formed in soap foam, present a great variety of sizes, ranging from 2 h^{-1} Mpc to 60 h^{-1} Mpc in diameter. In this section we review the main structures in our neighborhood. The reader interested in a more detailed description will find a colorful account of the nearby large-scale structures in the book by Fairall (1998). Fig. 1.27 shows some of these structures projected onto the sky. Fig. 1.28 shows a schematic representation of the main features (clusters, filaments, walls, and voids) surrounding the Milky Way up to a distance of 70 h^{-1} Mpc.

1.5.1 Groups and clusters of galaxies

Isolated galaxies are rarely found. Galaxies are usually found in groups or clusters, which typically form parts of still larger structures. Our Galaxy lies in the so-called Local Group. This is a gravitationally bound system of galaxies with

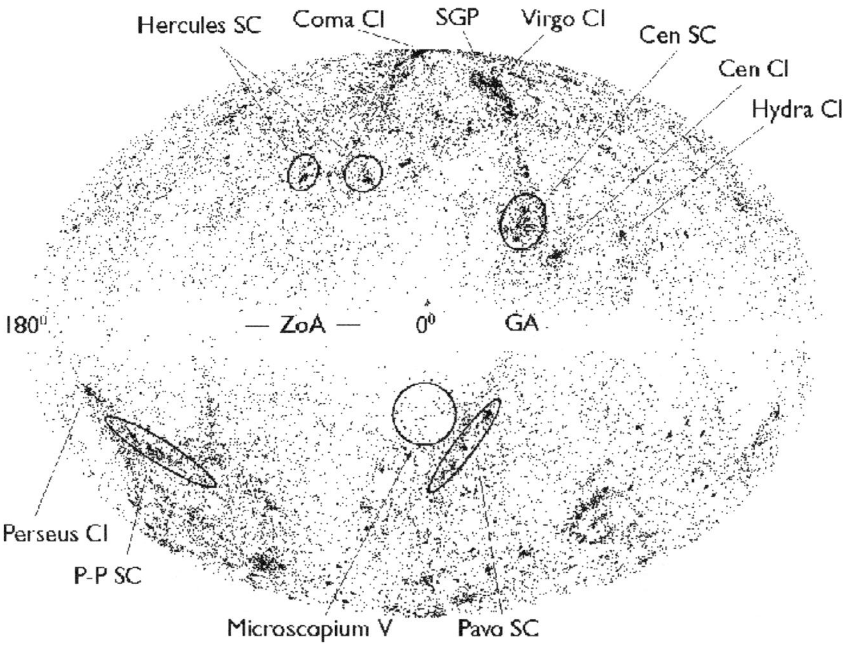

Figure 1.27 *Hammer–Aitoff projection of a compilation of redshifts performed by Christensen (1996) in which some of the features of the nearby distribution of galaxies can be located, as well as the zone of avoidance (ZoA), the location of the Great Attractor (GA) close to the galactic equator, and several clusters and superclusters. (Reproduced, with permission, from Christensen 1986, Compound Redshift Catalogues and Their Applications to Redshift Distortions of the Two-Point Correlation Function, University of Copenhagen.)*

about 30 members. Three spiral galaxies dominate the group: the Andromeda galaxy, M33, and the Milky Way. Nearly all the galaxies within the Local Group lie within a sphere of radius 1 Mpc. In general, groups of galaxies have fewer than 50 members, mostly spirals, except in very dense and compact groups in which a good fraction of ellipticals can also be found.

Clusters of galaxies are larger systems containing from 50 up to thousands of members within a sphere of radius up to 4 Mpc. They contain galaxies of all types. There are clusters with a well-defined spherical shape, but others show a more irregular and open structure. It is also possible to detect substructure within a large cluster. One example of an irregular cluster is the Virgo cluster (see Fig. 1.4). It has an average radial velocity of 1,100 km s^{-1}, and therefore the center of the cluster lies at a distance of 11 h^{-1} Mpc. From Fig. 1.27 we can see that the cluster covers a vast region of the sky (approximately $10° \times 10°$). Fig. 1.28 shows where

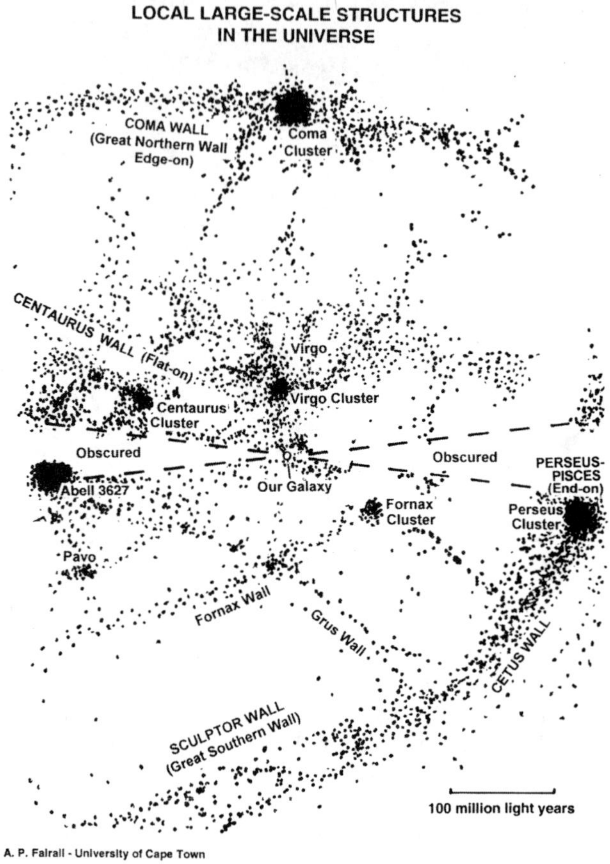

Figure 1.28 *A map showing the main large-scale features in our local universe. (Reproduced, with permission, from Fairall 1998, Large-Scale Structures in the Universe, John Wiley & Sons in association with Praxis Publishing.)*

this cluster lies in relation to our position. In fact, our Galaxy forms part of one of the branches growing outward from the cluster. The morphological segregation between ellipticals and spirals (see Section 1.4.3) is clearly evident in the spatial distribution of galaxies within the Virgo cluster. The central region of the cluster is populated mainly by elliptical galaxies. The biggest one, M87, is a giant E1 elliptical (see Fig. 1.4). Spirals are more or less distributed as a halo surrounding the cluster core.

The more famous regularly shaped rich cluster of galaxies is the Coma cluster, lying 5.4 times farther away from the Virgo cluster (see Fig. 1.28). Of the galaxies in the Coma cluster 85% are ellipticals and S0's. The diameter of the cluster is roughly 6 h^{-1} Mpc although, as a result of peculiar motions, it stretches into an elongated structure, resembling a "stick-man" in redshift space (see Fig. 1.12). Closer to the Earth and very similar to the Coma cluster, the cluster Abell 3627 was discovered later because, with a rather low galactic latitude, it is partially obscured by the dust of our own Galaxy. Another very big cluster is the Perseus cluster, which lies in the south galactic hemisphere. It is also partially obscured by the Milky Way.

1.5.2 Catalogs of clusters of galaxies

Clusters of galaxies are the largest and more massive virialized objects in the universe. As well-defined cosmological entities, clusters are used to probe the large-scale distribution of matter in the universe and their statistical properties complement and are related to those of the galaxy distribution. Similar to the galaxy catalogs described earlier, cluster catalogs also have been compiled, although some uncertainties regarding their statistical analysis have arisen from the loose definition of what a cluster is or from subjective visual identification of a cluster in projected maps (Sutherland and Efstathiou 1991). Projection effects can easily contaminate the detection of a cluster (van Haarlem 1996). For example, a filament along a line-of-sight could easily appear to be a rich cluster on a projected map.

Lately, automatic detection of clusters has been applied using more objective criteria and scanning machines. There is still a more powerful and less ambiguous way to detect these massive concentrations of galaxies. The hot intra-cluster gas produces X-ray emission by bremsstrahlung radiation. The X-ray luminosity is correlated with the cluster mass, and therefore can be used to infer the cluster richness, i.e., the number of galaxies in the cluster (see Reiprich and Börhinger 1999). In this way, galaxy clusters can be objectively detected by present-day X-ray satellites. In particular, the ROSAT satellite has provided more than 1,000 positions of clusters within a huge volume of the universe (Borgani and Guzzo 2001).

In this section we describe some of the cluster catalogs used as tracers of the large-scale structures of the universe. Statistical study of cluster catalogs is performed using tools similar to those used in the study of galaxy catalogs and explained elsewhere in this book.

The Abell cluster catalog

By visual inspection of the Palomar Observatory Sky Survey (POSS), Abell (1958) identified 2,712 rich clusters of galaxies. This catalog was the first well-defined and statistically useful sample of clusters of galaxies. The criteria used to select a cluster from the plates were the following. A cluster must contain at least 50 members with apparent magnitude in the range $[m_3, m_3 + 2]$, where m_3 is the third brightest member of the cluster. All cluster members must lie within a radius of 1.5 h^{-1} Mpc (Abell radius) from the center of the cluster. The survey contains clusters lying north of declination $\delta = -27°$. Abell assigned a richness class to each cluster depending on the number of members according to the previous criteria. Six richness classes were established, from $R = 0$ to $R = 5$. The number of cluster members of each class varies according to the following table.

Richness class	Number of cluster members
0	30–49
1	50–79
2	80–129
3	130–199
4	200–299
5	300 or more

A statistically homogeneous sample was selected by Abell covering a fraction of the sky of 4.26 steradians. It contains 1,682 clusters with richness $R \geq 1$ up to a nominal redshift $z = 2$. The early statistical analysis performed by Abell on his survey proved that clusters were indeed important tracers of the matter distribution in the universe. Stronger clumpiness of the cluster distribution is due to the fact that clusters trace high density peaks of the matter distribution (Kaiser 1984). In cosmology, this effect is known as "biasing," which relates the distribution of a given class of objects, galaxies or clusters, to the matter density field (see Section 7.9).

The ACO cluster catalog

The extension of the Abell catalog to southern declinations was published by Abell, Corwin, and Olowin (1989). This is the so-called ACO cluster catalog. Together with the Abell catalog, this extension gives an all-sky cluster survey containing 4,076 clusters with $R \geq 0$ nominally up to the redshift $z = 2$ and complete for high latitudes. Clusters were included when the apparent magnitude of the tenth brightest galaxy in the cluster, m_{10}, was less than or equal to 19.6. The ACO catalog has a north declination limit of $\delta = -17°$; therefore, within the region $-27° \leq \delta \leq -17°$ both catalogs overlap. This region was very important to calibrate the systematic differences between the catalogs and reduce the data to the same magnitude system. Many authors have contributed to the measurement of the redshifts of the Abell and ACO clusters. For example, Postman, Huchra, and Geller (1992) produced a redshift survey with 351 clusters.

The APM cluster catalog

In Section 1.4.1, we have already introduced the APM galaxy catalog. From this database, a cluster catalog was constructed based on automated algorithms (Dalton et al. 1997) that mimic the Abell selection criteria with smaller counting radius (half of the Abell radius). This survey covers an area of 4,300 square degrees (see Fig. 1.23) and contains 228 clusters up to a redshift limit of $z \simeq 1$.

The REFLEX cluster catalog

As we have already stated, the X-ray emission of clusters of galaxies can be used to identify clusters less ambiguously than optically. Once the clusters have been selected in a given region of the sky above a given flux limit, the goal is to measure their redshifts. The ROSAT-ESO flux limited X-ray (REFLEX) survey has been constructed in this way (Böhringer et al. 2001). The surveyed region of the sky is defined by the equatorial declination $\delta \leq 2.5°$, avoiding the region with low galactic latitude $|b| \leq 20°$ and two small patches around the Magellanic clouds. With these requirements, the covered area is 4.24 steradians (33.75 % of the sky). The survey contains 452 clusters with fluxes above $3 \cdot 10^{-12}$ erg s^{-1} cm^{-2}, up to the distance $\sim 1,000 h^{-1}$Mpc (see Fig. 1.29).

1.5.3 Superclusters: filaments and walls of galaxies

The very rich clusters mentioned in the previous sections are generally situated in larger structures known as superclusters, the clustering of clusters on a bigger scale. For example, the Virgo cluster dominates the Virgo supercluster, also known as the Local Supercluster, because it contains the Local Group. This structure was first recognized by G. de Vaucouleurs, who called it "Supergalaxy." It is the base of the supergalactic system of coordinates explained in Appendix A. Some of the superclusters have the shape of a wall, like the "Northern Great Wall" (Geller and Huchra 1989), which contains the Coma cluster (in Fig. 1.28 it is referred to as the Coma Wall). Its dimensions are 120 h^{-1} Mpc long (from east to west), 50 h^{-1} Mpc wide (north to south), and only 10 h^{-1} Mpc thick. The Sculptor Wall, a similar wall recognized by L. da Costa in the southern galactic hemisphere, presents a shape closer to a filament. It is about 200 h^{-1} Mpc long but only 30×10 h^{-1} Mpc of cross section. Other important superclusters include the Perseus–Pisces supercluster, studied in detail by Giovanelli and Haynes (1991), and the Hydra–Centaurus supercluster, which lies approximately in the opposite direction to the Perseus–Pisces supercluster.

1.5.4 Voids

The empty spaces surrounded by the filaments and walls are known as voids. The first very large region almost devoid of galaxies, detected by Kirshner et al.

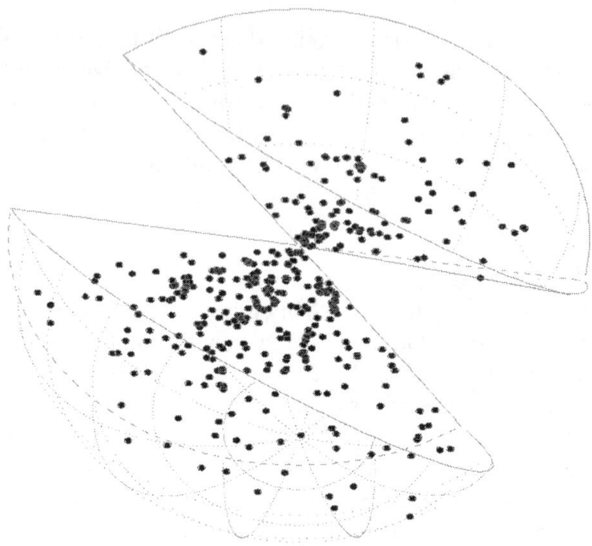

Figure 1.29 *The REFLEX cluster catalog. The zone of avoidance around the galactic plane ($-20^\circ < b < 20^\circ$) is not mapped. (Reproduced, with permission, from Borgani and Guzzo 2001, Nature, 409, 39–45. Macmillan Publishers Ltd.)*

(1981) and known as the Boötes void, has an overall size of 75 h^{-1} Mpc. This size is larger than the size of all other voids observed in the slice redshift surveys (like the CfA2 or Las Campanas). The maximum void size in these surveys is about 60 h^{-1} Mpc in diameter. In fact, the Boötes void was found to present some kind of rarefied structures in its interior. Voids in general are roughly spherical with some inlying galaxies pointing to their interior, although in many cases the core appears to be really empty. It is worth mentioning that the sizes of the voids do not increase with the depth of the redshift survey, as we can see in the deepest slice surveys constructed thus far, such as the Las Campanas redshift survey or the 2dF (see Figs. 1.25 and 1.26).

1.5.5 The texture of the galaxy distribution

We conclude this chapter by summarizing the overall view that the large-scale structures in the universe depict. Different expressions are often used to describe these large patterns: cellular structure, sponge-like topology,‡ foam-like bubbles

‡ A mathematical motivation for the use of "sponge-like" topology is given in Chapter 10.

surrounded by filaments and walls of galaxies, or labyrinth of interconnecting wall-like structures with big clusters lying in the intersections. All these expressions are more or less valid approximations to describe qualitatively the cosmic landscape provided by the distribution of galaxies. However, quantitative descriptors, being reliable, robust, unbiased, and physically interpretable, are needed to extract cosmological information from the data. The rest of this book reviews the statistical and mathematical tools used to quantitatively describe the cosmic web.

CHAPTER 2

The standard model of the universe

2.1 Introduction

In this chapter we present a brief introduction to basic cosmology. As long as we study nearby regions of space, the geometry can be considered Euclidean. But the largest contemporary data sets and, to a much greater extent, the planned future surveys, cover larger regions of space, where cosmological effects (curvature of the space, evolution of objects) must be taken into account.

We provide here only the basic facts and the formulae needed for statistical studies. For additional and much more detailed information we refer the reader to the textbooks by Peacock (1999), Coles and Lucchin (1995), and Peebles (1993).

While the basic theory we shall outline is well known, we also will describe some of the latest observational results, in order to show the present status of the standard cosmological models. These results are relatively recent and have not had the opportunity to withstand the test of time, so they must be considered as good examples, but examples only. Readers interested in timely information should search the physics e-print archives (*ArXiv*) on the Web. Because not all authors submit their papers as electronic preprints, it is also useful to search the NASA Astrophysics Data System site (*ADS*).

2.2 The Friedmann–Robertson–Walker universe

The mathematical basis of contemporary cosmology is Riemann geometry. A cosmological model consists of two ingredients, the geometry of the curved space (its metrics) and the assumptions about the physical content of the matter. The latter determines the evolution of the geometry.

The model geometry can be chosen either for its simplicity or for its concordance with observations. Historically, simplicity was the main reason; cosmology has become a truly observational science relatively recently, in the last 20 to 30 years.

The simplest geometry is, of course, Euclidean: a flat space. General relativity tells us that only a totally empty space can be Euclidean, so this geometry will not pass the simplest observational test of seeing stars out in the sky. The simplest curved-space models are based on the cosmological principle that supposes that there are no preferred locations and directions in the universe (the universe is homogeneous and isotropic). Homogeneity allows us to select a world coordinate system with a universal cosmological time, so that we can decompose the metric into its temporal and spatial parts, and homogeneity and isotropy together demand that the space-like sections must be spaces of constant curvature. (In fact, requiring isotropy for every observer implies homogeneity.) We know from observations that the universe is expanding, so we have to allow this curvature to change in time.

While the cosmological principle is certainly philosophical, its consequences and the principle itself can be tested, if we manage to observe a fairly large region of the universe. We shall describe the present status of these direct tests below.

The geometry of the four-dimensional space-time is described by its metric,

$$ds^2 = \sum_{i=0}^{3} \sum_{j=0}^{3} g_{ij}(\mathbf{x})\, dx^i dx^j ,$$

where ds is the distance between two nearby points (events) with coordinates $\mathbf{x}, \mathbf{x} + \mathbf{dx}$ and $g_{ij}(\mathbf{x})$ is the metric tensor. The cosmological principle leads us to the Robertson–Walker metric

$$ds^2 = c^2 dt^2 - R^2(t) \left(d\omega^2 + S_k^2(\omega)(d\theta^2 + \sin^2\theta \, d\phi) \right) , \tag{2.1}$$

where c is the speed of light, t is time, $R(t)$ is the scale factor (the radius of curvature of space-like sections), ω is a dimensionless radial coordinate, and θ and ϕ are the usual spherical coordinates. The function $S_k(\omega)$ has different forms for different types of space-like sections, marked by k ($k = 1$ describes a model of positive curvature, $k = 0$ means that the spatial sections are flat and the space is infinite, and $k = -1$ corresponds to a model of negative curvature):

$$S_k(\omega) = \begin{cases} \sin\omega, & k = 1, \\ \omega, & k = 0, \\ \sinh\omega, & k = -1. \end{cases} \tag{2.2}$$

The dimensionless scale factor $a(t)$ is defined by $R(t) = R_0 a(t)$, where R_0 is the value of the scale factor at present.

There are other possibilities for choosing the coordinates that lead to different forms of the metric for these spaces, but the form above is the simplest one. The spherical coordinates match best the observing practice (the observed positions of objects are ϕ and θ in some spherical coordinate system and a radial coordinate from which ω can be deduced).

The curvature of the space-like sections $k/R^2(t)$ and, hence, $S_k(\omega)$ and the evolution of the scale factor $R(t)$ in time, are determined by the matter content of the universe.

2.2.1 *Comoving and physical distances and volumes*

The coordinates ω, θ, ϕ are called comoving coordinates, and the scale factor $R(t)$ describes the general expansion (or contraction) of the space. Thus, distances between objects that have constant comoving coordinates change as $R(t)$; these distances (and volumes) are called "physical." It is useful to define comoving distances and volumes as the values of these quantities as they would be at the present moment, if all markers had retained their spatial coordinates.

Distances

Because the proper calculation of distances is important for spatial statistics, and years of working in an approximately Euclidean region of the space have introduced different distance definitions and "relativistic corrections," we shall consider this point in some detail here.

Usually we think of objects occupying a certain volume at a fixed moment in time and the distance between objects has the usual meaning here. If we speak of the spatial distribution of objects in cosmology, we have to take into account the fact that all observed objects lie along a "light cone," a 3-D section of 4-D space-time, and we see them at different moments. The farther away an object is from us, the earlier it has emitted the light we see.

In the nearby universe (in our Galaxy and between nearby galaxies, even between nearby clusters of galaxies) distances are so small that this difference of times does not matter. For deeper surveys it will matter.

Observations of distant objects in a universe described by the metric (2.1) occur as shown in Fig. 2.1, where we have suppressed two angular coordinates. We are located at a point marked O and the two luminous objects (galaxies) we observe (1 and 2) lie along our past light cone. The moments when they emitted their light (t_1, t_2) define the corresponding spatial sections and the dimensionless radial coordinates ω_1, ω_2 of the objects.

Because general relativity states that light follows null geodesics in space-time, the equation of the light cone is, from the metrics (2.1):

$$c^2 dt^2 = R^2(t) d\omega^2, \quad d\theta = d\phi = 0.$$

The observations we make lie along our past light cone, given, obviously, by

$$\frac{d\omega}{dt} = -\frac{c}{R(t)}. \tag{2.3}$$

Strictly speaking, all distances along a light cone are zero, by the definition of the metric in Riemann geometry. The distances that can be used are the time intervals (see Fig. 2.1), physical distances (lengths of the arcs from the time axis), and distances measured by dimensionless coordinates ω.

It would be meaningless to compare physical distances of galaxies because these depend on the time coordinate. The radial coordinate ω of an object at rest, on the contrary, does not change with time and can be used to find distances between objects. Such distances are called comoving distances. A comoving distance can be expressed either in dimensionless units of ω, or in the units of length $(R_0\omega)$. These distances can be thought of as distances between galaxies at the

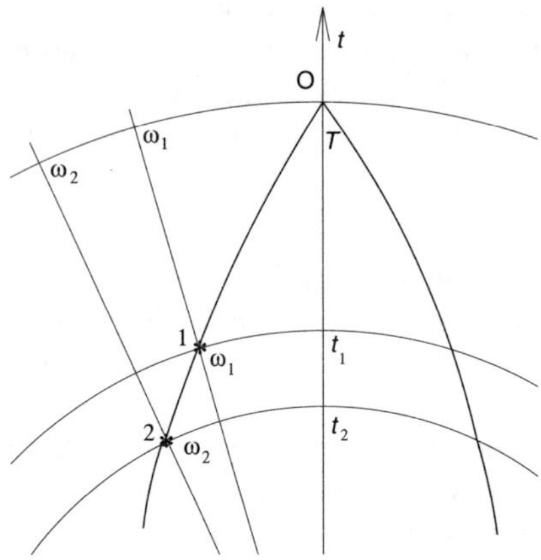

Figure 2.1 *Observations along a light cone (thick line). The location of the observer is marked by O and locations of two luminous objects (galaxies) are indicated by asterisks. Each galaxy and the observer define their instantaneous three-space (arcs in the figure). The time and dimensionless radial coordinate of the observer are T and 0, and for the two galaxies t_1, ω_1 and t_2, ω_2, respectively.*

present moment (along the space-like section defined by the age of the universe T), although we have observed them only in the past and cannot see them at their present location.

So, the first step in determining a distance between two galaxies is to find their comoving coordinates ω (comoving distances from the observer who has chosen his coordinates as $\omega = 0$). This depends on the assumed cosmological model and we shall describe it a little later. The second step is to find the distance between two galaxies in curved three-dimensional space.

Let us denote their coordinates as $\{\omega_i, \theta_i, \phi_i, i = 1, 2\}$ (the same used for the line element (2.1)). The coordinates θ and ϕ are the usual spherical coordinates in the sky (θ is the zenith distance and ϕ is the longitude). From a spherical triangle formed by the pole of the coordinates and by the points $\{\theta_i, \phi_i\}$ we can find the angular distance between the points on the sky η by the cosine theorem of spherical trigonometry:

$$\cos\eta = \cos\theta_1 \cos\theta_2 + \sin\theta_1 \sin\theta_2 \cos(\phi_1 - \phi_2). \tag{2.4}$$

Note that here θ_i are zenith distances, measured from the pole of the spherical coordinate system. Most astronomical coordinate systems use latitudes, measured from the equator; if this is the case with your catalog, you have to modify the above formula, replacing $\theta \to \pi - \theta$.

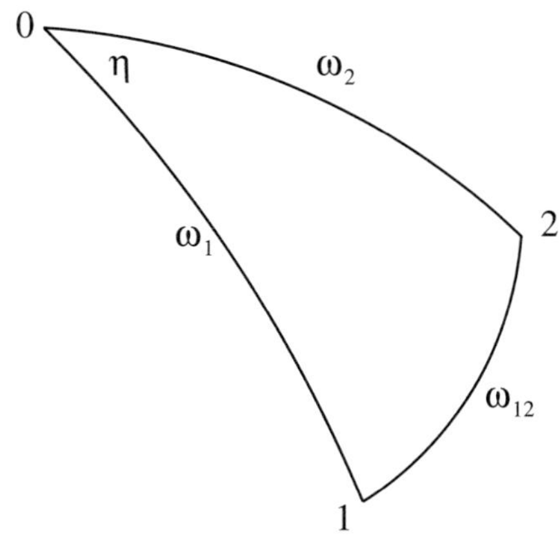

Figure 2.2 *Determination of the distance between two events. The spherical triangle is formed by our location (0) and by the two objects (1,2).*

Let us now move the pole of the celestial coordinates to the location of one of our objects. In the new coordinates

$$d\phi = 0, \qquad d\theta = d\eta,$$

and we can express the comoving distance element dl as

$$dl^2 = R_0^2(d\omega^2 + S_k^2(\omega)d\eta^2).$$

This formula tells us that in order to find the distance l between two objects we have to solve a triangle on an appropriate two-dimensional surface, either a sphere ($k = 1$), a plane ($k = 0$), or a pseudo-sphere ($k = -1$). The angular distance η is similar to the longitude ϕ and the dimensionless radius ω to the zenith distance θ of a usual spherical coordinate system. Fig. 2.2 shows determination of distance on a sphere.

For a flat three-space ($k = 0$) we have the cosine law:

$$\omega_{12}^2 = \omega_1^2 + \omega_2^2 - 2\omega_1\omega_2 \cos\eta. \tag{2.5}$$

For a closed hypersphere ($k = 1$) spherical trigonometry tells us that

$$\cos\omega_{12} = \cos\omega_1 + \cos\omega_2 + \sin\omega_1 \sin\omega_2 \cos\eta. \tag{2.6}$$

For the negative curvature, open universe ($k = -1$) there exists an analogous result in Lobatchevsky trigonometry:

$$\cosh\omega_{12} = \cosh\omega_1 + \cosh\omega_2 - \sinh\omega_1 \sinh\omega_2 \cos\eta. \tag{2.7}$$

These formulae give the dimensionless distances; in order to get the distance in units of length, we have to multiply the dimensionless distance by R_0:

$$r_{12} = R_0\omega_{12}. \tag{2.8}$$

We shall see below that for the Friedmann cosmological models in the case of a flat three-space ($k = 0$) R_0 and ω cannot be defined separately, but their product $r = R_0\omega$ can. Multiplying both sides of the formula (2.5) by R_0^2 transforms it into an equivalent formula for r.

Volumes

According to Riemann geometry, the spatial volume element dV can be written as

$$dV = \sqrt{-\det g_{\alpha\beta}}\, d^3x,$$

where $g_{\alpha\beta}$ is the spatial part of the metric tensor. For our metric it gives

$$dV = R^3(t) S_k^2(\omega) \sin\theta \, d\omega \, d\theta \, d\phi. \tag{2.9}$$

The corresponding comoving volume (the volume delimited by the same markers at the present epoch) is given by the same formula, if we substitute R_0 for $R(t)$.

The volumes of the observed galaxy samples are usually spherical sections. The comoving volume of a sphere with a radius ω is

$$\begin{aligned} V(\omega) &= 2\pi R_0^3(\omega - \sin\omega\cos\omega), & k &= 1, \\ V(\omega) &= \frac{4\pi}{3} R_0^3 \omega^3, & k &= 0, \\ V(\omega) &= 2\pi R_0^3(\sinh\omega\cosh\omega - \omega), & k &= -1. \end{aligned} \tag{2.10}$$

2.2.2 *Hubble's law and redshift*

Because the physical distance r between two nearby objects is proportional to their comoving radial separation, $r \sim R(t)\Delta\omega$, their mutual velocity

$$v = \frac{dr}{dt} = \dot{R}(t)\Delta\omega = \frac{\dot{R}(t)}{R(t)} r = H(t) r \tag{2.11}$$

is proportional to their distance. This relation is called Hubble's law — galaxies move away from us at velocities that are proportional to their distance from us. This law was established observationally by E. Hubble in 1929. For nearby galaxies and for the present epoch

$$v = H_0 r.$$

The ratio

$$H(t) = \frac{\dot{R}(t)}{R(t)} = \frac{\dot{a}(t)}{a(t)}$$

is Hubble's function. Its value at the present epoch H_0 is called Hubble's constant and must be found from observations.

We do not observe directly the recession velocity of a galaxy, but the changes in its spectra. Because of the Doppler effect, the light emitted from receding galaxies with frequency ν_e arrives to us with a lower frequency ν_0:

$$\nu_0 = \frac{\nu_e}{1 + v/c},$$

where v is the velocity of the emitter and c is the velocity of light (strictly speaking, this formula is valid only for small velocities, but we are using it for such a case). For small velocities (and small changes of frequency) we can write

$$\frac{d\nu}{\nu} = -\frac{v}{c} = -\frac{\dot{R}(t)}{R(t)}\frac{r}{c} = -\frac{\dot{R}(t)}{R(t)}dt = -\frac{dR}{R}.$$

The solution of this differential equation is

$$\nu \sim 1/R(t), \qquad \lambda = \frac{2\pi}{\nu} \sim R(t). \tag{2.12}$$

The Doppler interpretation of the change of the wavelength of radiation is useful for nearby objects and small velocities. For far-away galaxies the notion of their speed in respect to us has no clear meaning, because their light was emitted so long in the past. The formulae (2.12) remain correct in this case, too. In order to see that, we can imagine one standing electromagnetic wave (radiation of a single frequency) in an expanding universe. During expansion the number of nodes of this wave cannot change; thus, the wavelength of the wave has to grow at the same rate as the scale factor $R(t)$.

The observed shift of the frequency (or wavelength) of the spectra is described by a parameter z called redshift:

$$z = \frac{\lambda_0 - \lambda_e}{\lambda_e},$$

and this is the measure of distance we can find in observational catalogs. Here λ_e is the wavelength of a spectral line at the moment of emission (we know it on the basis of studies of spectra of nearby galaxies), and λ_0 is the wavelength we observe that line at here.

This definition, together with the formula (2.12) above, gives us

$$1 + z = \frac{R_0}{R(t)} = a^{-1}(t). \tag{2.13}$$

Thus, measuring the redshift of a luminous object, we can determine the scale factor $R(t)$ at the moment when the light was emitted. If we know the $R(t)$ dependence, we can find the cosmological time t, and from the equation of the light cone (2.3), the spatial distance to the object ω.

2.2.3 The Friedmann equations

The connection between the geometry and the physical content of the universe is given by the Einstein equations:

$$G_{ij} - \Lambda g_{ij} = \frac{8\pi G}{c^4} T_{ij},$$

where the Einstein tensor G_{ij} is formed from the metric tensor g_{ij}, its first and second derivatives, and Λ is the cosmological constant. The energy-impulse tensor T_{ij} on the right-hand side of the equation describes the physical content of the universe, the coefficient of proportionality given by a combination of the gravitational constant G and the velocity of light c.

For the Robertson–Walker line element (2.1) the above equations reduce to the Friedmann equations:

$$\frac{\ddot{R}}{R} = -\frac{4\pi G}{3}(\rho + 3p/c^2) + \frac{\Lambda c^2}{3}, \tag{2.14}$$

$$\frac{\dot{R}^2}{R^2} = \frac{8\pi G}{3}\rho + \frac{\Lambda c^2}{3} - \frac{kc^2}{R^2}, \tag{2.15}$$

$$\dot{\rho} = -3(\rho + p/c^2)\frac{\dot{R}}{R}, \tag{2.16}$$

which have to be supplemented by the equation of state, $p = p(\rho)$. Here R is the scale factor, ρ is the total energy density, p is the pressure and k is the curvature constant (see 2.2). These equations together with the Robertson–Walker metrics (2.1) define the "standard cosmological models." The three equations (2.14–2.16) are not independent; the first equation (2.14) can be derived, using the other two equations.

The main contributors to the total energy density ρ are baryonic matter, radiation, and nonbaryonic (dark) matter. These all have different equations of state and they dominate at different epochs of the evolution of the universe. For epochs when we can count objects and determine their statistics, the energy density of radiation is negligible compared with that of baryonic matter and dark matter, and for these constituents the pressure term p/c^2 in the Friedmann equations is much smaller than the density ρ. Thus the usual approximation is to set $p = 0$ and to consider the universe to be filled with pressureless "dust." In this case we get from the equation (2.16):

$$\rho \propto R^{-3}. \tag{2.17}$$

Now we are ready to introduce the standard parameterization of Friedmann cosmological models. Let us define the density parameter Ω_M as

$$\Omega_M = \frac{8\pi G \rho_0}{3H_0^2}, \tag{2.18}$$

where ρ_0 is the present average matter density, and the normalized cosmological constant Ω_Λ as

$$\Omega_\Lambda = \frac{\Lambda c^2}{3H_0^2}.$$

As the values of the parameters Ω_M, Ω_Λ determine the evolution of the cosmological model, they are called the cosmological parameters. The combination $(\Omega_M = 1, \Omega_\Lambda = 0)$ defines the Einstein–de Sitter model and $(\Omega_M = \Omega_\Lambda = 0)$ describes the Milne model, an empty expanding universe.

Another important parameter is the Hubble constant H_0, which determines the overall spatial and temporal scales of the universe.

Cosmological models can be also parameterized by the deceleration parameter q, which completely determines models without the cosmological constant:

$$q = -\frac{\ddot{R}R}{\dot{R}^2}.$$

For our parameterization we find from (2.14)

$$q = \frac{\Omega_M}{2} - \Omega_\Lambda.$$

Let us introduce now the normalized Hubble function $E(z)$ by

$$H(z) = H_0 E(z). \tag{2.19}$$

The best designation for that function would be $h(z)$, but h is already used in the same context to designate the dimensionless Hubble constant ($H_0 = h \cdot 100$ km/sec Mpc^{-1}). Using (2.17) and (2.13), we can rewrite the Friedmann equation (2.15) as

$$E^2(z) = \Omega_M(1+z)^3 + \Omega_\Lambda - \frac{kc^2}{H_0^2 R_0^2}(1+z)^2.$$

At the redshift $z = 0$ the function $E(0) = 1$, giving

$$\frac{kc^2}{H_0^2 R_0^2} = \Omega_M + \Omega_\Lambda - 1, \tag{2.20}$$

and we get, finally,

$$E^2(z) = \Omega_M(1+z)^3 + (1 - \Omega_M - \Omega_\Lambda)(1+z)^2 + \Omega_\Lambda. \tag{2.21}$$

This function allows us to describe the evolution of the cosmological model and to express all the observational relations in terms of an observed quantity, the redshift z. Some examples of that are given below.

We see also from (2.20) that the type of the curvature k is given by the sign of the difference $\Omega_M + \Omega_\Lambda - 1$. This combination of parameters is sometimes called "the curvature parameter" Ω_K and is defined as

$$\Omega_K = 1 - \Omega_M - \Omega_\Lambda.$$

Thus for negative curvature, $k = -1$, $\Omega_K > 0$, and for $k = 1$ $\Omega_K < 0$.

At any epoch z the cosmological model can be described by the current (instantaneous) cosmological parameters $\Omega_M(z), \Omega_\Lambda(z), H(z)$. The expression for $H(z)$ is given by the formula (2.19) and the formulae for the other parameters follow from the definition of Ω_M, Ω_Λ, above:

$$\Omega_M(z) = \frac{\Omega_M(1+z)^3}{E^2(z)}, \tag{2.22}$$

$$\Omega_\Lambda(z) = \frac{\Omega_\Lambda}{E^2(z)}. \tag{2.23}$$

These expressions show how a cosmological model "looks" at a particular epoch. To illustrate it, let us look at an open low-density $\Omega_\Lambda = 0$ model. For that model

$$\Omega_M(z) = \frac{\Omega_M(1+z)}{\Omega_M(1+z) + 1 - \Omega_M}$$

and we see that initially ($z \gg 1$) this model behaves as the Einstein–de Sitter model, $\Omega_M(z) \approx 1$. This behavior lasts until the redshift z_c given by $z_c + 1 \approx (1 - \Omega_M)/\Omega_M$, when the model starts expanding as a true open model. For a reasonable $\Omega_M \approx 0.3$ this happens rather late, at $z_c \approx 1.3$.

For a true $k = 0$ universe with the same density parameter $\Omega_M = 0.2$, $\Omega_\Lambda = 0.8$, the fast expansion starts much later, at $z_c \approx 0.6$. For the Einstein–de Sitter model the fast expansion stage never arrives.

2.2.4 *Cosmological time*

The expression for cosmological time can be derived from the definition of the function $g(z)$ (2.19) and the fact that $R(z) = R_0/(1+z)$:

$$H = H_0 E(z) = \frac{1}{R}\frac{dR}{dt} = -\frac{1}{1+z}\frac{dz}{dt}. \tag{2.24}$$

When this equation is integrated, the age of the universe at the epoch described by a redshift z is given as

$$t(z) = \frac{1}{H_0}\int_z^\infty \frac{dz'}{(1+z')E(z')}. \tag{2.25}$$

The time elapsed from the epoch z until now (the look-back time) is given by the same integral with the limits 0 and z, and we get the present age of the universe, if we integrate from 0 to ∞. This age is finite for all possible models.

For general combinations of Ω_M and Ω_Λ the integral has to be computed numerically. If $\Omega_\Lambda = 0$, there exist analytic expressions for it, different for different k, but these are very cumbersome and numerical integration is recommended in that case, too.

Simple analytic results exist for two models that can be used for quick estimates. For the Einstein–de Sitter model ($\Omega_M = 1, \Omega_\Lambda = 0$) we get

$$t(z) = \frac{2}{3H_0}\frac{1}{(1+z)^{3/2}}$$

with $T = 2/(3H_0)$ for the age of the universe. If we substitute here H_0 with the present mean density of the universe ρ_0 from (2.18) and note that $\rho_0(1+z)^3 = \rho$, we get the density behavior in the Einstein–de Sitter model:

$$\rho = \frac{1}{6\pi G t^2}. \tag{2.26}$$

This model is not only the simplest one, but it is also an excellent approximation for all other models for large redshifts. This is easy to understand if we look at the high-z asymptotics of the functions $\Omega_M(z)$ and $\Omega_\Lambda(z)$:

$$\begin{aligned}\Omega_M(z) &\approx 1 + (\Omega_M + \Omega\Lambda - 1)z^{-1}, \quad z \gg 1 \\ \Omega_\Lambda(z) &\approx \frac{\Omega_\Lambda}{\Omega_M} z^{-3}, \quad z \gg 1.\end{aligned}$$

For the empty Milne model $\Omega_M = \Omega_\Lambda = 0$

$$t(z) = \frac{1}{H_0}\frac{1}{1+z}$$

with $T = 1/H_0$ for the age of the universe. This model can be thought of as representing a low-density universe (the descriptive functions for $\Omega = 0.1$, for example, do not differ much from those for the empty model).

2.2.5 *The light cone equation*

In order to find the comoving radial distance of an observed event with the redshift z, we need to solve the light cone equation (2.3):

$$\frac{d\omega}{dt} = -\frac{c}{R(t)}.$$

Using the expression for dz/dt from (2.24), we can integrate this equation to find

$$\omega = \frac{c}{R_0 H_0}\int_0^z \frac{dz'}{E(z')}. \tag{2.27}$$

The coefficient before the integral depends only on Ω_M and Ω_Λ, by formula (2.20):

$$\frac{c}{R_0 H_0} = \sqrt{|\Omega_M + \Omega_\Lambda - 1|}.$$

In units of length, the comoving distance is

$$r = R_0\omega = \frac{c}{H_0}\int_0^z \frac{dz'}{E(z')}. \tag{2.28}$$

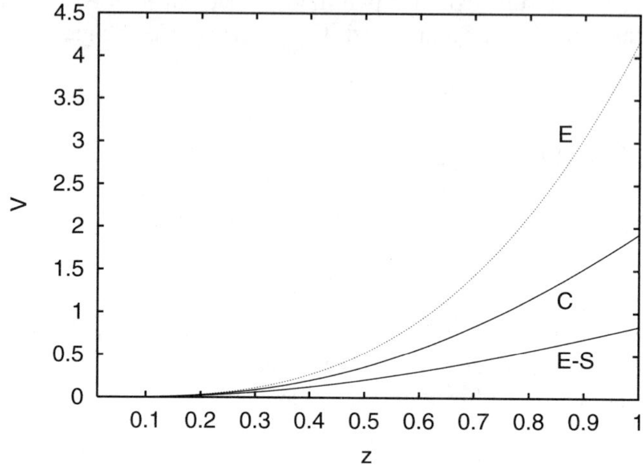

Figure 2.3 *The volume-redshift dependence for the Einstein–de Sitter model (E-S) and the concordance model (C). We also show the Euclidean behavior (E) for comparison. The volumes are given in the units of cubed Hubble length* $(c/H_0)^3$.

This last formula also can be used for $k = 0, R_0 = \infty$, while formula (2.27) evidently cannot.

The formula (2.28) can be integrated for $\Omega_\Lambda = 0$, giving different expressions for different k, but these are rather cumbersome to use, and numerical integration is the easy way out. For the two special models we have:

$$
\begin{aligned}
r &= \frac{2c}{H_0}\left(1 - \frac{1}{\sqrt{1+z}}\right), \quad && \Omega_M = 1,\ \Omega_\Lambda = 0, \\
r &= \frac{c}{H_0}\ln(1+z), && \Omega_M = 0,\ \Omega_\Lambda = 0.
\end{aligned}
$$

To illustrate this, we show in Fig. 2.3 the growth of the comoving volume of a sphere (formula 2.10) of a radius of redshift z in the Einstein–de Sitter cosmology and in another (concordance) model with $\Omega_M = 0.3, \Omega_\Lambda = 0.7$. The latter model is based on a comparison of the recent cosmic microwave background (CMB) data with data on the large-scale structure of the universe (Tegmark, Zaldarriaga, and Hamilton 2001). Although the three-space is flat in both those models, the space-time is curved, the $\omega(z)$ relation is nonlinear, and the volume does not grow as in the Euclidean case ($V \sim z^3$).

An important fact that can be read from formula (2.28) and the expression for $E(z)$ (2.21) is that the integral is finite for infinite z — there exists a maximum comoving distance $r_{max} = r(z \to \infty)$ (the only exception is the empty model). This is called the (particle) horizon, and it is caused by the finite lifetime of the universe; we can observe only those sources whose light has had enough time to reach us. Thus, we can, in principle, observe only a finite volume patch of the universe, even if the universe itself is infinite. This is a specific feature of the standard model we have described here; there are models for the early evolution of the universe that do not have particle horizons.

2.2.6 *Observational distances*

There are, by tradition, many distance definitions in observational cosmology. These are not really distances, but rather recipes for calculating different physical aspects of light sources.

Let us consider an extended object (galaxy) at an epoch z. Using the expression for metrics (2.1) and setting $d\omega = 0$, we can find the physical size of the galaxy Δl, knowing its angular diameter $\Delta\eta$ and comoving distance ω:

$$\Delta l = R(t) S_k(\omega) \Delta\eta,$$

In a Euclidean universe the corresponding relation would be written as

$$\Delta l = D\,\Delta\eta,$$

where D is the distance to the galaxy. The comparison of these two formulae leads to the definition of the angular diameter distance:

$$D_a = \frac{R_0}{1+z} S_k(\omega(z)).$$

So, if we insist on using Euclidean formulae for this procedure, an angular diameter distance is the function of z we have to use to calculate the angular diameter of an object. This is the recipe for other "distance" definitions, too.

Interestingly, the function $S_k(\omega(z))$ has the same rather simple form for all Ω_M in the case $\Omega_\Lambda = 0$, although the function $\omega(z)$ itself has to be calculated by three different formulae. If we define another "distance" $D_m(z)$ by

$$D_m(z) = R_0 S_k(\omega(z)), \tag{2.29}$$

then we can write

$$D_m(z) = \frac{2c}{H_0} \frac{\Omega_M z + (\Omega_M - 2)(\sqrt{1+\Omega_M z} - 1)}{\Omega_M^2 (1+z)} \tag{2.30}$$

This formula is known as the Mattig formula, and it sometimes is written in terms of the deceleration parameter $q = \Omega_M/2$. For small Ω_M it is better to use another form of it (Peacock 1999):

$$D_m(z) = \frac{c}{H_0} \frac{z(1+z+\sqrt{1+\Omega_M z})}{(1+z)(1+\Omega_M z/2 + \sqrt{1+\Omega_M z})}.$$

In terms of D_m, the angular diameter distance becomes

$$D_a(z) = D_m(z)/(1+z).$$

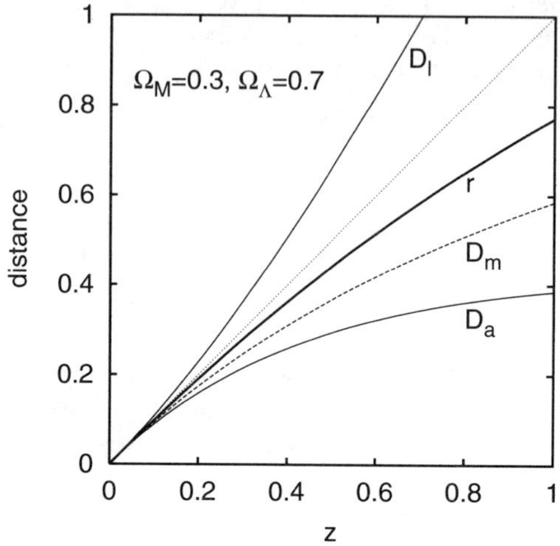

Figure 2.4 *Different distance definitions for a model universe with* $\Omega_M = 0.3, \Omega_\Lambda = 0.7$. *Here* r *is the true comoving distance,* D_a *is the angular diameter distance, and* D_L *is the luminosity distance. The curve labeled* D_m *shows the Mattig distance for* $q = 0.5$. *All distances are given in units of the Hubble length* c/H_0.

Another useful function is the luminosity distance D_L that is used to calculate the observed bolometric (total) flux density S of a source of bolometric luminosity L at a redshift z by the Euclidean formula

$$S = \frac{L}{4\pi D_L^2}. \tag{2.31}$$

This distance is given by

$$D_L(z) = (1+z)D_m(z). \tag{2.32}$$

Because the formula for $D_m(z)$ is so simple, it has been used extensively as a comoving distance in cosmological statistics of deep surveys. For smaller z the difference between $D_m(z)$ and the geometrical comoving distance $r(z) = R_0\omega(z)$ is not great. Note that in the case $k = 0$ (and $\Omega_\Lambda = 0$) both comoving distances coincide.

As an example, we show in Fig. 2.4 the distances $r(z), D_a(z)$, and $D_L(z)$ for the concordance model with parameters $\Omega_M = 0.3, \Omega_\Lambda = 0.7$. The Mattig distance is not defined for these models, but the usual practice is to use for all models the Mattig distance formula for $q = 0.5$. This curve is labeled D_m in the figure.

The deepest present galaxy surveys reach $z \approx 0.3$; at these distances the margin for error if we use the wrong distance definition is about 10%. New surveys will reach farther, and quasars have been observed up to redshifts $z \approx 5$, where the difference between the Mattig distance and the true comoving distance is about 30%. The effects we are trying to measure are much smaller; therefore, we must be careful when using distance-based statistics in cosmology.

We explained earlier in this chapter that the distance we need in spatial statistics is the comoving distance. It is the length of the geodesic of a space-like section, the minimal space-like distance possible, and it is the only counterpart for the Euclidean distance used in traditional statistics.

The method for calculating distances depends on the cosmological model that we assume. If this model has flat spatial sections ($k = 0$), as all popular models do, the calculation is simple. As the three-space is flat, we can continue using Cartesian coordinates. The only change is the mapping from redshift to comoving distance from the observer, given by Eq. 2.28. The distances between galaxies can then be found by the usual Euclidean formulae.

If the chosen model has curved spatial sections, the calculations are more complex. Although the spatial sections are homogeneous, coordinate systems that cover the whole spatial section are singular (as in the case of spherical coordinates), and the usual Euclidean coordinates cannot be used. First, we have to find the dimensionless comoving distances from the observer (2.27) for all galaxies. If we have to calculate distances between galaxy pairs, we have to find the line-of-sight angles for every pair by (2.4) and then use these angles to find the dimensionless comoving pair distances using (2.6) in the case of positive curvature or using (2.7) in the case of negative curvature. Finally, we have to use Eq. 2.8 to convert to comoving distances.

There is also the possibility of continuing to use Euclidean coordinates for galaxies, but in an artificial four-dimensional Euclidean (or Minkowski) space that embeds the spherical (or pseudo-spherical) three-dimensional spatial sections (Roukema 2001). The distances are then calculated as arc-lengths in four dimensions. The number of calculations is the same as for the method described above, but the concept could prove useful for future applications.

In most applications of spatial statistics in cosmology one of the goals is to estimate the cosmological parameters Ω_M and Ω_Λ. If we vary the values of these parameters, we should recalculate the galaxy positions and mutual distances for the initial catalog. Because this is an enormous amount of work, it is usually avoided. This approach can be justified for comparatively shallow catalogs, but the margin for error must be estimated.

2.3 Basic observational data

While the present cosmological models have been built on the basis of more or less general principles, the rapid development of observational techniques has allowed us to test directly most of these assumptions. We shall describe these assumptions briefly in this section, and some of them more fully in the respective statistics chapters.

Let us first give some numbers to get the feel of the characteristic scales. The natural unit of distance in cosmology is the megaparsec (Mpc), 1 Mpc $=$ 10^6 pc $= 3.086 \cdot 10^{24}$ cm. Average distances between galaxies and sizes of galaxy clusters are of the order of a few Mpc. As one parsec is 3.26 light-years, a step of one Mpc in space along the light cone takes us another 3.26 million years into the past.

We have seen above that the age and the radius of the curvature of the universe are determined by Hubble's constant. The values obtained for it vary rather widely, but the recent measurements seem to converge in the interval 60–70 km sec^{-1} Mpc^{-1}. In theoretical papers H_0 is usually expressed as $H_0 = 100 \cdot h\,\mathrm{km\,sec^{-1}\,Mpc^{-1}}$, giving the explicit dependence of the results on h. For further estimates we shall use the value $H_0 = 65\,\mathrm{km\,sec^{-1}\,Mpc^{-1}}$ ($h = 0.65$).

The age of the universe (the Hubble age) will then be about $T = 1/H_0 = 15 \cdot 10^9$ years, and its typical size (the Hubble radius) will be about $c/H_0 = 4{,}600$ Mpc. The relative amplitude of the curvature effects is of the order of the redshift z, or, if expressed in recession velocities, of the order of the ratio v/c ($c \approx 3 \cdot 10^5$ km/sec is the velocity of light). Another typical length, the curvature radius R_0, given by

$$R_0 = \frac{c}{H_0\sqrt{|\Omega_M + \Omega_\Lambda - 1|}},$$

is difficult to estimate if we do not know the cosmological parameters Ω_M and Ω_Λ, and it describes only the curvature effects inside the space-like sections; space-time is well curved even in the case $k = 0$, $R_0 = \infty$.

Because all distances and the look-back time t_l are linear functions of z for small z, the constants found above can be used to estimate them:

$$\begin{aligned} d &\approx \frac{c}{H_0} z \approx 4{,}600 \cdot z \text{ Mpc}, && z \ll 1, \\ t_l &\approx \frac{1}{H_0} z \approx 15 \cdot 10^9 \cdot z \text{ yr}, && z \ll 1. \end{aligned}$$

The present mean density of the universe is easy to calculate, using the parameterization formula (2.18) from above:

$$\rho = 1.9 \cdot 10^{-29} \Omega_M h^2 \mathrm{g/cm^3} = 2.8 \cdot 10^{11} \Omega_M h^2 M_\odot/\mathrm{Mpc}^3, \tag{2.33}$$

where $M_\odot$ is the mass of the sun.

The objects of the cosmological statistics are usually galaxies or their clusters. The various properties of galaxies were described in Chapter 1; for a cosmological model we need to know only the average properties of galaxies and their spatial distribution. As a rule of thumb, we may take the mean density of galaxies to be of the order of 0.01 Mpc^{-3}, a galaxy per 100 cubic megaparsec. Keep in mind that this estimate is very uncertain, because galaxies have widely different luminosities and galaxy samples are chosen usually on the basis of their observed light intensity. Thus, many galaxies, especially those farther away from us, remain unobserved.

All matter is not in galaxies. Most of it (95–99%) is dark and has been detected only by its dynamical effects. Its physical nature is not yet clear; because it determines largely the evolution of the structure in the universe, the hypothesis about the dark matter is frequently considered an extra "cosmological parameter."

There are also brighter objects that can be observed for larger distances, quasars and galaxy clusters. A (rich) galaxy cluster contains from a few hundred to a thousand galaxies and its characteristic size is a few Mpc. The spatial density of galaxy clusters is about 10^{-5} Mpc^{-3}.

2.3.1 Olber's paradox and the microwave background

One of the basic observational facts about the universe is that the sky we look at is mostly dark. We know that the observed flux density S of a luminous object (star or galaxy) from the distance r is

$$S = \frac{L}{4\pi r^2},$$

where L is the luminosity of the object. In an infinite Euclidean universe, populated with luminous objects of an average luminosity L with the number density n, the total observed flux density S_{tot} would be given by the integral

$$S_{tot} = \int_0^\infty \frac{nL}{4\pi r^2} r^2 dr,$$

which evidently diverges. In fact, we would not see an infinitely bright sky, as foreground stars would screen those farther away, but there would be a star shining at any point in the sky. This result is known as Olber's paradox.

Contemporary cosmological models do not predict an infinite luminosity of the sky for several reasons. First, the existence of the particle horizon means that the volume we observe is finite. Second, it is reasonable to suppose that in an evolving universe stars and galaxies form and die, and the spatial density of light sources will not be constant on the light cone. Third, the expansion of the universe causes the photons to lose their energy:

$$E_\nu \propto \nu \propto 1/R(t),$$

where E_ν is the energy of a photon, ν its frequency, and $R(t)$ the scale factor.

These factors all work to diminish the brightness of the sky, but they cannot reduce it to zero. The photons emitted by atoms during the evolution of the universe form the background light, and observations of its intensity and spectra (or corresponding observational limits) can be used to constrain theories of formation of structure in the universe.

Apart from radiation from discrete objects, there is a component of the background that once really made the whole sky shine uniformly. Because this happened at redshifts before $z \approx 1{,}300$, this light by now has been redshifted into the microwave region of the spectrum ($\lambda \approx 1$mm), and it is called the cosmic microwave background (CMB) radiation.

It can be shown that in an expanding universe the temperatures of matter T_m and radiation T_γ depend on the value of the scale factor as

$$\begin{aligned} T_m \propto & \ 1/R^2(t) \propto & (1+z)^2, \\ T_\gamma \propto & \ 1/R(t) \propto & (1+z). \end{aligned}$$

This means that the early universe ($z \gg 1$) must have been extremely hot. In order to determine how hot we have to know the present temperatures. The present mean temperature of matter is difficult to determine because it has been affected by many factors other than the cosmological expansion. As the energy density of radiation $\rho_\gamma \propto T^4 \propto (1+z)^4$ and that of matter $\rho \propto (1+z)^3$, it was the energy density of radiation that determined the overall evolution of the universe in early times. Although the present CMB temperature $T_\gamma = 2.74$ K is very small, at the redshifts $z > 1{,}300$ the temperature and the energy density of radiation were high enough to ionize all matter and to keep the matter and radiation temperatures equal.

As the universe expanded, temperatures dropped and at $z \approx 1{,}300$ recombination started, neutralizing the ionized matter and releasing the radiation. Because the only change that the radiation has undergone is the change of the wavelength, it has safeguarded information about the properties of the universe before recombination (recall that matter and radiation were then in close interaction).

2.3.2 *Isotropy of the matter distribution*

Isotropy of the spatial distribution of matter is one of the ingredients of the cosmological principle used to construct the standard cosmological models. While it is usually understood that the cosmological principle implies isotropy with respect to any point in the space, we can test it, naturally, only with respect to our location.

A straightforward test for isotropy would be the study of the distribution of galaxies. Because we are in a preferred point in space (on a planet in a galaxy), we cannot expect the nearby distribution of galaxies to be isotropic. It means that in order to check for isotropy we have to use rather deep surveys and to be able to discard nearby structures.

The fact that we live in a plane of a spiral galaxy makes testing difficult, because the galactic dust, concentrated in the plane, obscures a substantial region of the sky (the "zone of avoidance"). This means that it is impossible to make an optical galaxy survey that covers the whole sky, and an isotropy test will inevitably contain assumptions about the distribution of galaxies in the zone of

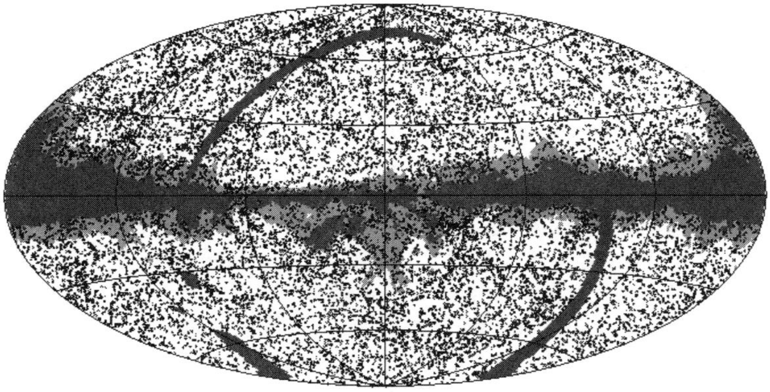

Figure 2.5 *Positions of the 15,431 PSCz survey galaxies in the galactic coordinates. The galactic longitude grows from $l = 0°$ at the right to $l = 360°$ at the left; the galactic north pole is at the top. The black masked regions either have not been observed or were excluded from the survey due to a too high infrared sky background. The gray masked regions have a high galactic extinction. (Reproduced, with permission, from Saunders et al. 2000, Mon. Not. R. Astr. Soc., 317, 55–64. Blackwell Science Ltd.)*

avoidance. Because dust is more transparent to infrared radiation, infrared surveys can achieve a better coverage of the sky.

For example, we show in Fig. 2.5 the sky distribution of 15,431 galaxies with measured redshifts from the PSCz survey (Saunders et al. 2000), based on the IRAS satellite survey of infrared objects. This is as complete sky coverage of galaxy positions as possible for the present time, and even here you can see the traces of the galactic plane and unobserved tracks (it covers 84% of the sky). Although foreground structures can be seen in places, the overall distribution of galaxies is fairly isotropic, showing no preferred directions.

Better results can be obtained using radio sources that are usually more distant and are not affected by the galactic absorption. But the best limits on possible anisotropy can be obtained from the measurements of the background radiation in different regions of spectra. Thus, as described in more detail by Peebles (1993), the measurements of the X-ray brightness of the sky limit a possible large-scale anisotropy by 10^{-3}. Even better limits come from the measurement of the cosmic microwave radiation background, where the temperature fluctuations at scales larger than $4°$ have been found to be about $\Delta T/T \approx 10^{-5}$ (Bennett et al. 1996). To be exact, there is one large-scale feature in the temperature distribution, a dipole moment, but that is due to our movement relative to the microwave background (the Doppler effect).

Thus, we can consider the assumption of the isotropy of the matter distribution to be well confirmed by observations.

2.3.3 Homogeneity of the matter distribution

Isotropy implies homogeneity, if we accept a Copernican attitude and declare that there are no preferred points in space. Homogeneity in and of itself does not demand isotropy — we can construct homogeneous cosmological models that have different velocities of expansion in different directions.

Observationally, however, homogeneity is more difficult to check. The basic method is to count galaxies for different distances from us (number counts).

For redshift catalogs we can directly count the number of galaxies $N(z)$ up to a redshift z and compare it with the expected Euclidean behavior $N(z) \propto z^3$ (for the nearby region of the universe, of course). Usually, however, catalogs contain all galaxies in a given region of sky up to a limiting apparent magnitude. This means that the larger the distance to a galaxy, the brighter (intrinsically) the galaxy has to be in order to be included in the survey, and the observed spatial density at large distances is smaller.

One solution is to use volume-limited samples: select the maximum limiting radius of the sample $D_{\rm lim}$, find the limiting absolute magnitude $M_{\rm lim}$ of a galaxy that corresponds to the apparent magnitude limit $m_{\rm lim}$ of the survey, and discard all intrinsically fainter galaxies. The cosmological magnitude-distance relation that connects the three values is similar to that given by (1.4):

$$m = M + 5\log_{10}(D_L/1\,\text{Mpc}) + 25. \tag{2.34}$$

The only difference is that we have to replace the distance by the cosmological luminosity distance D_L (2.32). Of course, the larger the volume is, the brighter and fewer galaxies we have to use.

As an example, Fig. 2.6 shows the number-redshift relation for the very deep (up to $z \approx 0.3$) ESO Slice Project redshift catalog, from Scaramella et al. (1998). Although there are 3342 galaxies in the catalog, the numbers for the volume-limited samples are much smaller, ranging from 815 to 43 (see the legends). The number counts are approximated by power laws, $N \sim R^n$, and the exponents n are shown in the figure. We see here all the problems of number counts. For smaller distances, up to $z \approx 0.1$, the local large-scale structure is distorting the picture, but at larger distances cosmological effects come into play.

In order to transform from redshifts to distances, we should know the parameters of the cosmological model, which are not well fixed at the moment. Strictly speaking, at these redshifts one should use the volume-redshift relation, and not look for a power law.

The second problem is the K-correction, described in Chapter 1. As we see in Fig. 2.6, different assumptions for that correction lead to widely different results about the homogeneity of the galaxy distribution. While the results in the left panel of the figure seem more sensible (after all, one should use cosmological distances and K-corrections), Joyce et al. (1999) have shown that the main factor at these distances is the form of the K-correction used and it is probably not

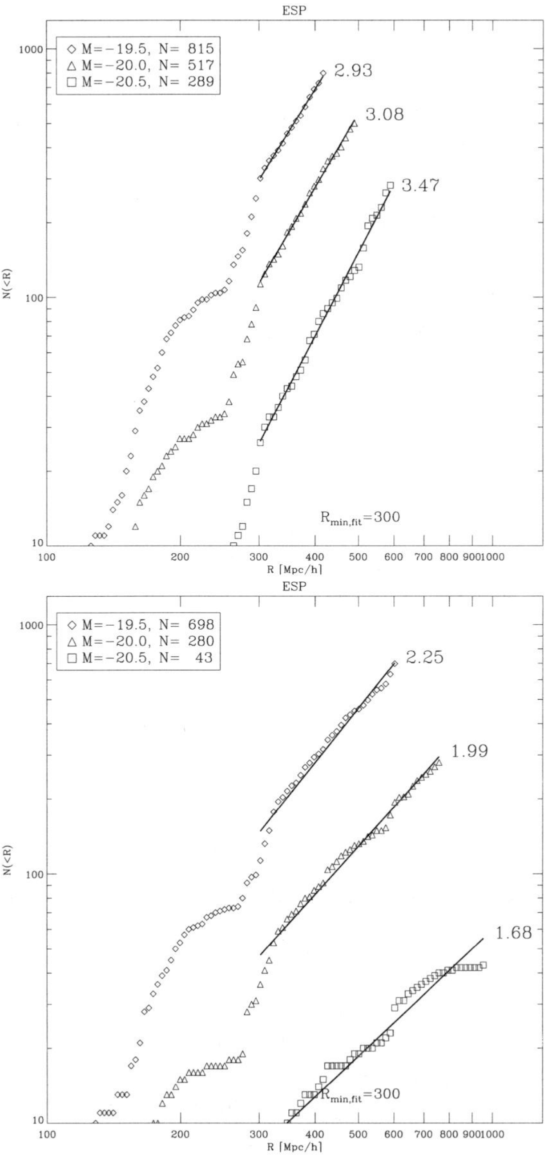

Figure 2.6 *The number-redshift relation for a deep (ESP) redshift catalog for different volume-limited subsamples, described in the legend. Cosmological ($\Omega_M = 1, \Omega_\Lambda = 0$) distances and the K-correction are used for the upper panel; Euclidean distances and no K-correction are used for the lower panel. The numbers at the straight lines show the exponents n for the number-distance relation $N \sim R^n$. (Reproduced, with permission, from Scaramella et al. 1998, Astr. Astrophys., 334, 404–408.)*

correct. Note also that the formulae of cosmological distances are based on the assumption of homogeneity that we are checking for.

It is possible to use another form of number counts, up to a limiting apparent magnitude. This method does not require difficult redshift measurements, and it was used already by Hubble in 1926.

We know that the observed flux density S from a galaxy of intrinsic luminosity L at a distance r is (in a Euclidean universe)

$$S = \frac{L}{4\pi r^2}. \tag{2.35}$$

All other galaxies of luminosity L and closer than r will look brighter than S, and their number $N(S' > S)$ is

$$N(S' > S|L) = n(L)V(r) = \frac{n(L)}{\sqrt{4\pi}}\left(\frac{L}{S}\right)^{3/2},$$

where we have expressed r by S, using (2.35). In order to get the number of all galaxies brighter than S we have to integrate the above expression over all luminosities L. If the galaxy luminosity function $n(L)$ does not depend on the distance r (and, thus, on S), it gives only a normalization constant and we find, finally

$$N(S' > S) \propto S^{-3/2}.$$

Here, of course, galaxies of different luminosity probe different volumes. If we change over from flux densities to apparent magnitudes, using the relation

$$m = -2.5\log_{10} S + \text{const}$$

from Chapter 1, we get

$$N(m' < m) \propto 10^{0.6m}.$$

Astronomers usually use differential number counts. These have to follow the same power law (with a different normalization):

$$\frac{dN(m)}{dm} \propto 10^{0.6m}.$$

In Fig. 2.7 we show the number count–apparent magnitude relation in the near-infrared region of the spectra (McCracken et al. 2000). For smaller magnitudes the classical 2/3 slope is well seen, but for the faint end (larger apparent magnitudes) the slope has to be modeled. These surveys extend to rather large redshifts ($z \approx$ 1–2) where the cosmological effects become important. In addition to the different volume-redshift dependence and the K-correction we also must consider the evolution of galaxies, which changes the galaxy luminosity function.

Thus, homogeneity of the spatial distribution of galaxies is not easy to establish observationally. It is difficult to disprove it, as well, because all the present cosmological machinery is based on the cosmological principle, and discarding a part of it would force one to rebuild the entire edifice from the start. Thus the

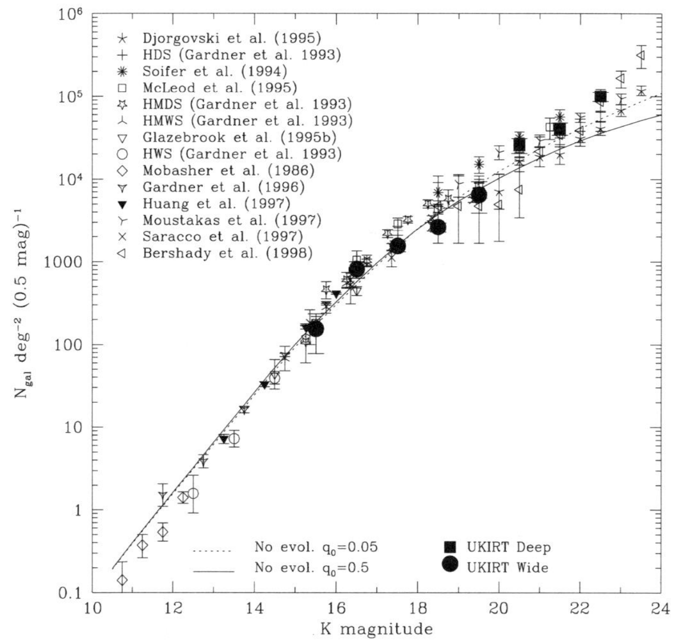

Figure 2.7 *Compilation of differential galaxy number counts in the K-filter, compared to model predictions. (Reproduced, with permission, from McCracken et al. 2000, Mon. Not. R. Astr. Soc., 311, 707–718. Blackwell Science Ltd.)*

main argument for homogeneity is indirect, adding up bit by bit from the success of all other achievements in cosmology.

2.3.4 Light element abundances

One of the best-studied areas in modern cosmology is the origin of the present abundances (number densities of atoms compared to that of hydrogen) of light elements. We have seen above that the universe was very hot in the early epochs, and there were times when it was hot enough for thermonuclear reactions to occur. Similar reactions also occur in stars, including the sun, but in the universe they occurred in difficult conditions of rapid expansion and diminishing densities. The interplay between the typical reaction times and the expansion (Hubble) time determines the final outcome of the reactions. In fact, a hot universe was first proposed by G. Gamow in 1948 to explain the cosmic abundance of elements. The CMB radiation itself (the proof that the universe had really been hot) was discovered by A.A. Penzias and R.W. Wilson only in 1965.

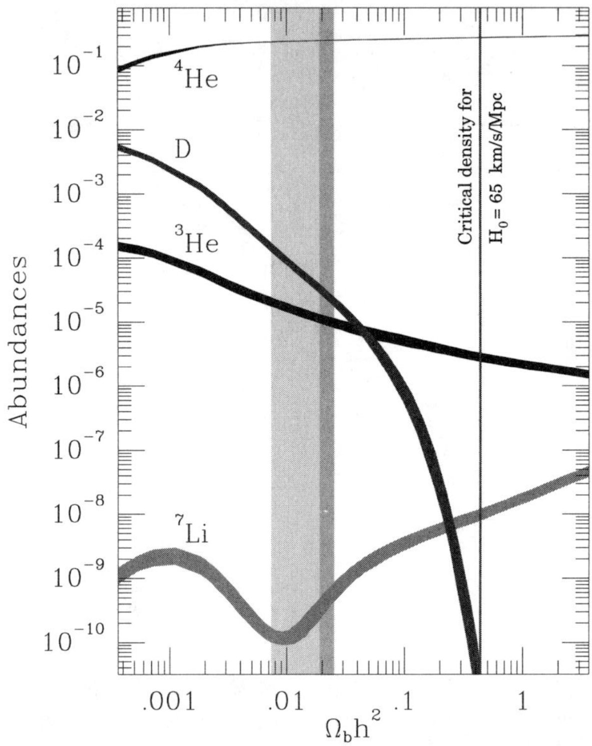

Figure 2.8 *Expected abundances of light elements for different values of the mean baryon density; the vertical shaded region shows the values of the density that are compatible with observed abundances. (Reproduced, with permission, from Schramm and Turner 1998, Rev. Mod. Phys., 70, 303–318. Copyright 1998 by the American Physical Society.)*

This is also one of the few cases in cosmology where we know the essential parameters of the cosmological model. In the early epochs the energy density of the matter and the role of the cosmological constant were both negligible compared to the energy density of radiation, and that is well known, determined by the present temperature of the CMB radiation.

The yields of light elements, except that of helium, depend strongly on the mean baryon density Ω_B in the universe and comparison with observations allows us to pinpoint it well, as shown in Fig. 2.8 (Schramm and Turner 1998). The main source of uncertainty is in estimating how primordial the observed abundances are. In general, the calculated abundances are in good agreement with observations, and thus support the overall picture of standard cosmology.

The inferred baryon density Ω_B is about twice that for luminous matter, which is natural, because, for example, there is a lot of mass in intra-cluster gas. An important point is that this parameter is much less than the estimates we have for the total $\Omega_M \approx 0.2$–1.0, showing that the bulk of the matter (dark matter) must be nonbaryonic.

2.3.5 *The cosmological parameters*

We have seen in the preceding sections how important it is to know the values of the cosmological parameters. The quest for these parameters started with the birth of the models themselves and is one of the most important problems in cosmology. Developments in observational techniques and in theory continually influence opinions about which methods are the best. We conclude this chapter with the description of the best estimates we have at present.

The first method used is the classical magnitude-redshift relation (2.34); we shall write it here once again, including explicitly the dependence on the cosmological parameters, the K-correction and the galactic extinction A_G:

$$m = M + 5\log_{10}(D_L(\Omega_M, \Omega_\Lambda, H_0, z)/1\ \text{Mpc}) + K(z) + A_G + 25, \quad (2.36)$$

where the luminosity distance $D_L(z)$ can be computed using the formulae (2.32, 2.29, 2.27 and 2.21) above. The galactic extinction accounts for the absorption of light in our Galaxy.

If we can find a class of objects with the same luminosity (the same absolute magnitude M, "standard candles"), then measuring the apparent magnitudes for these "candles" for different redshifts, we can fit the observed $m(z)$ relation to the formula (2.36) and find the parameters. This has been tried before, using high-luminosity galaxies in clusters and even quasars, but it was realized quickly that these objects evolve rapidly, and the $m(z)$ relation can tell us more about their luminosity evolution than about the cosmological model. The redshifts observed then were small, too, and the $D_L(z)$ relation does not depend much on the cosmological parameters for $z \ll 1$.

This method has been resurrected, using a new class of standard candles — distant supernovae. Supernovae are stellar explosions that do not depend on their environment, as galaxies do. They are also very bright and can be observed at high redshifts, but their main advantage is that their luminosity changes with the time-scales of weeks to months, allowing better selection of standard supernovae and better estimation of their intrinsic luminosities.

There are two groups currently observing distant supernovae: "The Supernova Cosmology Project," led by S. Perlmutter (Lawrence Berkeley Laboratory, U.S.A.) and the "High-Z Supernova Search Team," led by B. Schmidt (Mt. Stromlo and Siding Spring Observatories, Australia). Both groups are highly international and have discovered about 50 distant supernovae to date; that number is, however, less than the total membership of both groups. As you can see from the references, the number of authors of observational papers in cosmology is large, reflecting the

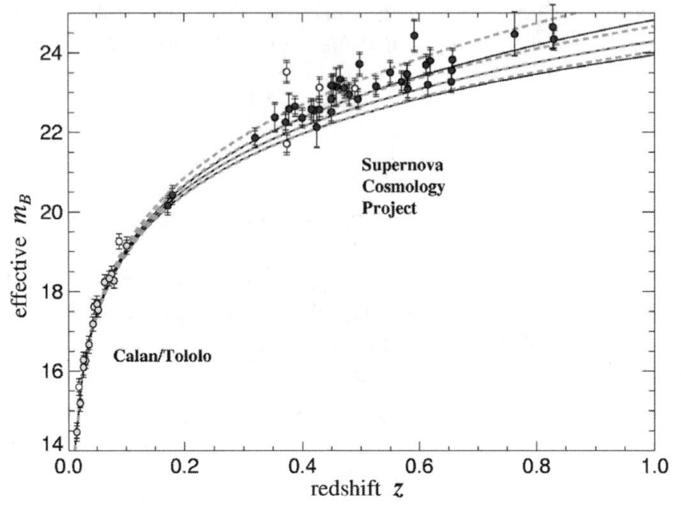

Figure 2.9 *The dependence of the apparent magnitude m_B of distant supernovae on their redshift z. (Reproduced, with permission, from Perlmutter et al. 1999, Astrophys. J., 516, 565–586. AAS.)*

fact that the observations are very difficult undertakings that require the cooperation of many observatories.

Fig. 2.9 shows the typical redshift-magnitude relation (for 42 distant supernovae) from the first group (Perlmutter et al. 1999). Different curves show the expected relation for different cosmologies, and we see that this difference is rather small even up to $z = 1$. The best parameter estimate obtained from supernovae thus far is that of the Hubble constant $H_0 = 65 \pm 2\,\mathrm{km\,sec^{-1}\,Mpc^{-1}}$. It is not easy to get good individual estimates of the remaining parameters from the magnitude-redshift relation, because both Ω_M and Ω_Λ influence the luminosity distance in similar ways and thus the estimates are highly correlated.

Another observationally very active branch of cosmology at the moment is measurement of the fluctuation spectra of the CMB radiation. Exact modeling of the recombination process predicts that for certain characteristic angular distances these fluctuations have an enhanced amplitude, and these distances and the amplitudes (oscillations) depend strongly on the cosmological model. Fig. 2.10 shows a compilation of observational data with a best-fit model (Hu 2001). Most of the data points in this figure correspond to a whole experiment.

This fit also allows us to estimate the cosmological parameters, and the estimates it gives are also well correlated. A good point is that the correlations are different from those arising from the $m(z)$ test, and combining the two tests can give us for the first time a rather well-defined estimate of the parameters. There have

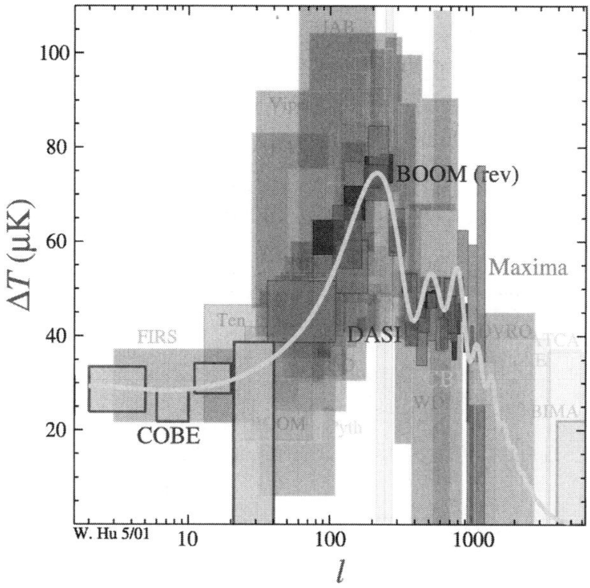

Figure 2.10 *Compilation of the CMB temperature fluctuation amplitudes* $\Delta T/T$ *and the effective multipole values* l *for recent experiments (courtesy of W. Hu). The gray boxes show the bandwidth–rms error boxes for different experiments; the light line shows the best model fit.*

been several such determinations recently with similar results. Fig. 2.11 shows the confidence regions for the parameters from the Boomerang measurements of the CMB anisotropy and supernovae data fits.

As we see, the different methods limit the possible region of parameters effectively, and this region also includes the line of the flat $k = 0$ models that are preferred by theories of the extremely early universe (inflationary models). We also see that the observations definitely require a cosmological constant. Up to now many cosmologists have tried to ignore this need because there is not much physical reason to include the cosmological term in the Einstein equations.

Another problem is the explanation of the numerical values of the parameters. Because these values have an extremely different behavior in time, the fact that they have comparable values at the present moment is rather mysterious. One possibility is to introduce a new physical field that mimics the role of the cosmological constant in the Friedmann equations, but changes slowly in time and can thus more easily have $\Omega_\Lambda \approx \Omega_M$. To the four main constituents of the universe (radiation, baryonic matter, and hot and cold dark matter), we add the new, yet completely hypothetical fifth field that has been termed "quintessence."

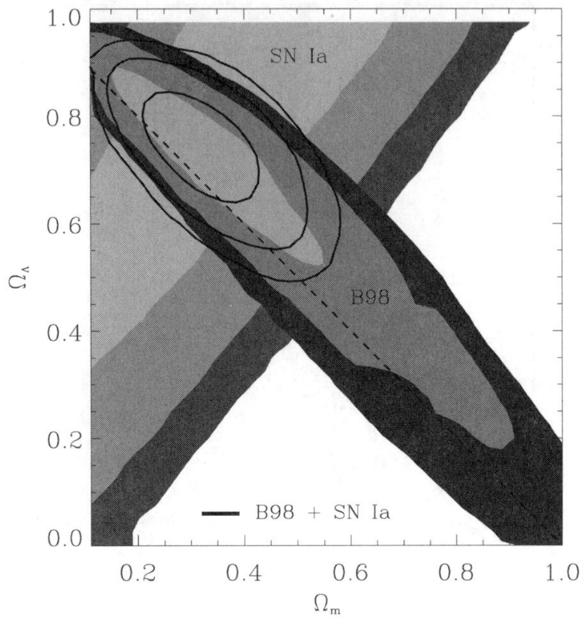

Figure 2.11 *Estimates of the cosmological parameters Ω_M and Ω_Λ, based on the Boomerang measurement of the CMB anisotropy, denoted by B98, and on the $m(z)$ relation for distant supernovae, denoted by SN 1a, The shaded regions correspond to the 68.3, 95.4, and 99.7% confidence levels. The line contours show the combined likelihood for the same confidence levels. The dotted line represents the flat $k = 0$ model. (Reproduced, with permission, from de Bernardis et al., Astrophys. J., in press. AAS.)*

Although the estimates of the parameters described above are among the best at the moment, they should be used with caution. Cosmological observations are difficult. Interpretation relies on a rather advanced theoretical machinery, and the more complex a machine is, the more easily it can break down. We should bear in mind a well-known fact from the 70-year history of contemporary cosmology — the best estimate of Hubble's constant has changed during these years from 480 km sec^{-1} Mpc^{-1}, as found by E. Hubble in 1929, to 65 km sec^{-1} Mpc^{-1}, as determined from the recent observations of distant supernovae. This means that our mental picture of the universe, our model universe, is almost ten times larger and older now than it was 70 years ago.

CHAPTER 3

Cosmological point processes

3.1 Introduction

In Section 1.2.1 we saw how redshift surveys provide a set of positions $\{\mathbf{x}_i\}_{i=1}^N$ of N galaxies in a portion of the universe with volume V. This can be regarded as a realization of a random point process. In this sense, the discrete distribution of galaxies, each one considered as a point of the process, is a good field of application of the statistical techniques developed in spatial statistics. However, in cosmology, a different approach is more often employed. The spatial distribution of matter in the universe can be considered, both now and in the past, as a continuous function of spatial locations, that is, a continuous random field. Both random fields and point processes are indeed examples of stochastic processes. This chapter deals with the discrete version, with point processes, while Chapter 7 discusses cosmological random fields.

We will introduce in depth most of the conventional methods to describe the discrete galaxy distribution. First, we will focus on discussing the relation between quantities defined for the situation of continuous density fields in the theory of random fields. In particular, we will see how the density field (intensity function) can be estimated from the galaxy distribution; other reconstruction methods will be presented in Chapters 5 and 9.

We then discuss the important concept of the two-point correlation function. How do we get the best estimates of the spatial correlation function from a set of data that has been selected in some specified way? This section is followed by a discussion of the higher-order correlation functions and their importance in estimating errors of the two-point correlation function. While the measurement of higher-order correlation functions becomes very cumbersome, a very important and related alternative is to study the moments and probability distribution functions of galaxy counts in cells. Both developed techniques and results of observations and simulations are discussed. Another interesting related statistic is the probability that a sphere of a certain radius is empty, the so-called void probability function. We conclude with a discussion of possible new methods of galaxy point statistics as suggested by new techniques developed in the context of spatial statistics (Babu and Feigelson 1996b). In particular, distance estimators are analyzed in some detail.

3.2 Point processes

In this section we introduce the basics of the theory of random point processes. For more formal and detailed descriptions of this theory we refer the reader to the books by Cox and Isham (1980), Ripley (1981, 1988), Diggle (1983), Daley and Vere-Jones (1988), Cressie (1991), Stoyan and Stoyan (1994), Stoyan, Kendall, and Mecke (1995), and the review articles by Babu and Feigelson (1996a) and Stoyan (2000).

A point process in $\mathbb{R}^d$ is mathematically defined as a random variable taking values in a measurable space formed by the family of all sequences Φ of points of $\mathbb{R}^d$ verifying two conditions:

1. *Local finiteness* — Any bounded set of $\mathbb{R}^d$ contains a finite number of points of Φ.
2. *Simplicity* — There are no multiple points.

The term "point process" and the symbol Φ are used with a double meaning; in fact, they are two different aspects of the same well-defined mathematical concept:

(i) The point process is considered a random sequence of discrete points, $\Phi = \{x_1, x_2, \ldots\}$, so the notation $x \in \Phi$ emphasizes that point x belongs to the point process.

(ii) The point process is a random measure counting the number of points lying in a given region $A \subset \mathbb{R}^d$, $\Phi(A) = n$.

The number of points lying in A can be expressed as

$$\Phi(A) = \sum_{x \in \Phi} \mathbb{1}_A(x),$$

where $\mathbb{1}_A(x)$ is the indicator function of the set A,

$$\mathbb{1}_A(x) = \begin{cases} 1 & \text{for } x \in A, \\ 0 & \text{for } x \notin A. \end{cases} \tag{3.1}$$

We shall use the terms point processes and point fields interchangeably. The term "process" was originally introduced because it was associated with time series. In our context, however, it would be more appropriate to use the term "field" as Stoyan and Stoyan (1994) have proposed, but we shall also retain the classical term "process." We shall typically deal with point fields in $\mathbb{R}^3$, and the bounded region where a given realization of the point field lies will be called the *window* and will be denoted by W. The volume of a region $A \subset \mathbb{R}^3$ will be denoted by $V(A)$ and the probability that k points of the process lie in A by $P(\Phi(A) = k)$.

Although the expressions in the following sections are given for point fields in the Euclidean space $\mathbb{R}^3$, it is important to keep in mind what was stated in Chapter 2 regarding the geometry of the space depending on the adopted cosmological model. Distances and volumes have to be calculated according to such a model.

3.2.1 *Intensity functions*

The statistical description of point processes is usually made by means of the intensity functions. Let us denote by dV the volume of an infinitesimal region that contains the point $\mathbf{x}$. The first-order intensity function or simply the intensity function of the process is defined by

$$\lambda(\mathbf{x}) = \lim_{dV \to 0} \frac{\langle \Phi(dV) \rangle}{dV}, \tag{3.2}$$

where the bracket $\langle ... \rangle$ denotes expected value of the random variable. For simplicity, we represent the infinitesimal region and its volume with the same notation, dV.

A point process is called stationary or homogeneous if its statistical properties are invariant under translations; if the invariance holds under rotations the process is called isotropic, and in that case there is no privileged direction.

For a homogeneous process the intensity function takes a constant value $\lambda(\mathbf{x}) = \lambda$, which is the mean number of points per unit volume. This value is called number density in cosmology, and it is often denoted by $\bar{n}$.

We can similarly define the second-order intensity function

$$\lambda_2(\mathbf{x}_1, \mathbf{x}_2) = \lim_{dV_1, dV_2 \to 0} \frac{\langle \Phi(dV_1) \Phi(dV_2) \rangle}{dV_1 dV_2}. \tag{3.3}$$

Again, if the process is isotropic and homogeneous this quantity depends only on the distance $|\mathbf{x} - \mathbf{y}| = r$

$$\lambda_2(r) = \lambda_2(\mathbf{x}, \mathbf{y}).$$

Other related second-order characteristics of the point fields will be introduced in Section 3.4.

3.2.2 *The binomial random field*

This is the kind of point distribution we get if we place N random points uniformly distributed within the window W, a compact set in $\mathbb{R}^3$. It is obviously a homogeneous and isotropic point field and therefore its intensity is just

$$\lambda = \frac{N}{V(W)}.$$

The expected number of points in a given region $A \subset W$ is then

$$\langle \Phi(A) \rangle = \lambda V(A).$$

The name of the process is quite natural if we consider that the form of the probability function of finding exactly k points in A is just

$$P(\Phi(A) = k) = \binom{N}{k} p_A^k (1 - p_A)^{N-k}, \qquad k = 0, 1, 2, \ldots, N,$$

where

$$p_A = \frac{V(A)}{V(W)}.$$

This kind of process is very often used in cosmology; in particular, auxiliary random catalogs, which are binomial fields, are used in the estimation of the two-point correlation function (see Section 3.4.1). A binomial random field is typically referred to as a Poisson catalog, random distribution, or Monte Carlo sample with N points. There is, however, a subtle difference between a homogeneous Poisson field and a binomial field. Strictly speaking, as we shall see in the next section, in a homogeneous Poisson field the number of points is also random.

3.2.3 Poisson processes

A fundamental property of a Poisson field is that if we consider a finite number k of disjoint regions A_i, the random variables $\{\Phi(A_i)\}_{i=1}^k$, providing the number of points of the process lying in A_i, are stochastically independent (Stoyan and Stoyan 1994). We can distinguish two kinds of Poisson processes, called, respectively, homogeneous or stationary and inhomogeneous or general.

Homogeneous Poisson processes

In this kind of process $\lambda(\mathbf{x})$ does not depend on position and therefore has a constant value $\lambda(\mathbf{x}) = \lambda$.

The name of the process is quite natural if we take into account that the number of events of the point process lying in a region A of space with volume $V(A)$ is a random variable, $\Phi(A)$, having a Poisson distribution with parameter $\lambda V(A)$,

$$P(\Phi(A) = k) = \frac{[\lambda V(A)]^k}{k!} e^{-\lambda V(A)}, \qquad k = 0, 1, \ldots$$

Khintchine (1955) refers to this kind of process in one dimension as "a simple stream of uniform events." Because the number of points of a homogeneous Poisson field is also random, the simulation of this process in a window W involves two steps (Stoyan and Stoyan 1994). First, we have to generate a random number N as a sample of a random variable following a Poisson distribution with parameter $\lambda V(W)$, and then we have to generate a binomial process with N points within W.

Inhomogeneous Poisson processes

For these general Poisson processes, the intensity function varies from one position to another, and then the value of $\lambda(\mathbf{x})$ can change when we vary the location $\mathbf{x}$, showing fluctuations in density. The probability distribution of the number of points lying in a bounded region A will be given now by

$$P(\Phi(A) = k) = \frac{[\Lambda(A)]^k}{k!} e^{-\Lambda(A)}, \qquad k = 0, 1, \ldots,$$

where $\Lambda(A)$ is the intensity measure defined as

$$\Lambda(A) = \int_A \lambda(x) dx.$$

It is clear from these formulae that, in general, the stationarity property is lost. For example, in Fig. 3.1 we show a simulation of an inhomogeneous Poisson process for which the intensity function follows a Plummer's model, a polytrope of index 5:

$$\lambda(\mathbf{x}) \propto \left(1 + \frac{x^2}{R^2}\right)^{-\frac{5}{2}},$$

where $x = \sqrt{x_1^2 + x_2^2 + x_3^2}$. The density decreases radially from the center of the cube following the curve showed in the bottom panel of Fig. 3.1.

In Chapter 5 we will study several models of point fields that have been used in different contexts of the description of the large-scale distribution of matter in the universe.

3.3 The relation between discrete and continuous distributions

As we have stated in the introduction to this chapter, the distribution of mass in the universe can be regarded as a continuous density field. This will be formalized in Chapter 7. The luminosity, however, is concentrated into galaxies, forming a discrete point process. In this section we discuss the connection between these two approaches, in particular how the density field can be estimated from the galaxy distribution.

The three-dimensional distribution of galaxies can be regarded as a point field, where $\{\mathbf{x}_i\}_{i=1}^N$ represents the coordinates of the positions. One way, although not the only one, to establish a connection between the discrete galaxy distribution and the underlying continuous density field is by means of an inhomogeneous Poisson point process with density $n(\mathbf{x})$ described formally as a sum of Dirac delta functions $\delta_D(\mathbf{x})$:

$$n(\mathbf{x}) = \sum_{i=1}^{N} m_i \delta_D(\mathbf{x} - \mathbf{x_i}) \tag{3.4}$$

where m_i is the mass of object at position $\mathbf{x_i}$.

If we consider that all masses are equal and normalize them to 1, ($m_i = 1$), the intensity of the point process $\lambda(\mathbf{x}) = \langle n(\mathbf{x}) \rangle$ (angle brackets denote expectation values over the ensemble of the point processes) is defined by a realization of the

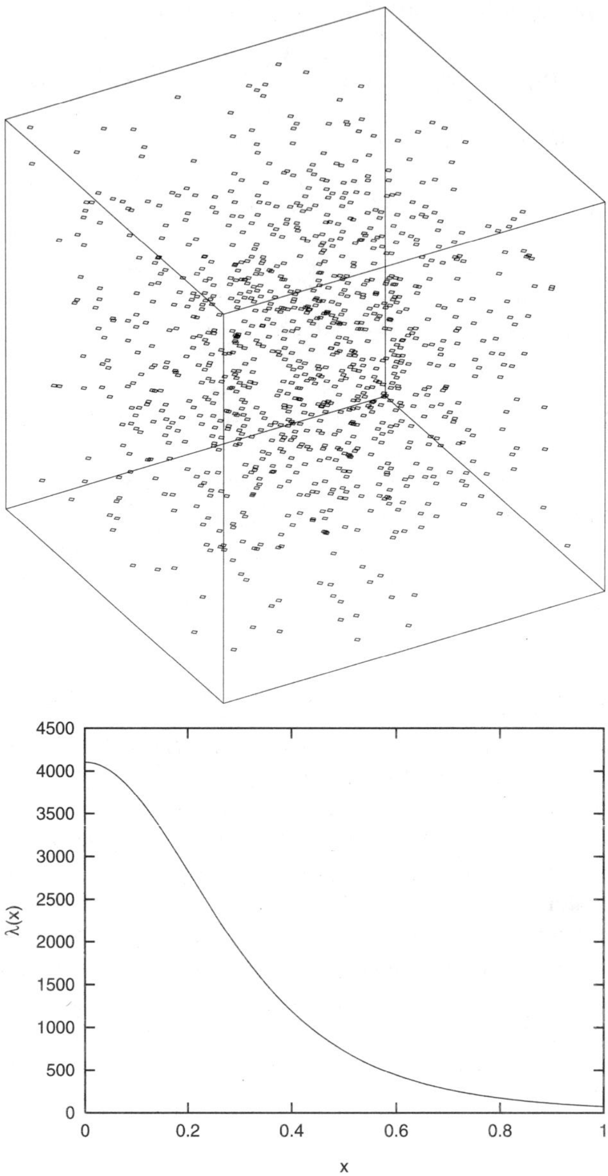

Figure 3.1 *Top: A 3-D simulation of an inhomogeneous Poisson process having a polytrope of index 5 as intensity function with* $R = 0.5$*; the center of the box is the origin of coordinates and the sidelength of the cube is 1. There are 1000 points in the simulation. Bottom: The intensity function of that process.*

random density field:

$$\lambda(\mathbf{x}) = \bar{n}\,[1 + \delta(\mathbf{x})]\,.$$

Here $\bar{n} = E\{\lambda(\mathbf{x})\}$ is the expectation value of the intensity over all realizations of the density field. We have assumed that density is a homogeneous random field; then by ergodicity $\bar{n}$ can be estimated by the mean number density over the whole sample volume. The random field $\delta(\mathbf{x})$ is the density contrast (a zero-mean dimensionless field).

Now, if we place randomly a volume dV at position $\mathbf{x}$, the probability that it contains a point is

$$\delta P = \lambda(\mathbf{x})dV.$$

This provides the basis of what is called in cosmology the Poisson model, introduced by Layzer (1956) (see also Peebles 1980, § 33 and Fry 1985 for a more detailed presentation). Using the statistical terminology, this is a double-stochastic point process,* because there are two different levels of randomness: the random field itself and the Poisson sampling.

3.3.1 Estimators of the density field: intensity functions

If galaxies are considered Poisson tracers of a continuous density field, we could be interested in estimating the underlying density field from the point field. The usual way of doing this is to convolve the spatial galaxy distribution with a window function in order to get a smooth field. Of course, under this hypothesis we are considering that galaxies trace mass, but this may not necessarily be true, and different biasing schemes can be adopted, once the density field has been estimated by considering that the galaxy formation probability is not just proportional to the local density, but instead follows more complicated dependence of the density environment (Coles 1993; Dekel and Lahav 1999). Biasing schemes are introduced in Section 7.9.

When cosmologists estimate the density field from the discrete galaxy distribution, they are making a kernel estimation of the intensity function (Saunders et al. 1991).

Let $\{\mathbf{x}_1, \mathbf{x}_2, \ldots, \mathbf{x}_N\}$ be the positions of N galaxies in a bounded region W. The estimator of the intensity function with kernel κ_ω (also called filter function) and smoothing radius (bandwidth) $\omega > 0$ is

$$\widehat{\lambda}_\omega(\mathbf{x}) = \sum_{i=1}^{N} \kappa_\omega(\mathbf{x} - \mathbf{x}_i), \qquad \mathbf{x} \in W, \tag{3.5}$$

where κ_ω is a symmetric density probability function. In the field of point processes the selection of the kernel function is not considered as relevant as the

* This process is also referred to as a Cox process. A more complete treatment is given in Chapter 5.

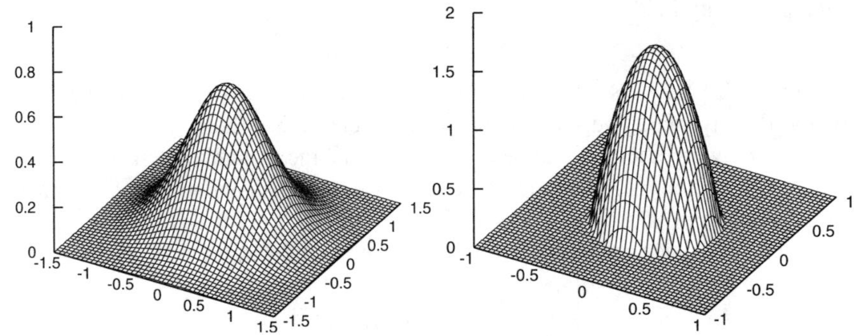

Figure 3.2 *The Gaussian kernel (left) and the Epanechnikov kernel (right) for the same value of the bandwidth,* $\omega = 0.5$.

choice of the value for the smoothing radius ω. The reconstructed density field strongly depends on the adopted value for ω.

The standard kernel function used in cosmology is the Gaussian filter

$$\kappa_\omega(\mathbf{y}) = \frac{1}{(2\pi)^{3/2}\omega^3} \exp\left(-\frac{|\mathbf{y}|^2}{2\omega^2}\right).$$

In the field of spatial statistics it is rather common to use the Epanechnikov kernel

$$k_\omega(\mathbf{x}) = \begin{cases} \frac{3}{4\omega}\left(1 - \frac{\mathbf{x}^2}{\omega^2}\right) & \text{for } |\mathbf{x}| \leq \omega \\ 0 & \text{otherwise} \end{cases}. \tag{3.6}$$

In Fig. 3.2 we show a two-dimensional representation of these two kernel functions, and in Fig. 3.3 we show a planar cluster point process and the kernel smoothed density field reconstructed by means of a Gaussian filter with different bandwidths.

An optimal selection of ω when one wants to estimate the intensity function from a point process is a well-studied problem in the mainstream of spatial statistics (Diggle 1983; Diggle 1985; Silverman 1986; Wand and Jones 1995) and it is a clear example of how the collaboration between astronomers and statisticians can be useful. For example, using adaptive intensity estimators would help in regions where the density varies quite abruptly (Stein 1997). A simple generalization of the Gaussian kernel with this goal is

$$\kappa_A(\mathbf{y}) = \frac{1}{(2\pi)^{3/2}\det(\mathbf{A})^{1/2}} \exp\left(-\frac{1}{2}\mathbf{y}^T\mathbf{A}\mathbf{y}\right).$$

where $\mathbf{A}$ is a matrix dependent on the position $\mathbf{y}$.

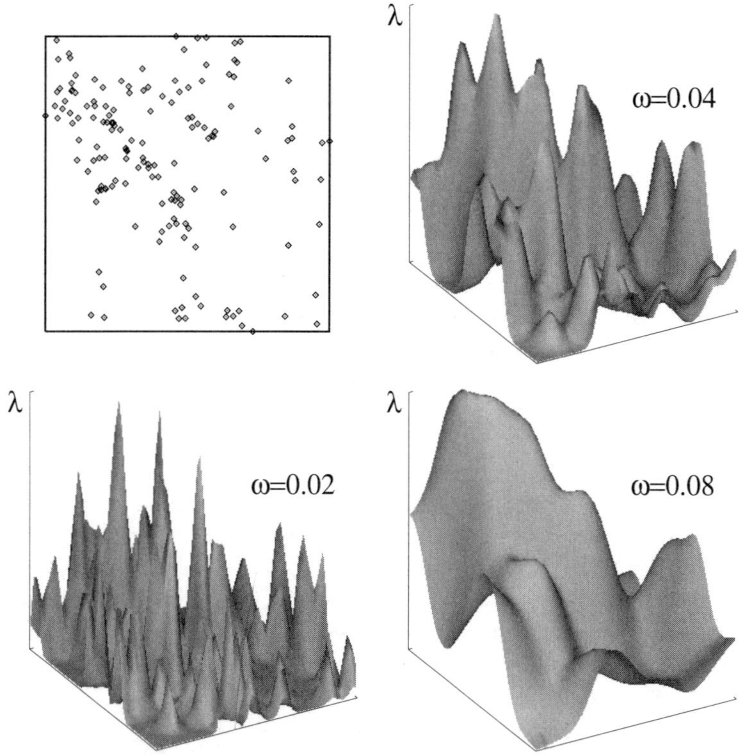

Figure 3.3 *A clustered planar point field and its reconstructed density field by means of a Gaussian filter with three values of the smoothing radius.*

3.4 The two-point correlation function

In Section 3.2.1 we defined the *second-order intensity function* for a point process, $\lambda_2(\mathbf{x}_1, \mathbf{x}_2)$. The infinitesimal interpretation of this quantity is the common one used in cosmology (Peebles 1973, 1980). This interpretation reads as follows: Let us consider two infinitesimally small spheres centered in $\mathbf{x}_1$ and $\mathbf{x}_2$ with volumes dV_1 and dV_2. The joint probability that in each of the spheres lies a point of the point process is

$$dP_{12} = \lambda_2(\mathbf{x}_1, \mathbf{x}_2) dV_1 dV_2. \tag{3.7}$$

In spatial statistics, this interpretation usually follows a more general treatment based on moments of the point processes that are measures [see Stoyan, Kendall, and Mecke (1995) for this approach]. $\lambda_2(\mathbf{x}_1, \mathbf{x}_2)$ is known as the second-order product density. This is also known as the Lebesgue density of the second factorial moment measure. In cosmology the correlation function $\xi(\mathbf{r})$ of a homogeneous point process is defined by

$$dP_{12} = \bar{n}^2 \left[1 + \xi(\mathbf{r})\right] dV_1 dV_2, \tag{3.8}$$

where $\mathbf{r}$ is the separation vector between the points $\mathbf{x}_1$ and $\mathbf{x}_2$.

For a Poisson process the second-order intensity function can be written as

$$\lambda_2(\mathbf{x}_1, \mathbf{x}_2) = \lambda(\mathbf{x}_1)\lambda(\mathbf{x}_2) + \delta_D(\mathbf{x}_1 - \mathbf{x}_2)\lambda(\mathbf{x}_1). \tag{3.9}$$

The first term comes from the statistical independence of the process in separate volumes and the second term comes from self-pairs. It can be derived, dividing the volumes dV_1, dV_2 into very small cells, so that the occupation numbers n_i of these cells are either 0 or 1 (Peebles 1980, § 36). Then the average

$$\langle n^2(\mathbf{x})\rangle\, dV = \langle \sum_i n_i^2 \rangle\, dV = \lambda(\mathbf{x}) dV.$$

Usually the correlation functions of point process are calculated by explicitly avoiding self-pairs; then the second term in (3.9) has to be omitted. When estimating Fourier power spectra (see Chapter 8), self-pairs are implicitly included and the full form (3.9) has to be used.

The galaxy distribution is very often modeled with Cox processes, inhomogeneous processes where the intensity measure is a random function (see Chapter 5). Taking the expectation value of $\lambda_2(\mathbf{x}_1, \mathbf{x}_2)$ over the full ensemble of realizations of the random field $\delta(\mathbf{x})$, we get

$$E\left\{\lambda_2(\mathbf{x}_1, \mathbf{x}_2)\right\} = \bar{n}^2 \left[1 + E\left\{\delta(\mathbf{x}_1)\delta(\mathbf{x}_2)\right\}\right] + \delta_D(\mathbf{x}_1 - \mathbf{x}_2)\bar{n}. \tag{3.10}$$

Omitting the last term, as we have explained above, and comparing (3.10) with the formula for the correlation function (3.8), we get

$$\xi(\mathbf{r}) = R_\delta(\mathbf{r}),$$

where $R_\delta(\mathbf{r})$ is the covariance function of the density contrast. Thus $\xi(\mathbf{r})$ can be used to estimate the covariance $R_\delta(\mathbf{r})$ and the covariance can be used to model $\xi(\mathbf{r})$.

If the point field is homogeneous, the second-order intensity function, $\lambda_2(\mathbf{x}_1, \mathbf{x}_2)$, depends only on the distance $r = |\mathbf{x}_1 - \mathbf{x}_2|$ and the direction of the line passing through $\mathbf{x}_1$ and $\mathbf{x}_2$. If, in addition, the process is isotropic, the direction is not relevant and the function depends only on r, $\lambda_2(r)$. The reduced two-point cumulant correlation function commonly used in cosmology (Eq. 3.8), $\xi(r)$, is related in this case to $\lambda_2(r)$ by

$$\xi(r) = \frac{\lambda_2(r)}{\bar{n}^2} - 1, \tag{3.11}$$

although in other fields of physics and statistics, the pair correlation function, also called the radial distribution function or the structure function, is just $g(r) = \xi(r) + 1$.

3.4.1 Measuring the two-point galaxy correlation function

For a statistical estimation of $\xi(r)$, N points are given inside a window W of observation, which is a three-dimensional body of volume $V(W)$. All the estimators are based on some kind of average of the counts of neighbors of galaxies at a given scale, or more precisely, within a narrow interval of scales. The problem that arises immediately is the fact that for galaxies close to the boundary the number of neighbors is obviously underestimated. One way to overcome this problem is to consider as centers for counting neighbors only galaxies lying within an inner window $W_{\rm in}$; then we can average

$$\widehat{\xi}_{\rm min}(r) = \frac{V(W)}{NN_{\rm in}} \sum_{i=1}^{N_{\rm in}} \frac{n_i(r)}{V_{\rm sh}} - 1, \tag{3.12}$$

where the quantity $\widehat{\xi}_{\rm min}(r)$ is the minus-estimator of the correlation function and $V_{\rm sh}$ is the volume of the shell of width dr,

$$V_{\rm sh} = \frac{4\pi}{3}[(r+dr)^3 - r^3],$$

which can be approximated by $4\pi r^2 dr$ if dr is small, whereas $n_i(r)$ is the number of points of the process within the whole window W lying at a distance between r and $r + dr$ from point i, which lies in $W_{\rm in}$. As we can see in Fig. 3.4, the scale $r + dr$ appearing in (3.12) has to be less than or equal to $r_{\rm max}$, the distance between the boundary of the inner window $W_{\rm in}$ and the edge of W. One can then choose between two possibilities: to fix $r_{\rm max}$ as the maximum scale at which we want to compute $\xi(r)$, or to shrink the window $W_{\rm in}$ as r increases, in such a way that $r_{\rm max}$ varies, being for a given scale $r_{\rm max} = r + dr$. The advantage of this estimator is that it can be applied safely to nonhomogeneous point fields, because no assumptions are made about the distribution of points outside W. One drawback is that this estimator does eliminate some of the information contained in the data. Moreover, at large distances only a small fraction of the galaxies are considered as centers, increasing the variance of the estimator (Kerscher, Szapudi, and Szalay 2000). If one wants to make full use of the data contained in the catalog, an edge correction has to be applied.

Several edge-corrected estimators of ξ are commonly used in cosmology. The simplest edge correction is incorporated in the so-called natural estimator, first used for the study of the angular correlation function by Peebles and Hauser (1974). An auxiliary random sample containing $N_{\rm rd}$ points must be generated in W (i.e., a binomial point process with $N_{\rm rd}$ points), and the estimator for the correlation function is defined as

$$\widehat{\xi}_{\rm PH} = \left(\frac{N_{\rm rd}}{N}\right)^2 \frac{DD(r)}{RR(r)} - 1, \tag{3.13}$$

where $DD(r)$ is the number of pairs in the catalog (within the window W) inside

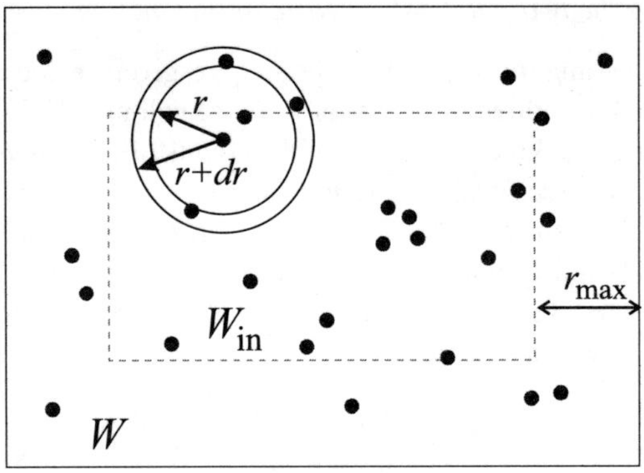

Figure 3.4 *The inner window used in the minus-estimator of the correlation function. Only points within $W_{\rm in}$ are considered as centers for counting neighbors.*

the interval $[r, r + dr]$, and $RR(r)$ is the number of pairs in the random catalog with separation in the interval mentioned above. The use of random samples in this and other estimators replaces the calculation of partial shell volumes; it is, in fact, the Monte Carlo integration for volumes. This also means that the number of random points must be much larger than the size of the data sample; ten times is usually considered sufficient, but more could be needed for the DR pairs (see below). The normalization factor in Eq. 3.13 $(N_{\rm rd}/N)^2$ can be replaced by $(N_{\rm rd}(N_{\rm rd} - 1))/(N(N - 1))$. In this case, the estimator is unbiased for a Poisson process. This choice is made by some authors and applies to all other estimators below that include this normalization constant.

However, the natural estimator has been shown to suffer from insufficient correction for edge effects. A better and extensively used estimator is that of Davis and Peebles (1983), sometimes called the standard estimator:

$$\widehat{\xi}_{\rm DP}(r) = \frac{N_{\rm rd}}{N}\frac{DD(r)}{DR(r)} - 1, \tag{3.14}$$

where $DR(r)$ is the number of pairs between the data and the binomial random sample with separation in the same interval.

Hamilton (1993b) calculated the bias of the above two estimators and found that the expectation value for the standard estimator (3.14) can be written as

$$E\{\widehat{\xi}_{\rm DP}(r)\} = \frac{\widehat{\xi}(r) + \psi(r) - \bar{\delta} - \psi(r)\bar{\delta}}{[1 + \psi(r)](1 + \bar{\delta})},$$

where $\widehat{\xi}(r)$ is the unbiased estimator of the correlation function, the mean overdensity $\bar{\delta}$ measures the error we make when we use for the mean density, $\bar{n}$, the estimate $n_{\rm est}$ found from our sample, $n_{\rm est} = \bar{n}(1 + \bar{\delta})$, and $\psi(r)$ is the galaxy–

random sample correlation function $\psi(r) = DR(r)/RR(r) - 1$. The quantity $\psi(r)$ can be interpreted as the large-scale variance we are missing in our finite sample. As our samples are limited in volume, the errors $\bar{\delta}$ and $\psi(r)$ are both usually nonzero.

The terms that are linear in $\bar{\delta}$ and $\psi(r)$ are of the same order and can be larger than the uncertainty in $\widehat{\xi}(r)$, especially for small correlation amplitudes (at large scales), introducing a substantial bias in the estimator. Hamilton (1993b) also calculated the bias for the natural estimator (3.13) and found that it is about twice that of the standard estimator.

This led Hamilton (1993b) to propose the estimator

$$\widehat{\xi}_{\rm HAM}(r) = \frac{DD(r) \cdot RR(r)}{[DR(r)]^2} - 1, \tag{3.15}$$

which has only a second-order bias, caused by the finite sample effects:

$$E\{\widehat{\xi}_{\rm HAM}(r)\} = \frac{\widehat{\xi}(r) - \psi^2(r)}{[1+\psi(r)]^2}.$$

The $DR(r)$ term in the above estimator (3.15) may introduce numerical noise at small distances. To eliminate that we must either choose larger random samples (Pons-Bordería et al. 1999) or integrate the angular part of the pair integral analytically, as Hamilton (1993b) does.

Another estimator, proposed almost simultaneously by Landy and Szalay (1993), has similar properties:

$$\widehat{\xi}_{\rm LS}(r) = 1 + \left(\frac{N_{\rm rd}}{N}\right)^2 \frac{DD(r)}{RR(r)} - 2\frac{N_{\rm rd}}{N}\frac{DR(r)}{RR(r)}. \tag{3.16}$$

The relation between the natural and Landy–Szalay estimators can be easily deduced from their definitions given in Eqs. 3.13 and 3.16:

$$\widehat{\xi}_{\rm LS} = \widehat{\xi}_{\rm PH} + 2 - 2\frac{N_{\rm rd}}{N}\frac{DR(r)}{RR(r)}. \tag{3.17}$$

The Landy–Szalay estimator has also a second-order bias, if the mean density is defined from the sample data, as in (3.16) above:

$$E\{\widehat{\xi}_{\rm LS}(r)\} = \frac{\widehat{\xi}(r) - 2\bar{\delta}\psi(r) + \bar{\delta}^2}{[(1+\bar{\delta}]^2}. \tag{3.18}$$

The authors of this estimator, however, propose to estimate the mean density by independent means (e.g., by a maximum likelihood method). Then we must substitute the mean overdensity in (3.18) with the fractional error in the mean density ϵ, which could be much less than $\bar{\delta}$, further reducing the bias.

Recently, Szapudi and Szalay (1998) generalized the LS-estimator for higher-order correlation functions. They argue that it is the most natural estimator; if we

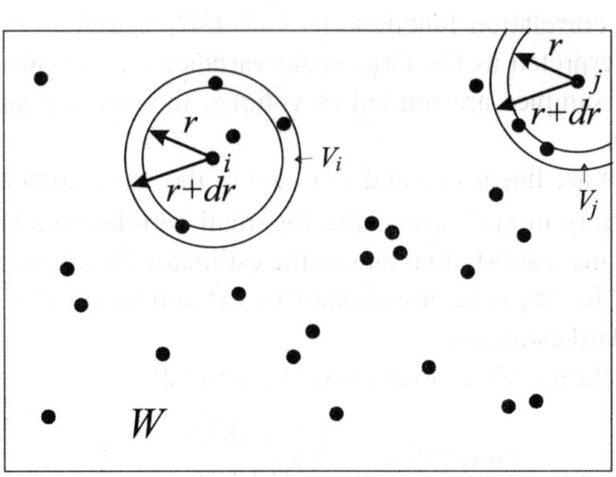

Figure 3.5 *In this illustration we see the volumes entering the definition of the Rivolo estimator for $\xi(r)$. If the shell touches the boundary of the window, only its intersection with W is considered.*

write the natural estimator (3.13) symbolically as

$$\widehat{\xi}_{\mathrm{PH}}(r) = \frac{\widetilde{DD}}{\widetilde{RR}} - 1 = \langle(1+\delta_1)(1+\delta_2)\rangle - 1 = \langle\delta_1\delta_2 + \delta_1 + \delta_2\rangle, \qquad (3.19)$$

where $\widetilde{DD}$ and $\widetilde{RR}$ are normalized pair counts, δ is the overdensity, and the angle brackets $\langle\cdot\rangle$ denote sample averages, we see that it is unbiased (for infinite samples), but the linear terms in (3.19) will generate additional terms in the sample variance of ξ. The symbolic presentation for the LS-estimator is

$$\widehat{\xi}_{\mathrm{LS}}(r) = \frac{(\widetilde{D}_1 - \widetilde{R})(\widetilde{D}_2 - \widetilde{R})}{\widetilde{RR}} = \langle\delta_1\delta_2\rangle,$$

which has only the necessary term for the correlation function, leading to an efficient (minimum-variance) estimator. This can be proved exactly; we shall describe that below.

If we have simple sample volumes, we can find the partial shell volumes without using the Monte Carlo trick of comparison with random samples. The simplest estimator of that kind is that of Rivolo (1986):

$$\widehat{\xi}_{\mathrm{RIV}}(r) = \frac{V(W)}{N^2} \sum_{i=1}^{N} \frac{n_i(r)}{V_i(r)} - 1, \qquad (3.20)$$

where $n_i(r)$ is the number of neighbors at distance in the interval $[r, r+dr]$ from galaxy i and $V_i(r)$ is the volume of the intersection with W of the shell centered at the ith galaxy and with radii r and $r+dr$. This is illustrated in 2-D in Fig. 3.5.

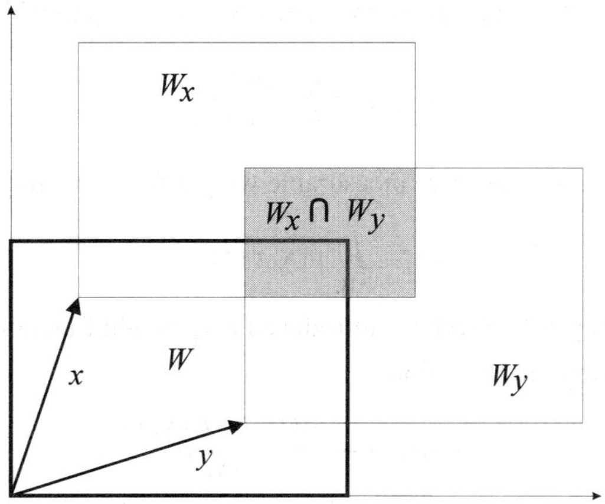

Figure 3.6 *A 2-D illustration of the denominator in the OS estimator of the correlation function, $W_{\mathbf{x}} \cap W_{\mathbf{y}}$. (Reproduced, with permission, from Pons-Bordería et al. 1999, Astrophys. J., 523, 480–491. AAS.)*

When W is a cube, an analytic expression for V_i is provided in Baddeley et al. (1993). This edge correction was introduced by Ripley (1976).

The estimator introduced by Ohser and Stoyan (1981), which is suitable for the case of homogeneous (not necessarily isotropic) point fields, reads:

$$\widehat{\xi}_{\mathrm{OS}}(r) = \frac{V(W)^2}{N^2 4\pi r^2} \sum_{i=1}^{n} \sum_{\substack{j=1 \\ j \neq i}}^{n} \frac{k(r - |\mathbf{x}_i - \mathbf{x}_j|)}{V(W \cap W_{\mathbf{x}_i - \mathbf{x}_j})} - 1, \tag{3.21}$$

where $k(x)$ is a kernel smoothing function (see also Fiksel 1988). The Epanechnikov kernel (3.6) is typically used in point field statistics.

Here $W_{\mathbf{y}}$ denotes the window W shifted by the vector $\mathbf{y}$, $W_{\mathbf{y}} = W + \mathbf{y} = \{\mathbf{x} : \mathbf{x} = \mathbf{z} + \mathbf{y}, \mathbf{z} \in W\}$. The denominator is the volume of the window intersected with a version of the window that has been shifted by the vector $\mathbf{x}_i - \mathbf{x}_j$. It also can be written as $W_{\mathbf{x}_i} \cap W_{\mathbf{x}_j}$ (see Fig. 3.6). The set covariance function is defined as $\gamma_W(\mathbf{x}) = V(W \cap W_{\mathbf{x}})$ (Stoyan, Kendall, and Mecke 1995).

We can still improve the OS estimator by replacing $4\pi r^2$ in the denominator of (3.21) by the quantity $4\pi|\mathbf{x}_i - \mathbf{x}_j|^2$. In particular, Stoyan and Stoyan (2000) recommend this replacement for small r and large value of the bandwidth ω of the kernel function. They also have shown that a geometric version of the Hamilton estimator can be obtained by improving the density estimator entering in (3.21). In fact, the square of the intensity is just estimated as

$$\bar{\lambda}^2 = \left(\frac{N}{V(W)}\right)^2,$$

while the improved estimators for the intensity have the general form

$$\widetilde{\lambda}_{\mathrm{P}}(r) = \sum_{i=1}^{N} \frac{p(\mathbf{x}_i, r)}{C(r)},$$

where $p(\mathbf{x}, r)$ is a non-negative measurable weight function and

$$C(r) = \int_W p(\mathbf{x}, r) dx < \infty.$$

Stoyan and Stoyan (2000) have introduced two weight functions:

1. The surface weighted function

$$p_S(\mathbf{x}, r) = \frac{A(W \cap \partial b(\mathbf{x}, r))}{4\pi r^2}, \tag{3.22}$$

   where A stands for the area, and $b(\mathbf{x}, r)$ is the ball centered at point $\mathbf{x}$ with radius r (and $\partial b(\mathbf{x}, r)$ is its boundary).

2. The volume-weighted function

$$p_V(\mathbf{x}, r) = \frac{V(W \cap b(\mathbf{x}, r))}{\frac{4\pi}{3} r^3}. \tag{3.23}$$

These two functions allow us to define the corresponding weighted improved intensity estimators $\widetilde{\lambda}_{\mathrm{S}}(r)$ and $\widetilde{\lambda}_{\mathrm{V}}(r)$.

The Stoyan estimator for the correlation function is then

$$\widehat{\xi}_{\mathrm{STO}}(r) = \frac{1}{\widetilde{\lambda}_{\mathrm{S}}(r) 4\pi r^2} \sum_{i=1}^{n} \sum_{\substack{j=1 \\ j \neq i}}^{n} \frac{k(r - |\mathbf{x}_i - \mathbf{x}_j|)}{V(W \cap W_{\mathbf{x}_i - \mathbf{x}_j})} - 1. \tag{3.24}$$

Stoyan and Stoyan (2000) have derived an approximation formula for the variance of the estimator valid for Poisson processes. On the basis of this formula and also considering simulations, they recommend the use of the rectangular kernel

$$k_\omega^r(r) = \frac{1}{2\omega} \mathbb{1}_{[-\omega, \omega]}(r)$$

because it has smaller variance than the Epanechnikov kernel.

Biases for all estimators listed above were calculated by Kerscher (1999). He also showed that the natural estimator $\widehat{\xi}_{\mathrm{PH}}$ (3.13) is nothing other than the isotropized Monte Carlo counterpart of $\widehat{\xi}_{\mathrm{OS}}$ in which the smoothing kernel has been substituted with the standard count of pairs $DD(r)$.

The estimators acting on a galaxy sample

To illustrate how different estimators act over a galaxy sample, we have chosen the complete volume-limited sample extracted from the Perseus–Pisces sample introduced in Section 1.4.2.

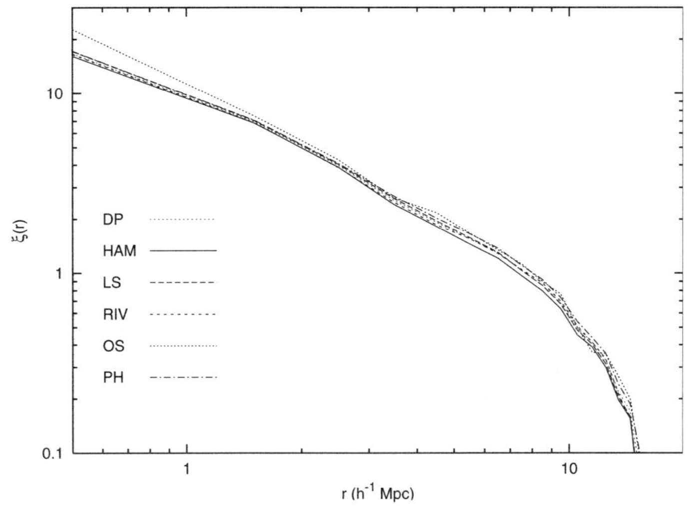

Figure 3.7 *The correlation function of a volume-limited sample of the Perseus–Pisces catalog calculated by means of six estimators described in the text.*

As we can see in Fig. 3.7, there is not much difference between the estimates of $\xi(r)$ provided by the formulae above when applied to the sample. We should mention only the departure of the OS estimator at small separations, taking higher values than the others, probably as a consequence of its lesser sensitivity to local anisotropies due to peculiar motions.

Several attempts have been made to compare different estimators, using real data samples, N-body simulations, and point processes with known correlation functions (Pons-Bordería et al. 1999, Kerscher, Szapudi, and Szalay 2000; Stoyan and Stoyan 2000). The results are the same; the differences between the estimates are more relevant at large scales where $\xi(r) \ll 1$ and fluctuations in the mean density affect the estimators more strongly. Pons-Bordería et al. (1999) find that at large scales the Hamilton and Landy–Szalay estimators provide the best results. Kerscher, Szapudi, and Szalay (2000) recommend the Landy–Szalay estimator because it is easier than the Hamilton estimator to calculate numerically.

Errors and variances in $\xi(r)$

The theoretical formula for the variance of the two-point correlation function depends on the three- and four-point correlation function (Peebles 1973, Hamilton 1993b). We shall follow Hamilton (1993b), and write the unbiased estimator $\widehat{\xi}(r)$

for the continuous overdensity field δ as

$$\widehat{\xi}(\mathbf{r}_{12}) = \xi_{12} = \frac{\langle W_{12}\delta_1\delta_2\rangle}{\langle W_{12}\rangle}, \tag{3.25}$$

where the window function W_{12} selects the proper coordinate shift $\mathbf{r}_1 - \mathbf{r}_2 = \mathbf{r}_{12}$, being nonzero for $\mathbf{r}_1 - \mathbf{r}_2 \in (\mathbf{r}_{12} - \mathbf{dr}, \mathbf{r}_{12} + \mathbf{dr})$, and the angle brackets denote sample averages:

$$\langle W_{12}\delta_1\delta_2\rangle = \frac{1}{V^2}\int\int W(\mathbf{r}_{12})\delta_1\delta_2\, dV_1 dV_2.$$

The finite width $2|\mathbf{dr}|$ of the window function represents the usual way we measure the correlation function, binning it into separate intervals; this can also be thought of as averaging over that interval. The window function can be written as

$$W_{12} = w_{12}\Phi_1\Phi_2, \tag{3.26}$$

where w_{12} is the pairwise weighting function, and Φ_i is the catalog selection function at the point $\mathbf{x}_i$.

The covariance of the estimator (3.25) for two different coordinate intervals $\mathbf{r}_{12}$ and $\mathbf{r}_{34}$ is

$$\langle\Delta\widehat{\xi}_{12}\Delta\widehat{\xi}_{34}\rangle = \frac{\langle W_{12}W_{34}(\delta_1\delta_2\delta_3\delta_4 - \langle\delta_1\delta_2\rangle\langle\delta_3\delta_4\rangle)\rangle}{\langle W_{12}\rangle\langle W_{34}\rangle}.$$

Its expectation value over similar samples is (see Section 7.2)

$$\langle\langle\Delta\widehat{\xi}_{12}\Delta\widehat{\xi}_{34}\rangle\rangle = \frac{\langle W_{12}W_{34}(\eta_{1234} + \xi_{13}\xi_{24} + \xi_{14}\xi_{23})\rangle}{\langle W_{12}\rangle\langle W_{34}\rangle}. \tag{3.27}$$

We usually use an isotropized correlation function

$$\widetilde{\xi}(r) = \int f(\mathbf{r})\widehat{\xi}(\mathbf{r})dV,$$

where $f(\mathbf{r})$ is a properly selected kernel function. Integrating the covariance (3.27) as above, we get the expression for the variance of $\widetilde{\xi}$:

$$\langle\langle\Delta^2\widetilde{\xi}\rangle\rangle = \int\int\int\int f_{12}f_{34}\widetilde{W}_{12}\widetilde{W}_{34}(\xi_{13}\xi_{24} + \xi_{14}\xi_{23} + \eta_{1234})\, dV_1 dV_2 dV_3 dV_4, \tag{3.28}$$

where $\xi_{12} \equiv \xi(r_{12})$ is the true (population) two-point correlation function, η_{1234} is the population four-point correlation function, and $\widetilde{W}$ denotes normalized pair windows,

$$\widetilde{W}_{12} = \frac{W_{12}}{\int W_{12}dV_1}.$$

We shall also introduce the normalized pair weight $\widetilde{w}_{12}$, which is defined in a similar way.

There are δ-function terms in the integral (3.28), where some of the points 1–4 coincide. Separating these, we arrive finally at

$$\begin{aligned}\operatorname{Var}\widetilde{\xi} &= \iiiint f_{12}f_{34}\widetilde{w}_{34}\Phi_3\Phi_4(\eta_{1234}+\xi_{13}\xi_{24}+\xi_{14}\xi_{23})\,dV_1dV_2dV_3dV_4 \\ &+\frac{4}{\bar{n}}\iiint f_{12}f_{13}\widetilde{w}_{13}\Phi_3(\zeta_{123}+\xi_{23})\,dV_1dV_2dV_3 \\ &+\frac{2}{\bar{n}^2}\iint f_{12}^2\widetilde{w}_{12}(1+\xi_{12})\,dV_1dV_2,\end{aligned} \tag{3.29}$$

where $\bar{n}$ is the mean number density and ζ_{123} is the three-point correlation function of the population. The first member in the sum (3.29) results from disjoint pairs 12 and 34, the second member comes from the configurations where the pairs share one point, and the last term comes from the case where the two pairs coincide. As we see, this variance includes both four- and three-point correlation functions, and is thus rather difficult to calculate.

The variance (3.29) depends on the pair weight function. Minimizing it in respect to the pair weighting gives an approximate result for the best weight function (Hamilton 1993b):

$$w_{12} \approx \frac{1}{[1+4\pi\bar{n}\Phi J_3(r)]^2},$$

where Φ is the selection function at the location of the pair and $J_3(r)$ is the integral of the correlation function (Peebles 1980):

$$J_3(r) = \frac{1}{4\pi}\int \xi(r)dV = \int_0^r \xi(s)s^2ds. \tag{3.30}$$

Usually a separable pair weighting $w_{ij} = w_i w_j$ is used, with

$$w_i = \frac{1}{1+4\pi\bar{n}\Phi_i J_3(r)}. \tag{3.31}$$

As the integral $J_3(r)$ depends on the pair separation, different weights should be chosen for different separations. This weighting is similar to the FKP weighting used in determination of power spectra (see Section 8.2). For dense regions, the weighting is inversely proportional to the local density and volume elements are weighted equally; for sparsely sampled regions, the weight function approaches unity, giving equal weight to separate galaxies.

The quantity $4\pi\bar{n}J_3(r)$ in (3.30) gives the average number of neighbors in excess over a random distribution up to a distance r for any sample point. It is usually determined from theoretical considerations. If we try to determine it on the basis of a point sample of size N, the value of that quantity for the whole sample volume will be -1, as every point has then $N-1$ neighbors, one less than expected for a random distribution (Peebles 1980, § 32). This integral constraint will cause the decrease of $J_3(r)$ at large separations. If we determine $4\pi\bar{n}J_3(r)$ from our point sample, we should stop when it reaches the maximum at a separation $r_{\max}$ and use this maximum value for larger separations $r \geq r_{\max}$.

Apart from the variance caused by the correlation structure of the data, the other sources of variance are the finite volume of the sample, insufficient edge correction, and the discreteness noise. These errors are usually coupled together as "cosmic variance" (Szapudi and Colombi 1996), although the term is more appropriate for the errors caused by the finite size of the sample. We have seen above how the different finite volume effects affect the bias of the estimator. These terms also contribute also to the variance; the smaller the volume of our sample, the larger the difference between our estimate of the correlation function and the actual correlation function.

In the case of a homogeneous point process, if we sample a large enough volume, we expect to obtain statistically reliable results, which approximate well the correlation function of the entire universe of galaxies. This is the "fair sample hypothesis" (Peebles 1980), which encompasses the cosmological principle, and the ergodic hypothesis introduced in Section 1.3. The size of a fair sample and the cosmic error that results from the finite size of the sample are difficult to estimate. Because these estimates require information about scales that have not yet been directly observed, we can find these estimates only by using theoretical assumptions.

The variance that is caused by insufficient edge corrections and by the discreteness noise is easier to estimate, in principle, especially for certain point processes. Ripley (1988) has discussed extensively the variance of second-order estimators for Poisson and binomial processes. The variance of these estimators contains both the Poisson pair term that is proportional to $\bar{n}^{-2}$ and the much larger $O(\bar{n}^{-1})$ terms. The latter terms are caused by insufficient correction for edge effects. Analytical expressions of the intrinsic variances of several estimators are also given in Hamilton (1993b) and Landy and Szalay (1993).

As noted above, Szapudi and Szalay (1998) proposed the L-S type estimators for correlation functions of any order. Their estimators can be written symbolically as

$$\widehat{\xi}_N = (\widetilde{D} - \widetilde{R})^N,$$

where $\widetilde{D}$ and $\widetilde{R}$ denote the normalized data and random reference processes. Expanding the power, we can also write

$$\widehat{\xi}_N = \frac{1}{S} \sum_i \binom{N}{i} (-1)^{N-i} \left(\frac{D}{\lambda}\right)^i \left(\frac{R}{\rho}\right)^{N-i}, \tag{3.32}$$

where λ and ρ are the intensities of data and the random reference process, respectively. Each factor in the sum (3.32) has to be calculated at a different point (N-point correlation functions depend on N arguments). Note that the overall normalization factor S is the number of Poisson N-tuples:

$$S = \int \Phi(\mathbf{x})\mu_N,$$

where μ_N is the N-dimensional Lebesgue measure and $\Phi(x_1, \ldots, x_N)$ is a function that selects a specific configuration of N points (e.g., for the two-point correlation function $\Phi(x_1, x_2)$ equals one when the distance between the points x_1 and x_2 falls into a given distance bin, and is zero otherwise).

Using the technique of factorial moment measures, Szapudi (2000) showed that the average of the estimator is

$$\langle \widehat{\xi}_N \rangle = \frac{1}{S} \int \Phi(\mathbf{x})\xi_N(\mathbf{x})\mu_N,$$

where ξ_N is the true (population) N-point correlation function. Szapudi derived a general expression for covariance of correlation function estimators of any order. While the general formula is rather complex, he found that the covariance of the estimator of the two-point correlation function coincides with the formula (3.29) above. Szapudi also gives an explicit formula for the covariance of the estimator (3.32) of the three-point correlation function.

Due to the specific structure of the estimators (3.32), the covariances are rather simple; most of the terms, which are due to the edge and discreteness effects, cancel. For a Poisson process their estimator gives the minimal variance

$$\mathrm{Var}\,\xi = \frac{1}{N_p},$$

where N_p is the number of distinct pairs for a pair distance bin.

Practical ways to estimate errors of the correlation function

Because even the minimum-variance formula (3.29) is extremely complicated to use, requiring knowledge of higher-order correlations, simpler methods to estimate errors of correlation functions are used in practice. There are at least six different ways of estimating the errors of the two-point correlation function. We summarize these methods as follows.

1. Poisson errors. They are derived assuming that the count of pairs $DD(r)$ follows a Poisson distribution (Peebles 1973). This implies that the standard deviation in the counts of pairs is just the square root of the count,

$$\sigma_{DD}(r) = \sqrt{DD(r)}.$$

Since similar expressions are valid for $DR(r)$ and $RR(r)$, it is quite straightforward to obtain an approximate expression of the Poisson errors associated with the estimators $\widehat{\xi}_{\mathrm{PH}}$, $\widehat{\xi}_{\mathrm{DP}}$, $\widehat{\xi}_{\mathrm{HAM}}$, and $\widehat{\xi}_{\mathrm{LS}}$:

$$\sigma_\xi(r) \simeq \frac{1 + \xi(r)}{\sqrt{DD(r)}}, \tag{3.33}$$

where it has been considered that the terms $1/DR(r)$ and $1/RR(r)$ vanish when choosing a large enough random sample.

Alternatively, we can estimate the Poisson errors by using in (3.33) instead of the number of data pairs $DD(r)$ the number of data–random pairs $\widetilde{DR}(r)$ or the number of random–random pairs $\widetilde{RR}(r)$, which are normalized to the total number of objects in the catalog:

$$\widetilde{DR}(r) = \frac{N}{N_{\text{rd}}} DR(r), \qquad \widetilde{RR}(r) = \left(\frac{N}{N_{\text{rd}}}\right)^2 RR(r).$$

These pair numbers give a better approximation of the Poisson error, especially for smaller scales with more data pairs.

For geometric estimators, Poisson errors have similar expressions. For example, for the Rivolo estimator

$$\sigma_\xi^{\text{RIV}}(r) = \sqrt{\frac{2V(W)(1+\xi(r))}{N\sum_{i=1}^N V_i(r)}},$$

while for the Stoyan estimator with the rectangular kernel the approximate error has the form

$$\sigma_\xi^{\text{STO}}(r) = \sqrt{\frac{1+\xi(r)}{4\pi r^2 \omega \widetilde{\lambda}_{\text{S}}^2(r)\bar{\gamma}_W(r)}},$$

where

$$\bar{\gamma}_W(r) = \frac{1}{4\pi}\int_0^{2\pi}\int_0^{\pi} \sin(\theta)\gamma_W(\mathbf{x}(r,\theta,\phi))d\theta d\phi\,,$$

and $\gamma_W(r)$ is the isotropized set covariance function of W.

For cluster processes, the Poisson method underestimates the errors. We can obtain this result if we use only part (the last term) of the minimum variance expression (3.29). Real estimators have additional errors caused by the edge effects; these, too, are ignored here. A rough rule is that Poisson errors give an order-of-magnitude estimate of the sample errors. Thus, frequently the errors given by (3.33) are multiplied by an empirical coefficient in the range 1–2 (Mo, Jing, and Börner 1992; Martínez et al. 1993).

2. Poisson enhanced errors. The assumption made previously about the randomness of the point distribution is in general not true for the clustered galaxy distribution. Peebles (1980) and Kaiser (1986) proposed estimating the variance of the correlation function on the basis of the "cluster model." This model, explained in more detail in Peebles (1980, § 40), assumes that all points are contained in isolated clusters. The mean number of points per cluster can be found as the mean number of neighbors for any point. The definition of the correlation function allows to write the probability $dP(r)$ of finding a neighbor for a point at the distance r in the volume element dV as

$$dP(r) = \bar{n}\big[1+\xi(r)\big]dV. \tag{3.34}$$

Integrating it, we get the mean number of neighbors

$$N_c = \bar{n} \int_V \xi(r) dV \approx 4\pi\bar{n} J_3(\infty) = 4\pi\bar{n} \int_0^\infty \xi(r) r^2 dr,$$

where the last equality assumes that the integral exists. Usually correlation functions fall off rapidly enough for that.

The integral J_3 has a meaning only in the context of the cluster model, where the pair separation range in the integral is thought to extend only to a typical cluster size. In this case $4\pi J_3 = N_c - 1$, where N_c is the number of points in a cluster, and averaging it over all clusters with different numbers of points gives (Peebles 1980, § 31)

$$4\pi\bar{n} J_3 = \frac{\langle N_c(N_c - 1)\rangle}{\langle N_c\rangle}$$

For pair separations larger than the typical cluster size the fluctuations in the correlation function are Poissonian in the number of clusters $N' = N/N_c$, where N is the total number of galaxies in the sample. This leads to the "enhanced Poisson errors," given by

$$\sigma_\xi(r) = \frac{1 + 4\pi\bar{n} J_3}{\sqrt{DD(r)}}. \tag{3.35}$$

3. Bootstrap errors. The bootstrap resampling estimation of error introduced in cosmology by Barrow, Sonoda, and Bhavsar (1984) is a well-known technique in statistics (see Efron and Tibshirani 1993). The idea is quite clear: we use the survey data, the best information about the parent population, to generate an ensemble of statistically equivalent samples. The method consists of generating a number M of bootstrap samples by choosing from the original sample randomly, with replacement, the same number of points as in the original sample. Consequently, some of the original positions may appear several times (there will be multiple points), whereas others will not be represented in a given bootstrap sample. Bootstrap errors are computed as follows:

$$\sigma_\xi^{\mathrm{BOO}}(r) = \sqrt{\sum_{i=1}^{M} \frac{(\xi_i(r) - \bar{\xi}(r))^2}{M - 1}}, \tag{3.36}$$

where $\xi_i(r)$ is the two-point correlation function calculated on the bootstrap sample labeled by i and

$$\bar{\xi}(r) = \frac{1}{M} \sum_{i=1}^{M} \xi_i(r).$$

It is clear that $\bar{\xi}(r)$ is not a good approximation of the two-point correlation function of the sample of galaxies, since a bias is introduced by the use of multiple points. Although this method has become rather popular for estimating errors of correlation functions in cosmology, it is clear that it is an incorrect application of a good idea. About one third of points in a generated sample, selected with replacements, are multiple. This is certainly not a good representation of the parent population, where there are no multiple points. Another possibility is to discard multiple points; in this case, the generated sample has about two thirds of the density of the original sample (and the parent population). Thus, there is only one possibility to generate a sample with the same correlation structure that the original sample. We have to select the same number of points without replacement, getting back the original sample. Kerscher, Szapudi, and Szalay (2000) express that idea, saying that in spatial statistics, a *whole* sample takes the role of one point in the bootstrap procedure. Moreover, Snethlage (1999) has shown that the bootstrap variance of the correlation function can be calculated analytically, without resorting to the bootstrap procedure, and that it differs completely from the real variance.

4. Disjoint subsampling. If the sample volume is large enough, we can measure the correlation function in different disjoint subregions and take as a measure of the error the dispersion of ξ for the different subregions (Maddox et al. 1990b, Buchert and Martínez 1993). The problem with this method is that it can be used only for large samples and the accuracy of the variances is inversely correlated with the size of the subregions. Maddox et al. (1990b) determined the (angular) correlation function of the APM survey at large separations, and used only four subregions to estimate variances of the correlation function.

5. Subregion fluctuations. This method was proposed by Hamilton (1993b). It estimates the variance of the correlation function estimator on the basis of the sample data. As Hamilton argues, it automatically includes all statistical sources of error.

This method can be derived by considering the correlation function and its estimator as functions of the pair window, $\xi(r) = \xi(W_{ij}(r))$, where the indices ij refer to all pairs of small volume elements. The error of the estimator is then caused by the difference of the sample windows W and the unknown population windows W_{pop}; assuming that this difference is small, we can write

$$\Delta\xi = \xi(W) - \xi(W_{\mathrm{pop}}) = \sum_{ij} (W_{ij} - W_{ij,\mathrm{pop}}) \left.\frac{\partial \xi}{\partial W_{ij}}\right|_{\mathrm{pop}} . \tag{3.37}$$

In this sum and below we suppose that the indices i and j are distinct. We also omit the correlation function argument r, when it is the same throughout a formula. The Taylor expansion (3.37) is valid if both windows are consistent (normalized). The normalization condition for sample windows can be expressed requiring that ξ does not depend on rescaling of the window functions:

$$\frac{\partial \xi(\lambda W)}{\partial \lambda} = \sum_{ij} W_{ij} \frac{\partial \xi}{\partial W_{ij}} = 0. \tag{3.38}$$

This eliminates the first sum in (3.37). Defining the fluctuation of the correlation function for the pair ij as

$$\Delta\xi_{ij} = W_{ij} \left.\frac{\partial\xi}{\partial W_{ij}}\right|_{\text{pop}}, \tag{3.39}$$

and reversing the sign of the error, we get from (3.37)

$$\Delta\xi = \sum_{ij} \Delta\xi_{ij}. \tag{3.40}$$

The population derivative in (3.39) is not known, but we can approximate it by the sample derivative, writing

$$\Delta\xi_{ij} \approx W_{ij} \frac{\partial\xi}{\partial W_{ij}}.$$

As Hamilton notes, this approximation can be done for any single pair, but not for their sum (3.40), where it will give zero by the normalization condition (3.38).

If we assign the fluctuations $\Delta\xi_{ij}$ caused by pairs to their volume elements by

$$\Delta\xi_i = \frac{1}{2}\sum_j \Delta\xi_{ij},$$

we can change the sum over pairs into the sum over volume elements,

$$\Delta\xi = \sum_i \Delta\xi_i.$$

In the usual case when the pair window is separable, $W_{ij} = W_i W_j$, the fluctuation for the volume element is

$$\Delta\xi_i = \frac{1}{2} W_i \frac{\partial\xi}{\partial W_i}.$$

The covariance of the estimator is

$$\langle \Delta\xi(r_1)\Delta\xi(r_2)\rangle = \sum_{ij}\sum_{kl} \Delta\xi_{ij}(r_1)\Delta\xi_{kl}(r_2).$$

This sum is similar to the integral (3.28) above. Choosing the same value for the pair distance r and replacing summation over pairs by summation over volume elements we get the expression for the variance

$$\operatorname{Var}\widehat{\xi} = \langle(\Delta\xi)^2\rangle = \sum_{ik} \Delta\xi_i \Delta\xi_k. \tag{3.41}$$

The expressions for $\Delta\xi_i$ depend on the form of the specific estimator. Hamilton (1993b) lists these for the standard, Hamilton, and Landy–Szalay estimators. For the Hamilton estimator the fluctuation for a volume element is

$$\Delta\xi_i = (1+\xi_{\text{est}})\left(\frac{D_i D}{DD} - \frac{D_i R}{DR} - \frac{DR_i}{DR} + \frac{R_i R}{RR}\right),$$

where D_iD is the pair number between the subvolume i and the full sample for the pair distance interval $(r, r + dr)$. Other pair numbers have a similar meaning.

In practice, the tiny subvolumes are substituted by finite, but small subregions. For the correlation analysis of the IRAS 2 Jy survey Hamilton (1993a) split the survey volume into 528 subregions; for estimating the variance of the power spectrum by the same method (see Section 8.5), Hamilton and Tegmark (2000) split the PSCz survey into 220 subregions.

The difficulty of this method is the integral constraint; the fluctuations sum to zero. This also leads to the zero sum for the variance. Hamilton (1993b) proposed to overcome this by first ordering the subvolume pairs in (3.41) by their mutual separation and by gradually including into the sum pairs with growing separation, until the sum stops increasing. This gives an underestimate of the true variance, but Hamilton (1993b) has shown that for real galaxy samples and for scales that are not close to the sample size the error is small. We note that the fluctuation analysis has to be carried out separately for all values of the argument r (argument intervals) of the correlation function $\xi(r)$.

6. Artificial samples. Since ideally we would like to have a large number of independent samples in order to measure the dispersion of the observed correlation function, one possibility is to construct artificial samples with correlation function similar to the observed one. The variance of the correlation function can then be found by comparing the correlation functions of artificial samples.

This can be done by using N-body simulations resembling the real distribution of galaxies (Fisher et al. 1993) or by means of realizations of stochastic models with known analytical expressions for ξ. For example, Cox processes, which will be explained in Chapter 5, are suitable for this purpose (Pons-Bordería et al. 1999). The results, however, must be interpreted with caution, because the higher-order correlations of artificial samples may differ from the higher-order correlations of the actual galaxy distribution, substantially changing the variance of the two-point correlation function. As we have seen above, at least the three- and four-point correlation functions must be well modeled to estimate the variance of the two-point correlation function.

Selection function

The selection function can enter the correlation function estimator explicitly, but usually it is included indirectly, via the random sample. This sample is generated to have not only the same geometry, but also the same selection function as the galaxy sample. Thus errors of the selection function generate additional errors in the correlation function estimator and, consequently, the selection function must be determined carefully.

The main method of finding the selection function is the maximum likelihood estimation of the luminosity function, given by an analytic form. This method is described in Chapter 1. There are also two other popular methods. In the "stepwise likelihood method" (Efstathiou, Ellis, and Peterson 1988) the luminosity function is represented as a step function. This method is thus essentially nonparametric. Saunders et al. (1990) use both methods and estimate the goodness of the parametric result by the χ^2 test, comparing the parametric and nonparametric luminosity functions.

Another possibility is to use the maximum likelihood method directly to find the selection function. This method was proposed by Saunders et al. (1990) and is similar to the determination of the luminosity function in Section 1.3.2.

The above methods give the selection functions for the space determined by the distance measure; it is usually redshift space. If we want to estimate the real space correlation function, we also have to transform the selection function to real space. This is rather complex; for details, see Hamilton (1998).

3.4.2 The angular two-point correlation function

Similar to the definition of the correlation function given in Eq. 3.34, we can define the (2-D) angular correlation function $w(\theta)$ in such a way that the conditional probability of finding an object in the solid angle $d\Omega$ at angular separation θ of an arbitrarily chosen galaxy is

$$dP = \mathcal{N}[1 + w(\theta)]d\Omega, \tag{3.42}$$

where $\mathcal{N}$ is the mean density of galaxies in the projected sky.

All the estimators of $\xi(r)$ previously introduced have their corresponding counterparts for $w(\theta)$. For example, the DP estimator for $w(\theta)$ is

$$\widehat{w}_{\text{DP}}(\theta) = F\frac{DD(\theta)}{DR(\theta)} - 1, \tag{3.43}$$

where $DD(\theta)$ is the number of galaxy–galaxy pairs in the catalog with separations in the interval $[\theta - d\theta/2, \theta + d\theta/2]$, and $DR(\theta)$ is the number of pairs between the data and the random sample with separation in the same range; F is the quotient of the density in the random catalog and the density in the galaxy catalog.

Angular correlation functions were measured on projected galaxy catalogs much earlier then the spatial correlation function on redshift surveys (Peebles and Hauser 1974; Groth and Peebles 1977). Recent results on the angular two-point correlation function from the observed sky maps were obtained by Maddox et al. (1996) for the APM survey described in Section 1.4.1. In Fig. 3.8, we can see these results for the estimator (3.43). The filled circles show the mean of $\widehat{w}_{\text{DP}}(\theta)$ for 185 Schmidt fields. In this case the density used to normalize the estimate has been calculated using the counts in each individual plate. When a global estimate of the mean surface density of galaxies is used, the estimated value of the correlation function increases at larger angular separations (see open circles in Fig. 3.8). The reason for this bias when estimating $w(\theta)$ is the integral constraint, which

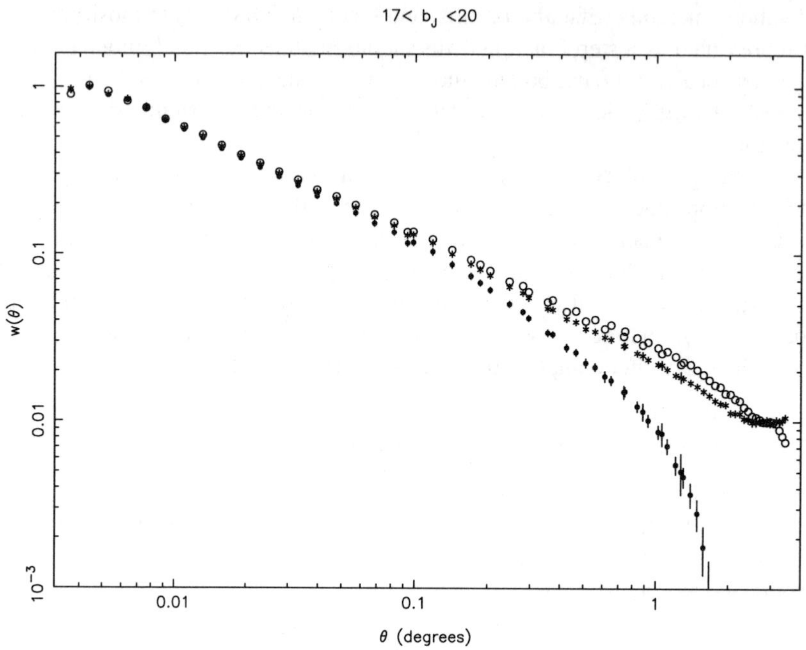

Figure 3.8 *The angular correlation function of the APM catalog calculated by the DP estimator (solid circles). At large angular scales the estimate is low biased due to the integral constraint. The open circles and the stars represent, respectively, the correction for that bias performed by using a global density in the estimator or by adding a certain constant that accounts for the power lacking at large scales. (Reproduced, with permission, from Maddox et al. 1996, Mon. Not. R. Astr. Soc., 283, 1227–1263. Blackwell Science Ltd.)*

will be explained in Section 8.2. In Fig. 3.8, Maddox et al. (1996) show how adding a constant to the estimate $\widehat{w}_{\mathrm{DP}}$ makes it possible to correct for the bias. This constant is estimated from the variance in the field number counts.

3.4.3 The correlation integral

The expected number of points within a distance r from an arbitrarily chosen galaxy is

$$\langle N \rangle_r = \int_0^r 4\pi n s^2 (1+\xi(s))ds = \frac{4\pi}{n}\int_0^r s^2 \lambda_2(s) ds. \tag{3.44}$$

The last expression may also be referred to as the correlation integral $C(r)$. $K(r) = C(r)/n$ is called the Ripley's K-function and is used extensively in the literature of point fields. In this context n is the first-order intensity function λ of the point field, which is constant for homogeneous processes.

$K(r)$ is a second-order cumulative function and is defined so that the expected number of neighbors a given galaxy will have at a distance less than r is $nK(r)$. Therefore its relation with the two-point correlation function is

$$K(r) = \int_0^r 4\pi s^2(1 + \xi(s))ds. \tag{3.45}$$

Its main advantage is that no binning of the data is necessary for the estimation. For a homogeneous Poisson process this function is just

$$K_{\mathrm{Pois}}(r) = \frac{4\pi}{3}r^3. \tag{3.46}$$

For this reason, statisticians often plot the so-called L-function (Ripley 1981) defined as

$$L(r) = \left(\frac{3K(r)}{4\pi}\right)^{1/3}$$

which increases linearly with r for a Poisson process.

Relation with other cumulant quantities

Peebles (1980) introduced the moments of the counts of neighbors $\langle N \rangle_r$ as the mean count of objects in balls of radius r excluding the central one. By definition $\langle N \rangle_r$ is the correlation integral $C(r)$ used by Martínez et al. (1995) in the analysis of the multiscaling properties of the matter distribution and is related to $K(r)$ simply by $\langle N \rangle_r = nK(r) = C(r)$.

Taking into account the shape of the K function for a random object distribution (Poisson process), we can just consider the difference between $K(r)$ and $K_{\mathrm{Pois}}(r)$ which leads to another cumulant quantity commonly used in the statistical description of the galaxy clustering, $4\pi J_3(r) = K(r) - K_{\mathrm{Pois}}(r)$, where $J_3(r)$ has been defined in (3.30) (Peebles 1980, 1993).

Finally, if we consider the quotient instead of the difference, we get the integral quantity used by Coleman and Pietronero (1992):

$$\Gamma^*(r) = \frac{nK(r)}{K_{\mathrm{Pois}}(r)}; \tag{3.47}$$

The methods used by Pietronero and collaborators will be discussed in Chapter 4 (see also Cappi et al. (1998)).

The advantages of the use of $K(r)$ with respect to other cumulant quantities are the following:

1. $K(r)$ is well normalized. We can compare directly the K function of different samples with different number density and within different volumes without additional normalization.

2. $K(r)$ is a well-known quantity in the field of spatial statistics and several analytical results regarding its shape and variance are already available for a variety of point processes.
3. It is very important to estimate the quantity $K(r)$ directly from the data and not through numerical integration of $1+\xi(r)$, which introduces artificial smoothing of the results. Several edge-corrected unbiased estimators are available for $K(r)$. In the context of the present application, the most desirable properties an estimator must have are to have little variance and not to introduce spurious homogeneity by means of the edge correction. In the next subsection we comment on different estimators for $K(r)$.

Estimators

We shall make the assumption that the process under consideration is homogeneous and isotropic.

Several estimators exist for $K(r)$. A comparison of some of them can be found in Doguwa and Upton (1989). From the definition of K and ignoring the edge effects one could consider the following naive estimator

$$\widehat{K}_{\mathrm{N}}(r) = \frac{V}{N^2} \sum_{i=1}^{N} \sum_{\substack{j=1 \\ j\neq i}}^{N} \theta(r - |\mathbf{x}_i - \mathbf{x}_j|), \tag{3.48}$$

where θ is Heaviside's step function, whose value is 1 when the argument is positive and 0 otherwise. Obviously, for a finite sample this estimator will provide values for K smaller than the true values since neighbors outside the boundaries are not considered. One possible solution is to use the counterpart for an integral quantity of the minus-estimator for $\xi(r)$: Let us consider only points in an inner region as centers of the balls for counting neighbors. The points lying in the outer region, a buffer zone (Upton and Fingleton 1985; Buchert and Martínez 1993), take part in the estimator just as points that could be seen as neighbors at a given distance r of the points in the inner region. The inner region might shrink as r increases. However, this solution leads to biases (the sample is not uniformly selected), wastes a lot of data, and obviously increases the variance of the estimator (Doguwa and Upton 1989). The standard solution adopted in the statistical studies of the large-scale structure is to account for the unseen neighbors outside the sample window by means of the following edge-corrected estimator,

$$\widehat{K}_{\mathrm{DU}}(r) = \frac{V}{N^2} \sum_{i=1}^{N} \sum_{\substack{j=1 \\ j\neq i}}^{N} \frac{\theta(r - |\mathbf{x}_i - \mathbf{x}_j|)}{f_i(r)}, \tag{3.49}$$

where $f_i(r)$ is the fraction of the volume of the sphere of radius r centered on the object i that falls within the boundaries of the sample. This estimator was introduced by Doguwa and Upton (1989) in the field of spatial statistics.

The most commonly used unbiased edge-corrected estimator in the analysis of point processes is Ripley's estimator, which according to our hypotheses reads (Baddeley et al. 1993):

$$\widehat{K}_{\mathrm{R}}(r) = \frac{V}{N^2} \sum_{i=1}^{N} \sum_{\substack{j=1 \\ j \neq i}}^{N} \frac{\theta(r - |\mathbf{x}_i - \mathbf{x}_j|)}{\omega_{ij}}, \tag{3.50}$$

where the weight ω_{ij} is an edge correction equal to the proportion of the *area* of the sphere centered at $\mathbf{x}_i$ and passing through $\mathbf{x}_j$ that is contained in W; in other words, ω_{ij} is the conditional probability that the point j is observed given that it is at a distance r from the point i. This correction is illustrated in Fig. 3.9.

Baddeley et al. (1993) give an analytic expression for ω_{ij} when W is a cube. Note, however, that using this estimator we introduce a certain bias when we estimate n through N/V but, on the other hand, we make full use of all sample points. For all the estimators mentioned it is possible to build the corresponding versions for flux-limited samples by simply adding a weighting factor representing the selection function.

Stoyan and Stoyan (2000) have proposed an estimator for the K function with a corrected (scale-dependent) intensity estimator:

$$\widehat{K}_{\mathrm{STO}}(r) = \frac{1}{\widetilde{\lambda}_{\mathrm{V}}^2(r)} \sum_{i=1}^{N} \sum_{\substack{j=1 \\ j \neq i}}^{N} \frac{\theta(r - |\mathbf{x}_i - \mathbf{x}_j|)}{V(W_{\mathbf{x}} \cap W_{\mathbf{y}})}, \tag{3.51}$$

Stein (1993) has introduced an estimator of $K(r)$ with similar properties to those of the Landy and Szalay estimator for $\xi(r)$. It is still possible that the best estimator depends on the kind of point process to be studied (clustered or regular) and even on the particular scale range.

3.5 N-point correlation functions

A generalization of the definition of the two-point correlation function can easily be made to higher orders. For example, the probability of finding three points within the infinitesimal volume elements dV_1, dV_2, and dV_3 placed at the vertices of a triangle with sides r_{12}, r_{23}, and r_{31} is just

$$dP = n^3 dV_1 dV_2 dV_3 [1 + \xi(r_{12}) + \xi(r_{23}) + \xi(r_{31}) + \zeta(r_{12}, r_{23}, r_{31})]. \tag{3.52}$$

where $\zeta(r_{12}, r_{23}, r_{31})$ is the reduced or connected† three-point correlation function, while the quantity in square brackets is the full three-point correlation function. The terms $\xi(r_{ij})$ account for the excess of triples found as a consequence of having more pairs than in a random distribution. Under the assumption of stationarity and isotropy ζ is symmetric in its arguments. Peebles (2001) gives a list

† The term *connected* comes from their analogy with the Green's functions in particle physics.

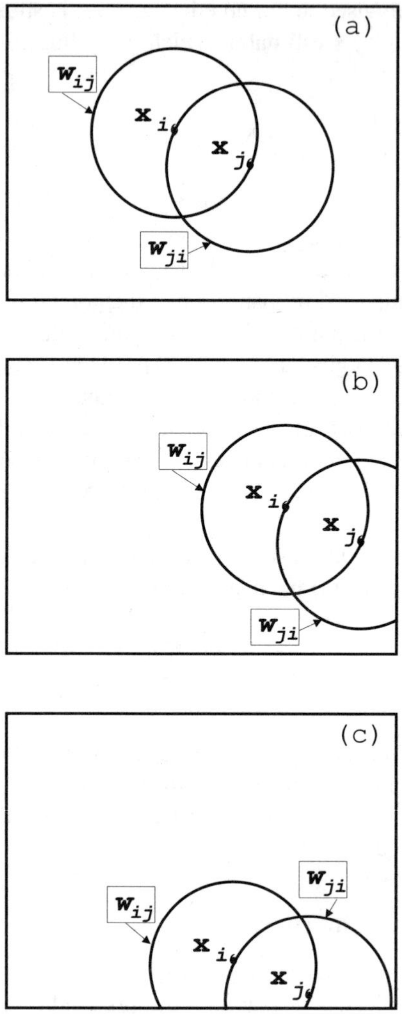

Figure 3.9 *An illustration of the weights used in the estimator of K (Eq. 3.50) in two dimensions. The rectangle represents the boundary of the sample. In this case, w_{ij} is the proportion of the circumference of the circle centered at $\mathbf{x}_i$, passing through $\mathbf{x}_j$, lying within the boundary of the sample. Depending on the relative positions of the galaxies with respect to the boundary, different cases are illustrated: (a) $w_{ij} = w_{ji} = 1$; (b) $w_{ij} = 1$, $w_{ji} < 1$; (c) $w_{ij} < 1$, $w_{ji} < 1$. It is clear from the plot that we weight the observed neighbor $\mathbf{x}_j$ of the galaxy $\mathbf{x}_i$ lying at a distance r (the radius of the circle) from it by the inverse of the probability that such a neighbor would be observed. (Reproduced, with permission, from Martínez et al. 1998, Mon. Not. R. Astr. Soc., 298, 1212–1222. Blackwell Science Ltd.)*

of the reasons that motivated people to use the three-point correlation function in the analysis of the galaxy distribution. Eq. 3.52 is just a generalization to order 3 of Eq. 3.7. In fact, this can be done to any order n, by considering n infinitesimal disjoint volumes dV_1, dV_2, $\cdots$, dV_n centered at the points $\mathbf{x}_1, \mathbf{x}_2, \ldots, \mathbf{x}_n$, then the probability of finding a point of the point process in each of the volumes dV_1, dV_2, $\cdots$, dV_n is

$$dP = \lambda_n(\mathbf{x}_1, \mathbf{x}_2, \ldots, \mathbf{x}_n) dV_1 dV_2 \cdots dV_n \tag{3.53}$$

where λ_n is the product density of the n-th factorial moment measure (Stoyan, Kendall, and Mecke 1995). If we adopt the notation $\xi_{(3)} = \zeta$ for the reduced three-point correlation function, Eq. 3.52 can be generalized to any order n, where $\xi_{(4)} = \eta$ is the reduced cumulant four-point correlation function and $\xi_{(n)}$ is the reduced n-point correlation function. For a Poisson process $\xi_{(n)} = 0$ for $n \geq 2$. Thus far, the analysis of the galaxy catalogs by means of higher-order correlation functions has been applied mainly to two-dimensional projected catalogs. The counterpart of Eq. 3.52 for a catalog listing angular positions of galaxies is the three-point angular correlation function, which is a generalization of Eq. 3.42,

$$dP = \mathcal{N}^3 d\Omega_1 d\Omega_2 d\Omega_3 [1 + w(\theta_{12}) + w(\theta_{23}) + w(\theta_{31}) + z(\theta_{12}, \theta_{23}, \theta_{31})], \tag{3.54}$$

where $\mathcal{N}$ is the mean surface number density.

Nevertheless, it is already possible to get fairly good estimates of the spatial higher-order correlation functions for the new generation of redshift surveys (Bonometto and Sharp 1980; Gaztañaga 1992). In fact, Jing and Börner (1998) and Szapudi et al. (2000) have presented three-dimensional analyses of the three-point correlation function for the Las Campanas redshift catalog and for the KPNO survey, respectively.

3.5.1 Hierarchical models for higher-order correlations

Higher-order correlation functions involve all distances separating n points. They are functions depending on $3n$ coordinates, minus 3 rotations and 3 translations, so their interpretation is rather complex. A model that simplifies enormously this task is the so-called hierarchical model, in which the n-function, $\xi_{(n)}$ is related to the $\xi_{(n-1)}$ through a scaling relation (Pebbles 1980, Balian and Schaeffer 1989). For example, the three-point correlation function, which is a function of three variables, can be written as products of two-point functions, which are functions of one variable,

$$\zeta = Q[\xi(r_{12})\xi(r_{23}) + \xi(r_{23})\xi(r_{31}) + \xi(r_{31})\xi(r_{12})], \tag{3.55}$$

where Q is a constant. This model has been applied very successfully to galaxy redshift catalogs. It fits the observed three-point correlation function with a value of $Q \simeq 1$ within the range $0.1 < r < 5\, h^{-1}$ Mpc.

Eq. 3.55 can be generalized to any order n by the expression

$$\xi_n(r_1, \ldots, r_n) = \sum_{t=1}^{T(n)} Q_{n,t} \sum_{L_{n,t}} \prod^{n-1} \xi(r_{ij}), \tag{3.56}$$

where $Q_{n,t}$ are structure constants. In Eq. 3.56, there is a product of $n - 1$ two-point correlation functions. Every $\xi(r_{ij})$ corresponds to an edge $r_{ij} = |\mathbf{x}_i - \mathbf{x}_j|$ linking the n galaxies. These links are arranged in a tree structure. For every tree, we have a product. There are $T(n)$ distinct trees (or trees with different topologies). For example, $T(2) = 1, T(3) = 1, T(4) = 2, T(5) = 3, T(6) = 6$, etc. (see Fig. 3.10). For each tree $t = 1, \ldots, T(n)$ there are $L_{n,t}$ possible relabelings, and there is a summation of all of them. It is intriguing why this model fits well with the observations at least up to $n = 8$ in the highly nonlinear regime (strong clustering).

A simplification of Eq. 3.56 is possible if we accept that the parameters $Q_{n,t}$ do not depend on the shape of the tree,

$$\xi_n(r_1, \ldots, r_n) = Q_n \sum_{t=1, L_{n,t}}^{T(n)} \xi(r_{ij}). \tag{3.57}$$

A more general model that verifies Eq. 3.56 is the scale-invariant model proposed by Balian and Schaeffer (1989),

$$\xi_n(kr_1, \ldots, kr_n) = k^{-(m-1)\gamma} \xi_n(r_1, \ldots, r_n). \tag{3.58}$$

3.6 Moments and counts in cells

The probability that a randomly placed cell A of volume $V(A)$ contains exactly N objects of the point process is denoted by $P(N, V(A))$. For a Poisson process with intensity λ, these quantities are completely known,

$$P(N, V(A)) = \frac{(\lambda V(A))^N}{N!} \exp(-\lambda V(A)). \tag{3.59}$$

The moments of order n of the counts are (Stoyan, Kendall, and Mecke 1995)

$$\mu^{(n)}(A \times \ldots \times A) = E(\Phi(A)^n) = \sum_{N=0}^{\infty} N^n P(N, V(A)), \tag{3.60}$$

i.e., these are the moments of order n of the random variable $\Phi(A)$, which provides the number of points in A.

The factorial moments of order n are defined as

$$\begin{aligned} \alpha_n(A \times \ldots \times A) &= E[\Phi(A)(\Phi(A) - 1) \cdot \ldots \cdot (\Phi(A) - n + 1)] \\ &= \sum_{N=0}^{\infty} N(N-1) \cdot \ldots \cdot (N - n + 1) P(N, V(A)) \end{aligned}$$

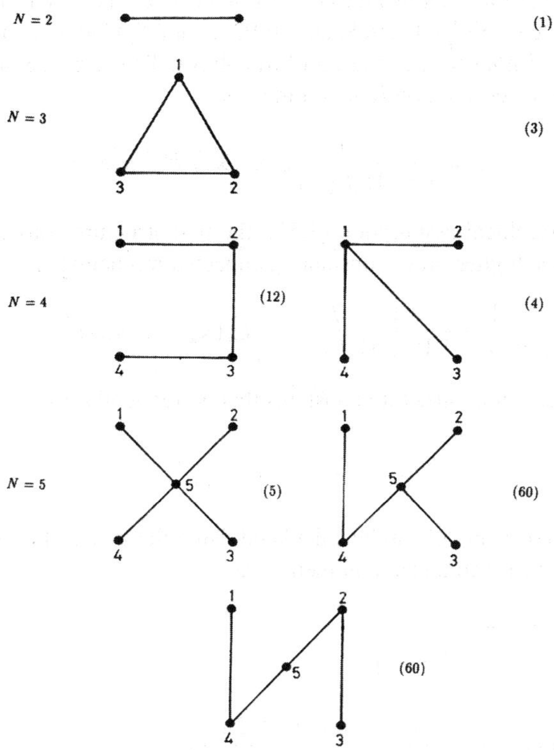

Figure 3.10 *The different shapes of trees connecting n points. The number of possible relabelings is indicated in parentheses. (Reproduced, with permission, from Coles and Luchin 1995, Cosmology. The Origin and Evolution of Cosmic Structure. ©John Wiley & Sons).*

$$= \int_A d\mathbf{x}_1 \ldots \int_A d\mathbf{x}_n \lambda_n(\mathbf{x}_1, \mathbf{x}_2, \ldots, \mathbf{x}_n), \tag{3.61}$$

i.e., the expression above is the factorial moment of order n of the random variable $\Phi(A)$. Eq. 3.61 is read in spatial statistics by stating that the density of the factorial moment α_n with respect to the Lebesgue measure is the product density of order n λ_n. In fact, Eq. 3.61 encapsulates the relation between the n-point correlation functions and the count in cell probabilities (White 1979).

As high-order correlation functions depend on a large number of arguments, they are rather difficult to estimate. Counts in cells are easy to estimate; Szapudi (1998) has developed an ingenious exact algorithm for the computation of counts in cells. The expected count distributions can be predicted theoretically, thus they are frequently used to study higher correlations in the galaxy distribution.

For that we have to establish the connection between the counts and correlation functions (see, e.g., Szapudi and Szalay 1993; Szapudi, Colombi, and Bernardeau 1999). This is similar to the relation (3.61) above. The variance of the counts is the average of the correlation function in a cell:

$$\bar{\xi} = \frac{1}{V^2(A)} \int_{A \otimes A} \xi(\mathbf{x}_1, \mathbf{x}_2)\, d^3x_1 d^3x_2 r.$$

Using the hierarchical assumption (3.57), the first structure constants are $Q_1 = Q_2 = 1$, and the higher-order constants (connected moments) are

$$Q_n = \frac{1}{n^{n-2}\bar{\xi}^{n-1}} \frac{1}{V^n(A)} \int_{A \otimes \ldots \otimes A} \xi_n(\mathbf{x}_1, \ldots, \mathbf{x}_n)\, d^3x_1 \ldots d^3x_n.$$

There is another set of structure constants that is frequently used:

$$S_n = n^{n-2} Q_n = \frac{\bar{\xi}_n}{\bar{\xi}^{n-1}}.$$

The structure constants Q can be calculated using the factorial moments defined above. The first four connected moments are:

$$\begin{aligned}
\bar{n} &= \alpha_1, \\
\bar{\xi} &= \frac{\alpha_2}{\alpha_1^2} - 1, \\
Q_3 &= \frac{\alpha_1(\alpha_3 - 3\alpha_1\alpha_2 + 2\alpha_1^3)}{3(\alpha_2 - \alpha_1^2)^2}, \\
Q_4 &= \frac{\alpha_1^2(\alpha_4 - 4\alpha_3\alpha_1 - 3\alpha_2^2 + 12\alpha_2\alpha_1^2 - 6\alpha_1^4)}{16(\alpha_2 - \alpha_1^2)^3}.
\end{aligned}$$

The skewness and kurtosis are usually described by the parameters S_3 and S_4.

The sample errors of these parameters are thoroughly analyzed in Szapudi, Colombi, and Bernardeau (1999); a program package (FORCE) for calculating the errors can be found at I. Szapudi's Web page (*Szapudi*). As the parameters Q are given by models of the correlation hierarchy, all these models can be tested using counts in cells.

The formalism has been mainly applied to photometric catalogs (projected data). Gaztañaga (1994) estimated the parameters S and Q up to the ninth order, using the APM galaxy catalog. Because there are far fewer galaxies in redshift catalogs, the only moments estimated using three-dimensional counts are skewness and kurtosis; the sample errors for higher moments are too large. Fig. 3.11 shows an example of the skewness and kurtosis of the PSCz redshift survey, found by Szapudi et al. (2000). As we see, the skewness and kurtosis practically do not depend on scale (the depth of a subsample), and their values are $S_3 = 1.9 \pm 0.6$ and $S_4 = 7.0 \pm 4.1$, in agreement with previous studies.

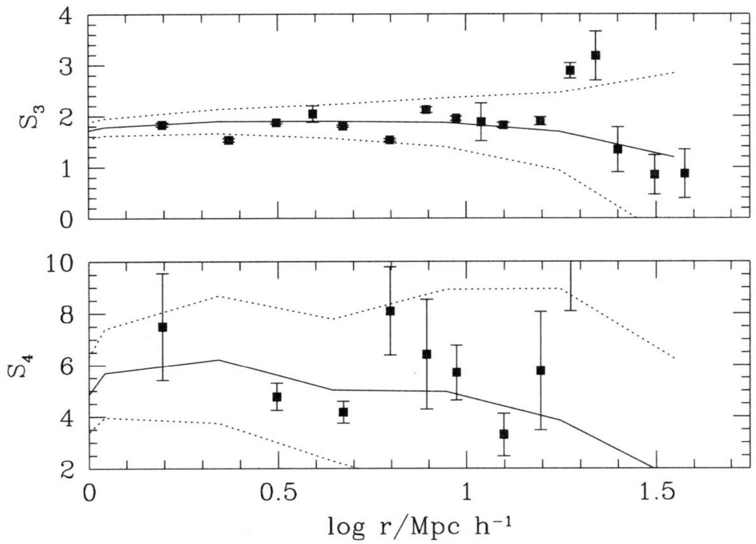

Figure 3.11 *Skewness (upper panel) and kurtosis (lower panel) for the PSCz redshift survey. Different points describe different volume-limited subsamples. The lines show the predictions of a galaxy formation model proposed by Benson et al. 2000; solid lines show the mean and dotted lines show the variance. (Reproduced, with permission, from Szapudi et al. 2000, Mon. Not. R. Astr. Soc. 318, L45–L50. Blackwell Science Ltd.)*

3.7 The void probability function

A special case of the counts-in-cells probabilities $P(N, V)$ is the so-called void probability function, which corresponds to $N = 0$; therefore, $P(0, V(A))$ indicates the probability that a randomly placed cell A with volume $V(A)$ is empty. In spatial statistics this quantity is known as the emptiness probability (Stoyan and Stoyan 1994). Typically, spherical cells are used for the definition of the void probability function and therefore the notation $P_0(r) = P(0, B_r)$ is also used where B_r is a ball or radius r.

Let us consider the random variable X, which represents the distance between a given random test point in $\mathbb{R}^3$ and its nearest neighbor in the point process (a galaxy). Cressie (1991) proposed calling the points from the point process (or from a given realization of the process) "events" to distinguish them from other arbitrary points of $\mathbb{R}^3$.

The distribution function of X is related to the void probability function by $P(0, B_r)$ by

$$F(r) = P(X \leq r) = 1 - P(0, B_r)$$

$F(r)$ (also referred as to $H_s(r)$ in spatial statistics) is the spherical contact distribution function. For a homogeneous Poisson process of intensity λ, it is given by

$$F_{\text{Pois}}(r) = 1 - \exp\left(-\frac{4}{3}\pi r^3 \lambda\right).$$

In cosmology $P(0, B_r)$ has been typically estimated (Maurogordato and Lachièze-Rey 1987) by randomly placing N_b balls of radius r, B_r, within the sample window (the balls do not cross the boundaries of the window), and counting the number of such balls having no galaxies $N_{\text{emp}}(r)$. The estimate of $P(0, B_r)$ is

$$\widehat{P}(0, B_r) = \frac{N_{\text{emp}}(r)}{N_b}$$

This is basically equivalent to the minus-sampling estimator (Baddeley et al. 1993) of the spherical contact distribution function, which reads

$$\widehat{F}_-(r) = \frac{V\{\mathbf{p} \in A : S(\mathbf{p}, r) \subset A \wedge \min_i |\mathbf{p} - \mathbf{x}_i| \leq r\}}{V\{\mathbf{p} \in A : S(\mathbf{p}, r) \subset A\}}, \qquad (3.62)$$

where $\{\mathbf{x}_i\}_{i=1}^N$ are the observed points (galaxies) and $S(\mathbf{p}, r)$ is the sphere centered at random test point $\mathbf{p}$ and having radius r. This estimator is unbiased. Fig. 3.12 shows the function $F(r)$ estimated for a volume-limited sample extracted from the CfA1 catalog with 80 h^{-1} Mpc depth, together with the same function for a binomial process having the same number of points.

If we arbitrarily place a point within the boundaries of a realization of a clustered point process, it is likely that it lies farther away from an event than it would in the case of placing it over on a homogeneous Poisson process. Therefore, for a clustered pattern at scale r, we should expect to obtain $F(r) < F_{\text{Pois}}(r)$.

Note that Eq. 3.62 provides an estimator of $1 - P_0(x)$, i.e., 1 minus the void probability function. White (1979) has shown how the void probability function serves as a generating functional for all the volume probabilities $P(N, V(A))$, and hence it is related to the n-order correlation functions.

3.8 Nearest neighbor distances

The n-point correlation function and related statistical measures are the most commonly used statistical measures in the cosmological literature. Other statistical measures, developed in spatial statistics and based on nearest neighbor distances, are extremely useful for revealing some aspects hidden in the correlation functions (Ripley 1981).

We have seen that the spherical contact distribution function $F(r)$ is the probability that the distance from an arbitrarily chosen point in $\mathbb{R}^3$ to its nearest event is less than or equal to r (Cressie 1991). Similarly, we can define the distribution

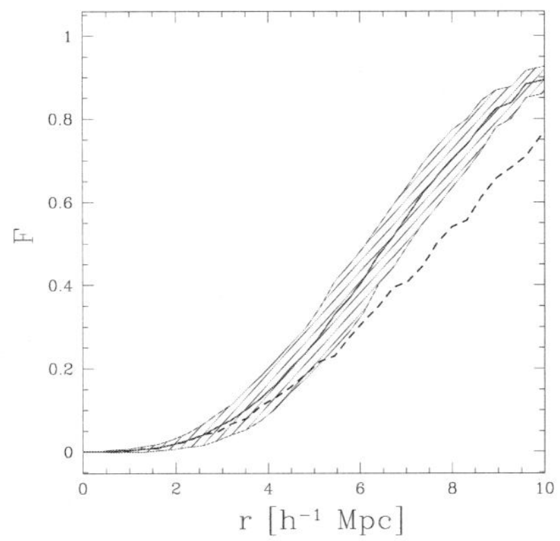

Figure 3.12 *The F-function for a galaxy sample (dashed line). The shaded area corresponds to the rms fluctuations of the same function calculated over 15 realizations of a binomial process with the same number of points. (Reproduced, with permission, from Kerscher and Martínez 1998, Bull. Int. Statist. Inst., 57-2, 363–366.)*

function $G(r)$ of the distance between events of the process (galaxies in our case), i.e., $G(r)$ is the probability that the distance between a randomly chosen galaxy and its nearest neighbor (event) is less than or equal to r.

An indication of clustering will be that $G(r) > G_{\rm Pois}(r)$, because the distance from clustered points to their nearest neighbor usually will be shorter than in a homogeneous Poisson distribution. Likewise, for a regular pattern $G(r) < G_{\rm Pois}(r)$, because a point from such a process is, on average, farther away from its nearest neighbor than in a Poisson process. This is illustrated in Fig. 3.13.

It is clear from the definitions that for a homogeneous Poisson process $G(r) = F(r)$. In general, however, $G(r)$ does not necessarily verify any continuity property (Baddeley 1999). The main drawback of nearest neighbor statistics is that they provide information of the clustering pattern within a rather restricted range of scales (Cressie 1991). Nevertheless, within these scales they are extremely useful. In fact, the quotient

$$J(r) = \frac{1 - G(r)}{1 - F(r)}, \tag{3.63}$$

suggested by van Lieshout and Baddeley (1996), has provided rather interesting results when applied to galaxy data and mock galaxy catalogs drawn from N-body simulations (Kersher et al. 1999). By means of this quotient we can compare the region around an event (galaxy) of the point process with the events lying within the neighborhood of a randomly chosen point. For a homogeneous Poisson

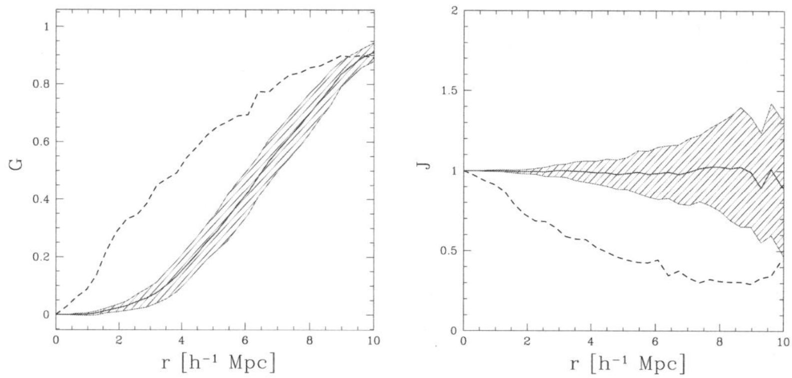

Figure 3.13 *The nearest neighbor distribution function and the J-function for a galaxy sample (dashed line). The shaded area corresponds to the rms fluctuations of the same functions calculated over 15 realizations of a binomial process with the same number of points. (Reproduced, with permission, from Kerscher and Martínez 1998, Bull. Int. Statist. Inst., 57-2, 363–366.)*

process, it is straightforward to see that $J_{\rm Pois}(r) = 1$, while clustered patterns indicate $J(r) < 1$ and regular point distributions show $J(r) > 1$.

In Fig. 3.13, we show the functions $G(r)$ and $J(r)$ for the CfA1 volume-limited sample used in the previous section. The corresponding functions for a binomial process with the same number of points also are displayed.

Improved estimators for the functions F(r), $G(r)$, and $J(r)$ have been developed in the field of spatial statistics (see, for example, Baddeley et al. 2000).

3.9 Galaxy distribution as a marked point field

Galaxies have different properties: mass, luminosity, morphology, color, size, etc. In modern cosmological databases, such as the SLOAN survey, many of these properties are listed together with the position of the galaxy. Working with these huge multidimensional databases represents an important challenge for statisticians and astronomers. In this section, we will study clustering properties of the galaxy distribution as a function of their intrinsic properties. Up to now, redshift surveys contained information on two additional properties of galaxies: luminosity and morphological type. Many studies have been performed to study the dependence of clustering on luminosity or morphology. This has given rise to the concept of segregation, which means that different galaxy types trace the matter distribution in different ways, and therefore different bias schemes can be tested from the observed galaxy distribution.

Measurements of the luminosity or morphological segregation have been attempted by applying statistical descriptors, such as the two-point correlation function, the nearest neighbor distribution functions, or multifractal measures to subsamples of the survey corresponding to populations of a given type, for example, all elliptical galaxies, or all galaxies brighter than a particular luminosity cutoff (Hamilton 1988; Davis et al. 1988; Salzer, Hanson, and Gavazzi 1990; Domínguez-Tenreiro, Gómez-Flechoso, and Martínez 1994; Loveday et al. 1995; Hermit et al. 1996; Guzzo et al. 1997). When the different populations provide different results for the statistical descriptors, showing different clustering properties at given scales, we can interpret this result as the fingerprint of segregation.

Other studies have been based on the cross-correlation function of subsamples (Alimi, Valls-Gabaud, and Blanchard 1988). The cross-correlation function between two types of objects, namely A and B, is defined similarly to Eq. 3.7 by considering the joint probability of finding an object of type A in the volume element dV_A, and an object of type B in the volume element dV_B, both volumes separated by a distance r:

$$dP = \bar{n}_A \bar{n}_B \left(1 + \xi_{AB}(r)\right) dV_A dV_B,$$

where $\bar{n}_A$ and $\bar{n}_B$ are the mean number densities of the two types of objects (Peebles 1980, § 44; Mo, Peacock, and Xia 1993). Ratios of the cross-correlation function of the two types of galaxies to the autocorrelation function of a single type are often used as a measure of segregation. Cross-correlation analysis between low and high redshift galaxies can be used to accurately estimate cosmological parameters (Benítez and Sanz 1999). Possible evolution effects in the correlations (see Section 8.7.2) must be considered.

In the literature of spatial statistics, when a characteristic is attached to all points of a process, the process is called a marked point field (Stoyan, Kendall, and Mecke 1995). For the galaxy distribution, we can use the notation $X^M = \{(\mathbf{x}_i, m_i)\}$, where the mark m_i of the galaxy at position $\mathbf{x}_i$ is an intrinsic property of the galaxy, for example, its luminosity, $m_i = L_i$. In this case, the mark is a continuous variable. However, discrete marks also can be considered, e.g., the morphological type. Marks can be of a complex structure, and many marks can be associated with a point process, giving rise to multivariate point processes (Diggle 1983). Vectors can be used as marks; for example, we could study directional correlations of angular momenta, orientations, or ellipticities. This latter mark would be closely related to the cosmic shear (see Section 9.6.3).

Beisbart and Kerscher (2000) have demonstrated the use of some statistical measures to quantify the mark correlation properties of point processes in the analysis of the galaxy distribution. Here we describe one of them: the normalized mark correlation function.

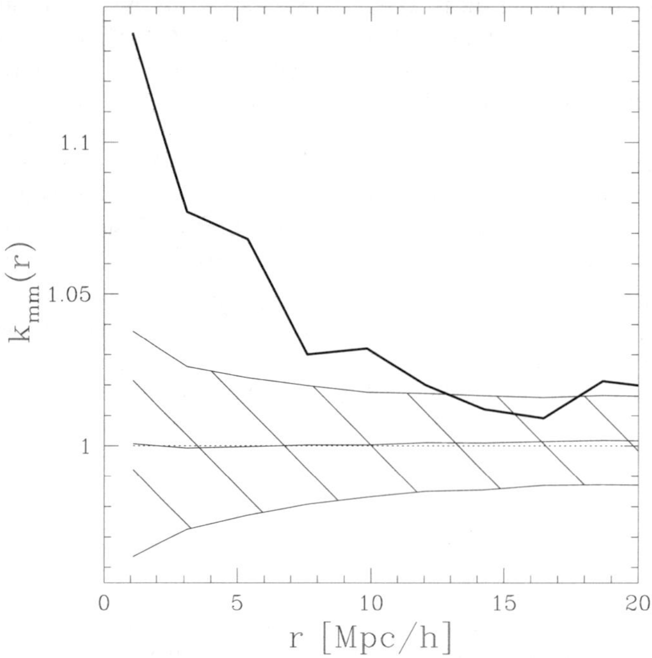

Figure 3.14 *The normalized mark correlation function for a volume-limited subsample of the SSRS2 catalog. The shaded area shows the rms fluctuations calculated over 1000 realizations of randomly shuffling the marks. (Reproduced, with permission, Beisbart and Kerscher 2000, Astrophys. J., 545, 6–25. AAS.)*

3.9.1 The normalized mark correlation function

Stoyan (1984a) applied second-order moment measures to study systematically correlations of marked point processes. A brief and practical summary can be found in Stoyan and Stoyan (1994). The starting point is to define a suitable weighting function $f(m_1, m_2)$ depending only on the marks of two points of the process, namely $\mathbf{x}_1$ and $\mathbf{x}_2$. As an example, we can define $f(m_1, m_2) = m_1 m_2$. If the marks are, for example, the masses of the galaxies, larger values of f will correspond to galaxies at position $\mathbf{x}_1$ and $\mathbf{x}_2$, both being very massive. Other functions f can be used, depending on the kind of marks we are considering.

For continuous real-valued marks, it makes sense to consider the density function of the mark $\rho(m)$, which can be used to define the mean mark

$$\bar{m} = \int m\rho(m)dm.$$

Analogously to the definition of the second-order product density, $\lambda_2(\mathbf{x}_1, \mathbf{x}_2)$, in Eq. 3.7, we can define the quantity

$$\lambda_2^M((\mathbf{x}_1, m_1), (\mathbf{x}_2, m_2))dV_1 dm_1 dV_2 dm_2$$

as the joint probability that in the volume element dV_1 lies a galaxy with the mark in the range $[m_1, m_1 + dm_1]$ and another galaxy lies in dV_2 with the mark in $[m_2, m_2 + dm_2]$. Then the Stoyan mark correlation function for an isotropic and homogeneous marked point process is calculated as

$$k_{mm}(r) = \frac{1}{\bar{m}^2 \lambda_2(r)} \int\int m_1 m_2 \lambda_2^M((\mathbf{x}_1, m_1), (\mathbf{x}_2, m_2)) dm_1 dm_2, \quad (3.64)$$

for $\lambda_2(r) \neq 0$.

Intuitively $k_{mm}(r)$ can be considered the squared geometric mean of the marks normalized by the overall mean mark squared, under the condition that the pairs of galaxies are separated by r. $k_{mm}(r) < 1$ represents inhibition of the marks at the scale r. For example, in forests it is typically found that trees with larger stem diameter (mark) tend to be isolated. Using luminosity as the mark, the opposite effect has been found for the galaxy distribution, i.e., $k_{mm}(r) > 1$ at small scales (Beisbart and Kerscher 2000), implying stronger clustering of brighter galaxies at small separations (see Fig. 3.14).

Other related measures such as the mark covariance function (Stoyan 1984b; Cressie 1991) and the mark variogram (Wälder and Stoyan 1996) can be used to analyze a marked point process. More details and practical formulae for the estimators for both continuous and discrete marks can be found in Stoyan and Stoyan (1994), § 15.4.4, and Beisbart and Kerscher (2000).

CHAPTER 4

Fractal properties of the galaxy distribution

4.1 Introduction

Already in the eighteenth century, Immanuel Kant and Johann Lambert conceived a hierarchical universe of stars clustered into larger systems, which today we call galaxies, and which, in turn, were clustered into larger systems and so on. This hierarchical view of the universe was proposed by John Herschel and Richard Proctor in the nineteenth century as a solution to Olber's paradox and was defended vigorously at the beginning of the twentieth century by several astronomers, including Carl Charlier and Fournier d'Albe (for a historical approach to the hierarchical universe, see Harrison 2000). More recently, Gerard de Vaucouleurs (1970) found observational evidence supporting this idea in the distribution of galaxies in clusters and superclusters.

With the introduction of the fractal concept by Mandelbrot (1982), hierarchical clustering has been reinterpreted in terms of a self-similar or scale-invariant fractal distribution of galaxies (Peebles 1978; Efstathiou, Fall, and Hogan 1979). In the past two decades, the mathematical concept of the fractal has greatly influenced a range of scientific disciplines.

Mandelbrot's book (1982) has presented many unresolved questions to the scientific community, because it was written by the author "without attempting completeness." This fact, together with the interdisciplinary character of the books, has prompted many scientists to explore the exciting fractal world. Other important readings on the application of fractals in different fields are the books by Barnsley (1988), Feder (1988), Takayasu (1989), Falconer (1990), Heck and Perdang (1991), and Bunde and Havlin (1994).

It seems clearly established that, at small distances ($r < 10\,h^{-1}$ Mpc), the galaxy clustering is fractal. If fractality were to extend to larger scales, our cosmological models would be called into question, because one of their fundamental tenets is the cosmological principle introduced originally by Einstein (although the term was coined by E.A. Milne in 1933).

The cosmological principle is the assumption that the large-scale universe is spatially homogeneous and isotropic (see Section 2.2), and according to this principle, the distribution of galaxies should break scale invariance at a certain distance, showing a clear transition to homogeneity at large scales.

The strongest observational evidence supporting the validity of the cosmological principle is the isotropy of the cosmic microwave background radiation. Other observations supporting this hypothesis include the angular distribution of radio sources, the analysis of the X-ray background, and the distribution of quasars and Lyman-α clouds. The distribution of γ-ray bursts also supports the homogeneity picture. These are all essentially two-dimensional tests because they analyze objects or radiation as seen on the celestial sphere. Additional observational evidence of this kind comes from the analysis of the angular correlation function for the Lick and the APM galaxy surveys. This function scales with magnitude as predicted for a homogeneous universe (Peebles 1993). In this chapter we analyze these tests.

The fractal or scaling approach is mathematically rigorous and applicable to measuring the galaxy clustering (Martínez 1991). In this chapter we explain how to measure fractal dimensions for the galaxy catalogs, how these quantities are related to other statistical descriptors, and the physical implications of the fractal prescription. Given that cosmic clustering has evolved under the influence of gravitation only, there is no physical motivation to think about preferred scales in the galaxy formation process. In this context multifractal measures are described as the most natural scaling behavior of the moments of the counts in cells introduced in Section 3.6. Recent views related to fractal aspects of galaxy clustering as the multiscaling prescription are discussed in this chapter.

4.2 Fractal models for the universe

Several fractal constructions have been proposed thus far as artificial universes trying to mimic some properties of the real galaxy distribution. Mandelbrot (1975) proposed a model based on a Rayleigh–Lévy flight, where galaxies are placed at the steps of a random walk. The direction of each jump is isotropically chosen at random and its length follows a power-law probability distribution function.

4.2.1 Rayleigh–Lévy dust

This model, based on the Rayleigh–Lévy flight, is built as follows. Choose a starting point of the simulation. Following a random walk, jump over to a second point, and so on. The power-law constraint affects only the distribution of the jump lengths, while the direction of each jump is taken isotropically at random. If X denotes the random variable that contains the values of the jump lengths, its probability distribution function must be a power law

$$P(X > \epsilon) = \left(\frac{\epsilon}{\epsilon_0}\right)^{-D}. \tag{4.1}$$

If we place a galaxy at each jump the output appears as shown in Fig. 4.1. The density of this distribution of points within a sphere of radius R varies as R^{D-3}. This

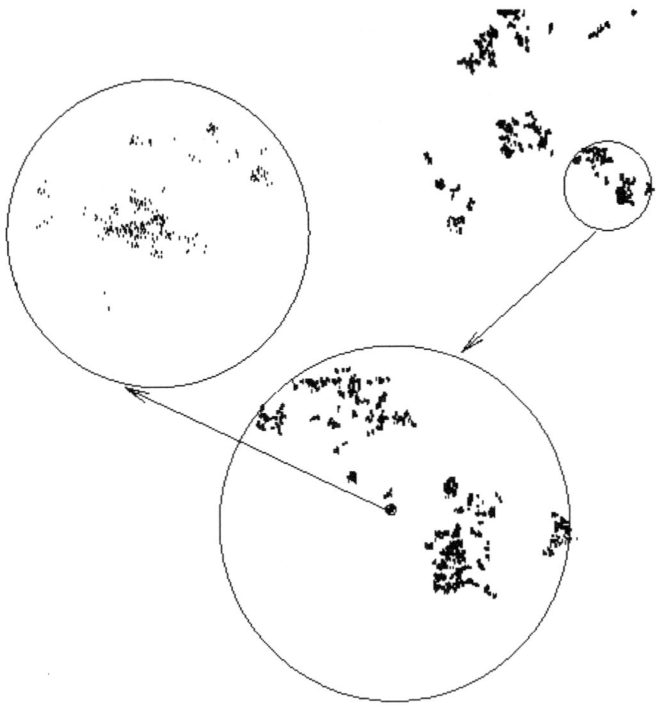

Figure 4.1 *Simulation of a Rayleigh–Lévy flight in two dimensions. Several zooms are shown to illustrate the self-similarity of this construction. (Reproduced, with permission, from Nusser and Lahav 2000, Mon. Not. R. Astr. Soc., 313, L39–L42. Blackwell Science Ltd.)*

behavior illustrates one of the definitions of fractal dimension. In a fractal object the *mass* (in our case the number of points) scales with the radius as $M(R) \propto R^D$. Note that for a homogeneous distribution on a line we get $M(R) \propto R^1$, and if the distribution is placed uniformly on a surface, the behavior is $M(R) \propto R^2$, etc. In a fractal object the exponent D is smaller than the dimension of the Euclidean space d where the object is embedded; therefore, the density scales as $\rho(R) \propto R^{D-d}$.

Mandelbrot used this kind of fractal to define the concept of *lacunarity* (Mandelbrot 1982). Lacunarity is related to the presence of large empty regions in the fractal. This fact agrees qualitatively with the observational evidence of the existence of very large voids almost devoid of galaxies (Kirshner et al. 1981, 1987), but the sizes of the empty regions in the model are much larger than in the observations. Thus, the lacunarity of the model is too high to be acceptable. Lacunarity

Figure 4.2 *This diagram illustrates how the Soneira and Peebles fractal model is built in two dimensions. The left panel shows the construction when overlapping of the spheres in each level is avoided. The right panel shows the opposite case when overlapping is allowed.*

will be studied in detail, both qualitative and quantitatively, in the last section of this chapter.

4.2.2 Soneira and Peebles fractal model

Soneira and Peebles (1978) introduced a model based on the superposition of fractal clumps, which reproduced not only the appearance of the projected distribution of galaxies as seen in the Lick maps (Seldner et al. 1977), but also the angular two- and three-point correlation function. Each fractal clump is built as follows: in a sphere of radius R, we randomly place η spheres of radius r/λ, with $\lambda > 1$. Within each of these spheres, we again place η new spheres of radius R/λ^2. The process is repeated L times and the last generation of η^L centers are considered the galaxies of a clustering hierarchy or a bounded fractal with dimension $D = \log \eta / \log \lambda$.

Fig. 4.2 shows several steps of the construction of this fractal pattern in two dimensions.

What does this fractal clump look like when projected onto the sky? Peebles (1998) showed a series of projections of the particles lying in concentric shells as seen by an observer situated at a given point close to the center of the first sphere. A similar set of equal-area Hammer–Aitoff projections is shown in Fig. 4.3. This is a fractal clump with $D = 2$, ($\eta = 2$, $\lambda \simeq 1.41$, and $L = 18$). We have scaled the fraction of particles plotted as a function of $1/r^2$, with r as the distance of a particle from the center. The width of a given shell is always twice the width of the previous one. The anisotropy of the point distribution provided by this model is remarkable. The inhomogeneity remains even in the larger shells far away from

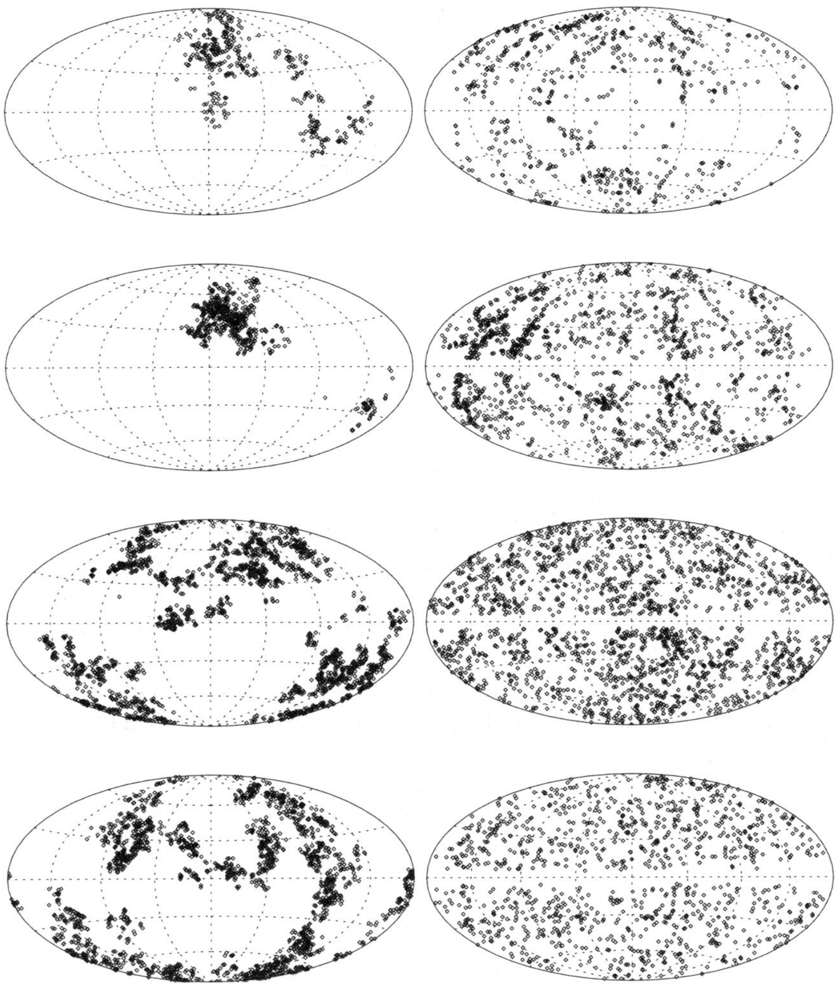

Figure 4.3 *Left panel: Equal-area Hammer–Aitoff projection of a single Soneira and Peebles fractal clump as seen by an observer situated at a point of the model close to the center of the first sphere. In each panel, from top to bottom, we have projected the points lying in concentric shells of increasing radius and width:* $1.25 < r \leq 2.5$, $2.5 < r \leq 5$, $5 < r \leq 10$, $10 < r \leq 20$, *in arbitrary units. Right panel: The same projections but for the IRAS 1.2Jy redshift survey. The division in shells is the same as before, but now the units are thousands of km/s in the radial velocity of each galaxy.*

the center. Davis (1997) performed the same kind of projection for a real flux-limited redshift survey, the IRAS 1.2Jy catalog. The right panels of Fig. 4.3 show the IRAS galaxies lying in concentric shells of increasing size projected onto the sky. The radial distribution of shells is the same as that shown for the Soneira and Peebles model (left panels in Fig. 4.3), but it is slightly different from that shown in Davis (1997). We can see that for the more distant and larger shells the distribution appears more homogeneous. This tendency toward homogeneity detected visually in the redshift catalogs will be quantified in the next sections.

4.3 Tests on projected data

The nearby distribution of galaxies is fairly inhomogeneous and can be considered fractal, but farther away the distribution appears to become more homogeneous. As described in Chapter 2, direct tests for homogeneity based on galaxy redshift catalogs do not yet provide definite answers. The problem lies in the unknown luminosity evolution of galaxies that can change selection limits and affect catalog densities at large distances.

A stronger result can be obtained by galaxy number counts (see also Chapter 2). Their dependence on the magnitude limits is also affected by galaxy luminosity and number density evolution, but only at larger distances. At closer distances the number counts behave exactly as they should for a uniform $D = 3$ density distribution (see the review in Peebles 1993).

Further evidence for a nonfractal $D = 3$ distribution of galaxies is provided by the scaling of the angular correlation function of galaxies $w(\theta)$ (see Chapter 8). The scaling predicted by a nonfractal distribution,

$$w(\theta) = \frac{1}{D} W(\theta D),$$

where D is the depth of a sample, is confirmed by observations. Peebles (1993) argues that it is strong evidence against a fractal distribution of galaxies.

Galaxy distribution at larger distances is presently known primarily from projected data. In principle, an imprint of a fractal distribution of matter should be clearly seen — relative density fluctuations for a fractal do not depend on scale and should remain similar to those observed nearby, at least about unity. Projected data do not confirm that; a thorough review of the current situation can be found in Wu, Lahav, and Rees (1999).

One class of sources at large distances are radio galaxies. Their typical redshifts are $z \approx 1$, while the depths of optical catalogs reach only $z \approx 0.1$. As shown in Fig. 4.4, their distribution in the sky is isotropic to a high degree. However, their luminosity functions are very wide, so projection effects tend to even out inhomogeneities in their spatial distribution.

Another class of distant sources produce the X-ray background (XRB). Their distribution is also isotropic, but projection effects could play again an important

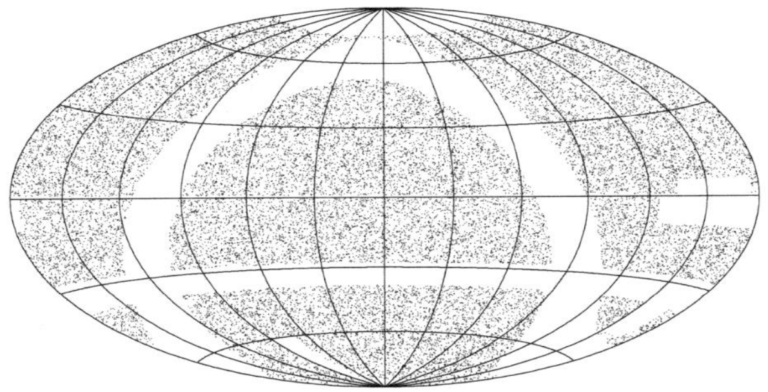

Figure 4.4 *Distribution of radio sources. (Reproduced, with permission, from Baleisis et al. 1998, Mon. Not. R. Astr. Soc., 297, 545–558. Blackwell Science Ltd.)*

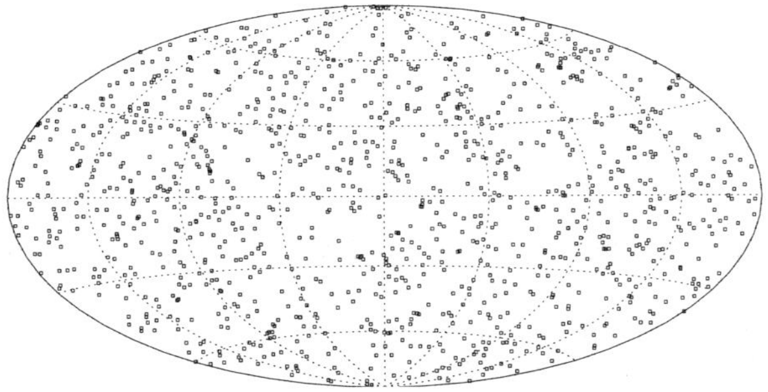

Figure 4.5 *Distribution of γ-ray bursts (courtesy of NASA Marshall's Space Flight Center, Space Sciences Laboratory).*

role. Fig. 4.5 shows the distribution of directions to gamma-ray bursts. This is also isotropic, but as yet nothing is known about the distances to their sources.

The maps listed above are isotropic. Although it is difficult to generate isotropic fractal maps (typical maps are shown in the previous section), it could be possible. Durrer et al. (1997) present fractal constructions that are more or less isotropic in projection. These are, however, rather special fractals and it is not clear if similar constructions could be used for random fractals.

The spatial distribution of optical and radio galaxies, and X-ray and gamma-ray sources could also be modified by spatially fluctuating biasing. The most isotropic background of all, the cosmic microwave background (CMB), measures, according to the present cosmological theory, fluctuations of total matter density at the moment of recombination. The typical CMB temperature fluctuations $\Delta T/T \approx 10^{-5}$ translate into fluctuations of the same amplitude for $\delta\rho/\rho$

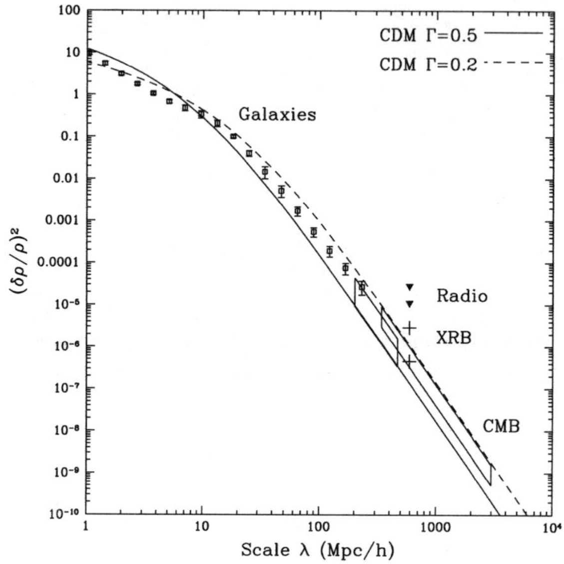

Figure 4.6 *Density fluctuations as a function of the scale for different models and different observations, including galaxy surveys, radio sources, X-ray background, and cosmic microwave background data. (Reproduced, with permission, from Wu, Lahav, and Rees 1999, Nature, 397, 225–235. Macmillan Publishers Ltd.)*

at scales of the horizon size. This value is impossible to reconcile with the fractal picture.

Present knowledge about density fluctuations on different scales is summarized in Fig. 4.6. The rms density fluctuation amplitudes are small already on the scales of a few tens of Mpc and diminish rapidly toward larger scales. Thus the fractal picture is probably useful only for smaller scales. The density distribution at large scales is fairly homogeneous. Although a totally fractal universe would be an enormous challenge both for observers to map out and for theoreticians to explain, the more modest task of discovering how the locally fractal distribution changes into a homogeneous one is also very interesting.

4.4 Fractal dimensions

4.4.1 Hausdorff dimension

Let us consider a nonempty subset $\mathcal{A}$ of the Euclidean space $\mathbb{R}^d$. For $r > 0$, an r-cover of $\mathcal{A}$ is a countable collection of sets with diameter less than or equal to

r that covers $\mathcal{A}$. We can consider the family of all possible r-covers of $\mathcal{A}$

$$\Upsilon^r_{\mathcal{A}} = \{\{B_i\}_{i=1}^{\infty} \mid \mathcal{A} \subset \cup_{i=1}^{\infty} B_i \mid r_i \leq r\}, \tag{4.2}$$

where r_i is the diameter of the covering set B_i (belonging to an r-cover of $\mathcal{A}$). Then, for each $\beta \geq 0$ the β-dimensional Hausdorff measure of $\mathcal{A}$ is defined as

$$H^\beta(\mathcal{A}) = \lim_{r \to 0} \inf_{\Upsilon^r_{\mathcal{A}}} \sum_{i=1}^{\infty} r_i^\beta. \tag{4.3}$$

It can be shown that this is a Borel measure. The definition of the Hausdorff dimension is contained in the following lemma (for a proof see Falconer 1985): For every subset $\mathcal{A}$ of $\mathbb{R}^d$, there exists a unique number $D_H(\mathcal{A})$, the so-called Hausdorff dimension, which verifies

$$H^\beta(\mathcal{A}) = \infty \quad \text{if} \quad \beta < D_H(\mathcal{A}) \tag{4.4}$$

and

$$H^\beta(\mathcal{A}) = 0 \quad \text{if} \quad \beta > D_H(\mathcal{A}). \tag{4.5}$$

Therefore, the Hausdorff measure has a critical behavior at the Hausdorff dimension, "jumping" from infinity to zero.

Mandelbrot defined a fractal set as one for which the Hausdorff dimension strictly exceeds its topological dimension.

It is rather straightforward to show that the Hausdorff dimension of any countable set is zero (Falconer 1990); therefore, this measure really is not useful for application to the galaxy distribution. It has been attempted, however, on the basis of the following assumption: If galaxies were good tracers of mass and the mass distribution had a Hausdorff dimension less than 3, could we estimate this value just from the analysis of the galaxy distribution?

Let us put this question in a different way using an example from the field of complex systems. We know that the Lorenz attractor is a fractal structure in 3-D with the Hausdorff dimension $D_H \simeq 2.1$. The previous question, in this context, should read as follows: If we know the coordinates of N points distributed randomly on the three-dimensional trajectory of the attractor, should we be able to calculate the attractor Hausdorff dimension just from the point distribution? This has been done by Domínguez-Tenreiro, Roy, and Martínez (1992) by means of the minimal spanning tree (MST). This is a graph-theoretical construction linking all points of the process without closed loops and with minimal total length (van de Weygaert, Jones, and Martínez 1992). In Chapter 10, we provide a more formal definition and see how this construction has been used as a measure of the structure of the galaxy distribution. Here, the MST plays the role of the minimal covering, implicitly given by the infimum requirement in the definition of the Hausdorff dimension (4.3). The method works as follows: Given N points sampling the attractor we randomly choose N_R points and calculate the lengths of the $m = N_R - 1$ branches of the MST connecting all those points $\{\ell_i\}_{i=1}^m$ (see

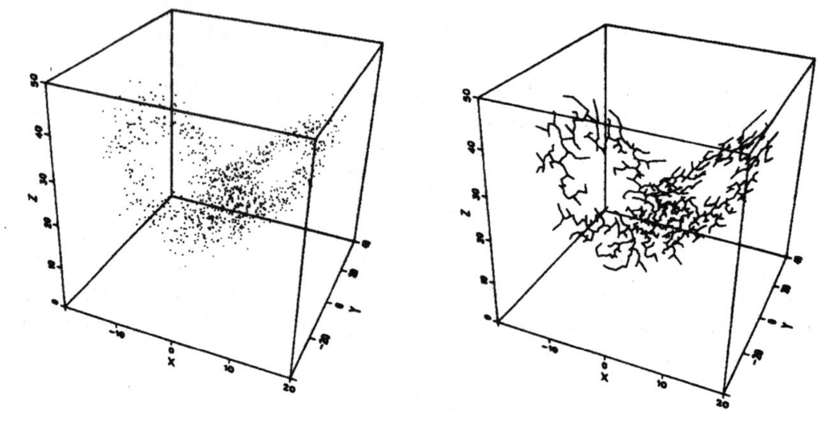

Figure 4.7 *The left panel shows 1000 points sampling the trajectory on the Lorenz attractor; the right panel shows its MST. (Reproduced, with permission, from Domínguez-Tenreiro, Roy, and Martínez 1992, Prog. Theor. Phys., 87, 1107–1118.)*

Fig. 4.7). We define the partition sum

$$S^\beta(m) = \frac{1}{m}\sum_{i=1}^{m} \ell_i^\beta .$$

If this function presents a scaling behavior as m varies when we perform different selections of N_R random points such as

$$S^\beta(m) = K(\beta) m^{-\beta/h(\beta)},$$

the fixed point of the function $h(\beta)$ provides a good estimator of the Hausdorff dimension of the attractor. Using this method Domínguez-Tenreiro, Roy, and Martínez (1992) estimated the Lorenz attractor Hausdorff dimension to be $D_H = 2.155 \pm 0.005$. A similar algorithm was applied by Martínez and Jones (1990) for the galaxy distribution. The value obtained by them for the CfA1 catalog was 2.1.

4.4.2 Box-counting dimension

In practice, one of the most widely measured fractal dimensions is the so-called box-counting or capacity dimension. It is rather easy to implement an empirical calculation of this quantity, and part of its popularity probably lies in this fact. For a window A in $\mathbb{R}^d$, we count the number of d-cubes of side ϵ that intersect A, $N(\epsilon)$ when we cover the set by an ϵ-mesh (a collection of disjoint and contiguous cubes of side ϵ covering the region where A lies). The capacity dimension is defined as

$$D_C = \lim_{\epsilon \to 0} \frac{\log N(\epsilon)}{\log(1/\epsilon)}. \tag{4.6}$$

For a point process this dimension is calculated performing a disjoint partition of the volume where the points lie in cells of equal size ϵ and counting the number of nonempty cells $N(\epsilon)$. Since in a discrete set the limit when $\epsilon \to 0$ has no meaning, this requirement in the previous definition is usually circumvented by estimating D_C by means of a logarithmic derivative

$$D_C \simeq \frac{d \log N(\epsilon)}{d \log(1/\epsilon)} \tag{4.7}$$

Direct measures of the capacity dimension by means of box-counting are not very efficient. In fact, for $\mathbb{R}^d$ with $d > 2$, it has been shown that box-counting algorithms provide rather bad performance.

VPF and box-counting dimension

The number of occupied cells can also be expressed in terms of the void probability function $P_0(r)$ or the empty space function $F(r) = 1 - P_0(r)$,

$$N(\epsilon) = N_{\text{cel}}(\epsilon) F\left(\sqrt[3]{\frac{3}{4\pi}}\epsilon\right), \tag{4.8}$$

where $N_{\text{cel}}(\epsilon) = V(A)/\epsilon^3$ is the number of cuboidal cells of side ϵ that can be put inside A.

In this way we are provided with a new procedure to estimate D_0 by replacing the above expression for $N(\epsilon)$v in (4.6) or in (4.7). For a homogeneous Poisson process with intensity λ, the situation is the following:

$$N(\epsilon) = \frac{V(A)}{\epsilon^3}\left[1 - \exp(-\lambda\epsilon^3)\right]. \tag{4.9}$$

Obviously, for a Poisson process, at large scales $N(\epsilon) \propto l^{-3}$ but, if λ is small, the exponential term will dominate at small scales. So, although the method introduced is correct, it can only be applied to samples with high enough number density and at long enough distances.

4.4.3 *Correlation dimension*

One of the most useful statistical measures to study fractality on large scales is the integral of the correlation function:

$$N(< r) = \bar{n} \int_0^r 4\pi s^2 (1 + \xi(s)) ds. \tag{4.10}$$

This function measures the number of neighbors on average that a given galaxy has within a distance r. A point pattern is said to be fractal with the correlation dimension D_2 when

$$N(<r) \propto r^{D_2}. \tag{4.11}$$

The correlation dimension is evaluated as:

$$D_2 = \frac{d(\log N(<r))}{d(\log r)} = 3 + \frac{d(\log \widehat{g}(r))}{d(\log r)}, \tag{4.12}$$

where $\widehat{g}(r)$ is the average of the structure function

$$\widehat{g}(r) \doteq \frac{1}{V}\int_0^r g(r')dV,$$

and the structure function $g(r) = 1 + \xi(r)$. As we see, $\widehat{g}(r)$ is the volume mean of Ripley's K-function, introduced in Chapter 3.

The two-point correlation function for a fractal pattern should vary as $1 + \xi(r) \propto r^{3-D_2}$. However, there are problems when we try to apply standard measures of correlation to fractal distributions. As Pietronero (1987) and Coleman and Pietronero (1992) have stressed, estimators of the correlation functions are based on the assumption of the homogeneity of the parent populations, which is not true for fractals. They introduce another statistic, the conditional density $\Gamma(r)$:

$$\Gamma(\mathbf{r}) = \frac{\langle n(\mathbf{x})n(\mathbf{x}+\mathbf{r})\rangle}{\bar{n}}, \tag{4.13}$$

where the angle brackets denote averages over a sample and $\bar{n}$ is the mean number density of points (galaxies) in the sample. The conditional density describes the distribution of mass around mass points and is usually assumed to be isotropic. For a simple fractal of dimension D the conditional density is a power law of r with the exponent $\gamma = D - 3$. The usual correlation function, which can be written as

$$\xi(r) = \frac{\Gamma(r)}{\bar{n}} - 1,$$

depends, if measured for a fractal, on the sample size (for a fractal of dimension D the mean density of a sample $\bar{n} \sim R^{D-3}$, where R is the depth of the sample).

If we try to verify the hypothesis of fractality of the galaxy distribution, we cannot use directly the usual border correction methods, as these assume the homogeneity of the parent population. Thus, the estimators for the conditional density $\Gamma(r)$ have to be chosen among minus-estimators, which use only those galaxies around a galaxy that are closer to that specific galaxy than the borders of the sample. This eliminates most of the sample points as centers for larger distances r and results in large sample variance. In any case, if edge-corrected estimators must be applied, caution has to be exercised to avoid the introduction of spurious homogeneity. We recommend applying the estimators to specific fractal point patterns to test this possibility. We return to this point in Section 4.4.5.

The average conditional density

$$\Gamma^\star(r) = \frac{3}{r^3}\int_0^r \Gamma(s)s^2 ds \tag{4.14}$$

is less noisy, at the expense of smoothing the possible details in $\Gamma(r)$. For a simple fractal, however, it is also a power law, $\Gamma^\star(r) \sim R^{3-D}$.

Pietronero and his colleagues applied these relations to galaxy catalogs in various papers (for a review see Sylos Labini, Montuori, and Pietronero 1988). They have always found a perfect power law for $\Gamma(r)$ and $\Gamma^\star(r)$ for the full argument range, confirming the fractal nature of the galaxy distribution. The left panel of Fig. 4.8 shows a typical result for the Stromlo–APM galaxy redshift survey. This can be compared with a later careful re-analysis of the same survey, using the same methods (Hatton 1999; the right panel of Fig. 4.8). The depths of the volume-limited samples, selected from the survey, are somewhat different, but the main difference is in the scale range chosen to determine the fractal dimension. Hatton showed that the results at small scales are severely contaminated by the Poisson noise. Eliminating these scales from consideration gives the fractal dimension that approaches the uniform $D = 3$, as the depths of the samples grow.

Crossover to homogeneity also has been found in other studies (Martínez, in press). In Fig. 4.9, we show the function $1 + \xi(r)$, estimated with proper care, for several deep galaxy redshift surveys (Guzzo et al. 1991; Bonometto et al. 1993; Martínez 1999). The fractal behavior at small scales disappears at larger distances, providing evidence for a gradual transition to homogeneity (Einasto and Gramann 1993). It is remarkable that the break of the scale-invariant power law appears at the same scale, approximately 10 to 15 h^{-1} Mpc, for the four samples. Martínez et al. (1998) have studied the correlation dimension by means of Eq. 4.11 for the Stromlo–APM survey, and have found a scale-dependent behavior of D_2. This quantity reaches a value of 2.8 at the largest scales probed by the sample in agreement with the results by Hatton (1999).

A scale-dependent correlation dimension was also found by Amendola and Palladino (1999) estimating the function $\widehat{g}(r)$ from the Las Campanas survey. These authors have developed an interesting method to reliably measure $\widehat{g}(r)$ at large distances based on radial cells, reporting values of D_2 around the homogeneity value 3 for the largest analyzed scales. In a recent paper, Pan and Coles (2000) studied the PSCz catalog by means of the multifractal algorithms. They obtained a value of $D_2 = 2.99$ at scales larger than $30\,h^{-1}$ Mpc, while for r below $\sim 10\,h^{-1}$ Mpc, they obtained $D_2 = 2.16$, a value comparable with that obtained by Martínez and Coles (1994) for the QDOT-IRAS galaxy redshift survey. The transition to homogeneity is also apparent in the plots of $\Gamma^\star(r)$ obtained by the proponents of the unbounded fractal universe. In Sylos Labini, Montuori, and Pietronero (1998), the plots of $\Gamma^\star(r)$ corresponding to samples such as the Stromlo–APM or the IRAS 2Jy show an unambiguous deviation from a power

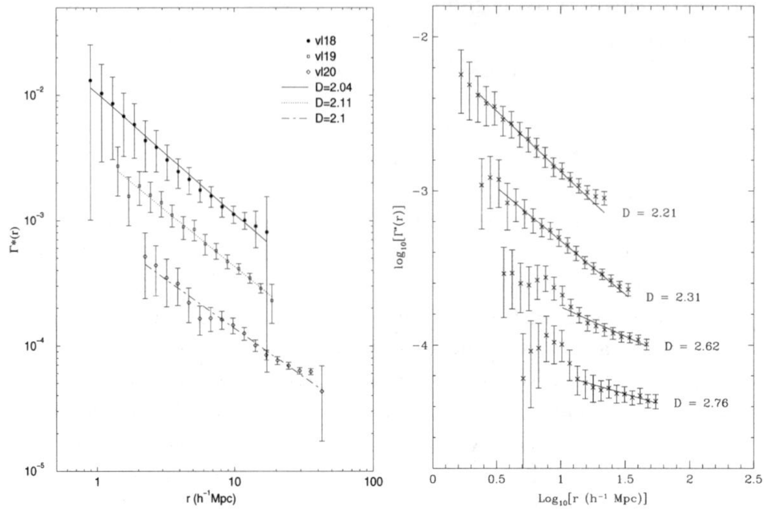

Figure 4.8 *The conditional average densities for volume-limited samples of different depth from the Stromlo–APM redshift survey. Observational results are denoted by dots and the respective power laws by lines (see the legends). Lower lines correspond to deeper samples, and the fractal dimensions for different samples are denoted by D. Errors are estimated by a bootstrap procedure. The left panel shows the results of Sylos Labini and Montuori (1998), and the right panel shows the results of Hatton (1999). (Reproduced, with permission, from Sylos Labini and Montuori 1998, Astr. Astrophys., 331, 809–814 and from Hatton (1999), Mon. Not. R. Astr. Soc., 310, 1128–1136. Blackwell Science Ltd.)*

law at large scales. Despite the fact that the authors interpreted these results in a different way, they could likely indicate the transition to homogeneity.

4.4.4 Correlation length and fractal behavior

The correlation length, r_0, is defined as the scale at which the correlation function reaches the value of 1, i.e., $\xi(r_0) = 1$. It is clear from the definition of the correlation function that at a distance r_0 of a given galaxy the density is, on average, twice the mean number density. The correlation length can be interpreted as the scale at which the density fluctuations change from the strongly nonlinear regime at short distances to the nearly linear regime at large scales.

One of the strong predictions of a fractal universe is that the correlation length must increase linearly with the radius of the sample (Pietronero 1987; Guzzo 1997). As the mean density of a fractal distribution decreases with sample depth, the correlation function of a fractal of dimension D in a spherical sample of radius

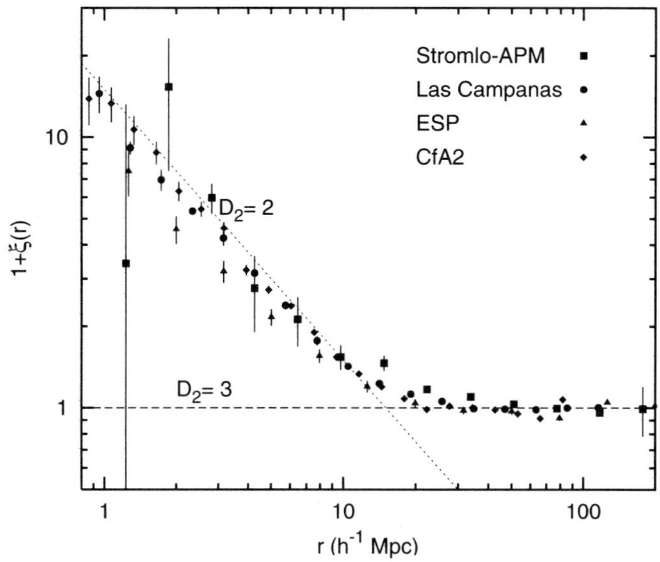

Figure 4.9 *The function* $1 + \xi(r)$ *for several redshift surveys (see the legend). Two reference lines have been plotted, one corresponding to a fractal with dimension* $D_2 = 2$ *and the other corresponding to a homogeneous distribution (*$D_2 = 3$*). We can appreciate how all data show a gradual transition from the fractal regime at short scales to a more homogeneous distribution at large scales.*

R_s centered on a point of the distribution is

$$\xi(r) = \frac{D}{3}\left(\frac{r}{R_s}\right)^{D-3} - 1,$$

and the correlation length

$$r_0 = \frac{D}{6}^{1/(3-D)} R_s$$

is proportional to the depth of the sample R_s.

This prediction has been tested on simple fractals, such as the one introduced in the previous section (Paredes, Jones, and Martínez 1995; Martínez, López-Martí, and Pons-Bordería 2001). Fig. 4.10 shows the behavior of r_0 with sample radius for spherical sample of a fractal clump generated by means of the Soneira and Peebles model introduced in Section 4.2.2.

We show here how this prediction is not supported by the observed galaxy distribution (Martínez et al. 1993; see Martínez, López-Martí, and Pons-Bordería 2001 for details). The analyses have been performed on the CfA2 redshift catalog (Geller and Huchra 1989; Park et al. 1994). From the CfA2 north survey, we have

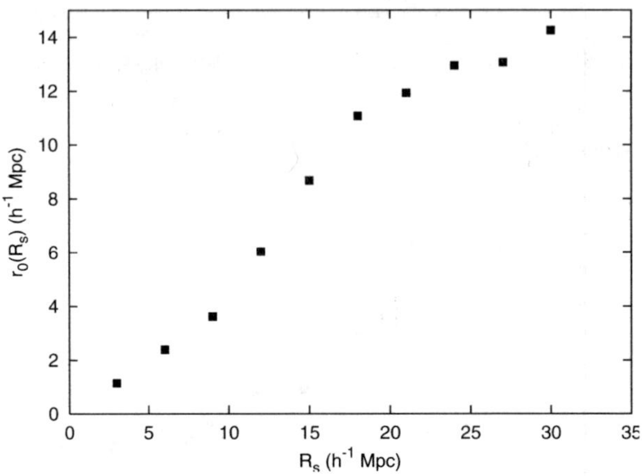

Figure 4.10 *The correlation length as a function of the sample radius for a fractal point distribution. (Reproduced, with permission, from Martínez, López-Martí, and Pons-Bordería 2001, Astrophys. J. Lett., 554, L5–L8. AAS.)*

extracted several volume-limited samples lying within the angular limits $8^h \leq \alpha \leq 16^h$ and $8.5° \leq \delta \leq 44.5°$, and with increasing depth 60, 70, 78, 85, and 101 h^{-1} Mpc, corresponding, respectively, to absolute magnitude limits of -18.49, -18.85, -19.10, -19.29, and -19.70 (omitting the term $+5\log h$).

Fig. 4.11 shows the three-dimensional diagrams of these samples. We have measured the correlation function of all these samples and estimated the value of the correlation length of each one by fitting a power-law function within the range 3–10 h^{-1} Mpc, weighting with Poisson errors. The values of the correlation length are given in the same figure (see also Fig. 4.12 with results also from the CfA southern hemisphere). These numbers show that the expected behavior for a fractal pattern is not observed; on the contrary, all samples present rather similar values of r_0, except the closest and smallest one. The nearly constant value of $r_0 \simeq 6.7\,h^{-1}$ Mpc in redshift space argues strongly against the unbounded fractal interpretation of galaxy clustering. A similar result was found by Cappi et al. (1998) analyzing the Southern Sky Redshift Survey 2. This result is also supported by the correlation length obtained for the deepest available redshift surveys analyzed to date: the Stromlo–APM, the Las Campanas redshift survey, and the ESP redshift survey. All of them provide a value $r_0 \simeq 6\,h^{-1}$ Mpc in redshift space (Loveday et al. 1995; Tucker et al. 1997; Guzzo et al. 2000). Moreover, recent studies of other deep samples have shown that the correlation length in real space is also very stable with redshift.

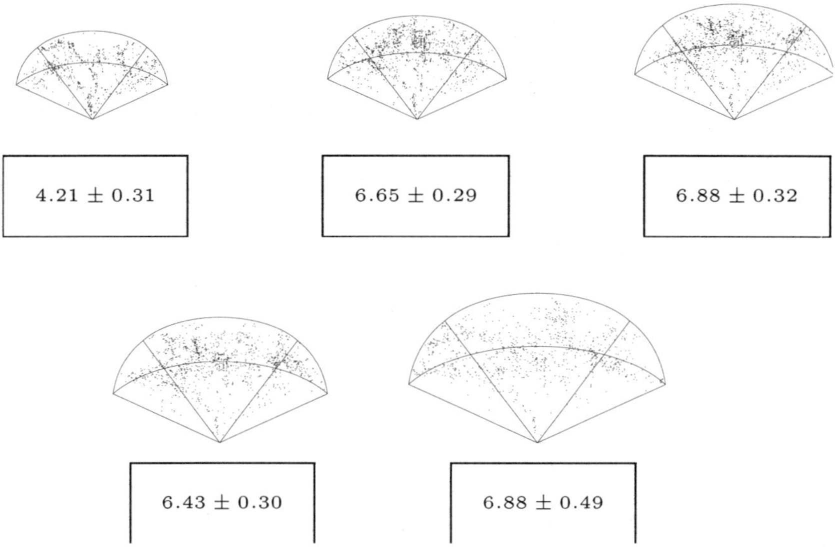

Figure 4.11 *Volume-limited samples extracted from the CfA2 north survey. From left to right and top to bottom the depth of the samples increases. The size of the diagrams* [illegible] *according to the real volume of the sample. The number of points in each one is, in the same order (left to right, above), 736, 1113, 1159, 1134, and 905. The correlation length* r_0 *of each sample is shown together with the standard deviation in the weighted log–log linear regression.*

4.4.5 Estimators, edge effects, and possible homogenization

Different estimators have been used in order to apply the statistical measures described above to the real galaxy distribution. Although it might not be very evident, the nub of the controversy lies in this matter. The problem arises because these estimates are affected by edge effects, which result from the inability to count neighbors outside the sample boundaries. Pietronero and co-workers avoid the edge correction by using the minus-estimators, which can be applied only up to the radius of the largest sphere enclosed within the sample volume. These estimators do not include as centers for counting neighbors at a given scale r those galaxies lying at a distance less than r of the sample boundary. The estimators used by most of the cosmologists assume implicitly that the point process is homogeneous (invariant under translations) and isotropic (invariant under rotations) and an edge correction is applied according to this assumption. For example, the denominator of Ripley's estimator for $K(r)$ (see Eq. 3.50), ω_{ij}, provides the edge correction, by measuring the conditional probability that the jth point is observed given that it is at a distance r from the ith point. For a homogeneous and isotropic

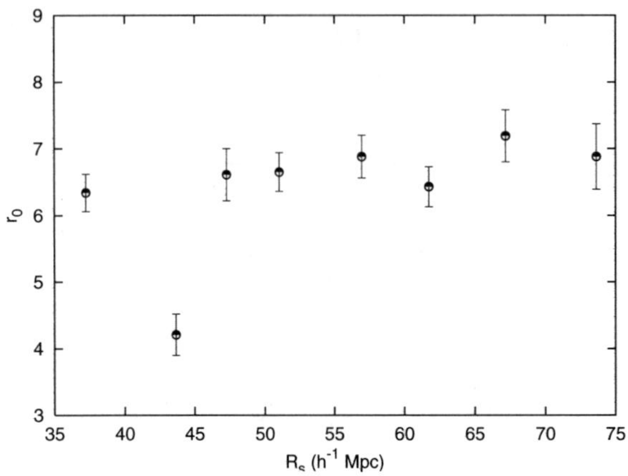

Figure 4.12 *The correlation length as a function of the sample size for several volume-limited subsamples drawn from the CfA2 redshift catalog. (Reproduced, with permission, from Martínez, López-Martí, and Pons-Bordería 2001, Astrophys. J. Lett., 554, L5–L8. AAS.)*

process it is rather straightforward to calculate this probability by geometrical methods. Similar corrections are applied to estimators of the correlation function $\xi(r)$, although typically, instead of geometrical corrections, the edge effects are taken into account by normalizing the number of galaxy pairs at a given distance with the corresponding quantities measured on Poisson samples. These samples are generated within the same volumes as the real ones and mimic their selection properties.

It has been shown, however, that on small scales all estimators provide comparable results (Kerscher 1999; Pons-Bordería et al. 1999), but at large scales the minus-estimator provides very poor performance (Kerscher, Szapudi, and Szalay 2000). Moreover, several authors (Martínez et al. 1990; Lemson and Sanders 1991; Provenzale, Guzzo, and Murante 1994) have simulated fractal point distributions generated within finite boundaries mocking the shape of the real galaxy catalogs. When applying the standard estimators for the correlation function to these fractal sets, these authors recover the expected theoretical power-law decaying behavior without finding the plateau observed in Fig. 4.12. This result indicates that the artificial homogenization supposedly introduced by the estimators is not playing an important role. Although this is a good test, strictly speaking, it only implies that if the galaxy distribution resembles such fractals, Fig. 4.12 should not present the observed flattening. Obviously, this does not resolve the issue unambiguously.

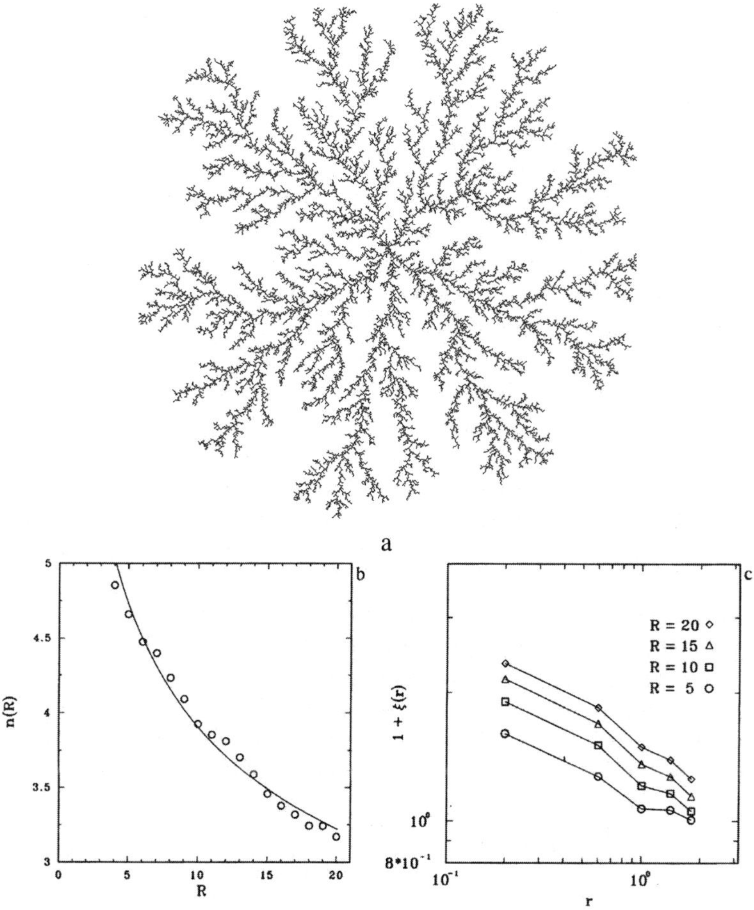

Figure 4.13 *(a) A realization of the diffusion-limited aggregation process. (b) The density within circles of radius R as a function of R. (c) The function $1 + \xi(r)$ calculated for the subsamples with different outer radius R. (Reproduced, with permission, from Paredes, Jones, and Martínez 1995, Mon. Not. R. Astr. Soc., 276, 1116–1130. Blackwell Science Ltd.)*

4.4.6 Mass–radius dimension

If, instead of an average as it appears in Eq. 4.10, we just look at the number of neighbors included in a sphere of radius r centered on the Earth, $M(r)$, we can define the "fractal dimension," D_M, as the exponent of the relation $M(r) \propto r^{D_M}$

(mass–radius relation) and therefore the radial density decreases as

$$n(R) \propto R^{D_M - d}$$

for a set in $\mathbb{R}^d$. Although this relation is less accurate than the one used in the definition of the correlation dimension, in which we average considering all the galaxies in the sample as possible centers, it has the advantage that it permits us to extend the measure of the dimension to much larger scales, due to the geometry of redshift surveys typically centered at the observer. We can illustrate the use of this dimension with a typical aggregation model studied in growth phenomena. It is the well-known diffusion-limited aggregation (DLA) introduced by Witten and Sander (1981). This process is simulated by setting a seed particle at the origin of a lattice. A new point moves around randomly until it visits a site in contact with the seed. Then its location is fixed and the particle becomes part of the cluster. Subsequent particles are introduced in the same way until they join the cluster.

The fractal cluster shown in Fig. 4.13 is the output of one DLA realization. It is obvious from the picture that the density in concentric circles of radius R is a decreasing function. In the figure we show how the density varies, falling off as a power law $n(R) \propto R^{D_M - 2}$ where D_M is the mass–radius dimension. Performing the calculation of $\xi(r)$ in several discs with different radius, we can see how the variation of density affects the amplitude of the correlation function. Fig. 4.13 shows the power-law shape of the structure function $(1 + \xi(r)) \propto r^{-\gamma}$ for the DLA subsamples marked with increasing radius. We can see how the smaller the density, the larger the amplitude of the structure function, while the slope always remains the same, $\gamma = 2 - D_M$. As explained in Section 4.4.4, this effect is a consequence of the density decrease with volume of the fractal pattern.

4.5 Multifractal measures

Multifractals have been applied to the description of chaotic dynamical systems in recent years. Multifractals are fractal sets with an invariant measure characterized by a whole spectrum of singularities, instead of a single number as in homogeneous fractals. The theory of multifractals was developed based on the pioneering work of Rényi in the framework of the probability theory (Rényi 1970). In this context the Rényi dimensions have been used in the description of strange sets by Hentschel and Procaccia (1983). However, a different approach came from a generalization of the classical concept of Hausdorff measures. This way of introducing the multifractal measures in the context of the analysis of nonlinear dynamical systems has been employed by Halsey et al. (1986), who noticed that the Hausdorff generalized dimensions are not always equivalent to the Rényi dimensions.

To illustrate a multifractal measure, we introduce here the model described by Meakin (1987). The construction of this set is performed following a multiplicative cascade (see also Martínez et al. 1990). We start dividing a square into four square pieces. We assign a probability number $\{f_i\}_{i=1}^4$ to each piece ($\sum_{i=1}^4 f_i = 1$). Each of the small squares is again subdivided into four pieces, assigning to each of the subsquares one of the numbers f_i randomly permuted. The probability attached to each one of the new 16 squares is the product of this number and the number of its parent. We continue this construction so on and so forth, each time multiplying the corresponding number f_i from the random permutation by all its ancestors. If we perform the previous construction until $L = 8$ levels, we have at the end a 256^2 lattice with a measure associated with each pixel. As an illustration we have performed several realizations of this process for different choices of the initial parameters:

Model	f_1	f_2	f_3	f_4
I	1/3	1/3	1/3	0
II	0.448	0.3	0.2	0.052
III	0.5714	0.2857	0.1429	0
IV	0.22	0.24	0.26	0.28

These realizations are shown in Fig. 4.14.

To characterize this multifractal measure one can use the so-called generalized or Rényi dimensions. We can illustrate this approach using the previous examples. First, we define the partition sum as

$$Z(q,\epsilon) = \sum_{i=1}^{N(\epsilon)} \mu(B_i(\epsilon))^q \tag{4.15}$$

where $\mu(B_i(\epsilon))$ is the measure associated with the box of size ϵ, $B_i(\epsilon)$. The Rényi dimensions are defined by

$$D_q = \lim_{\epsilon \to 0} \frac{\log Z(q,\epsilon)}{(q-1)\log \epsilon} \tag{4.16}$$

for $q \neq 1$; $D_1 = \lim_{q\to 1} D_q$, an avoidable discontinuity applying the l'Hôpital rule, is the information dimension. D_0 is the box-counting or capacity dimension (4.6). In the previous examples of multifractal measures it is rather straightforward to obtain the analytical expression of D_q (Falconer 1990, Martínez et al. 1990):

$$D_q = (1-q)^{-1} \log_2 \left(\sum_{i=1; f_i \neq 0}^{4} f_i^q \right).$$

It is clear that when all $f_i \neq 0$, $D_0 = 2$, being the dimension of the unit square. Model I is a simple fractal for which $D_q = D_0 = \log 3/\log 2 \simeq 1.58$. Model II is a "fat" fractal with $D_0 = 2$ (the dimension of the Euclidean space where the set is embedded), but it is also a multifractal that presents a value of the correlation dimension $D_2 \simeq 1.58$. Model III is a multifractal on a fractal support

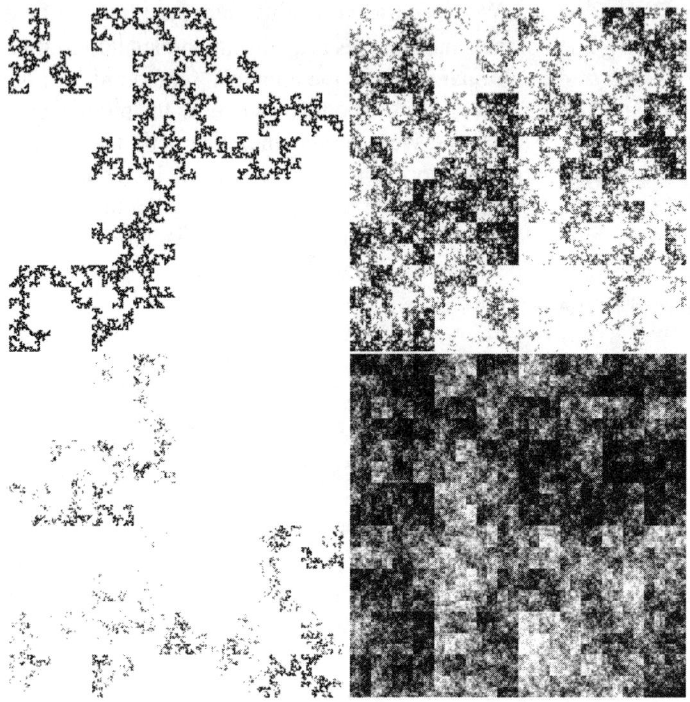

Figure 4.14 *Multifractal multiplicative cascades with parameters given in the text. Models are I, II, III, and IV, from top to bottom and left to right.*

with $D_0 \simeq 1.58$ and $D_2 \simeq 1.22$ and Model IV is, like Model II, a multifractal measure on a nonfractal support, the unit square. Fig. 4.15 shows the curves D_q for the four models. For the simple fractal, D_q is a straight line with zero slope. Multifractals present characteristic nontrivial shapes for D_q.

The previous approach is a generalization of the box-counting algorithm to define the function D_q. Likewise, it is often more efficient to generalize the correlation algorithm to obtain an expression for D_q. Formally the generalized dimensions defined in the following paragraphs are not completely equivalent, but we use the same notation as before for the dimensions, with the scale now represented by r. The scaling property used to define the correlation dimension $\langle N \rangle_r \propto r^{D_2}$ might be generalized to moments of any order if

$$Z(q,r) = \frac{1}{N} \sum_{i=1}^{N} n_i(r)^{q-1} \propto r^{\tau(q)}, \tag{4.17}$$

where $n_i(r)$ is the number of neighbors of point labeled by i within a sphere of

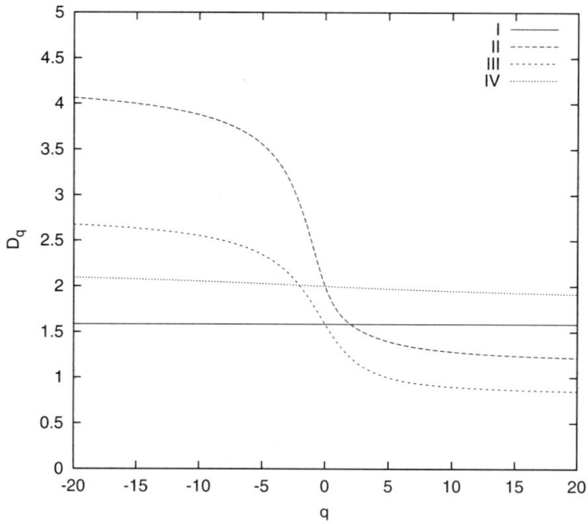

Figure 4.15 *The Rényi dimension for the multiplicative multifractal measures corresponding to Models I, II, III, and IV.*

radius r. A set is said to be a multifractal if the exponents or scaling indices $\tau(q)$ are well defined and independent of r in a suitable interval. Now we can define the generalized dimensions or Rényi dimensions as $D_q = \tau(q)/(q-1)$. It is obvious that for $q = 2$, we recover the scaling of (4.11), where $\tau(2) = D_2$. When the scaling relation (4.17) holds, we say that the point distribution has multifractal character. In a simple fractal $D_q =$ const for all q values, while for a multifractal set D_q is a decreasing function of q. The meaning of D_q is clear: when q is positive and large, the denser parts of the point distribution dominate the sums in (4.17), while for negative values of q the sums are dominated by the rarefied regions of the point set.

For $q < 2$, it is usually more convenient to obtain D_q through a different algorithm. Let us call $r_i(n)$ the radius of the smallest sphere centered at point i and enclosing n neighbors; in other words, $r_i(n)$ is the distance of point i to its nth neighbor. The exponents $\tau(q)$ are obtained through the relation

$$W(\tau, n) = \frac{1}{N} \sum_{i=1}^{N} r_i(n)^{-\tau} \propto n^{1-q}. \tag{4.18}$$

For $q = 0$, the previous equation provides a way for estimating the capacity dimension D_0. Galaxy samples such as the CfA1 provide a value of $D_0 \simeq 2.1$, while the correlation dimension of the same sample is $D_2 \simeq 1.3$ in the range of scales $1 \leq r \leq 10\,h^{-1}$ Mpc (Martínez et al. 1990; Jones, Coles, and Martínez

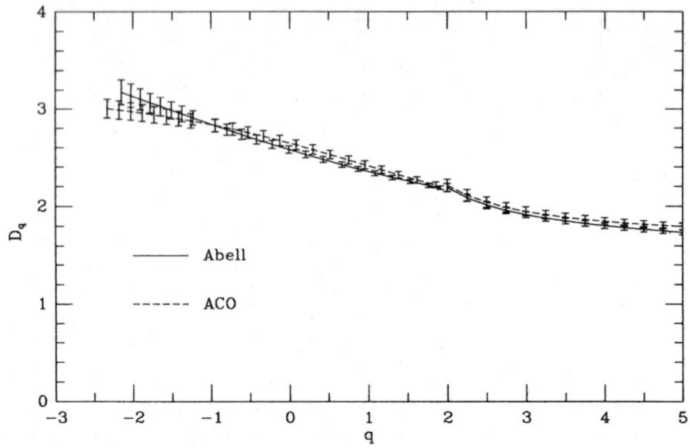

Figure 4.16 *The Rényi dimensions for the Abell and ACO clusters. (Reproduced, with permission, from Borgani et al. 1994, Astrophys. J., 435, 37–48. AAS.)*

1992). Eq. 4.17 is best suited for $q \geq 2$, while Eq. 4.18 works better for $q < 2$. Clusters of galaxies also present scaling behavior, although the D_q values are different from the D_q values for galaxies. Borgani et al. (1994) have shown that the correlation dimension turns out to be $D_2 \simeq 2$ for two different samples of clusters of galaxies within the range $15 \leq r \leq 60\,h^{-1}$ Mpc. The difference in the values of the correlation dimension for galaxies and for clusters of galaxies may be interpreted under the multiscaling hypothesis (Jensen, Paladin, and Vulpiani 1991), as shown in next section. The entire D_q function for clusters of galaxies is shown in Fig 4.16.

There is a clear connection between the multifractal approach and the moments of the counts in cells (Borgani 1993). In fact, the partition sum defined from the moments of the box-counting measures in (4.15) can be easily written in terms of the probabilities $P(N, V)$ introduced in Section 3.6:

$$Z(q, \epsilon) = N_{\rm cel}(\epsilon) \sum_{k=1}^{N_{\rm tot}} \left(\frac{k}{N_{\rm tot}} \right)^q P(N, \epsilon^3),$$

where $N_{\rm tot}$ is the total number of points and $N_{\rm cel}$ is the total number of cells. This approach has been exploited by Borgani (1993, 1995) who has obtained the D_q function for several point distributions and stochastic models with known expressions for the probabilities $P(N, V)$.

4.6 Multiscaling

Integral quantities such as $C(r)$ can still be estimated with enough reliability at large scales. For $r \geq 10\,h^{-1}$ Mpc, there is also information about the statistics of

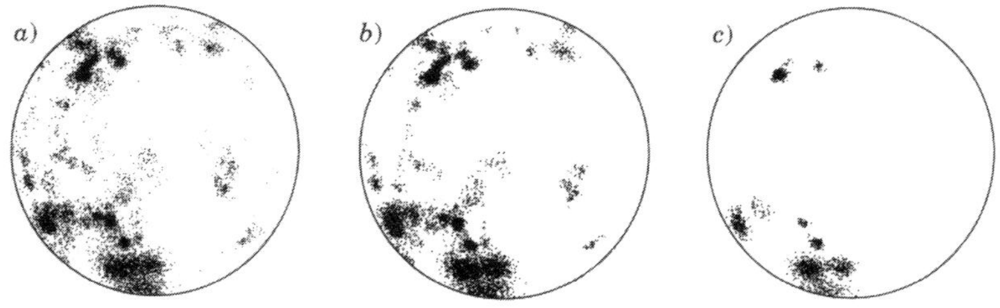

Figure 4.17 *The multifractal cascade model in 3-D (a), and after applying two different density thresholds (b and c). (Reproduced, with permission, from Martínez 1996, in Dark Matter in the Universe, IOS Press, Società Italiana di Fisica.)*

the distribution of clusters of galaxies. In this section, we will provide an explanation of the clustering of different objects such as galaxies or clusters within the same theoretical framework.

If the matter distribution is considered a continuous density field, we could think of galaxies as the peaks of the field above some given threshold. A larger threshold will correspond to clusters of galaxies. The higher the threshold, the richer the galaxy cluster. We can use the multiplicative model introduced in the previous section to illustrate this behavior. We build the model in 3-D by dividing a cube of side $100\,h^{-1}$ Mpc into eight equal cubic pieces of side $50\,h^{-1}$ Mpc. A probability number f_i is assigned to each part. The subdivision process is repeated L times. Jones, Coles, and Martínez (1992) have chosen $f_1 = f_2 = 0.07$, $f_3 = 0.32$, $f_4 = 0.54$, and $f_i = 0$ for $i = 5, \ldots, 8$. At the final stage we end with a lattice of $2^L \times 2^L \times 2^L$ cells, each with an assigned measure (density) $f_{i_1} \cdot f_{i_2} \cdots f_{i_L}$ resulting from the multiplicative cascade. The parameters of the model were chosen to provide values for $D_0 = 2.0$ and for $D_2 = 1.3$. Point distributions are generated on the cells of the lattice according to their densities. Small random shifts are applied to the position of the points to avoid regularity effects resulting from the lattice. By applying different density thresholds, which are quite naturally defined in this model, we obtain the distributions shown in Fig. 4.17.

The value of D_2 is approximately the slope of the log–log plot $Z(2,r)$ versus r, shown as a solid line in Fig. 4.18. After applying the threshold, $Z(2,r)$ still follows a power law, but with different slope (see the dotted and dashed lines in the figure). In the plot we can see that the higher the density threshold, the lower the value of D_2. Multiscaling is a scaling law where the exponent slowly varies with the length scale due to the presence of a threshold density defining the objects.

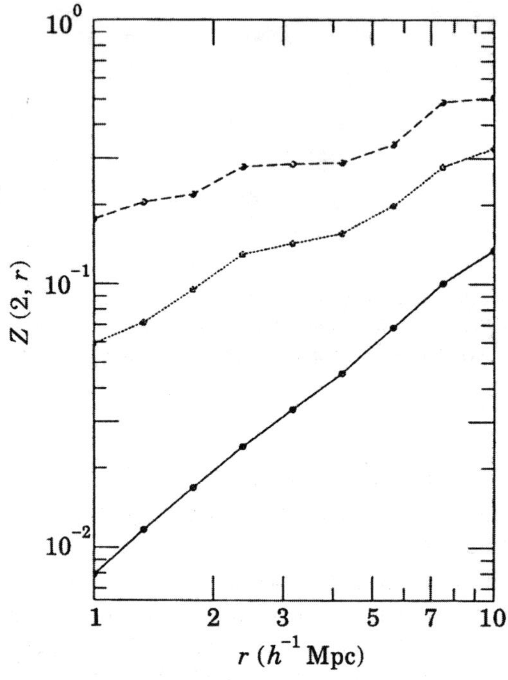

Figure 4.18 *The function $Z(2, r)$ for the multiplicative model. Note that $Z(2, r)$ is normalized here differently than in Eq. 4.17. Here $n_i(r)$ has been replaced with $n_i(r)/N_{\rm tot}$. Lower slopes are obtained when the density threshold is increased. (Reproduced, with permission, from Martínez 1996, in Dark Matter in the Universe, IOS Press, Società Italiana di Fisica.)*

We have seen that the observed matter distribution in the universe follows some sort of multiscaling behavior. If galaxies and clusters of galaxies with increasing richness are considered as different realizations of the selection of a density threshold in the mass distribution, the multiscaling argument implies that the corresponding values of the correlation dimension D_2 must decrease with increasing density.

We shall show the correlation integral for galaxy samples and for cluster samples in the range [10, 50] h^{-1} Mpc. In this range of scales $\xi_{gg}(r)$ does not follow a power-law shape, while $C(r)$ is nicely fitted to a power law. For galaxies we have analyzed the CfA1 sample, the Pisces–Perseus sample, and the QDOT-IRAS redshift survey. The cluster samples are the Abell and ACO catalogs, the Edinburgh–Durham redshift survey, the ROSAT X-ray-selected cluster sample, and the APM cluster catalog.

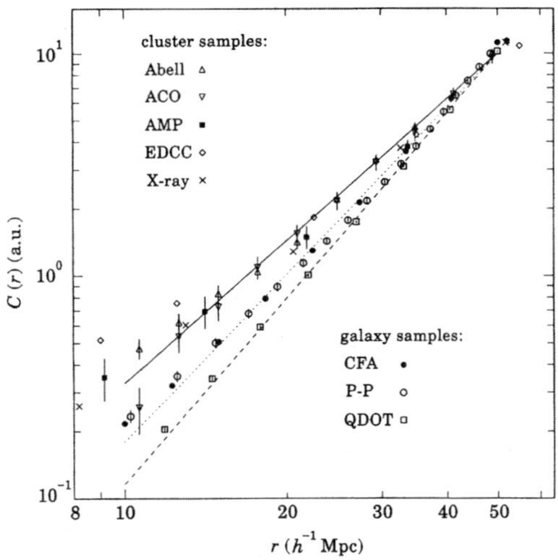

Figure 4.19 *The correlation integral for different galaxy and cluster samples in arbitrary units. Objects corresponding to higher peaks of the density field provide lower slopes, i.e., lower values of the correlation dimension. (Reproduced, with permission, from Martínez et al. 1995, Science, 269, 1245–1247; ©1995, American Association for the Advancement of Science.)*

Fig. 4.19 shows that three straight lines fit reasonably well the eight samples analyzed. All the cluster samples have a correlation integral well fitted by a power law with exponent $D_2 \simeq 2.1$. A value of $D_2 \simeq 2.5$ appears for the optical galaxy catalogs: the CfA1 volume-limited sample and the Pisces–Perseus survey within the range [10, 50] h^{-1} Mpc. Finally, a value of $D_2 \simeq 2.8$ is obtained for the QDOT-IRAS galaxies. Note that there is no contradiction between the value $D_2 \simeq 1.3$ for the range scale [1, 10] h^{-1} Mpc provided at the end of Section 4.5 for the CfA1 sample and the value reported here. In fact, as we have already noted in the discussion in Section 4.4.3, D_2 for the galaxy distribution is a scale-dependent quantity varying from 1.3 at small scales, where the approximation $D_2 \simeq 3 - \gamma$ is valid, to ~ 3 at large scales where the distributions are much more homogeneous. These results probe the multiscaling behavior of the matter distribution in the universe. The fact that D_2 for IRAS galaxies is larger than for optical samples indicates that IRAS galaxies are less correlated than optical galaxies; this is nicely interpreted if optical galaxies correspond to higher peaks of the density field. Clusters of galaxies have stronger correlations than single galaxies, corresponding to the highest peaks of the background matter density.

4.7 Lacunarity

Let us recall again the definition of the fractal mass–radius dimension D. If we have a ball of radius R, centered on the point belonging to a fractal structure, then the number of points in the ball (its mass) is given by

$$M(R) = FR^D, \tag{4.19}$$

where the prefactor F is a function (for deterministic fractals) or a random variable (for random fractals) that does not depend on R. Usually the main attention is focused on the determination of the fractal dimension (mass dimension), but that does not describe a fractal completely. Fractals of similar dimensions may have a completely different appearance, as illustrated in Fig. 4.20. We compare there two point distributions – one is an observed galaxy distribution in a thin slice from the Las Campanas survey (see Chapter 1) and another is its fractal model, the Rayleigh–Lévy flight, introduced in Section 4.2.1.

The two point distributions have the same fractal dimension, as shown by the comparison of the $\ln M$–$\ln R$ curves. But the overall appearance of the two distributions is different; they have different lacunarities. "Lacunarity" is a term that has been used loosely to describe the properties of the prefactor F. Blumenfeld and Mandelbrot (1997) proposed describing it by a series of variability factors

$$S_k = \frac{C_k}{C_1^k}, \tag{4.20}$$

where C_i is the i-th cumulant of the prefactor F. The simplest of these factors is the second-order variability factor

$$\Phi = \frac{E\{(F - \bar{F})^2\}}{\bar{F}^2}. \tag{4.21}$$

The distribution of the prefactor F for one-dimensional Lévy flight of dimension D is the Mittag–Leffler distribution with the density

$$f_D(x) = \frac{1}{\pi} \sum_{k=1}^{\infty} \frac{(-1)^{k-1}}{(k-1)!} \sin(\pi k D)\Gamma(kD)x^{k-1}$$

(Blumenfeld and Mandelbrot 1997). Using that distribution, Blumenfeld and Mandelbrot derived a simple expression for the variability factor:

$$\Phi(D) = \frac{2\Gamma^2(1+D)}{\Gamma(1+2D)} - 1.$$

For larger dimensions the distribution is not known, but the variability factor is easy to estimate from observations. The variability factors (lacunarities) Φ for our two distributions are also given in Fig. 4.20 (the lower right panel). For a strictly fractal distribution this factor should not depend on the radius of a ball R and any such dependence is evidence for nonfractality. Fig. 4.20 shows that

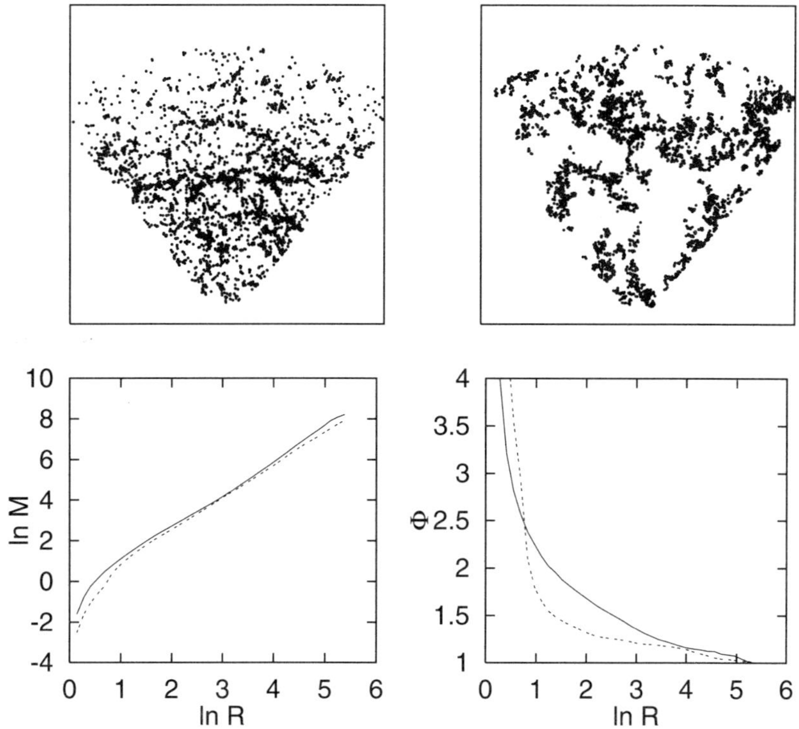

Figure 4.20 *Comparison of a Las Campanas survey slice (upper left panel) with its Lévy flight model (upper right panel). The fractal dimensions of both distributions coincide, as shown by the* $\ln M$–$\ln R$ *curves in the lower left panel, but the lacunarity curves (in the lower right panel) differ considerably. The solid lines describe the galaxy distribution, and the dotted lines show the model results. The monotonous decrease of the lacunarity curve for the LCRS slice indicates that the distribution is not as perfectly fractal as suggested by the lower left panel.*

both distributions are not strictly fractal. At small scales this is due to the Poisson noise, as in the Poisson distribution

$$\Phi \sim \frac{\lambda}{\lambda^2} = \frac{1}{\lambda} \sim \frac{1}{\bar{n}R^2},$$

where λ is the intensity and $\bar{n}$ is the mean density of the Poisson process. At larger scales the Lévy flight model soon approaches a constant value; the small deviations are probably due to the fact that this particular model is a sum of several (about 20) Lévy flights with random starting points. In contrast, the lacunarity curve for the galaxy distribution keeps diminishing with R, indicating that the observed sample is not such a perfect fractal as seems to be indicated by the almost perfect coincidence of the two $\ln M$–$\ln R$ curves. The fractal dimensions of the distributions, estimated for the interval 4–140 (h^{-1} Mpc), are $D = 1.58$ for the LCRS slice and $D = 1.59$ for the Lévy flight model.

Thus, the lacunarity analysis can complement the usual dimension-seeking fractal studies. It is easy to carry out; the only caveat is that we must use minus-estimators for cumulants in (4.20) and (4.21), as we cannot assume the characteristics of the point distributions outside the sample volume.

CHAPTER 5

Statistical and geometrical models of the galaxy distribution

5.1 Introduction

In the previous chapters we focused in particular on the development of statistical measures of the galaxy distribution. This chapter goes one step farther and introduces statistical models of the large-scale distribution of galaxies. While statistical models cannot claim to offer a full physical model of structure formation, unlike, e.g., N-body simulations, they do have the advantage of being more manipulative and having greater flexibility, two properties that are useful in studying the behavior of statistical measures under diverse circumstances and assessing the significance of these measures.

We introduce one of the first models that was used to describe the galaxy distribution, based on the Neyman–Scott prescription. The virtues and potential of this process and related applications are discussed. We show how different statistical descriptors of the point process function in the different stochastic models presented here. For example, in most cases, we provide the analytical expressions for the two-point correlation function and other statistical measures, although we do not intend to be exhaustive about this aspect. We only want to illustrate the use of the descriptor, providing additional information about the point process.

We next describe the cellular or sponge-like patterns in the galaxy distribution. We then discuss some ad-hoc models, and examine one of the most frequently used models, the Voronoi model, and its physical motivation, and give a mathematical introduction to the Voronoi tessellations. Using this as a basis we describe a kinematical model for clustering in the context of the Voronoi model. In connection with this we discuss how to simulate galaxy surveys in Voronoi models, and the nature of clustering properties of galaxies that would be clustered in the various elements of a Voronoi tessellation. In particular, we emphasize the clustering properties of the vertices in a Voronoi tessellation and how they relate naturally to the clustering of clusters.

The last part of the chapter deals with models for the galaxy clustering based on a phenomenologically motivated or physically motivated selection of the one-point distribution function of the density field. Two models are analyzed in some detail, one based on the lognormal distribution and the other one based on the gravitational quasi-equilibrium distribution proposed by Saslaw (2000).

5.2 The Neyman–Scott process and related models

In Chapter 3, we presented different examples of point processes: uniform Poisson, binomial, and inhomogeneous Poisson. In addition, in Chapter 4, we generated several fractal point patterns. In this chapter we introduce point field models that have been used as particular models of galaxy clustering in both cosmology and spatial statistics.

5.2.1 Cox fields

This is a generalization of the inhomogeneous Poisson process. The intensity measure is now a random function with a given distribution function. For this reason Cox processes are also known as doubly stochastic Poisson processes. They are simulated following two steps (Stoyan and Stoyan 1994):

1. An intensity function $\lambda(\mathbf{x})$ is chosen according to a given distribution function. More formally speaking, the intensity measure is generated from a driving random measure distribution (Stoyan, Kendall, and Mecke 1995), M; therefore, it is also said that the Cox process is directed by the random measure M (Cressie 1991).

2. An inhomogeneous Poisson process with that intensity function is generated.

It is obvious that, from a unique sample, a Cox process is indistinguishable from an inhomogeneous Poisson field. Having several samples of the Cox process, however, it should be possible, in principle, to infer the distribution function used to generate the different intensity functions.

In summary, in a Poisson process, the intensity λ is constant, in an inhomogeneous Poisson process the intensity function $\lambda(\mathbf{x})$ varies from point to point, and in a Cox process the intensity function is itself a random function.

Example: Segment Cox process

This point process was recently introduced in cosmology because its simulation is rather straightforward and its two-point correlation function is known analytically (Martínez et al. 1998; Pons-Bordería et al. 1999). The point field is produced in the following way: Segments of length l are randomly distributed within a cube W. Points are then scattered on the system of segments (see Fig. 5.1). The length density of the system of segments is $L_V = \lambda_s l$, where λ_s is the mean number of segments per unit volume. The intensity of the point process λ is the product of the mean number of points on a segment per unit length, λ_l, by the length density, L_V,

$$\lambda = \lambda_l L_V = \lambda_l \lambda_s l\,.$$

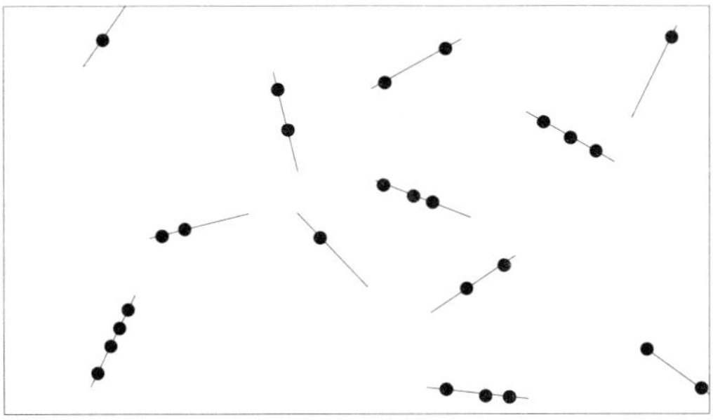

Figure 5.1 *An illustration of the generation of a planar segment Cox process.*

Fig. 5.2 shows a realization of a segment Cox process generated inside a cube of side length $L = 100\ h^{-1}$ Mpc. The chosen values for the parameters are $\lambda_s = 10^{-3}$, $\lambda_l = 0.6$, and $l = 10\ h^{-1}$ Mpc. Point patterns containing $N \simeq 6,000$ points are produced with this selection.

To calculate the correlation function of this point field we have to consider that the driving random measure of the point field is equal to the random length measure of the system of segments. Stoyan, Kendall, and Mecke (1995) showed that the two-point correlation function of the Cox point process equals the two-point correlation function of the system of segments, which reads

$$\xi_{\rm Cox}(r) = \frac{1}{2\pi r^2 L_V} - \frac{1}{2\pi r l L_V} \tag{5.1}$$

for $r \leq l$ and vanishes for larger r. It is worth noting that the expression (5.1) does not depend on the mean number of points per unit length of the segments.

The K function, the integral of the correlation function (see Eq. 3.45), is therefore:

$$K_{\rm Cox}(r) = \begin{cases} \frac{4\pi}{3} r^3 + \frac{r}{\lambda_s l}\left(2 - \frac{r}{l}\right), & r \leq l, \\ \frac{4\pi}{3} r^3 + \frac{1}{\lambda_s}, & r > l, \end{cases} \tag{5.2}$$

and the correlation dimension D_2 can be calculated analytically simply as:

$$D_2(r) = \frac{r}{K(r)} \frac{dK(r)}{dr}. \tag{5.3}$$

To illustrate these functions on a segment Cox processes, we have generated 10 realizations within a cube of side $100\ h^{-1}$ Mpc. The values for the parameters are $l = 20$, $\lambda_s = 4 \times 10^{-5}$, $\lambda_l = 1.88$. Fig. 5.3 shows the average values of the estimates of $K(r)$ obtained by means of the Ripley estimator (Eq. 3.50) together with the standard deviations (which are too small to be well appreciated).

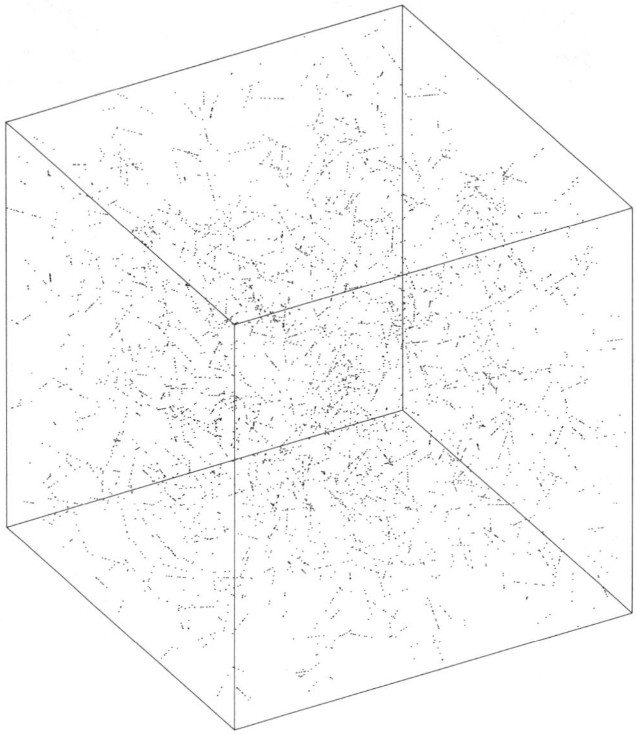

Figure 5.2 *Simulation of the segment Cox process with parameters* $\lambda_s = 10^{-3}, \lambda_l = 0.6$, *and* $l = 10\,h^{-1}$ *Mpc.* $N = 6,007$ *points have been plotted. (Reproduced, with permission, from Pons-Bordería et al. 1999, Astrophys. J., 523, 480–491. AAS.)*

In the plot the solid line of the bottom main panel represents the expected theoretical function given by Eq. 5.2. As we see, the empirical estimate of K (black dots) reproduces the expected theoretical behavior quite satisfactorily. The edge correction included in the estimator has not destroyed the goodness of the estimates; in particular, it has not introduced spurious homogenization. The estimator used (see Eq. 3.50) works well not only in the "easy" case of absence of structure, which represents Poisson processes (Martínez et al. 1998), but it has also been able to reproduce quite exactly the very precise value of K for a clustered Cox process. This test gives us enough confidence to believe that the K results obtained from the galaxy samples effectively reflect the structure existing there.

Models related to these kinds of Cox processes have been used in cosmology to mimic the galaxy distribution. For example, Buryak and Doroshkevich (1996) devised a model in which galaxies are randomly placed on straight lines and planes.

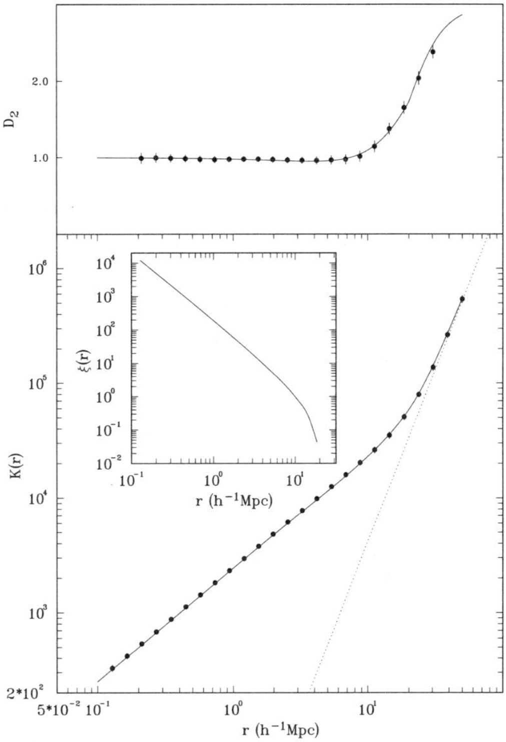

Figure 5.3 *Bottom panel: The average and rms error bars of the function $K(r)$ for 10 realizations of the Cox point processes (solid circles). The inset shows the two-point correlation function of this stochastic model. Top panel: The local correlation dimension D_2 with rms errors calculated by means of a five-point weighted log–log least square fit on the average of K. In both panels the solid line shows the theoretical values, while the dotted line in the bottom panel corresponds to $K_{\rm Pois}(r)$. (Reproduced, with permission, from Martínez et al. 1998, Mon. Not. R. Astr. Soc., 298, 1212–1222. Blackwell Science Ltd.)*

Caution with fractal interpretations

The Cox processes are also a good example with which to prevent the naive use of fractals in the analysis of point fields, as was anticipated by Stoyan (1994). The Fig. 5.3 inset shows the function $\xi(r)$ expected for this process. The function is the sum of two power laws: $\xi(r) = Ar^{-2} + Br^{-1}$ with $A = (2\pi\lambda_s l)^{-1}$ and $B = -(2\pi\lambda_s l^2)^{-1}$. At short distances the first power law dominates and therefore the function $\xi(r)$ can be nicely fitted to a power law $\xi(r) \propto r^{-\gamma}$ with $\gamma = 2$. This could lead us to interpret that the point set is a fractal when clearly it is not (Stoyan 1994; McCauley 1998). Because in this regime $\xi(r) \gg 1$, the same can be said for the function $1 + \xi(r)$. Looking at the top panel of Fig. 5.3, we can see the behavior of the empirical local correlation dimension D_2 calculated over the average of the 10 realizations of the Cox processes together with the rms deviations. The solid line represents the expected theoretical values (5.3). Again, we can see the reliability of the estimates. But what is more interesting in this example is the long plateau observed in the plot of the correlation dimension. The value $D_2 \simeq 3 - \gamma \simeq 1$ remains nearly constant for a broad range of scales, due to the particular behavior of the K function for this model. After the "fractal" behavior, a transition to homogeneity is clearly appreciated in both D_2 and $K(r)$. At this point we want to remark that, in the same way that the term "fractal" is not appropriate for the Cox process, even with a correlation function decaying as a power law at short scales (Stoyan 1994), the galaxy distribution, even holding a similar property, is not a fractal in a rigorous sense. However, in a looser use of the term fractal (Avnir et al. 1998), it could be appropriate to talk about a "fractal" regime to describe the range of scales where $K(r)$ follows a power law, bearing in mind that a real self-similar point pattern, for example the Soneira–Peebles model described in Section 4.2.2, satisfies other conditions (self-similarity) apart from a power-law decaying correlation function.

5.2.2 *Neyman–Scott fields*

A kind of point process very popular in the field of spatial statistics was originally introduced by Neyman and Scott (1958) in cosmology to model the distribution of galaxies. They are Poisson cluster processes because they are based on an initial homogeneous Poisson process with intensity λ whose events are called parent points. Around each parent point, a cluster of daughter points is scattered. The number of points per cluster is randomly generated according to a given discrete probability distribution function. The location of the daughter points with respect to their parent center is independently generated for each parent following a given density function. This law is the same for all parents. The final process is formed only by the offspring.

Example: Matérn process

In this subclass of the Neyman–Scott field proposed by Matérn (1960), each event of the parent Poisson process is surrounded by a sphere of radius R in which m points are distributed randomly (following a binomial process). The offspring of each parent point varies from center to center following a Poisson distribution with mean μ. Fig. 5.4 shows a sketch of a simulation of a Matérn process in two dimensions.

For a Matérn cluster process, Santaló (1976) derived an analytical expression for the two-point correlation function $\xi(r)$ (see also Stoyan, Kendall, and Mecke

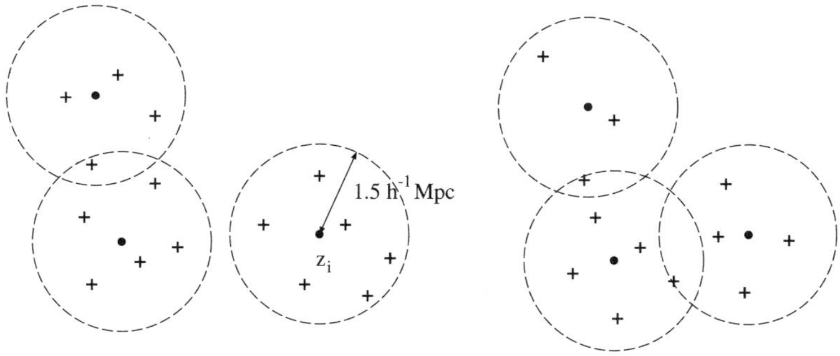

Figure 5.4 *A Matérn process is sketched here. The radius of the ball centered at each parent point is $R = 1.5\,h^{-1}$ Mpc and $\mu = 5$ is the mean number of points per cluster. (Reproduced, with permission, from Kerscher et al. 1999, Astrophys. J., 513, 543–548. AAS.)*

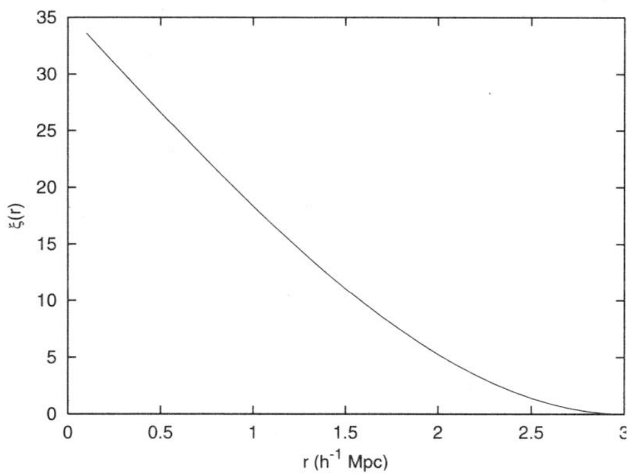

Figure 5.5 *The correlation function of a Matérn process with parameters $\lambda = 0.01$, $\mu = 5$ and $R = 1.5\,h^{-1}$ Mpc.*

1995). In $\mathbb{R}^3$ it has the form (see Fig. 5.5)

$$\xi(r) = \begin{cases} \dfrac{1}{\lambda\mu}\dfrac{3}{8\pi R^6}\left(R - \dfrac{r}{2}\right)^2\left(2R + \dfrac{r}{2}\right)\displaystyle\sum_{n=2}^{\infty}\dfrac{e^{-\mu}\mu^n}{(n-2)!}, & 0 \le r < 2R, \\ 0, & r > 2R. \end{cases}$$

Van Lieshout and Baddeley (1996) have derived the formula of the J-function corresponding to this point field, introduced in Section 3.8 In three dimensions the expression is the following (Kerscher et al. 1999):

$$J_{\mathrm{M}}(r) = \begin{cases} \dfrac{1}{\mathrm{Vol}(B_R)} \displaystyle\int_{B_R} e^{-\mu V(\mathbf{x},r,R)} d^3x, & 0 \le r \le 2R, \\ e^{-\mu}, & r > 2R, \end{cases} \tag{5.4}$$

where

$$V(\mathbf{x}, r, R) = \frac{\mathrm{Vol}(B_r(\mathbf{x}) \cap B_R)}{\mathrm{Vol}(B_R)}$$

denotes the ratio of the volume of the intersection of two balls to the volume of a single ball. Here $B_r(\mathbf{x})$ is a ball of radius r centered at the point $\mathbf{x}$, while B_R is a ball of radius R centered at the origin. This quantity can be calculated from basic geometric considerations (Stoyan and Stoyan 1994). In three dimensions, the expression is

$$V(x, r, R) = \begin{cases} c_3 x^3 + c_1 x + c_0 + c_{-1} x^{-1}, & 0 \le r < R, \quad R - r < x < R, \\ & \text{or } R \le r \le 2R, \quad r - R < x < r, \\ r^3/R^3, & 0 \le r < R, \quad 0 \le x \le R - r, \\ 1, & R \le r \le 2R, \quad 0 \le x \le r - R. \end{cases}$$

with $x = |\mathbf{x}|$ and

$$c_3 = \frac{1}{16R^3}, \qquad c_1 = -\frac{3}{8}\left(\frac{r^2}{R^3} + \frac{1}{R}\right),$$
$$c_0 = \frac{1}{2}\left(\frac{r^3}{R^3} + 1\right), \quad c_{-1} = \frac{3}{16}\left(\frac{2r^2}{R} - \frac{r^4}{R^3} - R\right).$$

Fig. 5.6 shows $J_{\mathrm{M}}(r)$ for $R = 1.5\, h^{-1}$ Mpc and several values of μ. Obviously, $J(r)$ discriminates between the varying richness classes of the Matérn cluster processes.

Matérn point processes can be easily modified by changing the distribution law for locating the points within each cluster. For example, if points are scattered within the balls following a symmetric normal distribution, we get a modified Thomas process. This process has a known correlation function (Cox and Isham 1980) following an exponential law.

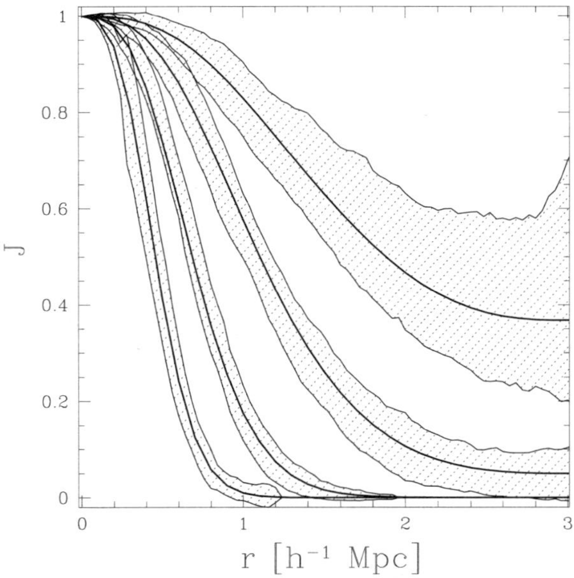

Figure 5.6 *The function* $J_{\rm M}$ *for Matérn processes with parameters* $R = 1.5\,h^{-1}$ *Mpc and* $\mu = 1, 3, 10, 30$ *(bending down successively). The solid lines correspond to the analytical expression (5.4) while the areas represent the rms fluctuations of the* J*-function calculated using 50 realizations of the point process. (Reproduced, with permission, from Kerscher et al. 1999, Astrophys. J., 513, 543–548. AAS.)*

Barlett (1964) has shown the formal equivalence between Cox processes and Neyman–Scott processes when, in this latter type, the number of points per cluster follows a Poisson distribution.

5.3 The Voronoi model

The Voronoi tessellation (Voronoi 1908) is a partitioning of space uniquely defined by a discrete point set. Each point of the set (nucleus) is surrounded by a Voronoi cell that encloses that part of space that is closer to its nucleus than to any other nucleus. The Voronoi tessellation is quite dependent on the initial distribution of nuclei. Depending on the degree of clustering or spatial correlation between points in the initial point set, the corresponding tessellation presents a complete different aspect. Fig. 5.7 shows the Voronoi tessellation corresponding to six different planar point distributions where different clustering patterns have been simulated, from regular (anticorrelated) distributions to distributions with a high degree of clustering, passing through a Poisson distribution. The regular point pattern for the nuclei produces cells in the Voronoi tessellation with similar shapes and areas. Increasing the clustering creates much more variability in the forms of the cells and in their areas as well. The visual differences between the Voronoi tessellations can be quantified by the study of Monte Carlo simulations. Van de Weygaert (1991) presents a detailed study of how several statistical quantities such as the area, the number of vertices, the edge lengths, the angles between neighboring edges, etc. are related to the cells of a tessellation. He shows how these quantities present quite different distributions as a function of the initial degree of clustering of the nuclei. In a sense, these quantities can be used as a measure of the clustering of the initial point pattern (van de Weygaert 1994).

In three-dimensional space, any realization of this process, called a "Voronoi foam" (Icke and van de Weygaert 1987), is built of three topologically distinct elements: walls, which are the bisecting planes between two neighboring nuclei; lines, where three walls intersect; and vertices, where four lines come together. The vertices are the centers of the circumscribing spheres of the Delaunay tetrahedra, whose packing is the Delaunay tessellation, the dual of the Voronoi tessellation. Each Delaunay tetrahedron consists of four nuclei whose circumscribing sphere does not contain any other nucleus. Fig. 5.8 shows a stereoplot of a packing of three Voronoi cells.

The Voronoi tessellation has acquired several alternative names: Dirichlet regions, Voronoi polygons or polyhedrons, Wigner-Seitz cells, Thiessen figures, and domains.

Icke and van de Weygaert (1987) and van de Weygaert and Icke (1989) used the Voronoi tessellation as a description of the geometric skeleton of the distribution of galaxies in space, based on the notion by Icke (1984) that underdense regions will expand with respect to the background in such a way that they become more and more spherical. The physical mechanism was called the Bubble theorem, in which the evolution of the low-density regions in a pressure-free self-gravitating collapse is followed. In this approach, one can see how voids expand while high-density regions collapse. The Voronoi walls corresponded to walls of galaxies, the lines to filaments of galaxies, the vertices to clusters of galaxies and the interior of the cells to the voids in the galaxy distribution (see Section 1.5). One of the most outstanding results was that the two-point correlation function of the Voronoi vertices with Poissonian nuclei has a power-law form on scales smaller than the average cell size with a slope $\simeq -2$ and an amplitude completely in accordance with the observational cluster–cluster correlation function determined by Bahcall and Soneira (1983), a result that nearly all proposed models of large-scale structure formation have failed to reproduce. This result, empirically found by van de Weygaert and Icke (1989), has been verified by the analytical results of Heinrich et al. (1998). These authors show that $\xi(r) \propto r^{-2}$ for the vertices process of a Voronoi tessellation with Poisson-distributed nuclei. Snethlage et al. (submitted) have shown that randomly shifting the positions of the vertices of the

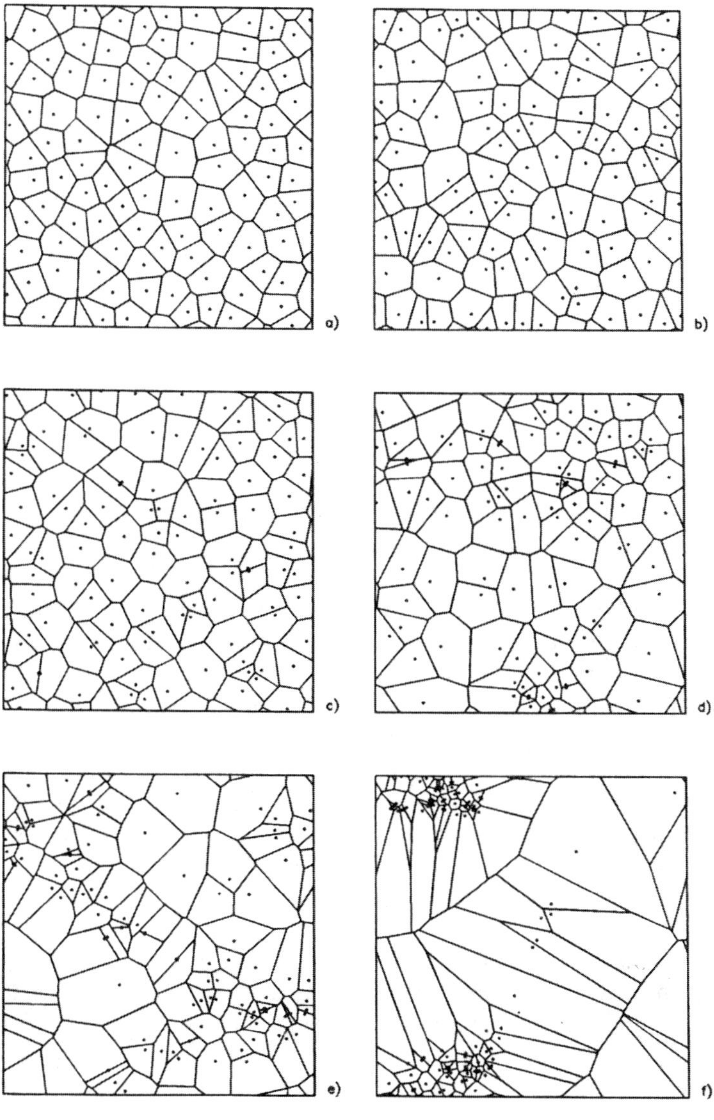

Figure 5.7 *The Voronoi tessellation corresponding to six different point patterns (as nuclei of the tessellation). The degree of clustering of the nuclei increases from top to bottom and left to right. The greater the clustering, the greater the variability of the shapes and sizes of the Voronoi cells. (Reproduced, with permission, from van de Weygaert 1991, Voids and the Geometry of the Large Scale Structure, Leiden University.)*

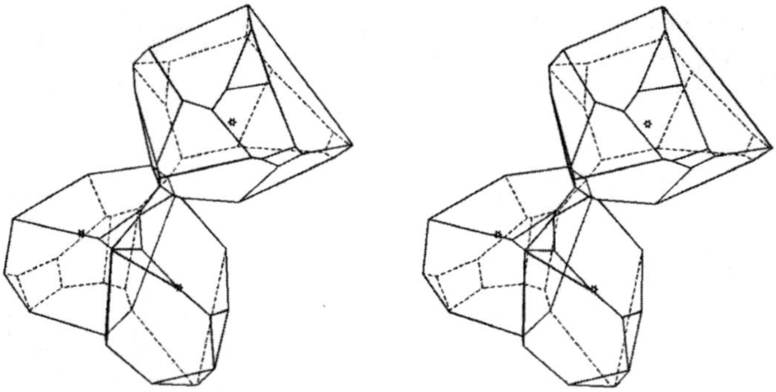

Figure 5.8 *Stereoscopic pair of three adjacent Voronoi cells. (Courtesy of Rien van de Weygaert.)*

tessellation produces a reduction of the value of γ. If the shifts follow a power-law probability distribution, one can get any desired value, $1 < \gamma < 2$, for the order of the singularity.

Details on Voronoi tessellations from the point of view of spatial statistics can be found in Okabe et al. (1999). In this section we use the Voronoi vertices as a paradigm for point clustering. Note that here we use the Voronoi vertex distribution purely as an example of a stochastic point process, without the physical background that underlies the use of Voronoi tessellations in the description of the large-scale structure in Icke and van de Weygaert (1987) and van de Weygaert and Icke (1989). The motivation for using the Voronoi vertices in this way is the convenient power-law clustering property as expressed in their two-point correlation function. The Voronoi vertex catalog used here consists of the 10,085 vertices of a Voronoi tessellation of 1,500 cells defined by 1,500 nuclei forming a binomial field. The 10,085 vertices are situated in a box 100 Mpc $\times$ 100 Mpc $\times$ 100 Mpc. Fig. 5.9 shows the point distribution, its projection onto the sky (only of those points within a sphere of radius 50 Mpc centered at the cube), and the two-point correlation function which follows a perfect power law $\xi(r) \propto r^{-2}$, at short scales (in the range $0.1 - 3$ Mpc). This property was exploited by van de Weygaert and Icke (1989) to model the distribution of clusters of galaxies by the vertices in the Voronoi foams, since the cluster–cluster correlation function follows a power law with exponent $\gamma \simeq 2$.

5.3.1 Simulating galaxy surveys

As an extension of the model for the distribution of clusters of galaxies, van de Weygaert (1991) introduced an elegant method to simulate galaxy samples. The

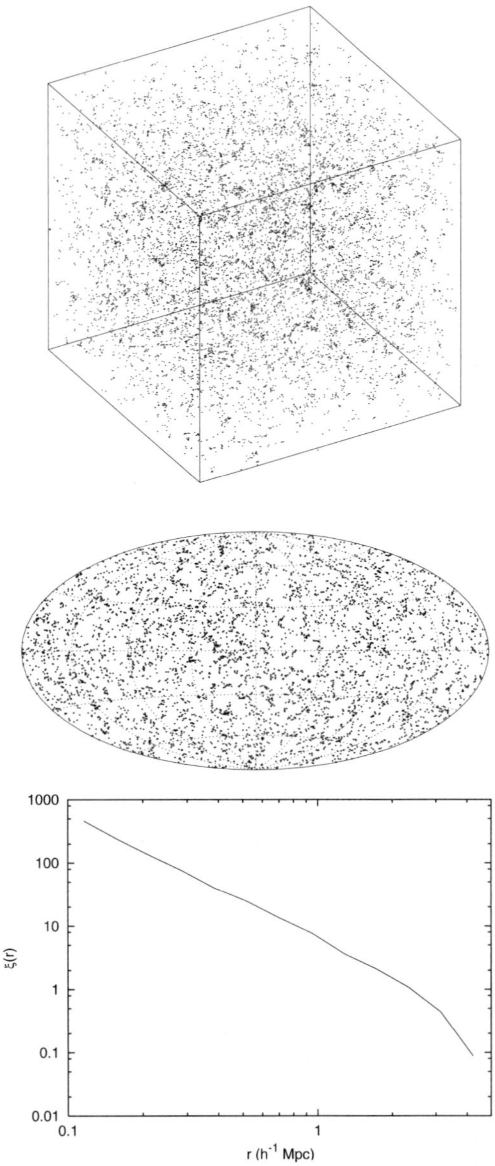

Figure 5.9 *A point field based on the Voronoi model. The top panel shows the distribution of vertices of a Voronoi tessellation built from 1,500 randomly distributed nuclei. A Hammer–Aitoff projection is shown at the center, while the bottom panel shows the two-point correlation function of this process, which follows a power law with exponent* -2.

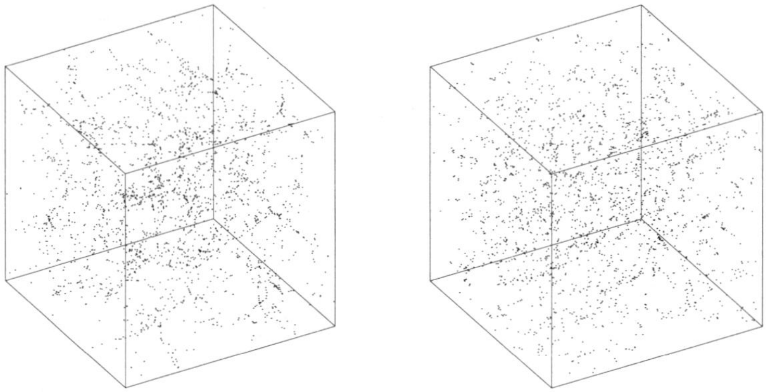

Figure 5.10 *Two realizations of point processes based on the scattering of points along filaments (left panel) or walls (right panel) of a Voronoi tessellation.*

idea comes from the observational fact, clearly noted in Chapter 1, that galaxies cluster around filaments and walls. Bearing this in mind, we can use the Voronoi foams as a skeleton with an evident cell-like structure, and place galaxies as desired:

1. At the edges of the tessellation (to populate filaments)

2. At the walls of the tessellation (to populate walls)

3. At the vertices of the tessellation (to populate clusters)

4. Or in any combination of the above elements with variable proportions

Different methods can be used to distribute the galaxies within a given Voronoi element. The distribution can be uniform or follow a given prescription such as a power-law density profile or a Gaussian distribution. Fig. 5.10 shows two realizations of this kind of Voronoi-based point process. In the left panel, points were placed following a Gaussian distribution along the edges of a Voronoi tessellation with Poisson-distributed nuclei, while in the right panel, points were placed populating the walls. This is obviously a highly interesting family of Voronoi-based point processes that merits further investigation both analytically and by Monte Carlo simulations.

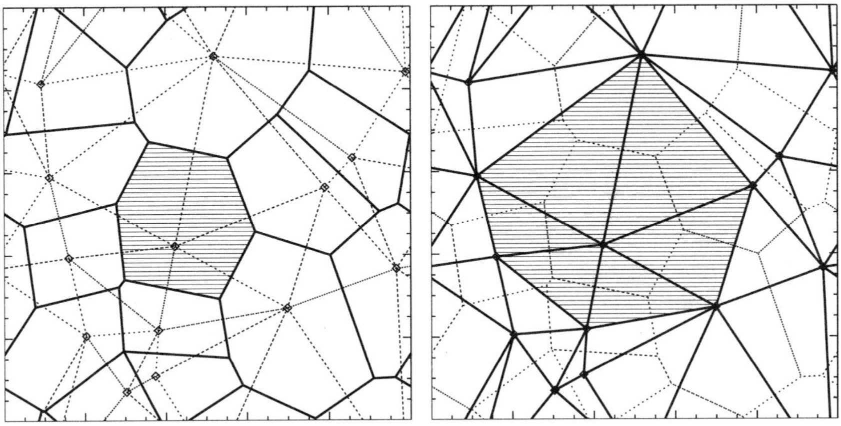

Figure 5.11 *A planar point process consisting of 20 points is depicted together with its Voronoi tessellation (left panel: solid lines) and its Delaunay tessellation (right panel: solid lines). In the left panel, the shaded region marks the Voronoi cell of a given point, while the right panel indicates the contiguous Voronoi cell associated with the same point. (Reproduced, with permission, from Schaap and van de Weygaert 2000, Astr. Astrophys., 363, L29–L32.)*

5.3.2 Spatial interpolation through Voronoi/Delaunay tessellations

One of the most promising cosmological applications of the Voronoi–Delaunay tessellations is the implementation of density reconstruction methods based in these kinds of geometric constructions. Schaap and van de Weygaert (2000) have introduced a fully self-adaptive method to estimate the intensity of a point process from the discrete realization. We saw in Section 3.3 that the estimation of the underlying density field from a Poisson sampling is usually done by smoothing the data by means of a kernel function. Schaap and van de Weygaert propose estimating the density at location $\mathbf{x}_i$ by the inverse of the volume of the contiguous Voronoi cell with the appropriate normalization, which guarantees the conservation of mass of the estimator. The contiguous Voronoi cell is the union of all Delaunay tetrahedra having point i as a vertex (see Fig. 5.11). Once the values of the intensity function are known for all points $\lambda(\mathbf{x}_i)$, a first-order interpolation procedure is applied to obtain the intensity at any point $\mathbf{x}$ (Bernardeau and van de Weygaert 1996) from the values of λ at the vertices of the Delaunay tetrahedron where $\mathbf{x}$ is included.

A very successful application of this reconstruction method on N-body simulations has been performed by Schaap and van de Weygaert (2001). In fact, the method is part of a more general approach to continuous field value estimation pioneered in astronomy by Bernardeau and van de Weygaert (1996) on rendering statistical distributions of discretely sampled velocity fields. The use of tessellations for spatial interpolation was anticipated by Sibson (1981). More recently, Braun and Sambridge (1995) have applied similar methods for solving partial differential equations.

5.4 Statistical models for the counts in cells

In Section 3.6 we introduced the statistics of the counts in cells, where $P(N, V)$ represented the probability that a randomly placed cell of volume V contains N galaxies. Several models of galaxy clustering are based on particular choices of this function, which are physically motivated. In some cases the physical motivation comes from the continuous version of the counts in cell statistics, the one-point distribution function of the density field. In other cases, as for the Saslaw distribution, more general physical principles are used to invoke such distributions.

5.4.1 The lognormal model

Hubble (1934), analyzing very preliminary projected catalogs of galaxies, was the first to notice that the distribution of galaxy counts in two-dimensional cells could be well approximated by a lognormal distribution.

The lognormal model for the density field in cosmology was introduced by Coles and Jones (1991), who found how this random field could be obtained from the gravitational evolution of an initially Gaussian field. They also present a model for the distribution of galaxies based on the lognormal one-point probability distributed function of the smoothed field. Fig. 5.12 shows different realizations of this model with different degrees of clustering obtained by increasing the variance of the underlying Gaussian field. It is worth mentioning that a similar cluster process was introduced much more recently in the field of spatial statistics by Møller, Syversveen, and Waagepetersen (1998). These authors termed this model the log Gaussian Cox process and they present more sophisticated ways of simulating point patterns of this model.

Assuming the Poisson model to extend the continuous lognormal distribution to provide a discrete distribution of galaxies, Coles and Jones (1991) derived an expression for the counts-in-cells probabilities

$$P(N = n) = \frac{1}{\sigma\sqrt{2\pi}}\frac{1}{n!}\int_o^\infty e^{-\lambda}\lambda^{n-1}\exp\left[-\frac{(\ln(\lambda/\beta) - \mu)^2}{2\sigma^2}\right]d\lambda,$$

where $\lambda = \beta\rho$ is the intensity of the Poisson process, ρ represents the mass density (the continuous density field), and β is a normalizing constant introduced to match the correct number density of objects. The correlation function of this model is presented in Section 7.5.

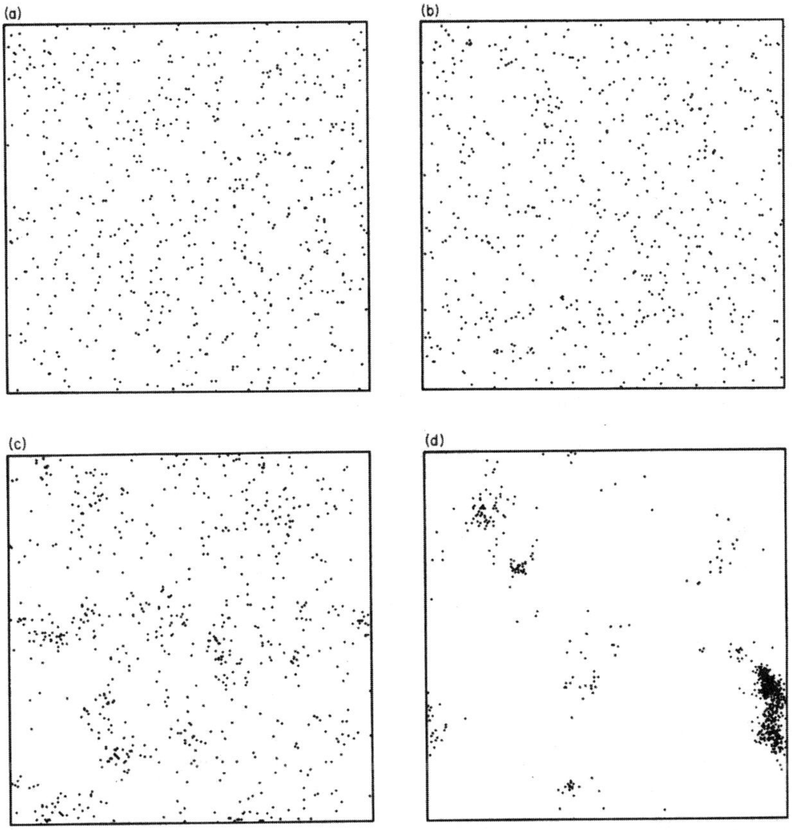

Figure 5.12 *Four realizations of a point process based on a lognormal random field with underlying Gaussian fluctuations having variances (a) 0.01, (b) 0.1, (c) 1.0, and (d) 10.0. (Reproduced, with permission, from Coles and Jones 1991, Mon. Not. R. Astr. Soc., 248, 1–13. Blackwell Science Ltd.)*

5.4.2 The Saslaw distribution function

The one-point galaxy distribution function has been approximated by different known distributions – the Poisson distribution, the Gaussian distribution, and the lognormal distribution. Each of these distributions has its own range of application; the motivation for selecting the distribution is based mainly on mathematical simplicity.

W. Saslaw proposed a different distribution function, based on physical (thermodynamical) arguments, which he calls the gravitational quasi-equilibrium distribution (GQED). Saslaw describes its properties thoroughly in his recent book

(Saslaw 2000). This function was derived assuming that a system of gravitating point masses in an expanding universe is in thermodynamical equilibrium. The probability of finding N galaxies in a randomly placed volume V is

$$P(N,V) = \frac{\bar{N}(1-b)}{N!}\left[\bar{N}(1-b)+Nb\right]^{N-1} e^{-[\bar{N}(1-b)+Nb]}. \qquad (5.5)$$

Here $\bar{N} = \bar{n}V$ is the expectation value of N ($\bar{n}$ is the mean galaxy number density), and the parameter b is defined as

$$b = -\frac{U}{2K},$$

where U is the average gravitational energy of density fluctuations and K is the average energy of peculiar velocities (fluctuations from the uniform Hubble expansion). As the distribution function is derived in a thermodynamical framework, an important quantity is the average temperature T of the point ensemble, defined by

$$K = \frac{mN\langle v^2\rangle}{2} = \frac{3}{2}NT,$$

where m is the mass of a particle and $\langle v^2\rangle$ is the dispersion of peculiar velocities. The energy ratio b can now be written as

$$b = \frac{2\pi G m^2 \bar{n}}{3T}\int_V \frac{\xi(\bar{n},T,r)}{r} r^2\, dr, \qquad (5.6)$$

where G is the gravitational constant and $\xi(r)$ is the correlation function of density fluctuations. Eq. (5.6) shows that the correlation function depends both on the mean number density and the temperature of the ensemble. Saslaw (2000) has adopted the dependence

$$b = \frac{b_0 \bar{n} T^{-3}}{1 + b_0 \bar{n} T^{-3}}.$$

The parameter b is either 0 for the case of negligible gravitational perturbations ($U = 0$) or 1 for gravitationally virialized systems where $2K + U = 0$. In the limit $b = 0$ the distribution (5.5) reduces to the Poisson distribution. The other limit $b = 1$ describes extremely large density fluctuations. This can be seen most easily from (5.5) for the case $N = 0$ (the void probability distribution):

$$P(0,V) = e^{-\bar{n}V(1-b)}.$$

This probability approaches 1 for $b \to 1$ (points concentrate to diminishingly small subvolumes). The dependence of b on the volume V can be found if we know the correlation function $\xi(r)$ or from the observed variance of number counts in cells (Saslaw 2000, § 28.4):

$$\langle(\Delta N)_V^2\rangle = \frac{\bar{N}}{\left(1-b(V)\right)^2}.$$

Assuming adiabatic (slow) evolution of structure, Saslaw derives the time evolution of b in cosmological context:

$$a(t) = a_\star \frac{b^{1/8}}{(1-b)^{7/8}},$$

where $a(t)$ is the dimensionless scale factor and $a_\star$ is a constant defined by the initial value of b at a certain epoch t.

Assuming that the kinetic energy of the fluctuations is proportional to their potential energy (gravitational quasi-equilibrium of the same degree on all scales) Saslaw obtains the peculiar velocity distribution density

$$f(u)du = \frac{2\alpha^2\beta(1-b)}{\Gamma(\alpha u^2+1)} \left[\alpha\beta(1-b)+\alpha b u^2\right]^{\alpha u^2-1} e^{-\alpha\beta(1-b)-\alpha\beta u^2} u\, du.$$

Here the normalized peculiar velocity $u = v(R/Gm)^{1/2}$, R is the size of the ensemble volume, Γ is the gamma function, $\beta = \langle u^2\rangle$ is the (normalized) peculiar velocity dispersion, and the density form factor $\alpha = \langle R/r\rangle_P \langle u^2\rangle^{-1}$, where $\langle R/r\rangle_P$ is the value of R/r averaged over a Poisson distribution with mean number density $\bar{n}$.

Specifying both the velocity and number density distributions gives a complete description of a gravitating system, but the velocity distribution also is important for statistical studies. We shall see in subsequent chapters that peculiar velocities can severely distort our estimates of the observed density distribution.

Borgani (1993) has shown that this model exhibits a very interesting multifractal behavior. From the expression of the $P(N,V)$ it is rather straightforward to show that the Rényi dimensions are

$$D(q) = \begin{cases} 3 - \dfrac{\gamma q}{1-q}, & q \le 1/2, \\ 3 - \gamma, & q > 1/2, \end{cases} \tag{5.7}$$

where γ is the exponent of the two-point correlation function.

CHAPTER 6

Formation of structure

6.1 Introduction

Because our observations are based on the properties of different physical fields, objects, and their populations (CMB, galaxies, etc.), we must understand how these objects form and evolve and how their properties are connected with those of the universe and with one another.

The present consensus in cosmology is that the observed structure developed from small initial perturbations of the physical fields (density, velocity, gravitational potential, etc.) resulting from the instability of the Friedmann models for small perturbations. The Newtonian theory of gravity also predicts the instability of a self-gravitating distribution of matter. While in the Newtonian theory the growth rate of perturbations is exponential in time, in cosmology the expansion of the universe slows the growth of perturbations and the growth rate is typically a power law. This slow growth continues until local densities become large enough to stop the expansion. After that, structure starts to evolve rapidly, forming the objects we observe now.

Below we delineate the minimal theory of formation of structure needed to understand the present statistical applications. A detailed representation of the theory can be found in Peebles (1980).

6.2 Dynamics of structure

Using the theory of the initial inflational stages of the universe, it can be shown that the present structure evolved from small fluctuations of the continuous matter density, velocity, and temperature fields. It can be also shown that when we are describing physics at spatial scales substantially less than the radius of the curvature of the universe, we can use the usual Eulerian hydrodynamics and Newtonian mechanics. The equations describing the evolution of the matter are then (see Peebles 1980; Peacock 1999)

$$\begin{aligned} &\frac{D\rho}{Dt} + \nabla \cdot (\rho \mathbf{w}) = 0, \\ &\frac{D\mathbf{w}}{Dt} = -\frac{1}{\rho}\nabla p - \nabla\Phi, \\ &\Delta\Phi = 4\pi G\rho, \end{aligned} \tag{6.1}$$

where $D/Dt = \partial/\partial t + \mathbf{w} \cdot \nabla$ is the convective derivative, ρ is the density of matter, p is the pressure, $\mathbf{w}$ is the velocity field, Φ is the gravitational potential, Δ is the Laplacian, and G is the gravitational constant. The first equation is the continuity equation, the second is the equation of motion, and the third is the Poisson equation. These equations have to be complemented by the equation of state $p = p(\rho)$ and, because this equation usually depends on temperature, an equation of energy, as well. During these stages of evolution when we need to use dynamics to predict statistical descriptors the pressure term in the equation of motion is negligible, and we shall omit it from further discussion. Pressure effects are important for the very early universe (radiation pressure) and at final stages of formation of objects when they counteract gravitational forces. There are no statistical applications for the first case yet and the pressure effects for the second case are accounted for in statistical theories in an approximate manner.

In cosmology it is more convenient to work in comoving coordinates $\mathbf{x}$ that are connected to the Eulerian coordinates $\mathbf{r}$ by

$$\mathbf{r} = a(t)\mathbf{x}, \tag{6.2}$$

where $a(t)$ is the dimensionless scale factor. Differentiating the above equation, we get

$$\partial\mathbf{r}/\partial t = \mathbf{w} = H(t)\mathbf{r} + a(t)\mathbf{u}, \tag{6.3}$$

where we have used (6.2) to replace $\mathbf{x}$ by $\mathbf{r}/a(t)$ and introduced the notation $\mathbf{u} = \partial\mathbf{x}/\partial t$ (recall that $H(t) = \dot{a}(t)/a(t)$). This equation tells us that the velocity in the Eulerian (physical) space is the sum of the Hubble expansion and the "peculiar" velocity $\mathbf{v} = a(t)\mathbf{u}$.

Since the continuity equation that expresses the conservation of matter has the same form in all coordinates, we get

$$\frac{D\rho}{Dt} + \nabla \cdot (\rho\mathbf{u}) = 0. \tag{6.4}$$

We can use the transformation (6.2, 6.3) to transform the Eulerian equation of motion into comoving coordinates, but it is easier to derive it as Peacock (1999) has done, differentiating $\mathbf{x}$ in (6.2) twice to get

$$\frac{D^2\mathbf{x}}{Dt^2} = a\frac{D\mathbf{u}}{Dt} + 2\dot{a}\mathbf{u} + \frac{\ddot{a}}{a}\mathbf{x} = -\nabla_r\Phi.$$

The last equality comes from the Eulerian equation of motion, and we have used the subscript r to show that this gradient is calculated with respect to the Eulerian coordinates. The transformation formula (6.2) says that $\nabla_r = (1/a)\nabla_x$. Using the Eulerian equation of motion together with the Poisson equation for the case of the smooth motion $\mathbf{u} = \mathbf{0}$, $\mathbf{w} = (\dot{a}/a)\mathbf{r}$, we get

$$\frac{D\mathbf{w}}{Dt} + \mathbf{w} \cdot \nabla\mathbf{w} = \frac{\ddot{a}}{a}\mathbf{x} = \nabla_r\bar{\Phi},$$

where $\bar{\Phi}$ is the solution of the equation

$$\Delta_r \bar{\Phi} = 4\pi G \bar{\rho}.$$

This solution diverges at infinity ($\bar{\Phi} \sim x^2$). Newtonian theory is not applicable there, but it is correct for most cases, and we use this solution as a tool to eliminate the term $(\ddot{a}/a)\mathbf{x}$ from our equation of motion and write

$$\frac{D\mathbf{u}}{Dt} + 2\frac{\dot{a}}{a}\mathbf{u} = -\frac{1}{a^2}\nabla_r \Phi',$$

where Φ' is the perturbation of the gravitational potential $\Phi' = \Phi - \bar{\Phi}$. As the Laplace operator is linear, we can write

$$\Delta_r \Phi' = 4\pi G(\rho - \bar{\rho}).$$

There is no trouble with this Poisson equation because the volume integral of the source term, the fluctuating density, is zero.

We shall drop below the coordinate index r and the prime and write the fluctuating potential as Φ. We will also use the density contrast δ instead of density:

$$\delta(\mathbf{x}) = \frac{\rho(\mathbf{x}) - \bar{\rho}}{\bar{\rho}}.$$

Note that $\bar{\rho} = \bar{\rho}_C$, if calculated in comoving coordinates, does not depend on time, and is related to the Eulerian $\bar{\rho}_E$ we have used thus far as $\bar{\rho}_E = a^{-3}\bar{\rho}_C$. The equations of dynamics in comoving coordinates now become

$$\frac{D\delta}{Dt} + \nabla \cdot [(1+\delta)\mathbf{u}] = 0, \tag{6.5}$$

$$\frac{D\mathbf{u}}{Dt} + 2\frac{\dot{a}}{a}\mathbf{u} = -\frac{1}{a^2}\nabla\Phi, \tag{6.6}$$

$$\Delta\Phi = 4\pi G a^{-1}\bar{\rho}\delta. \tag{6.7}$$

Using the comoving density simplified the derivation of the continuity equation. You will not see these equations often in the cosmological literature, however, because the Eulerian (physical) density field is preferred in the dynamical equations. Because the density contrast is defined as a normalized fractional amplitude, its value is the same in both cases. The mass of a volume element is

$$dM = \rho_E(1+\delta_E)\,dV_E = \rho_C(1+\delta_C)\,dV_C,$$

and because $\rho_E\,dV_E = \rho_L\,dV_L$, the density contrasts are really the same. The comoving velocity $\mathbf{u}$ is usually also replaced by the physically observable peculiar velocity $\mathbf{v} = a\mathbf{u}$. Using the relation $\rho_E = a^{-3}\rho_C$ and dropping the subscript E, we get another version of the equations of dynamics

$$\frac{\partial\delta}{\partial t} + \frac{1}{a}\nabla \cdot [(1+\delta)\mathbf{v}] = 0, \tag{6.8}$$

$$\frac{\partial\mathbf{v}}{\partial t} + \frac{1}{a}\mathbf{v}\cdot\nabla\mathbf{v} + \frac{\dot{a}}{a}\mathbf{v} = -\frac{1}{a}\nabla\Phi, \tag{6.9}$$

$$\Delta\Phi = 4\pi G a^2 \bar{\rho}\delta. \tag{6.10}$$

These equations are approximate only in the sense that we cannot use them for scales close to the curvature radius and in extremely strong fields (e.g., in black holes), but everywhere else they are perfectly applicable. In particular, we can use them to describe the large densities and velocities that occur in the observable universe.

6.3 The linear approximation

6.3.1 Density evolution

Although the equations of dynamics do not appear too complex, they are impossible to solve analytically, in general, but nonetheless many useful approximations have been found. In cosmology even the most simple linear approximation goes a long way — the amplitude of the density fluctuations after recombination, when the real growth of the structure starts, is about 10^{-3}. We shall also see later that at many observationally relevant scales the density contrast is still so small that it can be described by linear dynamics.

In the linear approximation we can replace convective derivatives by partial time derivatives, as $|\mathbf{u}| \ll 1$. We can also neglect the term $\delta\nabla\cdot\mathbf{u}$ in the continuity equation and write

$$\frac{\partial\delta}{\partial t} + \nabla\cdot\mathbf{u} = 0. \tag{6.11}$$

Taking appropriate derivatives, we obtain from (6.5, 6.6), respectively,

$$\frac{\partial^2\delta}{\partial t^2} + \nabla\cdot\frac{\partial\mathbf{u}}{\partial t} = 0, \tag{6.12}$$

$$\nabla\cdot\frac{\partial\mathbf{u}}{\partial t} + 2\frac{\dot{a}}{a}\nabla\cdot\mathbf{u} = -\frac{1}{a^2}\Delta\Phi. \tag{6.13}$$

Combining now equations (6.12, 6.13, 6.11) and (6.10) we get a second-order equation for the evolution of the density contrast:

$$\ddot{\delta} + 2\frac{\dot{a}}{a}\dot{\delta} - 4\pi G\bar{\rho}\delta = 0. \tag{6.14}$$

Because this equation does not depend explicitly on the spatial coordinates and thus is local, we have replaced partial time derivatives with ordinary ones. Although gravity is a nonlocal force, it enters in the linear approximation only through the Laplacian of the potential, which is local. This property disappears in subsequent approximations, which are nonlocal.

Equation (6.14) is a linear second-order ordinary differential equation, and its solution can be written as

$$\delta(\mathbf{r},t) = A(\mathbf{r})D_1(t) + B(\mathbf{r})D_2(t).$$

The partial solutions $D_1(t)$ and $D_2(t)$ are called, respectively, the growing and the decaying modes (obviously, we can order them in this way).

We shall illustrate the solutions in a simple case of the Einstein–de Sitter ($\Omega_M = 1, \Omega_\Lambda = 0$) universe. In this case $4\pi\bar{\rho} = 2/(3t^2)$, $\dot{a}/a = 2/(3t)$, and the density contrast equation is

$$\ddot{\delta} + \frac{4}{3t}\dot{\delta} - \frac{2}{3t^2}\delta = 0.$$

This equation is homogeneous in t; thus, the partial solutions are power laws, $\delta \sim t^s$. Substituting this form into the equation we get two solutions $s = 2/3$ and $s = -1$, the first for the growing mode $D_1(t) = t^{2/3}$ and the second for the decaying mode $D_2(t) = t^{-1}$. The decaying mode is usually discarded in theoretical studies; for longer periods of evolution its amplitude becomes negligible compared with that of the growing mode if we do not have very specific initial conditions.

As we see, the growing mode is proportional to $a(t)$ in this simple model. We see also that although the density contrast δ is growing with time, the physical overdensity $\widetilde{\rho} = \rho - \bar{\rho} = \bar{\rho}\delta$ actually gets smaller, $\widetilde{\rho} \sim a^{-2} \sim t^{-4/3}$; in the linear regime the infall of matter into objects is slower than the expansion of the universe. The formation of structure is due to the fact that at some limiting value of δ the process becomes nonlinear and causes a really fast growth of density.

Another simple case is that of the empty universe, when $a(t) \sim t$ and $\bar{\rho} = 0$. Of course, there can be no matter density perturbations without matter, but as before, we can consider this model as a good and simple approximation of a low-density universe. The linear density contrast equation is

$$\ddot{\delta} + \frac{2}{t}\dot{\delta} = 0.$$

This equation is easy to solve and it gives for the modes the results $D_1(t) = \mathrm{const}$ and $D_2(t) = t^{-1}$. If we recall the evolution of the instantaneous cosmological parameters in the case of an open model (it is close to the flat case until a certain redshift and then starts to expand faster), we can say that the growth of the density contrast "freezes" in an open model at a certain epoch.

In a general case the linear density contrast equation (6.14) can be solved almost completely analytically (Saar 1973). Changing the argument from time to redshift by

$$\frac{dz}{dt} = -H_0(1+z)E(z)$$

(Section 2.2) the density contrast equation transforms to

$$\delta'' + \left(\frac{E'}{E} - \frac{1}{x}\right)\delta' - \frac{3}{2}\Omega_M\frac{x}{E^2}\delta = 0.$$

Here $x = 1 + z$ is a redshift-type variable and primes denote differentiation with respect to x. Defining now a new function $\beta(x)$ by

$$\delta = E\beta$$

and substituting that into the above equation we get

$$E\beta'' + \left(3E' - \frac{E}{x}\right)\beta'' + \left(E'' + \frac{E'^2}{E} - \frac{E'}{x} - \frac{3}{2}\Omega_M \frac{x}{E}\right)\beta = 0.$$

It can be checked that the coefficient for β is zero for any $E^2(x)$ that is a cubic polynomial in x, and that this class also includes the E^2 for the canonical $(\Omega_M, \Omega_\Lambda)$ models; see (2.21). Thus we can write

$$E\beta'' + \left(3E' - \frac{E}{x}\right)\beta' = 0,$$

which gives

$$\beta' = A\frac{x}{E^3},$$

where A is a constant. So we get the full solution

$$\delta(z) = AD_1(z) + BD_2(z),$$

where the growing mode is

$$D_1(z) = E(z)\int_z^\infty \frac{(1+z')}{E^3(z')}dz'$$

and the decaying mode is

$$D_2(z) = E(z).$$

The integral for the growing mode can be taken analytically for all $\Omega_\Lambda = 0$ models (Sahni and Coles 1995):

$$D_1(z) = \frac{1 + 2\Omega_M + 3\Omega_M z}{|1 - \Omega_M|^2} + 3\Omega_M \frac{(1+z)\sqrt{(\Omega_m z + 1)}}{|1 - \Omega_M|^{5/2}} I(\Omega_M, z), \quad (6.15)$$

where

$$I(\Omega_M) = \begin{cases} -\dfrac{1}{2}\log\left[\dfrac{\sqrt{\Omega_M z + 1} + \sqrt{1 - \Omega_M}}{\sqrt{\Omega_M z + 1} - \sqrt{1 - \Omega_M}}\right], & \Omega_M < 1, \\ \arctan\sqrt{\dfrac{\Omega_M z + 1}{\Omega_M - 1}}, & \Omega_M > 1. \end{cases}$$

For the models with a cosmological constant the solutions have to be found numerically.

6.3.2 *Velocity evolution*

Although we had to eliminate the velocity field from the dynamical equations in order to get the equation for the density contrast, we must also describe it, as it is, in principle, directly observable.

The linear velocity equation is

$$\frac{\partial \mathbf{u}}{\partial t} + 2\frac{\dot{a}}{a}\mathbf{u} = \mathbf{g}, \quad (6.16)$$

where the gravitation acceleration $\mathbf{g}$ can be written, solving the Poisson equation, as

$$\mathbf{g}(\mathbf{x}) = -G\bar{\rho}\nabla \int \frac{\delta(\mathbf{x}')}{|\mathbf{x}' - \mathbf{x}|}. \tag{6.17}$$

The velocity field, as any vector field, can be written as the sum of a longitudinal (nonrotational) field and a transverse (rotational) field, $\mathbf{u} = \mathbf{u}_L + \mathbf{u}_T$, $\nabla \times \mathbf{u}_L = \mathbf{0}$, $\nabla \cdot \mathbf{u}_T = 0$. Because the nonhomogeneous term in (6.16) is a gradient of a scalar function, the linear velocity equation splits into two:

$$\frac{\partial \mathbf{u}_T}{\partial t} + 2\frac{\dot{a}}{a}\mathbf{u}_T = \mathbf{0},$$
$$\frac{\partial \mathbf{u}_L}{\partial t} + 2\frac{\dot{a}}{a}\mathbf{u}_L = \mathbf{g}$$

The first equation gives the homogeneous solution

$$\mathbf{u}_T = \mathbf{C}_T(\mathbf{x})a^{-2}(t),$$

where the constant field $\mathbf{C}_T$ is purely rotational; initial rotational motions decay rapidly.

The equation for the longitudinal (divergent) part can be solved, noting that both the velocity (6.19) and the acceleration (6.17) are proportional to the density contrast δ. Because this is a sum of two modes and our equations are linear, $\mathbf{u}_L$ also must be a sum of two modes, linear in density contrast. Comparing the velocity equation (6.16) and the density equation (6.14) it is easy to see that the velocity component resulting from the first density mode $D_1(t)$ can be written as

$$\mathbf{u} = \frac{\mathbf{g}}{4\pi G\bar{\rho}D_1}\frac{dD_1}{dt} \tag{6.18}$$

(the normalization of $\mathbf{g}$ is chosen to eliminate its dependence on time). Substituting (6.18) into the velocity equation (6.16) we get the equation for the density contrast that is solved by $D_1(t)$, so our solution (6.18) must be correct. Just as there are two modes for the density contrast, there are two modes for the longitudinal velocity: one growing, the other decaying. Recalling the results for $D_1(t)$, $D_2(t)$ from above, we can write for the standard Einstein–de Sitter model:

$$D_1(t) \sim t^{2/3}, \quad \mathbf{u}_{L1} \sim t^{-1/3}, \quad \mathbf{v}_{L1} \sim t^{1/3},$$
$$D_2(t) \sim t^{-1}, \quad \mathbf{u}_{L2} \sim -t^{-2}, \quad \mathbf{v}_{L2} \sim -t^{-4/3}.$$

We see that the comoving velocities diminish in time, and the word "growing mode" is justified only for peculiar velocities in the physical space. The minus sign for the second mode means that these velocities are directed against the gravitational acceleration, working to erase the initial density contrast.

For the second example, the empty universe ($\Omega_M = 0, \Omega_\Lambda = 0$), we get

$$D_1(t) = \text{const}, \quad \mathbf{u}_{L1} = \mathbf{0}, \quad \mathbf{v}_{L1} = \mathbf{0},$$
$$D_2(t) \sim t^{-1}, \quad \mathbf{u}_{L2} \sim -t^{-2}, \quad \mathbf{v}_{L2} \sim -t^{-4/3}$$

(zero velocities mean freezing). The growing and decaying modes for density and velocity are connected. The usual approach is to discard the decaying modes and use only the growing mode. In a generic situation the amplitudes of both modes should be similar at the start of the growth of structure (after recombination, at $z \approx 1300$). Because the ratio of the amplitudes $D_2/D_1 \sim t^{-5/3} \sim (1+z)^{-5/2}$ (in the Einstein–de Sitter universe), this ratio for the recent past is about 10^{-8}, so we can safely disregard the decaying mode.

6.3.3 Dimensionless growth rate

Using the fact that in the linear approximation the density contrast is a function of time only, we can write

$$\frac{d\delta}{dt} = \frac{da}{dt}\frac{d\delta}{da} = \dot{a}\frac{\delta}{a}\frac{d\log\delta}{d\log a} = H\delta\frac{d\log D_1}{d\log a}$$

(if we neglect the decaying mode, we know the time behavior of the density contrast, $\delta(a) \sim D_1(a)$). The continuity equation (6.11) now gives us

$$\delta = -\frac{1}{Hf(a)}\nabla\cdot\mathbf{u}. \tag{6.19}$$

For direct comparison with observations we would use the peculiar velocity $\mathbf{v}$ and compute the divergence in physical coordinates. The scaling (6.2) ensures that the above formula remains the same ($\nabla_r\cdot\mathbf{v} = \nabla_x\cdot\mathbf{u}$).

This expression is frequently used in applications. The function f is called the dimensionless growth rate. It can be computed, using the above expressions for D_1, but in most applications approximate expressions are used. For the present epoch and in the case of $\Omega_\Lambda = 0$ a good approximation is

$$f(\Omega_M, z=0) = \frac{d\log D_1}{d\log a}|_{z=0} \approx \Omega_M^{0.6}. \tag{6.20}$$

The approximation as a power of Ω_M is very good, with a margin of error of up to a few percent. Lahav et al. (1991) demonstrated that in a general case the function f almost does not depend on the cosmological constant:

$$f(\Omega_M, \Omega_\Lambda) \approx \Omega_M^{0.6} + \frac{\Omega_\Lambda}{70}\left(1+\frac{\Omega_M}{2}\right).$$

They also found that the approximation (6.20) can be used for all epochs z, if we use the instantaneous value of the parameter $\Omega_M(z)$ (formula 2.27):

$$f(\Omega_M, z) \approx \Omega_M^{0.6}(z).$$

The formula for the velocity (6.18) can be rewritten, replacing the time variable by a and using the definition of the function f (6.20):

$$\mathbf{v} = \frac{H_0 f}{4\pi G\bar{\rho}}\mathbf{g} = \frac{2f}{3H_0\Omega_M}\mathbf{g}. \tag{6.21}$$

The last equality follows from the definition of Ω_M (2.18).

6.3.4 The Zeldovich approximation

The above derivation was done in different Eulerian coordinates (the coordinate $\mathbf{r}$ referred to the physical space, the coordinate $\mathbf{x}$ to the comoving space). In both spaces mass points (objects) change their coordinates, if their velocities are not zero. An alternative description is the purely Lagrangian one, where the main coordinates are the labels of the particles (e.g., these labels could easily be their Eulerian positions at a freely chosen moment). The dynamics is now described by the mapping from the Lagrangian to the Eulerian coordinates, which defines all other physical fields.

The reader can find a strict mathematical representation of the cosmological Lagrangian dynamics in Buchert (1992) and a more pedestrian approach in Bouchet et al. (1995).

We shall start by taking another look at the formula connecting velocity and acceleration formula (6.21) in the linear regime. Besides showing how we can calculate velocity given the acceleration, it also implies that the direction of the velocity is constant; particles move along straight lines, otherwise the directions of the velocity and acceleration would differ.

If we know that the growing mode of the velocity is irrotational, we can introduce the initial velocity potential $\Psi(\mathbf{q})$ and write for the linear approximation in the Lagrangian approach

$$\mathbf{x}(t) = \mathbf{q} + D_1(t)\nabla_q\Psi(\mathbf{q}), \tag{6.22}$$

where $\mathbf{x}$ is the (Eulerian) comoving position of a particle, $\mathbf{q}$ are its Lagrangian coordinates, and $D_1(t)$ is the growing mode of the linear density contrast. In order to prove that the factor $D_1(t)$ is correct, we have to calculate the density. Because density is strictly constant in the Lagrangian space, the density in the comoving coordinates can be found as

$$\rho = \rho_0|x_{i,k}|^{-1} = \rho_0|\delta_{ik} + D_1(t)\Psi_{,ik}|^{-1},$$

where $|x_{i,k}|$ is the Jacobian of the mapping (6.22), and we use the notation $x_{,i} = \partial x/\partial q_i$ for partial derivatives. Let the eigenvalues of the deformation tensor $f_{ik} = \Psi_{,ik}$ be $\alpha_i, i = 1,2,3$ — these are the contraction (or expansion) coefficients along the three eigenvectors (f_{ik} is a symmetric tensor and can be easily diagonalized). The formula for density simplifies then to

$$1 + \delta = \rho/\rho_0 = \left[(1 + D_1(t)\alpha_1)(1 + D_1(t)\alpha_2)(1 + D_1(t)\alpha_3)\right]^{-1}. \tag{6.23}$$

For small perturbations ($D_1(t) \ll 1$) we get

$$\delta(t) = -D_1(t)(\alpha_1 + \alpha_2 + \alpha_3),$$

thus the density contrast grows as it should in the linear regime (wherever the sum of the eigenvalues of the deformation tensor is negative).

Differentiating the mapping formula (6.22) we get the comoving velocity

$$\mathbf{u} = \frac{dD_1(t)}{dt} \nabla \Psi(\mathbf{q}).$$

Comparing this formula with (6.18) and substituting there $\mathbf{g} = \nabla\Phi/a^2$, we find

$$\Phi = -4\pi G\bar{\rho}a^2 D_1 \Psi.$$

Replacing $4\pi G\bar{\rho}a^3$ with the model parameters, we get

$$\Phi = -\frac{3}{2} H_0^2 \Omega_M \left(\frac{D_1}{a}\right) \Psi.$$

Thus, the velocity potential and the gravitational potential are proportional and carry the same information. The label "inertial motion" that we used above to describe the linear motion is somewhat misleading — the initial velocities and the initial gravitational field are closely connected, thus we may also say that the motion is "purely gravitational." This is the result of having chosen the growing mode of fluctuations. In the Einstein–de Sitter model ($\Omega_M = 1, \Omega_\Lambda = 0$) the density contrast is proportional to the scale factor, the coefficient before the velocity potential is constant, and both potentials remain constant in the linear regime.

The Zeldovich approximation goes farther than the usual Eulerian linear approximation. As proposed by Zeldovich (1970), it can be applied after the linear regime, when the typical perturbation amplitude $D_1(t)\alpha_i \sim 1$ and the density contrast $\delta > 1$. While in the linear regime it was the sum of the eigenvalues of the deformation tensor that determined the amplitude of the density contrast, at other times that amplitude will be determined only by the most negative eigenvalue of the deformation tensor. If we write that eigenvalue as $-\alpha$ and designate the eigenvector to which it corresponds as $\mathbf{e}_1$, we see that the density contrast grows along $\mathbf{e}_1$ faster than in any other directions, $\delta \sim (1 - \alpha D_1 t)^{-1}$, and it reaches infinity at the moment $D_1(t) = \alpha_1$. Thus, the Zeldovich ansatz predicts the collapse of matter into planes (sheets). As the initial configuration becomes increasingly flatter (more sheetlike) during the density evolution, the approximation works increasingly better. The reason is that for the collapse of plane-parallel configurations of matter the acceleration is exactly constant — it is perpendicular to the plane of the symmetry and does not vary in time, because it is determined by the mass coordinate of a point (the Lagrangian $\mathbf{q}$ in our case) only.

Comparison with numerical simulations shows that the Zeldovich approximation works well beyond the linear regime. It breaks down, obviously, when the trajectories of the particles begin to cross one another, and it does not describe well spherical condensations, where the collapse is essentially three-dimensional.

The Zeldovich approximation is widely used in cosmological studies, both for analytical description of large-scale motions and for setting up initial conditions for numerical modeling — we need to generate only one scalar function $\Psi(\mathbf{q})$ to calculate all the necessary initial fields.

6.4 Exact solutions

6.4.1 Plane-parallel collapse

The Zeldovich approximation is also one of the three exact solutions describing the dynamical evolution of objects in the cosmological context. For an exact solution we have to postulate a plane-parallel density distribution ("sheet") of infinite extent in two other dimensions. This model can be identified locally with the "walls" we see in the distribution of galaxies, although the identification was easier at the times when we supposed that the way of formation of structure was "top-down," meaning that large-scale structures formed first and fragmented then into smaller structures and finally into galaxies. In the presently popular clustering picture, small objects form first and cluster then into objects of larger size, and it seems a bit unnatural to suppose that there exist conditions for formation of well-delineated sheets. However, fashions change with times, and perhaps the plane-parallel collapse will be useful once again.

The plane-parallel collapse is essentially one-dimensional, as all fields depend only on the coordinate perpendicular to the plane. We get the simplest example if we choose the Einstein–de Sitter cosmological model and study the evolution of a configuration given by a single-wave initial velocity potential $\Phi = L^2 \cos(kq)$, where $k = 2\pi/L$ and the potential is defined in the interval $[-L/2, L/2]$. We can also imagine, if we wish, that the potential is periodic with the period L. The formulae for the physical fields (the comoving coordinates, the velocity, and the density contrast) will then be

$$\begin{aligned} x &= q - (t/t_0)^{2/3} L^2 k \sin(kq), \\ u &= -\frac{2}{3t_0}(t/t_0)^{-1/3} L^2 k \sin(kq), \\ \delta &= \frac{1}{1 - (t/t_0)^{2/3} L^2 k^2 \cos(kq)}. \end{aligned}$$

This "sheet" will collapse at $t_c = t_0 (Lk)^{-3} = (2\pi)^{-3} t_0$.

We illustrate this simple solution in Fig. 6.1. We have chosen a moment rather close to the start of the collapse $t = 0.9t_c$. The left panel shows the velocity potential, the velocity, and the density contrast in the Lagrangian coordinates, and the right panel shows the same in the comoving (Eulerian) coordinates. The velocity and the potential are arbitrarily rescaled. While there is not much happening in the

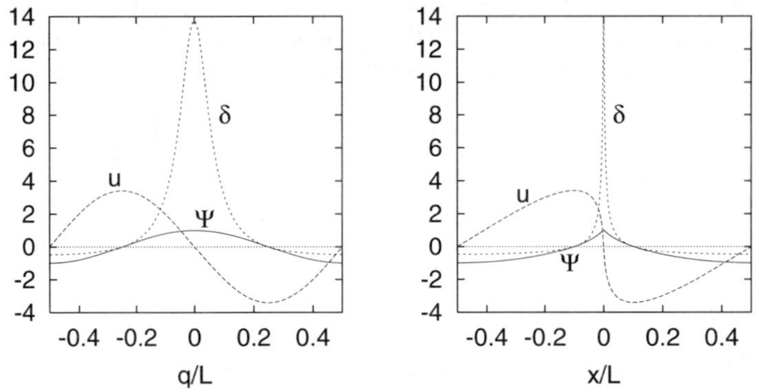

Figure 6.1 *Gravitational evolution of a plane wave in the Einstein–de Sitter universe. The physical fields are shown at the time* $t = 0.9t_c$*, where* t_c *is the collapse time; the velocity and the velocity potential have been rescaled. The left panel shows the coordinate profiles of the velocity potential* Ψ*, the comoving velocity* u*, and the density contrast* δ *in the Lagrangian coordinates; the right panel shows the same in the comoving Eulerian coordinates.*

Lagrangian space apart from the growth of density, the coordinated movement of matter is creating steep profiles in the Eulerian space. Note, however, that the amplitudes are the same in both panels. The high density contrasts that occur during the evolution of structure are typical for the gravitational instability picture.

Another typical feature of gravitational collapse is the way the density evolves. Fig. 6.2 shows the evolution of the central density of the collapsing plane (again in the Einstein–de Sitter cosmology). As shown, the density contrast evolves for a long time, almost up to $\delta \approx 1$, by the linear rule $\delta \sim t^{2/3}$. After that it starts growing extremely fast, becoming infinite in a finite time. In physics such a growth is called "explosive." This justifies the usual approximation of describing collapse by the linear and highly nonlinear stages. As density contrasts are very high in the nonlinear stage, the expansion of the universe no longer influences the dynamics and the nonlinear stage can be treated as Newtonian.

The Zeldovich approximation predicts that the particles cross the central plane of the sheet after $t = t_c$ with the same velocity at which they arrived there (the velocity changes very slowly, as $\dot{D}_1(t)$, compared to the extremely rapid growth of the density at collapse). Although this crossing may seem strange, it is possible and natural, as the main constituent of matter, the dark matter, can be thought of as dust, tiny particles that pass one another without colliding. This sort of matter will develop streams of different velocities that coexist at the same point of space. Thus far, the Zeldovich model is correct, but the accelerations and the velocities it predicts after collapse are incorrect, because it cannot account for the gravitational forces from the collapsed sheet itself. The further evolution of the sheet has

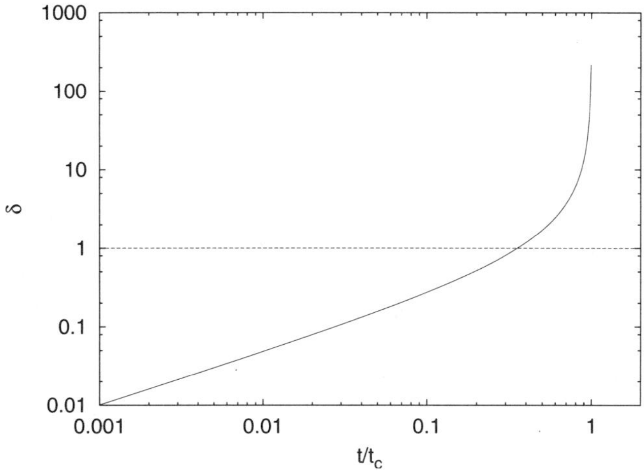

Figure 6.2 *The growth of the central density contrast of a plane wave in the Einstein–de Sitter universe. After a long linear growth period the density contrast δ becomes infinite in a finite time t_c.*

been followed numerically in many papers, partly because it is one of the easiest problems to understand and partly because the collapse of a plane is an excellent test of a numerical code.

Similar flows can develop in the baryon component of the matter, too, if it is condensed into compact objects. For example, stars in galaxies usually participate in different flows that cross one another easily. In the early universe, however, baryons are more likely to form a gas, and that gas will develop high pressure in the high density regions, high temperatures, and shock waves that accompany collapse.

6.4.2 Spherical collapse

There are two more exact solutions — the collapse of a constant density sphere and that of a similar ellipsoid. The ellipsoid solution is rather complicated and is not used much, but the spherical collapse is a well-loved workhorse, and that solution is especially popular in statistical applications. This is understandable, because the simplest approximation for an isolated object is a sphere.

In a matter-dominated $\Omega_\Lambda = 0$ universe the dynamics of a gravitating sphere of a mass M and a radius R can be described by the Newtonian equation (we shall work in physical coordinates for a while)

$$\frac{d^2R}{dt^2} = -\frac{GM}{R^2}.$$

Cosmology will enter here through initial conditions — a cosmological sphere will usually expand initially. Integrating the above equation we get

$$\left(\frac{dR}{dt}\right)^2 = \frac{2GM}{R} + C, \tag{6.24}$$

where C is the constant of integration that determines the total energy of the sphere (if $C > 0$, the sphere will expand forever; if $C < 0$, expansion will change to contraction at a certain R). As we see, this equation is practically the same as the Friedmann equation (2.15), because any spherical region in a universe expands at the same rate as the overall expansion factor. This allows us to fix the value of the constant C as $C = kc^2$, where c is the velocity of light.

We are interested in the formation of an object, so we shall choose the case $C < 0, k = -1$. If we change now to a new nondimensional variable η by

$$dt/d\eta = R/c, \tag{6.25}$$

we can write

$$\left(\frac{dR}{d\eta}\right)^2 = \frac{2R_c}{R} - R^2, \tag{6.26}$$

where $R_c = GM/c^2$.

As the reader can check, this equation has a (well-known) solution, the cycloid

$$R = R_c(1 - \cos\eta). \tag{6.27}$$

Integrating (6.25) we obtain the solution for the time

$$t = t_c(\eta - \sin\eta), \tag{6.28}$$

where

$$t_c = R_c/c. \tag{6.29}$$

For an expanding sphere the start of expansion corresponds to $\eta = 0$, and for the initial stage $\eta \ll 1$ we can expand (6.27, 6.28) in power series. For the time we get

$$t \approx \frac{t_c\eta^3}{6}\left(1 - \frac{\eta^2}{20}\right),$$

which gives

$$\frac{t_c\eta^3}{6} \approx t\left[1 + \frac{1}{20}\left(\frac{6t}{t_c}\right)^{2/3}\right]. \tag{6.30}$$

The formula for the radius gives

$$R \approx \frac{R_c\eta^2}{2}\left(1 - \frac{\eta^2}{12}\right),$$

and substituting here η^2 by t from (6.30) above, we get

$$R \approx \frac{R_c}{2}\left(\frac{6t}{t_c}\right)^{2/3}\left[1 - \frac{1}{20}\left(\frac{6t}{t_c}\right)\right].$$

The density of the sphere can be calculated as

$$\rho = \frac{3M}{4\pi R^3} \approx \frac{1}{6\pi G t^2}\left[1 + \frac{3}{20}\left(\frac{6t}{t_c}\right)^{2/3}\right]. \tag{6.31}$$

The factor before the parenthesis,

$$\rho_{ES} = \frac{1}{6\pi G t^2}, \tag{6.32}$$

is the density of the Einstein–de Sitter model (2.26), so the (linear) density contrast of our sphere in that model is

$$\delta = \frac{3}{20}\left(\frac{6t}{t_c}\right)^{2/3}. \tag{6.33}$$

As we know, all other models approach that model in the past, so their mean density also can be written as the density of the $\Omega_M = 1$ model plus a small density contrast, and the real density contrast of our sphere in any model is the difference of those two. Another model-dependent quantity is the growing mode of the density contrast $D_1(t)$. Because the final results do not depend much on the background model, we shall consider below only the simplest $\Omega_M = 1$ case, where (6.33) gives the true density contrast.

The meaning of the above exercise was to find a connection between the real dynamics of the sphere given by the cycloid formulae (6.27, 6.28) and the linear approximation.

The exact formulae tell us that the sphere starts to expand from a zero radius until a maximum radius, from where it contracts to the zero radius again. The first essential moment of the evolution is called "turnover" — until that the sphere expands together with the universe, and after that real collapse begins. This moment corresponds to $\eta = \pi$, the time is then $t_T = \pi t_c$ and the radius of the sphere is $R_T = 2R_c$. The density of the sphere is $\rho = 3M/4\pi R_T^3$, and the ratio of that to the background density is

$$\frac{\rho_T}{\rho_{ES}} = \frac{9\pi^2}{16} \approx 5.552$$

(to get that one has to use (6.29)). The density contrast at that moment is, accordingly, $\delta_T \approx 4.552$. It is also useful to know the density contrast for that moment that would be predicted by the linear theory. By (6.33) it is $\delta_T^{lin} = (3/20)(6\pi)^2 \approx 1.062$.

Table 6.1 *Evolution of density contrast during a spherical collapse*

Moment	δ	δ_{linear}
Turnover:	4.552	1.062
Collapse:	∞	1.686
Virialization:	177.7	1.686

Another important moment at $\eta = 2\pi$ is the final collapse of the sphere, $R_F = 0$, $t_F = 2\pi t_c$. The real density contrast at that moment is infinite, but the linear theory predicts $\delta_F^{lin} = (3/20)(12\pi)^{2/3} \approx 1.686$.

The final stage of collapse is called virialization. As in the plane collapse we discussed above, streams of dust particles (dark matter) pass one another. But in this case the Newtonian theory of gravitation predicts the future of our collapsed sphere, namely, the law of conservation of angular momentum demands that the equilibrium of a gravitating sphere should be described by the virial theorem

$$U + 2K = 0, \tag{6.34}$$

where U is the potential and K is the kinetic energy. At the point of maximum expansion (turnover) $K_T = 0$, and the total energy is conserved, thus $E = U_T = U_V + K_V$. From (6.34) we get the initial kinetic energy $K_V = -U_V/2$, leading to $U_T = U_V/2$. As the gravitational energy

$$U = 4\pi G \int_0^R \frac{m(r)\rho(r)}{r} r^2 dr,$$

($m(r)$ is the mass inside a sphere with a radius r) is inversely proportional to R, we get $R_V = R_T/2$. The density of the virialized object will thus be 8 times higher than at the turnover, and as the time until virialization is twice as long as that until turnover and $\rho_{ES} \sim t^{-2}$, the relative density of the virialized object will be $\rho_V/\rho_{ES} = 8 \times 4\rho_T/\rho_{ES} = 2 \times 9\pi^2 \approx 177.7$. The collapse density contrasts are summarized in Table 6.1.

The specific numbers in that table refer to the Einstein–de Sitter model. Peacock (1999) derives an approximate expression

$$1 + \delta_{vir} \approx 178\Omega_M^{-0.7}$$

for $\Omega_\Lambda = 0$ models, and Lahav et al. (1991) show that the value of Ω_Λ influences little the dynamics of the spherical collapse. The commonly used virial contrast values (real and linear) in cosmological statistics are those from Table 6.1.

Fig. 6.3 illustrates the spherical collapse. The left panel shows the evolution of the densities, both for the sphere and for the background. The density of the sphere follows rather closely the background density almost until the turnover. At the moment of collapse ($t = 2\pi t_c$) the density of the sphere becomes infinite, but

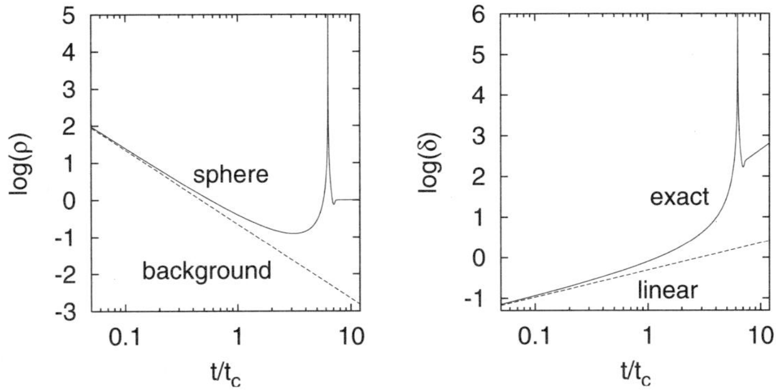

Figure 6.3 *Collapse of a sphere in the Einstein–de Sitter universe. The left panel shows the evolution of the density of the sphere and of the background density, the right panel shows the behavior of the exact and linear density contrasts. The time unit t_c is defined in the text; the density units are rescaled.*

stabilizes then at the level of 8 times the minimum (turnover) density. The density contrast, shown in the right panel, grows continuously, and follows the $\delta \sim t^2$ law after virialization (the density of the sphere is constant, but that of the universe decreases). We can also see the extremely fast density evolution at the late stages of collapse, similar to that in the case of the plane collapse above (faster, in fact).

This figure also shows why we searched for the linear theory connection. The point is that it can be used to specify the initial conditions for a collapse in the past. For example, let us suppose that a certain class of objects is so young that it must have formed in the immediate past. Setting then the present time for the collapse epoch, we can predict the amplitude of the density contrast as

$$\delta(t) = 1.686\frac{D_1(t)}{D_1(T)} = \frac{1.686}{1+z},$$

where T is the age of the universe and the last equality is true for the Einstein–de Sitter universe. For any other hypothesis for the collapse epoch t_c we have only to replace $D_1(T)$ by $D_1(t_c)$ in the above formula. As seen from the figure, this prediction is correct almost up to the present time.

The spherical collapse formulae above do not refer to the density of the sphere; the only variables are its radius and mass. Thus they can be used to describe the collapse of spherical distributions of varying density profile. As the mass of the sphere M is supposed to be constant, the different spherical shells should not overtake one another. In cosmological applications usually $d\rho/dR < 0$ and this condition is satisfied.

6.5 Numerical experiments

The two exact solutions we have described show that the evolution of structure splits roughly into two stages, the linear and the strongly nonlinear regimes. The linear solution is thus a very useful tool that can be applied for a wide redshift range. In contrast, various higher-order approximations, although useful, span much shorter periods, mostly because the transition from the linear to the fully nonlinear dynamics is rather sharp. Most of the observed structures are decidedly nonlinear with densities at least 10^3 times higher than the background density. Nonlinear dynamics demands use of numerical methods, and numerical modeling of structure is one of the theorist's most popular tools in cosmology.

Numerical models are used to connect cosmological hypotheses (the background model and the initial properties of the physical fields) with the observable structure. As cosmological observations are difficult and the influences of various processes do not factor out easily, numerical models are also frequently used as a testbed of new statistical methods.

6.5.1 Dynamics of dark matter

We mentioned above that most of the matter in the universe is noncollisional and interacts only gravitationally, and thus streams of different velocity can coexist at the same place. Thus even the simplest formulation of dynamics must work with the six-dimensional (one-particle) phase space $(\mathbf{x}, \mathbf{v})$. As this is extremely demanding computationally, the usual approach is to sample the initial phase space by discrete particles and then to follow the trajectories of these particles. Such models are called N-body models, for obvious reasons. A review of recent work on N-body modeling is given by Bertschinger (1998).

After choosing the coordinate volume and the number of particles (thus fixing a mass of a particle), the initial conditions for the calculation are set up by placing the particles uniformly in space with zero velocity. The initial small departures from uniformity are described by the linear theory.

The initial conditions are usually set up using the Zeldovich approximation. The initial velocity potential is determined as a realization of a random Gaussian field with a given power spectrum; we describe that process in Chapter 7. The chosen cosmological model gives the time-dependent factor $D_1(t)$ of the velocity of a particle and the gradient of the velocity potential describes the spatial dependence of velocity. Choosing the initial moment for the calculations so that the perturbations are small and the linear theory applies, we can find the initial velocities and the displacements for all particles.

The further evolution of the system is carried out, using the formula for the dynamics (6.6) that can be written for particles as

$$\dot{\mathbf{u}} + 2\frac{\dot{a}}{a}\mathbf{u} = \mathbf{g},$$

$$\dot{\mathbf{x}} = \mathbf{u}.$$

The most time-consuming part is the calculation of the gravitational acceleration $\mathbf{g}$, as its value depends on the positions of all particles.

The first models found accelerations using the pairwise force summation over all particles,

$$\mathbf{g}_i \sim \sum_{j=0}^{j=N} m_i \frac{\mathbf{x}_j - \mathbf{x}_i}{|\mathbf{x}_j - \mathbf{x}_i|^3}, \tag{6.35}$$

where the indices i, j label particles. (In practice the force law is slightly modified to account for the finite size of the particle; this is called "force softening.") It is evident, however, that this approach can be used only for rather small numbers of particles N, because the number of operations needed to find all accelerations scales as N^2.

The number of particles necessary to model a representative volume of space and to have a proper mass resolution is large. If we want to model a typical observed volume of $(100 \text{ Mpc})^3$, and we recall that the mean density of the universe is $2.8 \cdot 10^{11} \Omega_M h^2 M_\odot/\text{Mpc}^3$, we see that by choosing 10^6 particles the mass of a particle would be about $3 \cdot 10^{11} M_\odot$, typical for a galaxy. So that huge model of a million particles is still rather rough, able to explain formation of clusters and larger structures, but not of galaxies themselves.

Pairwise force summation for 10^6 particles gives the operation count of about 10^{13}. This is prohibitively large; thus, pairwise summation is not a useful method for cosmological simulations.

Different algorithms for building numerical models of structure differ mainly in the methods used to calculate gravitational forces. The first category is formed by "tree-codes" that divide the computational volume into a hierarchy of cells until the lowest level cells contain only one particle. When calculating the acceleration for a particle, it is first determined at what angle a cell is seen from the position of that particle. If that value is smaller than a given limit (the usual choice is $s/d < 1$, where d is the distance to the cell and s is the size of the cell), the sum of accelerations caused by the particles belonging to the cell is approximated by the acceleration from their total mass at the mass center of the cell. This trick reduces the total operations count to $O(N \log N)$. A good description of the tree code can be found in Hernquist (1987). Tree codes can be parallelized, and the record particle number used is over 16 million.

Another approach, the "particle-mesh" method (PM), calculates accelerations by first estimating the density on a fixed spatial grid (mesh), assuming the particle size and its density distribution. The density distribution is assumed to be periodic on the scale of the computational cube, allowing the Poisson equation to be solved on the grid by the Fast Fourier Transform method (FFT). This is the step that reduces the operation count to $O(N \log N)$ in this case, too. The accelerations are calculated then by interpolating the gravitational potential. This method is

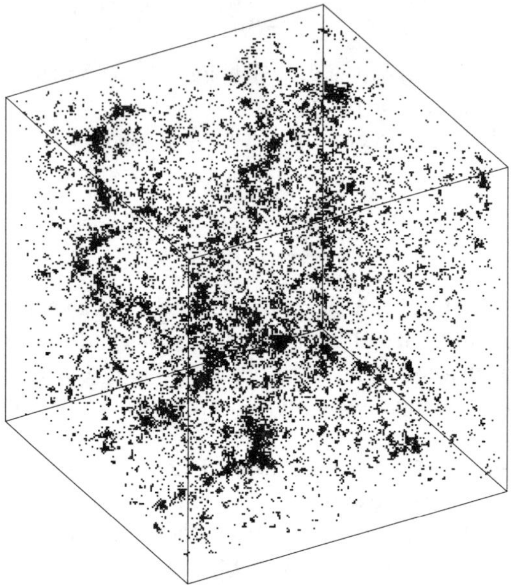

Figure 6.4 *Evolved density distribution from a N-body model. The point set has been uniformly diluted and only a few percent of the total $N = 64^3$ points are shown. (Reproduced, with permission, from Pons-Bordería et al. 1999, Astrophys. J., 523, 480–491. AAS.)*

well described in Hockney and Eastwood (1988). As interpolation damps small-scale forces, an improved method, called P^3M, uses the FFT solver for large-scale matter distribution and adds pairwise summed forces from neighboring points. The application of that method to cosmology is demonstrated in Efstathiou et al. (1985).

Fig. 6.4 shows an example of an evolved density distribution obtained this way. It was generated using a P^3M solver for 64^3 points (a diluted version of the point distribution is shown, otherwise the projection effects would destroy the picture). The structures shown are typical for dark matter simulations and resemble, to a large extent, the observed large-scale structure as well.

6.5.2 Gas and galaxies

The above codes model the evolution of dark matter. The objects we observe were formed from baryonic matter, which was originally in a gaseous state. Introducing that component into the dynamical equations yields a new set of physical fields — apart from the density and velocity of gas we also have its pressure and temperature, and the equations for the time evolution of those fields.

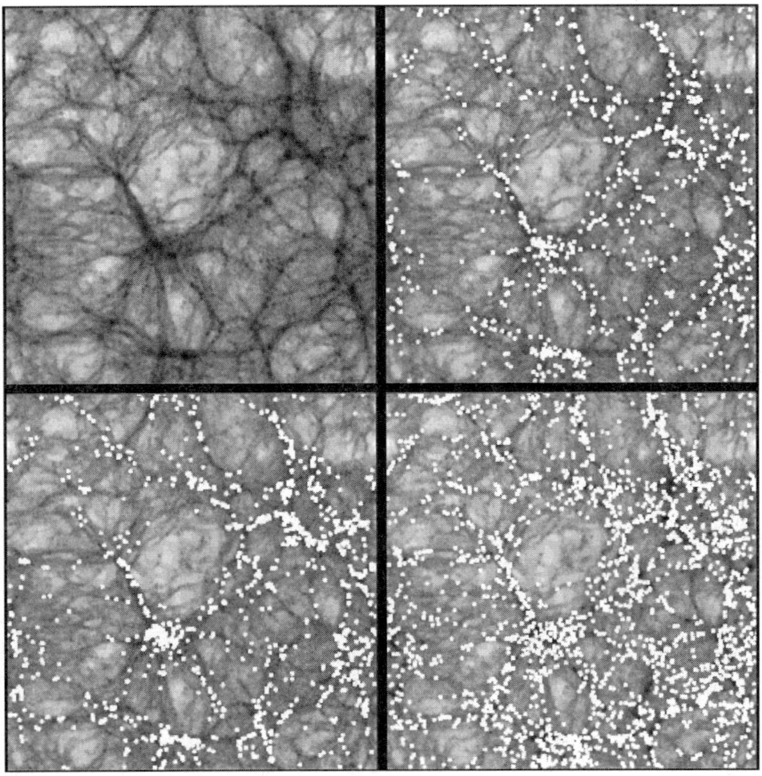

Figure 6.5 *Thin slices of the spatial distribution of dark matter and galaxies from a recent numerical simulation. The size of the region is 85 h^{-1} Mpc; the thickness of the slice is 8 h^{-1} Mpc. The top left panel shows the distribution of dark matter. In the other three panels, galaxies are added as white circles. The top right panel shows the locations of bright galaxies, and lower luminosity galaxies in the bottom panels are divided into two groups — those with a low star formation rate (bottom left panel) and those with a high star formation rate (bottom right panel). (Reproduced, with permission, from Kauffmann et al. 1999, Mon. Not. R. Astr. Soc., 303, 188–206. Blackwell Science Ltd.)*

The most popular method of incorporating gas dynamics into cosmological structure models is called "smoothed particle hydrodynamics" (SPH). Here the physical fields are defined for "gas particles," and the values of these fields at different points in space are obtained by interpolation. That approach is similar to the density calculations in PM, but the interpolation kernel here is not connected with the grid and is different. A good review of SPH can be found in Monaghan (1992). In recent years, SPH has been combined with (adaptive) P^3M to produce a state-of-the-art code (called "Hydra") for cosmological modeling (Couchman, Thomas, and Pearce 1995). "Adaptive" here means that the number of neighbors for pairwise force summation is kept approximately constant, however high the local density grows.

Most of the codes described above are freeware. Their availability has provided a strong impetus for the use of numerical models in cosmological studies. Both Hydra and the adaptive P^3M (without gas dynamics) can be obtained from the Hydra Consortium Web page (*Hydra*).

Fig. 6.5 shows slices from a recent simulation, an example of a typical numerical model of the structure. The basic simulation was done by the "Virgo Consortium," a group of astronomers pooling computational resources to create detailed structural models based on the parallel version of Hydra (Jenkins et al. 1998). This model has 256^3 particles, but no gas. The cosmology is specified by $\Omega_M = 1, \Omega_\Lambda = 0$, and it is supposed that galaxies started to form later than usual due to a neutrino background (the τCDM model). The size of the cube is $85h^{-1}$ Mpc, and the mass of a particle is $10^{10}h^{-1}M_\odot$. (Including gas dynamics slows the computation considerably, and the best gas–dark matter models computed thus far have 128^3 of both types of particles.) The dark matter distribution in a thin slice with thickness of $8h^{-1}$ Mpc (more than 1.5 million particles) is shown in the top left panel of Fig. 6.5.

In order to predict the formation of galaxies in a dark matter model, the so-called "semianalytical" methods are frequently used — recipes for the collapse of gaseous objects, of their further merging, of star generation, etc. Applying such methods to the above model, Kauffmann et al. (1999) generated a model galaxy population that can be directly compared with observations. The high-luminosity part of that population is shown in the top right panel of Fig. 6.5. Galaxies of lower luminosity are shown in the bottom panels, divided according to their star formation rate. Those with the low rate are shown in the left panel, and those with the high rate are shown in the right panel. We can see that generally galaxies follow the dark matter density distribution, but differences in the luminosity and star formation rate lead to different spatial distributions of galaxies.

Thus far the models that have been compared with observations have been mostly dark matter models. As we shall show in the later chapters (and as can be seen from the above model), galaxies probably do not follow exactly the dark matter distribution, so these comparisons cannot be considered entirely decisive. However, with the rapid evolution of computer hardware and programming methods both semianalytical methods and direct gas–dark matter simulations will soon provide us with decent models of galaxy distribution.

CHAPTER 7

Random fields in cosmology

7.1 Introduction

Many physical observables in cosmology can be treated as continuous functions of spatial coordinates and time. The theory of the early universe tells us that baryonic matter had to be distributed continuously before recombination and for a long time thereafter. The same can be said of dark matter; the most likely candidates are tiny elementary particles whose dynamics can best be described by continuous density and velocity fields.

It is of obvious interest to find out the properties of these fields, both present and past. It can be done, however, only in a statistical manner, both because the universe is huge and because we can observe only glimpses of distant space at different moments of time — the light cone limitation. Thus the theoretical machinery is that of random fields, and it provides the basis for the statistical models of the present-day galaxy distributions.

Becuse these fields are defined for a specific background geometry, they share the homogeneity and isotropy of cosmological models.

Exactly as for simple random variables, all statistics for random fields are meant to be measured averaging over ensembles of fields. Because we have only one universe to observe, we can never get our hands on an ensemble, and we have to suppose that our fields are ergodic — that is, ensemble averages are equal to spatial averages. For certain fields physical theories that explain how the fields are generated also can predict the ergodicity of the field (quantum fluctuations from inflation, for example).

For a detailed presentation of the theory of random fields and their realizations we refer the reader to the classical monographs by Adler (1981) and by Cramér and Leadbetter (1967).

7.2 Random fields

Random fields can be multidimensional, but in order to keep our formulae simple we shall talk below mainly about scalar random fields. All that can be easily generalized to higher dimensions (e.g., vector and tensor fields). A scalar random field (or a stochastic process) is a set of random variables $Y(\mathbf{x})$, $\mathbf{x} \in \mathbb{R}^N$, together with a collection of distribution functions

$$P(Y(\mathbf{x}_1) \leq y_1, \ldots, Y(\mathbf{x}_n) \leq y_n)$$

for any number of points n. Because we will later work with Fourier transforms, it is useful to think of the field Y as having complex values.

The properties of a random field are determined by its distribution functions. For example, homogeneous random fields are defined by the condition that all distribution functions have to be invariant under a common shift of their arguments:

$$P(Y(\mathbf{x}_1) \le y_1, \ldots, Y(\mathbf{x}_n) \le y_n) = P(Y(\mathbf{x}_1+\mathbf{u}) \le y_1, \ldots, Y(\mathbf{x}_n+\mathbf{u}) \le y_n)$$

for every vector $\mathbf{u}$ and for any number of points n.

Random fields begin to really differ from a simple collection of random variables if we start to apply the continuity conditions. This can be done in one of two ways. If we think of a random field as a collection of realizations, then we can speak about continuity in mean square. Alternatively, we can demand continuity of sample functions (realizations), which places additional conditions on the distribution functions. We shall describe these conditions below after introducing the appropriate machinery.

7.2.1 *Spatial correlations*

The simplest statistics for random fields are their first and second moments. The mean of a random field $Y(\mathbf{x})$ is

$$m(\mathbf{x}) = E\{Y(\mathbf{x})\}$$

(E denotes the expectation value). For a homogeneous random field $m(\mathbf{x})$ does not depend on $\mathbf{x}$ and must thus be a constant. This constant is usually taken to be zero, but that is not always the case (e.g., in the case of asymmetric one-point distribution functions).

The second moment is called the two-point noncentral covariance function $R_2(\mathbf{x}_1, \mathbf{x}_2)$ and is defined by

$$R_2(\mathbf{x}_1, \mathbf{x}_2) = E\left\{Y(\mathbf{x}_1)Y^\star(\mathbf{x}_2)\right\}.$$

The central two-point covariance function $\xi(\mathbf{x}_1, \mathbf{x}_2)$ is usually called simply the covariance function and is defined by

$$\xi(\mathbf{x}_1, \mathbf{x}_2) = E\left\{[Y(\mathbf{x}_1) - m(\mathbf{x}_1)]\,[Y(\mathbf{x}_2) - m(\mathbf{x}_2)]^\star\right\}$$

(a star denotes complex conjugation). The covariance function is positive definite: for any collection of points $\{\mathbf{x}_1, \ldots, \mathbf{x}_k\}$ and arbitrary complex numbers $\{z_1, \ldots, z_k\}$ the Hermitian form

$$\sum_i \sum_j \xi(\mathbf{x}_i, \mathbf{x}_j) z_i z_j^\star = E\left\{\left|\sum_i [Y(\mathbf{x}_i) - m(\mathbf{x}_i)]\, z_i\right|^2\right\} \ge 0$$

is always real and non-negative. This condition limits the class of possible correlation functions. It also allows us to represent the correlation function below as a Fourier integral.

For homogeneous fields the covariance function has to be the function of the difference $\mathbf{x}_1 - \mathbf{x}_2$ only. The value of the covariance function for zero lag $\xi(\mathbf{0})$ is the variance of the field Y and is, evidently, real.

Yet another possibility to limit the class of random fields is to define isotropic random fields by the requirement that the covariance function should not depend on the direction:

$$\xi(\mathbf{x}, \mathbf{x} + \mathbf{r}) = \xi(\mathbf{x}, r).$$

The most popular random fields in cosmology are both homogeneous and isotropic, and their covariance function $\xi(r)$ depends only on the distance between the two points $r = |\mathbf{r}|$.

The covariance function determines the (mean square) local properties of random fields. Namely, if the covariance function $\xi(\mathbf{r})$ of a homogeneous random field is continuous at the origin ($\mathbf{r} = \mathbf{0}$), then the field itself is continuous (in mean square) everywhere. And if the $2k$-th order partial derivative of the covariance function of a random field $\partial^{2k}\xi(\mathbf{x})/\partial x_1^2 \ldots x_k^2$ exists and is finite at $\mathbf{r} = \mathbf{0}$, then the field itself has partial derivatives of the order of k.

The covariance function is the simplest case in a hierarchy of n-point covariance functions

$$\mu_n(\mathbf{x}_1, \ldots, \mathbf{x}_n) = E\left\{[Y(\mathbf{x}_1) - m] \ldots [Y(\mathbf{x}_n) - m]\right\}. \tag{7.1}$$

The complete description of a random field demands knowledge of all distribution functions, or, alternatively, of all n-point covariance functions. These functions also contain all the higher moments of the one-point probability distributions (in the case of coinciding arguments).

Higher-order covariance functions have a special structure, which we demonstrate in the case of the noncentral three-point covariance function

$$R_3(\mathbf{x}_1, \mathbf{x}_2, \mathbf{x}_3) = E\left\{Y(\mathbf{x}_1)Y(\mathbf{x}_2)Y(\mathbf{x}_3)\right\}. \tag{7.2}$$

In the event that we move one point (say, $\mathbf{x}_3$) far away from the other two, the values of the field at that point become independent of those at the two other points and the expectation value in (7.2) reduces to

$$E\left\{Y(\mathbf{x}_1)Y(\mathbf{x}_2)Y(\mathbf{x}_3)\right\} = E\left\{Y(\mathbf{x}_1)Y(\mathbf{x}_2)\right\} E\left\{Y(\mathbf{x}_3)\right\} = mR_2(\mathbf{x}_1, \mathbf{x}_2),$$

where $R_2(\mathbf{x}_1, \mathbf{x}_2)$ is the two-point noncentral covariance. As the covariance functions are symmetric in their arguments, we can write

$$R_3(\mathbf{x}_1, \mathbf{x}_2, \mathbf{x}_3) = m\left[R_2(\mathbf{x}_1, \mathbf{x}_2) + R_2(\mathbf{x}_1, \mathbf{x}_3) + R_2(\mathbf{x}_2, \mathbf{x}_3)\right] + R_3^c(\mathbf{x}_1, \mathbf{x}_2, \mathbf{x}_3).$$

The last term in this sum depends on the coordinates of all three points and is called the irreducible three-point covariance function.

The expansion above is an example of a cluster expansion, already introduced in Section 3.5. Correlations in a configuration of n points are expressed by a sum of correlations between all possible disjoint subsets (clusters) of the configuration. In probability theory the moments of a random variable can be found from the series expansion of the characteristic function. In the theory of random fields one can use the moment-generating functional for the same purpose. This approach is well presented in E. Bertschinger's lectures (1992) and in Matarrese, Verde, and Heavens (1997). The cumulant expansion of moments carries then over into the cluster expansion of the covariance functions, and the irreducible covariance functions are the analogs of cumulants. These functions are also frequently called the connected covariance functions.

As in probability theory, the second- and third-order cumulants (two- and three-point irreducible covariance functions) coincide with the corresponding central moments.

In cosmology higher-order covariance functions are mainly used to describe the density field ρ. The first three (noncentral) density covariance functions can be written as

$$\begin{aligned} R_2(\mathbf{x}_1,\mathbf{x}_2) &= \bar{\rho}^2(1+\xi_{12}), \\ R_3(\mathbf{x}_1,\mathbf{x}_2,\mathbf{x}_3) &= \bar{\rho}^3(1+\xi_{12}+\xi_{23}+\xi_{23}+\zeta_{123}), \\ R_4(\mathbf{x}_1,\mathbf{x}_2,\mathbf{x}_3,\mathbf{x}_4) &= \bar{\rho}^4(1+\xi_{12}+\xi_{13}+\xi_{14}+\xi_{23}+\xi_{24}+\xi_{34}+ \\ &\quad +\xi_{12}\xi_{34}+\xi_{13}\xi_{24}+\xi_{14}\xi_{23}+ \\ &\quad +\zeta_{123}+\zeta_{124}+\zeta_{134}+\zeta_{234}+\eta_{1234}). \end{aligned} \tag{7.3}$$

Here ξ, ζ and η are the two-, three-, and four-point irreducible (connected) density correlation functions, respectively, and we have used the notation $\xi_{12} \equiv \xi(\mathbf{x}_1,\mathbf{x}_2)$.

Point processes and random fields share many common features. In the framework of the Poisson model (Peebles 1980), the cosmological point processes are interpreted as Cox processes (see Chapter 5), which can be thought as the result of Poisson sampling the mass density field, which is assumed to be a Gaussian random field. For this model, the probability of an object being placed within a volume dV is just $\rho(\mathbf{x})dV$. This model allows us to mirror many properties of the point processes onto those of random fields. In particular, the observed correlation functions of the galaxy distribution are given by the same expressions as the covariance functions of the random field (compare formulae (7.3) with formulae (3.8, 3.52) from Chapter 3). However, we want to make clear that this is just a model (see Kerscher 2001).

The covariance functions of the density contrast $\delta = (\rho - \bar{\rho})/\bar{\rho}$ are given by

$$\begin{aligned} \mu_2(\mathbf{x}_1,\mathbf{x}_2) &= \xi_{12}, \\ \mu_3(\mathbf{x}_1,\mathbf{x}_2,\mathbf{x}_3) &= \zeta_{123}, \\ \mu_4(\mathbf{x}_1,\mathbf{x}_2,\mathbf{x}_3,\mathbf{x}4) &= \xi_{12}\xi_{34}+\xi_{13}\xi_{24}+\xi_{14}\xi_{23}+\eta_{1234}. \end{aligned} \tag{7.4}$$

7.2.2 Fourier representation

An important class of random fields are those that can be represented by their Fourier transform. The spectral representation theorem states that every mean-square continuous, zero-mean, homogeneous random field $Y(\mathbf{x})$ can be represented as a mean-square integral

$$Y(\mathbf{x}) = \int_{\mathbb{R}^N} e^{-i\mathbf{x}\cdot\mathbf{k}}\, dZ(\mathbf{k}),$$

where the random field $Z(\mathbf{k})$ is a field with orthogonal increments. The latter means that different $dZ(\mathbf{k})$ can be thought of as being independent. The field $Z(\mathbf{k})$ does not have to be differentiable; if it is, we can write its density as $\widetilde{Y}(\mathbf{k})$, and the above formula reduces to the usual (inverse) Fourier transform formula

$$Y(\mathbf{x}) = \int_{\mathbb{R}^N} \widetilde{Y}(\mathbf{k}) e^{-i\mathbf{x}\cdot\mathbf{k}} \frac{d^N k}{(2\pi)^N}. \tag{7.5}$$

The field $Y(\mathbf{x})$ and its Fourier representation $\widetilde{Y}(\mathbf{k})$ form a Fourier transform pair, and the Fourier representation can be found from the direct Fourier transform

$$\widetilde{Y}(\mathbf{k}) = \int_{\mathbb{R}^N} Y(\mathbf{x}) e^{i\mathbf{k}\cdot\mathbf{x}}\, d^N x.$$

There are many different Fourier transform conventions; those used here are common in cosmological papers.

The properties of the field $Z(\mathbf{k})$ allow us to say that the Fourier amplitudes $\widetilde{Y}(\mathbf{k})$ are independently distributed, namely

$$E\left\{\widetilde{Y}(\mathbf{k}_1)\widetilde{Y}^\star(\mathbf{k}_2)\right\} = P(\mathbf{k}_1)(2\pi)^N \delta_D^N(\mathbf{k}_1 - \mathbf{k}_2), \tag{7.6}$$

where δ_D^N is the Dirac delta-function and $P(\mathbf{k})$ is the spectral density of the field. The factor $(2\pi)^N$ is due to the adopted Fourier transform convention. It is easy to remember, using $d^N k/(2\pi)^N$ for the Fourier space volume element. If the field $Y(\mathbf{x})$ is real, $\widetilde{Y}^\star(\mathbf{k}) = \widetilde{Y}(-\mathbf{k})$ and the above formula can be written as

$$E\left\{\widetilde{Y}(\mathbf{k}_1)\widetilde{Y}(\mathbf{k}_2)\right\} = P(\mathbf{k}_1)(2\pi)^N \delta_D^N(\mathbf{k}_1 + \mathbf{k}_2).$$

The statistical independence of the Fourier amplitudes of a homogeneous random field makes the study of such fields much easier.

Another possibility to describe random fields is the Hilbert space formalism, described by Hamilton (1998), where the real-space and Fourier components of a function are considered as components of the same vector in different bases.

7.2.3 Power spectrum

The covariance function of a mean-square continuous, zero-mean, homogeneous random field $Y(\mathbf{x})$ has an integral representation similar to the field itself:

$$\xi(\mathbf{r}) = \int_{\mathbb{R}^N} e^{-i\mathbf{r}\cdot\mathbf{k}}\, dF(\mathbf{k}),$$

where $F(\mathbf{k})$ is the spectral distribution function of $Y(\mathbf{r})$. It usually has a spectral density $P(\mathbf{k})$ (called power spectrum in astronomy) and the previous formula can be written as

$$\xi(\mathbf{r}) = \int_{\mathbb{R}^N} e^{-i\mathbf{r}\cdot\mathbf{k}} P(\mathbf{k}) \frac{d^N k}{(2\pi)^N}. \tag{7.7}$$

Because the correlation function is positive definite, the spectral density in (7.7) has to be positive in its entire argument range (Adler 1981). This condition conforms to the definition of spectral density in Eq. (7.6).

For an isotropic random field the spectral density must be a function of the norm $k = |\mathbf{k}|$ only. In this case the above formula can be simplified, integrating over the angles first. For two dimensions, we get

$$\xi(r) = 2\pi \int_0^\infty P(k) J_0(kr) \frac{k\, dk}{(2\pi)^2}$$

($J_0(x)$ is the 0-th order Bessel function). For three dimensions,

$$\xi(r) = 4\pi \int_0^\infty P(k) \frac{\sin(kr)}{kr} \frac{k^2\, dk}{(2\pi)^3}.$$

For completeness, in the one-dimensional case we have

$$\xi(r) = 2 \int P(k) \cos(kr) \frac{dk}{2\pi}.$$

In the Hilbert space formalism the covariance function and the power spectrum are the same function, only expressed in different bases. Application of this approach can be seen in papers by Hamilton (e.g., Hamilton 2000).

The variance of an isotropic random field can be written as

$$\sigma^2 = \xi(0) = 4\pi \int_0^\infty P(k) \frac{k^2\, dk}{(2\pi)^3} \tag{7.8}$$

and this has led to another frequently used form of the power spectrum in cosmology:

$$\Delta^2(k) = \frac{1}{2\pi^2} P(k) k^3. \tag{7.9}$$

Using the above definition, we can write the total variance of the field (7.8) as

$$\sigma^2 = \int_0^\infty \Delta^2(k)\, d(\ln k);$$

thus, $\Delta^2(k)$ gives the variance of the field per $\ln k$. Because the power spectrum $P(\mathbf{k})$ is a positive function of $\mathbf{k}$, Eq. (7.7) describes all possible covariance functions; any model covariance function should satisfy this relation. If we start by choosing a model power spectrum, we achieve this automatically. However, in cosmology we frequently use approximations for covariance functions that clearly do not conform with (7.7). A typical example is the popular approximation for the covariance function $\xi(r) = (r/r_0)^{-\gamma}$, which does not satisfy the above condition. In this case we implicitly assume that our model is defined for a limited argument interval.

As (7.6) shows, the power spectrum describes the second moment of a random field. The third moment of the field can be written as

$$E\left\{\widetilde{Y}(\mathbf{k}_1)\widetilde{Y}(\mathbf{k}_2)\widetilde{Y}(\mathbf{k}_3)\right\} = B(\mathbf{k}_1, \mathbf{k}_2, \mathbf{k}_3)(2\pi)^N \delta_D^N(\mathbf{k}_1 + \mathbf{k}_2 + \mathbf{k}_3). \quad (7.10)$$

The function $B(\mathbf{k}_1, \mathbf{k}_2, \mathbf{k}_3)$ is called the bispectrum. Eq. (7.10) says that the third moment is nonzero only for wavevectors that form closed triangles. This is the result of homogeneity of the field. Because the correlation functions depend only on the differences $\mathbf{x}_i - \mathbf{x}_j$, one of the volume integrals in the Fourier transform in (7.10) vanishes, except in the case $\mathbf{k}_1 + \mathbf{k}_2 + \mathbf{k}_3 = 0$ (see Peebles 1980, § 43). This leads to the definition of the bispectrum as the Fourier transform of the irreducible three-point correlation function:

$$B(\mathbf{k}_1, \mathbf{k}_2, \mathbf{k}_3) = \int e^{i(\mathbf{k}_2\mathbf{x}_{12} + \mathbf{k}_3\mathbf{x}_{13})} \zeta(\mathbf{x}_{12}, \mathbf{x}_{13}) d^3x_{12} d^3x_{13},$$

where the three wave vectors form a closed triangle, $\mathbf{x}_{ij} \equiv \mathbf{x}_i - \mathbf{x}_j$ and we have assumed that the field is homogeneous. The Fourier transform of the irreducible four-point correlation function is called the trispectrum.

As we have seen above, many local properties of random fields depend on the covariance functions and their derivatives. As the covariance function and the power spectrum form a Fourier transform pair, these properties also can be described by the spectral moments

$$\lambda_{i_1 i_2 \dots i_n} = \int_{\mathrm{I\!R}^N} k_{i_1} k_{i_2} \dots k_{i_n} P(\mathbf{k}) \frac{d^N k}{(2\pi)^N}. \quad (7.11)$$

The zeroth-order moment is

$$\lambda = \int_{\mathrm{I\!R}^N} P(\mathbf{k}) \frac{d^N k}{(2\pi)^N} = R(\mathbf{0}).$$

In the case of real-valued random fields the covariance function is real valued and the power spectrum is a symmetric function, $P(-\mathbf{k}) = P(\mathbf{k})$; hence, the odd-order spectral moments are zero. The second-order spectral moments

$$\lambda_{ij} = \int_{\mathrm{I\!R}^N} k_i k_j P(\mathbf{k}) \frac{d^N k}{(2\pi)^3}$$

also can be written as

$$\lambda_{ij} = \left.\frac{\partial^2 R(\mathbf{x},\mathbf{y})}{\partial x_i \partial y_j}\right|_{\mathbf{x}=\mathbf{y}} = -R_{,ij}(\mathbf{0})$$

(we have used the comma-notation for partial derivatives, e.g., $\partial f/\partial x_i \equiv f_{,i}$). Later on we shall need expressions for correlations between a Gaussian random field and its derivatives, which can be found from the formula:

$$E\left|\frac{\partial^{\alpha+\beta}Y(\mathbf{x})}{\partial^\alpha x_i \partial^\beta x_j} \cdot \frac{\partial^{\gamma+\delta}Y(\mathbf{x})}{\partial^\gamma x_l \partial^\delta x_m}\right| = \left.\frac{\partial^{\alpha+\beta+\gamma+\delta}}{\partial^\alpha x_i \partial^\beta x_j \partial^\gamma x_l \partial \delta x_m} R(\mathbf{x})\right|_{\mathbf{x}=\mathbf{0}} =$$
$$= (-1)^{\alpha+\beta} i^{\alpha+\beta+\gamma+\delta} \int_{\mathbb{R}^N} k_i^\alpha k_j^\beta k_l^\gamma k_m^\delta P(\mathbf{k}) \frac{d^N k}{(2\pi)^N}. \quad (7.12)$$

This formula can be easily derived, using the spectral representations of the random field and of its covariance function. With appropriate choices for $\alpha, \beta, \gamma, \delta$ we find that the field $Y(\mathbf{x})$ and its derivatives $Y_i(\mathbf{x})$ are uncorrelated ($\alpha = 1, \beta = \gamma = \delta = 0$), and the second and first derivatives are also uncorrelated ($\alpha = \gamma = \delta = 1, \beta = 0$).

7.3 Gaussian random fields

The definition of a general random field demands specification of all n-point probability distributions. The additional conditions we have used thus far (homogeneity, isotropy etc.) have allowed us to better describe these fields, but the next, practically unavoidable simplification is to define the n-point distributions as multivariate distributions of the same type. The simplest class of random fields defined in this way are the Gaussian random fields, where we demand the distribution functions to be multivariate Gaussian. A multivariate Gaussian distribution of a random vector $\mathbf{Y} = \{Y_1, \ldots, Y_n\}$ has the probability density

$$f(\mathbf{Y}) = \frac{1}{(2\pi)^{n/2}\sqrt{|\mathbf{S}|}} e^{\left[-\frac{1}{2}(\mathbf{Y}-\mathbf{m})\mathbf{C}^{-1}(\mathbf{Y}-\mathbf{m})^T\right]},$$

where $\mathbf{m} = E\{\mathbf{Y}\}$ and $\mathbf{C}$ is the covariance matrix

$$C_{ij} = E\left\{(Y_i - m_i)(Y_j - m_j)^\star\right\}.$$

The matrix $\mathbf{C}$ is symmetric and non-negative definite.

For a random field the covariance matrix is replaced by the covariance function:

$$C_{ij} = R(\mathbf{x_i}, \mathbf{x_j}).$$

As the covariance matrix of a multivariate Gaussian defines completely all the distribution functions (and, naturally, all the higher moments), the covariance function plays the same role for Gaussian random fields.

The same can be said about the power spectrum of a Gaussian random field. The representation of a Gaussian random field $Y(\mathbf{x})$ in the Fourier space $\widetilde{Y}(\mathbf{k})$

is especially simple. As the Fourier components of the field depend linearly on $Y(\mathbf{x})$, they also have a Gaussian distribution (with the variance given by the power spectrum $P(\mathbf{k})$), but they are statistically independent, in contrast to the values of the function $Y(\mathbf{x})$ at different points in real space.

The higher-order (three-point, four-point, etc.) irreducible correlation functions and spectra of a Gaussian field (bispectrum, trispectrum, etc.) are identically zero, so their determination from observations is a good test for Gaussianity of the field.

Gaussian random fields arise naturally in cosmology, where the initial small fluctuations of physical fields (e.g., gravitational potential, temperature, velocity, and density of matter, etc.) are due to quantum fluctuations that are Gaussian.

These basic cosmological random fields are real homogeneous isotropic (Gaussian) random fields, reflecting the structure of the space they live in, at least when their amplitudes are small. During the evolution of structure they evolve gradually away from Gaussianity. A typical example is that of the matter density field $\rho(\mathbf{x})$ that is close to Gaussian at the beginning, but that tends to be closer to lognormal at later stages (Coles and Jones 1991). The density contrast field $\delta(\mathbf{x}) = (\rho(\mathbf{x}) - \bar{\rho})/\bar{\rho}$ has a minimal value $\delta_{\mathrm{min}} = -1$ ($\rho_{\mathrm{min}} = 0$), but its maximum values are around 10–1000, depending on the smoothing scale used to estimate the density.

The velocity field is tightly connected with the density field and the knowledge about one can be used to predict the other. The velocity field is an example of a random vector field that can be described by its component fields, three values along the coordinate directions at every point. Such a field has to be described by a covariance tensor $R_{ij}(\mathbf{x}_1, \mathbf{x}_2), i, j = 1, 2, 3$ that is a function of a coordinate pair, with obvious simplifications for homogeneous and isotropic cases.

Gaussian random fields also have an extremely useful property of being ergodic. Gaussian random fields are ergodic if the spectral density is everywhere continuous, or, equivalently, if $R(\mathbf{r}) \to 0$ for $r \to \infty$ (Adler 1981). This condition is natural for physical fields and allows us to infer the statistics of cosmological fields from the study of the unique realization of our universe. Of course, we must always keep in mind that it is possible to have a universe with non-Gaussian initial conditions or to have a mixture of Gaussian and non-Gaussian fields.

7.3.1 *Filtered fields*

If we try to apply the Gaussian random field techniques to real objects, we have to take into account that objects have a finite size, and we are not interested usually in the behavior of the field at smaller scales. This can be formalized by defining a new random field that is the convolution of the original one with a specific filter $W(\mathbf{r}; R_F)$,

$$Y(\mathbf{x}; R_F) = \int W(\mathbf{x} - \mathbf{x}'; R_F) Y(\mathbf{x}')\, d^3x',$$

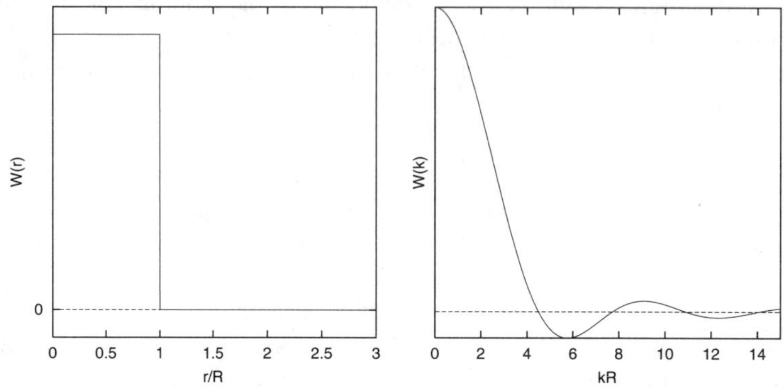

Figure 7.1 *The top-hat filter in the coordinate space (left panel) and in the Fourier space (right panel). Compare the widths of the filters.*

where R_F is the filter scale — the typical scale for a class of objects in which we are interested. The Fourier image of the filtered field is then

$$\widetilde{Y}(\mathbf{k}; R_F) = \widetilde{W}(\mathbf{k}; R_F)\widetilde{Y}(\mathbf{k})$$

and the power spectrum of the filtered field is

$$P(\mathbf{k}; R_F) = |\widetilde{W}(\mathbf{k}' R_F)|^2 P(\mathbf{k}),$$

where $P(\mathbf{k})$ is the power spectrum of the original field.

Two different filters are commonly used for smoothing cosmological random fields. The first is the top-hat filter

$$W_{TH}(r; R_{TH}) = \frac{1}{(4\pi/3)R_{TH}^3}\theta(R_{TH} - r),$$

where $\theta(x)$ is the theta-function. This filter smooths over a finite spherical volume of radius R_{TH} around a point, but its Fourier image is rather extended:

$$\widetilde{W}_{TH}(k; R_{TH}) = \frac{3}{k^3 R_{TH}^3}\big(\sin(kR_{TH}) - kR_{TH}\cos(kR_{TH})\big). \tag{7.13}$$

We show this filter in Fig. 7.1 along with its Fourier image.

The second filter is the Gaussian filter that is equally smooth both in the coordinate space and in the wavenumber space:

$$\begin{aligned} W_G(r; R_G) &= \frac{1}{\sqrt{(2\pi)^3}R_G^3}\exp(-r^2/2R_G^2), \\ \widetilde{W}_G(k; R_G) &= \exp(-k^2 R_G^2/2). \end{aligned}$$

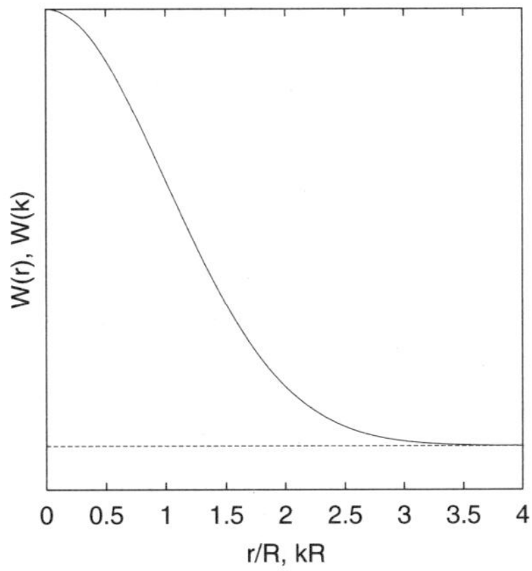

Figure 7.2 *The Gaussian filter, both in the coordinate space and in the Fourier space. The y-axis units are different for the two cases.*

The effective volume for the top-hat filter is

$$V_{TH} = \frac{4\pi}{3} R_{TH}^3. \tag{7.14}$$

It is more difficult to define the effective volume for the Gaussian filter. A popular convention is

$$V_G = \sqrt{(2\pi)^3} R_G^3. \tag{7.15}$$

These volumes are necessary when we wish to define objects by their masses, as is customary in cosmology (during evolution the radii of objects change considerably, but their masses remain largely the same).

Although we all have seen Gaussian curves, we illustrate the Gaussian filter in Fig. 7.2 so that you can compare it with the top-hat filter (see Fig. 7.1).

For a Gaussian filter the new field will be

$$Y(\mathbf{x}, R_F) = \frac{1}{(2\pi R_F^2)^{3/2}} \int e^{-|\mathbf{x}-\mathbf{x}'|^2/2R_F^2} Y(\mathbf{x}')\, d^3x',$$

giving a simple expression for the power spectrum of the filtered field:

$$P(k, R_F) = P(k) e^{-k^2 R_F^2}. \tag{7.16}$$

7.3.2 *Spectra of cosmological Gaussian random fields*

Many approximations in the study of the statistics of Gaussian random fields depend on the form of their power spectra. Although we shall discuss estimation of the spectra of cosmological fields later, we give here a brief description of a typical case.

The basic form is that of the density (contrast) spectrum. We know from the discussion of the dynamics of structure in Chapter 6 that growing modes of density, velocity, and gravitational potential are closely connected; thus, if we know the spectrum of one of them, we can easily derive the others.

Contemporary cosmology predicts that the inhomogeneities that grow later into the observed structure are generated during inflation. Cosmology predicts their spectra, as well. There are many theoretical possibilities, but because we as yet know little about the physics of inflation, most of these derivations have tried to conform with previous intuitive ideas.

Intuitively, the best choice is the initial power-law spectrum $P(k) \sim k^n$, as there are no preferred scales for gravitation (the exponent n is usually called the spectral index). In Chapter 6 we saw that in the comoving coordinates the density contrast δ and the fluctuating gravitational potential Φ are connected via the Poisson equation

$$\Delta\Phi(\mathbf{x}) = 4\pi G\bar{\rho}\delta(\mathbf{x})$$

($\bar{\rho}$ is the mean density of the universe). Thus, the Fourier components of these fields are connected by

$$\widetilde{\Phi} = -4\pi G\bar{\rho}\widetilde{\delta}(k)/k^2,$$

and the spectra

$$P_\Phi(k) \sim P(k)/k^4 \sim k^{n-4}.$$

The relative importance of the contributions of different wavenumbers is better shown by the Δ^2 descriptor (7.9):

$$\Delta^2_\Phi(k) \sim k^3 P_\Phi(k) \sim k^{n-1}.$$

This singles out the spectral index $n = 1$, because in this case there is equal power in the fluctuations at any scale, meaning that the geometry of the universe is equally wrinkled at any scale (the gravitational potential Φ is the metric perturbation from the viewpoint of general relativity).

The $n = 1$ or $P(k) \sim k$ spectrum is called the Harrison–Zeldovich spectrum and it is the most widely used form for the initial density contrast spectrum in cosmology.

This is not the whole story, however. The initial fluctuations are influenced by interactions between different species of particles in the early universe, by interaction of matter and radiation, etc. Because the initial perturbations are small, their dynamics are linear. It is easy to see that in the linear approximation the Fourier components of the density contrast $\widetilde{\delta}(k)$ evolve independently. Thus we can write

$$\widetilde{\delta}(k,t) = \frac{D(t)}{D(t_i)} T(k,t,t_i)\widetilde{\delta}(k,t_i),$$

where $D(t) \equiv D_1(t)$ is the growing mode of the density perturbation we derived in Chapter 6. The function $T(k,t,t_i)$ defined by the above equation describes the wavelength-dependent part of the evolution of the Fourier modes. It is called the transfer function and for a usual choice of the initial and current time moments it does not depend on t and t_i. The reason for that is that the physical processes that depend on the wavelength of the mode work only for a limited time interval, from the moment the mode enters the horizon until after recombination.

If we know the transfer function, it is easy to find the time-dependent spectrum of the perturbations:

$$P(k,t) = \left(\frac{D(t)}{D(t_i)}\right)^2 T^2(k) P(k,t_i).$$

For an initial power spectrum $P(k,t_i) = A(t_i)k^n$ the dependence on the initial moment is given by its amplitude $A(t_i)$, which is usually determined from observations (normalization of the spectrum). We shall describe that below.

Computation of the transfer function is rather difficult, but it is possible. It involves integrating the Boltzmann equation describing various interactions, but the perturbations are linear. The results of numerical studies are usually given by approximation formulae. One of the most widely used forms is that for the cold dark matter (CDM) universe, given by Bond and Efstathiou (1984):

$$T(k) = \left\{1 + \left[aq + (bq)^{3/2} + (cq)^2\right]^\nu\right\}^{-1/\nu}, \tag{7.17}$$

where $q = k/\Gamma$ describes the scaling of the wavenumber with a cosmological model. A good approximation for the shape parameter Γ is $\Gamma = \Omega_M h$. In fact, Γ also depends weakly on the baryon content of the universe Ω_B, but that dependence usually can be ignored. The number h above is the dimensionless Hubble constant defined by $H_0 = h \cdot 100$ km/sec/Mpc.

Exact computation of transfer functions has become easy lately thanks to the work of Seljak and Zaldarriaga (1996), who proposed an improved fast algorithm for that and distribute the program as freeware. The program, known as CMB-FAST, can be obtained from the Web site (*CMBFast*). Fitting formulae based on that program are given by Eisenstein and Hu (1999); they also explain physical mechanisms producing the transfer functions.

If we know the transfer function, the next step is to fix the overall amplitude (normalization) of the spectrum. This is done presently by two methods. The first method is the so-called "COBE normalization." The COBE satellite measured for the first time fluctuations in the cosmic background radiation (CMB) that cosmologists had been trying to detect for years without success. These fluctuations are the imprint of initial velocity fluctuations and are caused by the fluctuations of the initial gravitational potential (or density), so the fluctuations of the temperature of the CMB can be used to describe the density fluctuations at recombination. These measurements fix the amplitude of the density power spectrum at large scales ($L > 1000h^{-1}$ Mpc) that cannot yet be reached by galaxy catalogs. The main reference for this normalization is the paper by Bunn and White (1997). They describe the matter density power spectrum by

$$\Delta^2(k) = \delta_H^2 \left(\frac{ck}{H_0}\right)^{3+n} T^2(k), \tag{7.18}$$

where δ_H is the rms amplitude of density fluctuations at the scale of the present-day horizon c/H_0. Because the normalization depends on the cosmological model, they provide approximate expressions for δ_H:

$$\delta_H = \begin{cases} 1.94 \cdot 10^{-5} \Omega_M^{-0.785-0.05\ln\Omega_M} e^{-0.95\tilde{n}-0.169\tilde{n}^2}, & \Omega_M + \Omega_\Lambda = 1; \\ 1.95 \cdot 10^{-5} \Omega_M^{-0.35-0.19\ln\Omega_M} e^{-\tilde{n}-0.14\tilde{n}^2}, & \Omega_\Lambda = 0, \Omega_M < 1, \end{cases} \tag{7.19}$$

where $\widetilde{n} = n - 1$ is the "tilt" of the power spectrum, the difference of its initial slope from the Harrison–Zeldovich value $n = 1$. The rms error they quote is 7%.

Another method of normalization is by the value of the (linearly extrapolated) rms fluctuations of mass in spheres of certain radius. The radius adopted in cosmology for that is $8h^{-1}$ Mpc, which is considered to be approximately the smallest scale where the dynamics is still almost linear. This value is called σ_8 and is given by

$$\sigma_8^2 = 4\pi \int_0^\infty \widetilde{W}_{TH}^2(k, R_{TH} = 8h^{-1}\,\text{Mpc}) P(k) \frac{k^2\,dk}{(2\pi)^3}, \tag{7.20}$$

where $\widetilde{W}_{TH}$ is the Fourier image of the top-hat filter (7.13). This value does not coincide with the observed rms variation of galaxy counts, which differs from σ_8 by the bias factor and by the effects of nonlinear dynamical evolution. Thus, σ_8 should be treated as only a normalization constant.

Of course, once we know the COBE normalization and the cosmological parameters, we can calculate σ_8, so it is not really a free parameter. Using the σ_8 normalization means that we do not absolutely rely on known physical mechanisms of generating the initial power spectrum and allow for modifications that can change its slope and amplitude near $k = 2\pi/8\,h\,\text{Mpc}^{-1}$. Thus, when determining the σ_8 normalization, the shape parameter Γ in (7.17) is usually also treated as a free parameter.

As the window function $\widetilde{W}^2(k, R)$ drops sharply around $k \approx 1/R$, the main contribution to the $\sigma^2(R)$ comes from the same region because of the volume effect. Peacock (1999) gives a useful formula for quick estimation of σ_8 for CDM-like spectra:

$$\sigma_8^2 \approx \Delta^2(k_{eff}),$$
$$k_{eff} = \left(0.172 + 0.011\ln^2(\Gamma/0.34)\right) h\,\text{Mpc}^{-1}.$$

The number density of clusters of galaxies is sensitive to the overall amplitude of the power spectrum, so the σ_8 normalization is usually found on the basis of cluster mass–number density distribution. There have been many attempts to do

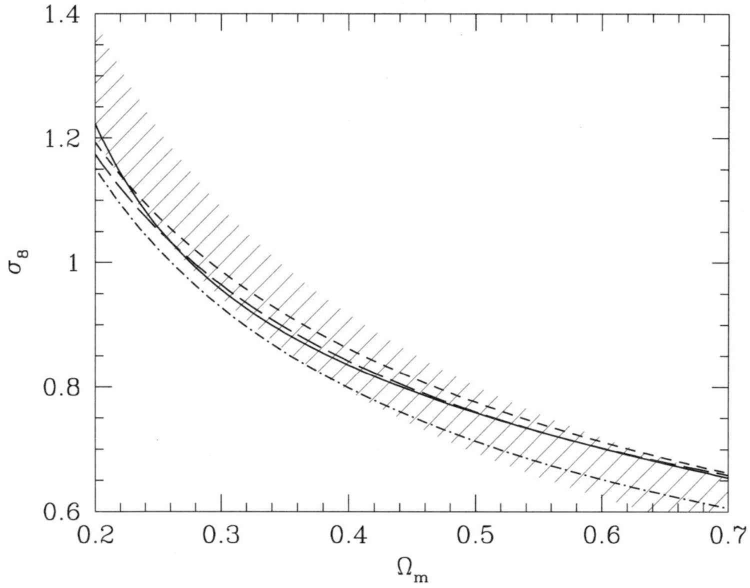

Figure 7.3 *The normalization factor σ_8 for flat cosmological models. The hatched region shows their result with rms error limits; lines show the results of different authors. (Reproduced, with permission, from Pierpaoli, Scott, and White 2001, Mon. Not. R. Astr. Soc., 325, 77–88. Blackwell Science Ltd.)*

this and, in general, the results agree. In recent years σ_8 is mainly found on the basis of the X-ray temperature distribution of rich clusters of galaxies (the X-ray temperature is determined by the mass of a cluster). We refer the reader to a careful analysis of methods used, their inherent uncertainties, and data problems in Pierpaoli, Scott, and White (2001). They give detailed fits for σ_8 for different cosmologies and shapes of the power spectra. For a flat cosmological model ($\Omega_M + \Omega_\Lambda = 1$) and typical power spectra ($n = 1$ and $\Gamma = 0.23$ in (7.17 above) the result is

$$\sigma_8 = 0.495\Omega_M^{-0.60}.$$

For comparison, Viana and Liddle (1999) find

$$\sigma_8 = 0.56\Omega^{-0.47}.$$

Cen (1998), using slightly different methods, gets

$$\sigma_8 = 0.5\Omega^{-0.43}.$$

The scatter in different normalizations is illustrated in Fig. 7.3. There is a general agreement, but the uncertainties are larger than for the COBE normalization. As we see, this normalization depends rather strongly on the cosmological model.

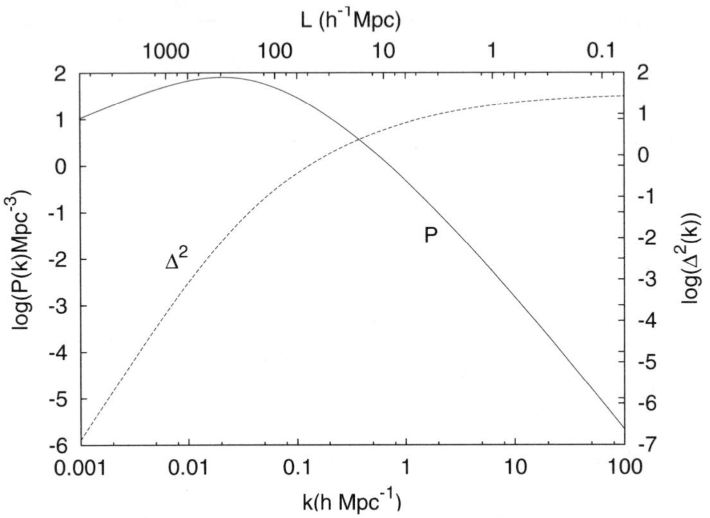

Figure 7.4 *The density contrast power spectrum in a CDM cosmology. The lower x-axis shows wavenumbers; the upper x-axis shows the corresponding spatial scales. Both the spectral density $P(k)$ and the spectral energy $\Delta^2(k)$ are in their real-life units.*

The reason for that is simple — in low-density models the growth of structure stops at late cosmological epochs, so the observed clusters must have formed earlier, and the power spectrum needs a larger amplitude.

Those two amplitudes (δ_H and σ_8) restrict the overall behavior of the density power spectrum rather well. For studies of galaxy clustering frequently only the σ_8 normalization is used, because it describes the power spectrum at typical clustering scales.

We show a typical density power spectrum in Fig. 7.4. It was calculated using the coefficients obtained by Jenkins et al. (1998) by comparing the exact transfer function with the form (7.17): $\gamma = 0.21$, $a = 6.4h^{-1}$Mpc, $b = 3.0h^{-1}$Mpc, $c = 1.7h^{-1}$Mpc, $\nu = 1.13$. The normalization was determined by requiring $\sigma_8 = 0.9$.

We see that the spectral density $P(k)$ has a maximum at rather large scales (around $200h^{-1}$Mpc), and drops rapidly at small wavelengths. We also see that the variance per logarithmic wavelength interval tends to a constant at small scales; thus, the initial density distribution is highly irregular.

This is usually referred to as a hierarchical picture. The spherical collapse solution in Chapter 6 gives the collapse time as $t_{coll} = 12\pi t_c = 12\pi R_c/c$, where R_c is the characteristic radius of the sphere — smaller objects collapse faster. Thus, the density distribution can be visualized as a hierarchy of clouds of different size and of different formation rate.

Such a hierarchy of clouds is typical for the cold dark matter models. In once popular hot dark matter models the transfer function decayed exponentially with the wavenumber, starting at the characteristic scale of superclusters. In these models huge superclusters formed first and the statistics of objects was that of fragmentation, completely different from the present-day approach.

The spectra described above and shown in Fig. 7.4 define the density distribution at redshifts $z \approx 1000$, at the start of the slow process of structure formation. By the present day the power spectra have been distorted by dynamical processes, but their overall form remains rather similar to the initial one.

Dynamical processes also can generate nonzero values of the bispectrum. In the linear approximation the physical fields remain Gaussian, but in the second-order approximation the bispectrum for density contrast is (Fry 1984)

$$B_{123} = B(\mathbf{k}_1, \mathbf{k}_2, \mathbf{k}_3) = Q_{12}P_1P_2 + Q_{13}P_1P_3 + Q_{23}P_2P_3,$$

where $P_i \equiv P(\mathbf{k}_i)$ and the coefficients Q_{ij} describe the triangle:

$$Q_{ij} = (1+\kappa) + \cos\theta_{ij}\left(\frac{k_i}{k_j} + \frac{k_j}{k_i}\right) + (1-\kappa)\cos^2\theta_{ij}.$$

Here θ_{ij} is the angle between the wavevectors $\mathbf{k}_i$, $\mathbf{k}_j$ and κ is a constant that depends slightly on the cosmological model ($\kappa = 3/7$ for the Einstein–de Sitter model).

Both the bispectrum and the power spectrum vary strongly with scale (and with time). The reduced bispectrum

$$Q_{123} = \frac{B_{123}}{P_1P_2 + P_1P_3 + P_2P_3}, \tag{7.21}$$

on the contrary, depends mainly only on the geometry of the triangle $\{123\}$. We shall describe recent determinations of the bispectrum in Chapter 8.

7.4 Realizations of random fields

An instance of a random field is called a realization of that field. All the observed fields are realizations of random fields. When building numerical models to compare with observations, we have to generate specific realizations of model fields. Such realizations are also widely used as initial data for numerical simulations. An overview of methods used for simulations of random variables, random fields, and point processes in statistics is given in Ripley (1987).

The random fields used to model the fluctuating physical fields (density, velocity, etc.) in cosmology are usually Gaussian fields or local transforms of Gaussian fields. As a realization of a random field is a set of field values $\mathbf{Y} \equiv \{Y(\mathbf{x}_1), \ldots, Y(\mathbf{x}_n)\}$ at n points, a realization of a Gaussian field can be considered as a multivariate Gaussian random variable of dimension n. Such a random variable is completely specified by its mean $\mathbf{m}$ and covariance matrix $\mathbf{C}$.

Two methods to generate multivariate Gaussian variables are described in Ripley (1987). The most common method is to represent the covariance matrix as $\mathbf{C} = \mathbf{S}\mathbf{S}^T$. The "square root" $\mathbf{S}$ of the matrix $\mathbf{C}$ can be found by Cholesky decomposition (Press et al. 1992). Next we generate n independent Gaussian random variables $\mathbf{Z} \equiv \{Z_1, \ldots, Z_n\}$ and calculate the sum

$$\mathbf{Y} = \mathbf{m} + \mathbf{S}\mathbf{Z}.$$

It is easy to see that the multivariate random variable $\mathbf{Y}$ is the realization we seek. This method, however, is suitable only for rather small n, because the decomposition of large matrices is computationally expensive.

Other methods for generating Gaussian random fields, listed in Ripley (1987), use the central limit theorem. One example is Matheron's "turning band" method, which is used to generate realizations of homogeneous and isotropic Gaussian random fields. This method starts by simulating a stationary one-dimensional Gaussian process $Y(x)$ with a known covariance function $\xi_1(r)$. Then we select N-dimensional random rotations O and carry the values of the process on the line over to the points in N-dimensional space covered by the rotated image of the line, $Y(Ox) = Y_1(x)$. The generated process is isotropic with the covariance function $\xi_N(r)$ (Ripley 1987):

$$\xi_N(r) = \frac{2\Gamma(N/2)}{\sqrt{\pi}\Gamma[(N-1)/2]} \int_0^1 \xi_1(vr)(1-v^2)^{(N-3)/2} dv.$$

The relation between the covariance functions is especially simple for $N = 3$:

$$\xi_1(r) = \frac{d}{dr}[r\xi_3(r)].$$

As the generated process is not Gaussian, we have to generate many realizations of this process and average them; by the central limit theorem the result is close to a Gaussian.

We have to confess that cosmologists do not use these classical methods; the methods used in cosmology are described below.

7.4.1 Fourier method

Numerical realizations of random fields have to be generated, naturally, for limited coordinate regions. Because in the case of cosmological fields these regions somehow have to represent an infinite (or very large) volume, the fields are usually assumed to be periodic.

Let us evaluate a realization of a real-valued random function $Y(\mathbf{x})$ for a cube of a size L, on a regular lattice of spacing $\Delta x = L/N$ (in one direction), as a matrix $Y(\mathbf{j}), \mathbf{j} = (j_1, j_2, j_3), j_k = 1, \ldots, N$, where the indices determine the

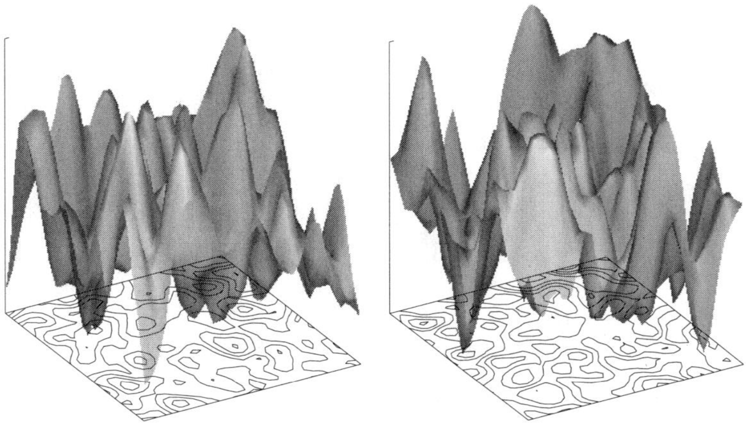

Figure 7.5 *Two different realizations of a Gaussian random field.*

spatial coordinate $\mathbf{x}(\mathbf{j})$ by

$$\mathbf{x}(\mathbf{j}) = \Delta x \sum_{l=1}^{l=3} j_l \mathbf{e}_l,$$

and $\mathbf{e}_l$ are the unit vectors in coordinate directions. Because of periodicity the function Y can be represented as a discrete Fourier transform (see, e.g., Press et al. 1992)

$$Y(\mathbf{j}) = \frac{1}{(2\pi)^3} \sum_{\mathbf{l}} e^{-i\mathbf{x}(\mathbf{j})\cdot\mathbf{k}(\mathbf{l})} \widetilde{y}(\mathbf{k}(\mathbf{l})), \tag{7.22}$$

where the Fourier amplitudes $\widetilde{y}(\mathbf{k})$ are also defined on a lattice $\mathbf{l} = (l_1, l_2, l_3), l_n = -N/2, \ldots, N/2$,

$$\mathbf{k}(\mathbf{l}) = \Delta k \sum_{n=1}^{n=3} l_n \mathbf{u}_n.$$

Here $\mathbf{u}_n$ are the unit vectors in the Fourier space, and the k-space lattice spacing is $\Delta k = 2\pi/L$. The finite number of lattice points in one direction limits the range of the Fourier sum by the Nyquist frequency $k_N = N\Delta k/2$. Comparing Eqs. (7.5 and 7.22) we see that

$$\widetilde{y}(\mathbf{k}) = (\Delta k)^3 \widetilde{Y}(\mathbf{k}).$$

So, in order to build the realization, we have to generate (complex) Gaussian-distributed Fourier amplitudes for every point of the grid $\mathbf{k}(\mathbf{l})$. The real and imaginary parts of the amplitude are both Gaussian with a zero mean and with a

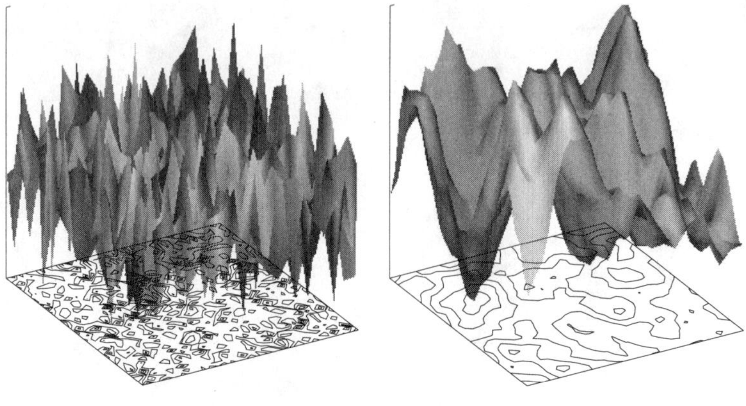

Figure 7.6 *Realizations of Gaussian random fields with different spectra,* $P(k) = k^2$ *(for the left panel) and* $P(k) = k^{-4}$ *(for the right panel).*

variance

$$\sigma^2 = \frac{(\Delta k)^3}{2} P(\mathbf{k}).$$

As our field is real, its Fourier amplitudes have to satisfy the relation

$$\widetilde{y}(-\mathbf{k}) = \widetilde{y}^\star(\mathbf{k}).$$

This means that we can choose freely only the Fourier amplitudes for a half-cube in the wavenumber space, assigning the amplitudes to the other half-cube using the relation above.

The calculation of the realization in coordinate space can be done easily using the Fast Fourier Transform (FFT, see Press et al. 1992). In Fig. 7.5 we show two different realizations of the same two-dimensional Gaussian random field with a simple power spectrum with a Gaussian cutoff,

$$P(k) = k^n \exp(-k^2 R^2/2), \tag{7.23}$$

with the exponent $n = -1$ and the smoothing radius $R = 0.3$, on a 32^2 grid.

It is clear that different spectra lead to different types of random fields. This is illustrated in Fig. 7.6, which compares realizations with pure power spectra, $P(k) = k^n$, but with the exponents $n = 2$ and $n = -4$. The larger exponent gives more weight to fluctuations (Fourier components are essentially plane waves) with higher wavenumbers $|\mathbf{k}|$ and with shorter wavelengths, making the realization more erratic (although both realizations are smooth). The smaller exponent, in contrast, weighs larger wavelengths more heavily, and leads to a rather smooth realization even without a high-wavenumber cutoff.

We can represent a complex Fourier amplitude also as $\widetilde{y} = |\widetilde{y}| \exp(i\phi)$, where $|\widetilde{y}|$ is its module and ϕ is its phase. The independence of the real and imaginary parts of $\widetilde{y}$ means that the phases ϕ are distributed uniformly (this is sometimes referred to as a "random phase distribution"). The modules $x = |\widetilde{y}|$ have the Rayleigh distribution with the probability density

$$f(x) = \frac{2x}{\sigma^2} e^{-x^2/\sigma^2}, \tag{7.24}$$

where

$$\sigma^2 = (\Delta k)^3 P(\mathbf{k}).$$

The integral probability function for (7.24) is

$$F(x) = 1 - \exp(-x^2/\sigma^2), \tag{7.25}$$

thus the function $\exp(-|\widetilde{y}|^2/\sigma^2)$ has a uniform distribution on the unit interval. This gives another means for generating the Fourier amplitudes $\widetilde{y}$: generate a random value from the uniform distribution and transform it to the module, using (7.25). Then select a random phase and use that to generate the real and imaginary parts of the complex amplitude by

$$\widetilde{y} = |\widetilde{y}| \exp(i\phi) = |\widetilde{y}| \cos\phi + i|\widetilde{y}| \sin\phi.$$

The above construction gives a periodic random function, not an isotropic one, which is needed in most cases. Of course, it is impossible to get a realization of an isotropic random field for a finite volume, so the periodicity of the field is not much of a drawback. The resulting fields will be nearly isotropic for larger wavenumbers (smaller separations).

For small wavenumbers the field will certainly be anisotropic, and there is another drawback to this picture. Because of periodicity the small wavenumber region is sampled badly; there will be only a few discrete Fourier amplitudes per wavenumber interval. If long-wavelength features of the generated field are important, one could use the FFTLog, a Fast Fourier Transform algorithm for a constant step in the logarithm of the wavenumber. A detailed description of that transform was given by Hamilton (2000), who has also written a public-domain implementation, available from the Web site (*FFTLog*).

Another possibility is to drop the periodicity assumption and to calculate the Fourier integrals by direct summation, choosing the grid in the wavenumber space at will (logarithmic grids are frequently used). Of course, direct summation is much more time-consuming than the FFT. Direct summation also can be used to improve the isotropy of the generated field, as done by Nusser and Dekel (1990), who choose for integration spherical coordinates in the k-space, select a grid for the values of radii k, and choose a number of random directions of $\mathbf{k}$ for every radial interval.

7.4.2 Noise convolution

Apart from the Fourier transform, another way to generate a Gaussian random field is to start with a collection of independent Gaussian random numbers of unit variance on the grid (a white noise field $Y_W(\mathbf{x})$). Given the spectral density $P(\mathbf{k})$ of the field we wish to generate, we can calculate the inverse Fourier transform of its square root

$$G(\mathbf{x}) = \int \sqrt{P(\mathbf{k})} e^{-i\mathbf{x}\cdot\mathbf{k}} \frac{d^3k}{(2\pi)^3}.$$

If we convolve the white noise field with the function $G(\mathbf{x})$, the resulting field

$$Y(\mathbf{x}) = \int G(\mathbf{x} - \mathbf{x}') Y_W(\mathbf{x}')\, d^3x'$$

is also Gaussian and has the spectral density $P(\mathbf{k})$, as necessary.

If we assume that the field is periodic and use FFT to calculate the convolution, we obtain the same result that we found with the Fourier method. Dropping the periodicity assumption and using direct summation in real space, we sample the small wavenumber region (large spatial scales) very well.

The difference is due to the periodicity requirement that is, in fact, extremely restrictive. It demands that the values of the field at any grid point be equal to their replicas in all periodic cubes and that creates a huge number of conditions to satisfy.

7.4.3 Erratic realizations

All the realizations we have built thus far are, in principle, smooth (differentiable). The rule is that realizations of Gaussian random fields are either smooth or extremely erratic; there is no intermediate behavior. For homogeneous random fields with finite $R(\mathbf{0})$ the realizations are mostly smooth, although there are limitations on the spectral density. As shown in Adler (1981), a real homogeneous Gaussian random field will have smooth sample functions with probability 1, if its spectral density $P(\mathbf{k})$ satisfies the condition

$$\int_{\mathbb{R}^N} |\ln(1 + |\mathbf{k}|)|^{1+\varepsilon} P(\mathbf{k})\, d^N k < \infty,$$

where $\varepsilon > 0$. As the variance of the field is given by a similar integral,

$$\sigma^2 = \int_{\mathbb{R}^N} P(\mathbf{k}) \frac{d^N k}{(2\pi)^N},$$

we see that the finiteness of the variance does not yet guarantee the smoothness of realizations. For an isotropic 3-D Gaussian random field the above condition limits the high-wavenumber behavior of the spectral density, demanding that

$$\lim_{k\to\infty} \ln(k) k^3 P(k) = 0.$$

For example, Gaussian random fields characterized by covariance functions

$$R(\mathbf{x}_1, \mathbf{x}_2) \propto \left(|\mathbf{x}_1|^{2\beta} + |\mathbf{x}_2|^{2\beta} - |\mathbf{x}_1 - \mathbf{x}_2|^{2\beta}\right)$$

(index-β fields) have realizations that are erratic (nondifferentiable) everywhere for $0 < \beta < 1$. These fields are not homogeneous, however, and so they are difficult to apply in a cosmological context.

The best-known example of such fields is the isotropic Brownian motion $B(\mathbf{x})$ with

$$R(\mathbf{x}_1, \mathbf{x}_2) = \frac{1}{2}\left(|\mathbf{x}_1| + |\mathbf{x}_2| - |\mathbf{x}_1 - \mathbf{x}_2|\right).$$

The realizations of such fields are more difficult to construct, although they are widespread; two-dimensional index-β fields serve as the basis for "fractal landscapes." These are usually built by fractal interpolation (the random midpoint displacement method). In this method the values of the field are generated initially at a coarse grid (the four corners of a square already represent a grid) and these are interpolated to successively finer grids, adding to the interpolated values random Gaussian displacements with the variance chosen to satisfy the correlation function. The construction of fractal landscapes is well described in the literature and there are many freeware programs to generate them. The best-known collection currently can be found at the Spanky Web site (*Spanky*). We also refer the interested reader to a volume edited by Peitgen and Saupe (1988). Fig. 7.7 shows an example of such a landscape, a realization of the two-dimensional isotropic Brownian motion.

As Fig. 7.7 shows, this surface is really extremely erratic. In fact, this is not a normal surface any more, because realizations of index-β fields are fractal with the fractal dimension $D = 3 - \beta$ (for two-dimensional fields). The dimension of the surface in Fig. 7.7 is $D = 2.5$, and this surface is halfway to filling the volume it occupies (Mollerach et al. 1999).

Thus, Gaussian random fields can be used to model fractal distributions of matter in the universe, should the need to do that arise.

7.5 Non-Gaussian fields

General non-Gaussian random fields are difficult to characterize. The fields that are used in cosmology are all derived from a Gaussian field, with a long list given by Coles and Barrow (1987). An example is a lognormal field $L(\mathbf{x}) = \exp(G(\mathbf{x}))$, where $G(\mathbf{x})$ is a Gaussian random field. This field has been described in detail by Coles and Jones (1991). Although the field $L(\mathbf{x})$ is homogeneous, if we start from a homogeneous Gaussian field, it is not wise to consider L as a zero-mean field, as its one-point distributions can be very asymmetric. There is a simple relation between the covariance functions:

$$R_L(r) = \exp(R_G(r)) - 1,$$

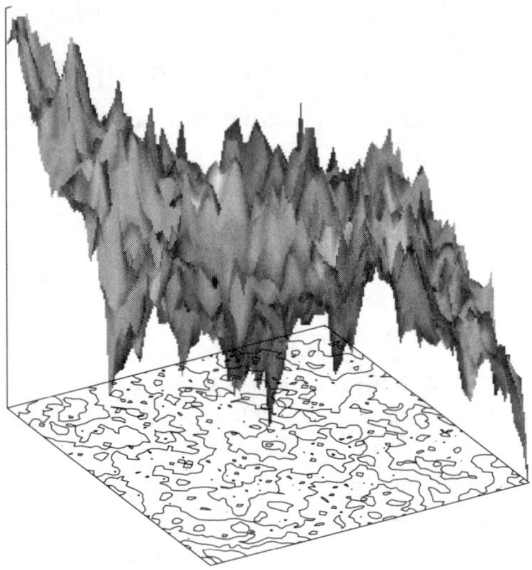

Figure 7.7 *A realization of the two-dimensional Brownian motion — a fractal landscape with the fractal dimension* $D = 2.5$.

where $R_L(r)$ is the covariance function of the lognormal field, $R_G(r)$ is the covariance function for the Gaussian field, and we have supposed that the Gaussian (hence also the lognormal) field is isotropic.

Another rather well-studied class of fields are the χ^2 fields, also described by Adler (1981). A parameter-n χ^2 field $Y_n(\mathbf{x})$ is defined as

$$Y_n(\mathbf{x}) = \sum_{i=1}^{n} G_i^2(\mathbf{x}),$$

where $G_i(\mathbf{x})$ are independent homogeneous (zero-mean) real Gaussian random fields with the same covariance function $R(\mathbf{r})$. A simple relation exists between the covariance functions in this case, too:

$$R_n(\mathbf{r}) = 2nR_G^2(\mathbf{r}),$$

where $R_n(\mathbf{r})$ is the covariance function of the parameter-n χ^2 field and $R_G(\mathbf{r})$ is the covariance function of the component Gaussian fields.

The Rayleigh fields are similar to the χ^2 fields and are defined as

$$RR_n(\mathbf{x}) = \sqrt{Y_n(\mathbf{x})}.$$

Several versions of these fields for specific n are described by Coles and Barrow (1987). Sheth (1995) gives the one-point distributions and correlation functions of these fields for the general case.

A non-Gaussian field that has become rather popular recently in cosmology is called the quadratic model. It is defined as

$$Q(\mathbf{x}) = G(\mathbf{x}) + \alpha \left(G^2(\mathbf{x}) - \langle G^2(\mathbf{x}) \rangle \right),$$

where $G(\mathbf{x})$ is a Gaussian random field and α is the coefficient of non-Gaussianity. This model can be studied analytically. It was proposed by Verde et al. (2000), who derived its bispectrum and studied the possibilities of observational determination of the degree of non-Gaussianity. Coles and Chiang (2000) explained how the bispectrum measures the quadratic phase coupling, specific to this model, and Shandarin (2001) derived the expressions for its Minkowski functionals, applied to the two-dimensional CMB maps.

7.6 Statistics of peaks in Gaussian random fields

As stated above, there are physical reasons to suppose that the initial density distribution in the universe can be described as a Gaussian (real, homogeneous, and isotropic) random field. This distribution is evolving in time, giving rise, finally, to the objects we observe, galaxies, and their clusters. The main force behind this evolution is gravitation that amplifies density fluctuations. The gravitational growth of fluctuations takes the fluctuation amplitude, finally, to a level after which rapid (nonlinear) formation of gravitationally bound objects begins. It is clear that the higher the initial density amplitude, the faster an object will form, and initial density maxima are the best candidates for present-day objects. Thus the peaks of Gaussian density fields give the most direct statistical model for galaxy clustering.

The first study of the statistics of Gaussian fields in cosmology was done by Doroshkevich (1970). Because this paper was published in Russian, many of his results were later rediscovered. The recognized milestone in this field, the "BBKS" article (Bardeen, Bond, Kaiser, and Szalay 1986), gives a detailed presentation of essential statistics and is required reading for anybody who is seriously interested in the topic.

7.6.1 Number density of peaks

The simplest statistics describing peaks in random fields is their number density. We can expand the field $Y(\mathbf{x})$ in the vicinity of a maximum at $\mathbf{x} = \mathbf{0}$ (if the field is homogeneous, we can always shift the coordinates) as

$$Y(\mathbf{x}) = Y(\mathbf{0}) + \frac{1}{2} \sum_{ij} Y_{,ij}\, x_i x_j,$$

where $i, j = 1, \ldots 3$ label the coordinates and $Y_{,ij} = \partial^2 Y / \partial x_i \partial x_j(\mathbf{0})$ are the second derivatives of Y at the maximum. The first derivatives Y_i are then

$$Y_{,i}(\mathbf{x}) = \sum_j Y_{,ij}\, x_j. \tag{7.26}$$

The probability of having an extremum of Y is $p(Y, Y_{,i} = 0, Y_{,ij})\, dY d^3 Y_{,i} \times d^6 Y_{,\{ij\}}$, where $p(Y, Y_{,i}, Y_{,ij})$ is the joint probability density of the field Y, its first and second derivatives. The volume element for the second derivatives is six-dimensional, as the matrix $Y_{,ij}$ is symmetric in indices. Because of (7.26) we can write $d^3 Y_{,i} = |Y_{,ij}|\, d^3 x$, where the bars denote the absolute value of the determinant of the matrix $Y_{,ij}$. The probability of having an extremum of Y in a volume $d^3 x$ is

$$dY d^3 x \int p(Y, Y_{,i} = 0, Y_{,ij})\, |Y_{,ij}|\, d^6 Y_{,\{ij\}} = n_{ext}(Y)\, dY d^3 x.$$

This gives us an expression for the peak number density n_p:

$$n_p(Y) = \int_{\mathcal{M}} p(Y, Y_{,i} = 0, Y_{,ij})\, |Y_{,ij}|\, d^6 Y_{,\{ij\}}\,, \tag{7.27}$$

where the integration is over the region $\mathcal{M}$ in the space of second derivatives that corresponds to the maxima of Y, namely, where the eigenvalues of $Y_{,ij}$ are negative. (Note that our notation for peak number densities differs from that of BBKS, who use N for differential and n for integrated peak number density.)

The joint probability density of the Gaussian field and its derivatives is also Gaussian:

$$p(Y, Y_{,i}, Y_{,ij}) = \frac{1}{(2\pi)^{N/2}\sqrt{|\mathbf{C}|}} e^{-\mathbf{y}\mathbf{C}^{-1}\mathbf{y}^T},$$

where the vector $\mathbf{y}$ stands for the collection of arguments $Y, Y_{,i}, Y_{,ij}$. Its dimension is 10, as there are only 6 independent second derivatives, and the matrix $\mathbf{C}$ is the 10×10 correlation matrix of $\mathbf{y}$. The components of the correlation matrix, the correlations between the field and its derivatives, are given by the correlation function and its derivatives or, alternatively, by the spectral moments, as we described above. For a real field the odd-ordered spectral moments are zero, which simplifies the matrix.

The correlation matrix can be found using the general formula (7.12). Using these amplitudes and the mapping $y_1 = Y$, $y_{i+1} = Y_{,i}$ $(i = 1, 2, 3)$, $y_{i+4} = Y_{,ii}$, $y_8 = Y_{,23}$, $y_9 = Y_{,13}$, $y_{10} = Y_{,12}$, the nonzero components of the correlation matrix can be written as

$$\begin{aligned} C_{00} &= \sigma_0^2, \\ C_{0i} &= -\frac{1}{3}\sigma_1^2, \quad i = 5, 6, 7, \\ C_{ii} &= \frac{1}{3}\sigma_1^2, \quad i = 2, 3, 4, \end{aligned} \tag{7.28}$$

$$\begin{aligned} C_{ii} &= \frac{1}{5}\sigma_2^2, \quad i = 5, 6, 7, \\ C_{ii} &= \frac{1}{15}\sigma_2^2, \quad i = 8, 9, 10, \\ C_{ij} &= \frac{1}{15}\sigma_2^2, \quad i, j = 5, 6, 7, \quad i \neq j, \end{aligned}$$

where σ_i are the spectral amplitudes. These are defined in cosmological literature as

$$\sigma_i^2 = \frac{1}{2\pi^2} \int_0^\infty k^{2i} P(k)\, dk. \tag{7.29}$$

These are usually called "spectral moments" and they differ from the moments used in the theory of random fields (see 7.11).

The next quantity we have to know to calculate the integral (7.27) is the volume element $d^6 Y_{,\{ij\}}$. Bardeen et al. (1986) show that this can be written as

$$d^6 Y_{,\{ij\}} = |(\lambda_1 - \lambda_2)(\lambda_2 - \lambda_3)(\lambda_1 - \lambda_3)|\; d\lambda_1 d\lambda_2 d\lambda_3 \frac{d\Omega}{6}, \tag{7.30}$$

where λ_i are the eigenvalues of the matrix $Y_{,ij}$ and $d\Omega$ is the volume element of a three-sphere of Eulerian angles that describe the orientation of the eigenvalues. This expression shows that the volume of the regions in the $Y_{,ij}$ space described by close eigenvalues λ_i (the probability of having close eigenvalues) is very small.

The integral (7.27) is rather difficult to calculate, but the integration can be carried out analytically almost all the way through, leaving only one numerical step (see Bardeen et al. 1986). The result can be written as

$$n_p(\nu) = \frac{1}{(2\pi)^2 R_\star^3} e^{-\nu^2/2} G(\gamma, \gamma\nu). \tag{7.31}$$

Here ν is the normalized height of the peak

$$\nu = \frac{Y}{\sigma_0},$$

the function G is defined as

$$G(\gamma, x_\star) = \int_0^\infty F(x; \gamma, x_\star)\, dx, \tag{7.32}$$

and

$$F(x; \gamma, x_\star) = \frac{1}{\sqrt{2\pi(1-\gamma^2)}} e^{-(x-x_\star)^2/2(1-\gamma^2)} f(x). \tag{7.33}$$

The parameters γ and $R_\star$ characterize the shape of the spectrum and are defined as

$$\gamma = \frac{\sigma_1^2}{\sigma_0 \sigma_2}, \tag{7.34}$$

$$R_\star = \sqrt{3}\frac{\sigma_1}{\sigma_0}. \tag{7.35}$$

In order to understand the meaning of these parameters, let us choose a simple power spectrum as an example. We start with the power spectrum $P(k) \sim k^n$ and filter it with a Gaussian filter of radius R_F:

$$P(k, R_F) = k^n \exp(-k^2 R_F^2).$$

The spectral parameters for this spectrum are:

$$R_\star = \sqrt{\frac{6}{n+5}} R_F,$$

and

$$\gamma^2 = \frac{n+3}{n+5}.$$

Thus $R_\star$ is proportional to the filter scale; it describes the correlation length of the field. The concentration parameter γ changes between zero for $n \approx -3$ (this is a limiting case, otherwise the integral over $P(k)$ would diverge at $k \to 0$) and one for large n, when the spectral density has a sharp maximum.

The function $f(x)$ in formula (7.33) describes the distribution of the second derivatives of the field:

$$\begin{aligned} f(x) &= \frac{(x^3 - 3x)}{2} \left[\text{erf}\left(\sqrt{\frac{5}{2}} x \right) + \text{erf}\left(\sqrt{\frac{5}{2}} \frac{x}{2} \right) \right] + \\ &+ \sqrt{\frac{2}{5\pi}} \left[\left(\frac{31x^2}{4} + \frac{8}{5} \right) e^{-5x^2/8} + \left(\frac{x^2}{2} - \frac{8}{5} \right) e^{-5x^2/2} \right]. \end{aligned}$$

Here x is the normalized Laplacian of the field:

$$x = -\nabla^2 Y / \sigma_2. \tag{7.36}$$

As shown by Mann, Heavens, and Peacock (1993), the formula (7.31) can be integrated analytically to get the total peak number density over a level ν

$$N_p(\nu) = \frac{1}{8\pi^2 R_\star^3} I(\nu), \tag{7.37}$$

where $I(\nu)$ is an integral

$$I(\nu) = \int_0^\infty f(x) e^{-x^2/2} \text{erfc}\left(\frac{\nu - \gamma x}{\sqrt{2(1-\gamma^2)}} \right) dx. \tag{7.38}$$

The peak number density $n_p(\nu)$ and the total density $N_p(\nu)$ (the total number density of peaks with height $\geq \nu$) are shown in Fig. 7.8 for various values of the concentration parameter of the spectrum γ. We see that the amplitudes of peaks can reach up to 5σ, and that there can be peaks with height much lower than the mean density of matter, down to -3.5σ. There are no low-density peaks only in the limiting case of a very sharp spectrum, $\gamma = 1$.

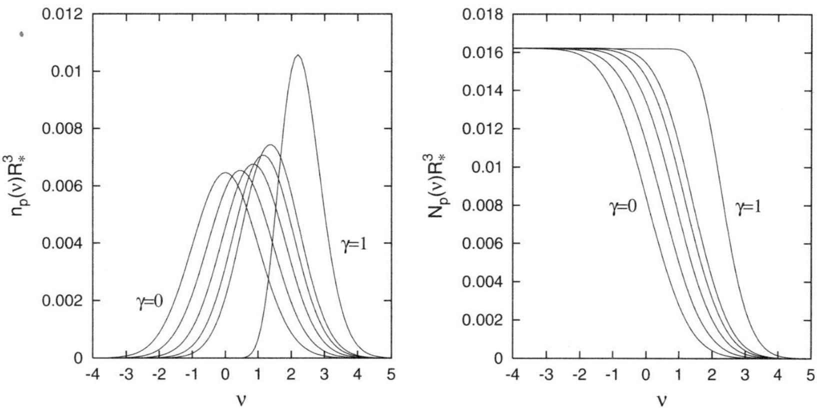

Figure 7.8 *Peak number densities for Gaussian fields with a different spectral parameter γ. The left panel shows differential number densities dependent on the normalized peak height ν. The different curves correspond to different γ, ranging between the limiting values $\gamma = 0$ and $\gamma = 1$ with the step 0.2. The right panel shows the total number density (for all peaks with height larger than ν) for the same choice of γ.*

We see also that the mean peak density grows with the concentration of the spectrum, ranging from 0 for $\gamma = 0$ up to about 2.2 for $\gamma = 1$. The right panel of Fig. 7.8 shows us that the total peak number density does not depend on the concentration, but only on the correlation length $R_\star$. This also can be seen from the formula (7.32) — the integral of $\exp(-\nu^2/2)G(\gamma, \gamma\nu)$ over all peak heights does not depend on γ.

The total peak number density can be found exactly:

$$N_p(-\infty) = \frac{29 - 6\sqrt{6}}{8\pi^2\sqrt{125}R_\star^3} \approx 0.00162 R_\star^{-3},$$

one peak per about $(9R_\star)^3$ volume. If we suppose that the object connected with the peak has a volume of the sphere of a radius of $R_\star$, the peaks occupy roughly $1/50$ of the total volume for any spectra.

In the high peak limit, the formulae for the number densities simplify:

$$n_p(\nu) = \frac{\gamma^3}{4\pi^2 R_\star^3}(\nu^3 - 3\nu)e^{-\nu^2/2}, \quad \nu \to \infty, \tag{7.39}$$

$$N_p(\nu) = \frac{\gamma^3}{4\pi^2 R_\star^3}(\nu^2 - 1)e^{-\nu^2/2}, \quad \nu \to \infty. \tag{7.40}$$

7.6.2 Structure of peaks in Gaussian random fields

In the neighborhood of a peak, the field Y can be approximated as an ellipsoid

$$Y(\mathbf{x}) = Y_0 - \sum \lambda_i x_i^2, \qquad Y_0 = Y(\mathbf{0}),$$

if we orient the coordinate system along the principal axes (the eigenvectors of $Y_{,ij}$). At a peak the eigenvalues λ_i are all positive and they define the semiaxes of the profile:

$$a_i = \left[\frac{2(Y_0 - Y_c)}{\lambda_i}\right]^{1/2},$$

where Y_c is a constant value of the field.

The steepness of the profile (its curvature at the center) is

$$\nabla^2\delta = -\sigma_i\lambda_i = -I_1,$$

where I_1 is the first invariant of the matrix $Y_{,ij}$. Let us order the eigenvalues by $\lambda_1 \geq \lambda_2 \geq \lambda_3$. The asymmetry of the profile is described then by its ellipticity (in the x_1, x_3 plane)

$$e = \frac{\lambda_1 - \lambda_3}{2I_1}$$

and by its prolateness

$$p = \frac{\lambda_1 - 2\lambda_2 + \lambda_3}{2I_1}.$$

The last parameter has values $p \in (-e, e)$; for prolate spheroids $p = e$ and for oblate spheroids $p = -e$.

Bardeen et al. (1986) derive the joint probability distributions of the height of the peak ν, its steepness I_1, and the asymmetry parameters e and p. The formulae are rather cumbersome and we will not repeat them here. The main result is that the most probable value of p is close to zero ($\lambda_2 \approx (\lambda_1+\lambda_3)/2$), but the ellipticity e is definitely larger than zero; the peaks are triaxial. The higher the peak, the more spherical it is — the most probable ellipticity is

$$e_{\max} \approx \frac{1}{\sqrt{5}\gamma\nu}. \tag{7.41}$$

Probably more important than the initial density profile is its evolution during the collapse. In the Zeldovich approximation this evolution is determined by the velocity potential $\Psi(\mathbf{x})$. Let us diagonalize the (symmetric) deformation tensor $\Psi_{,ij}$ and denote these eigenvalues again λ_i (they are different from those that describe the density profile) and order them. Doroshkevich (1970) found a rather simple expression for the joint distribution of the eigenvalues:

$$\begin{aligned} p(\lambda_1, \lambda_2, \lambda_3) &= \frac{15^3}{8\pi\sqrt{5}\sigma_0^6} \exp\left(-\frac{3I_1^2}{\sigma_0^2} + \frac{15I_2}{2\sigma_0^2}\right) \times \\ &\quad (\lambda_1 - \lambda_2)(\lambda_1 - \lambda_3)(\lambda_2 - \lambda_3), \end{aligned}$$

where I_1 and $I_2 = \lambda_1\lambda_2+\lambda_1\lambda_3+\lambda_2\lambda_3$ are the first and the second invariants of the deformation tensor. This formula shows that spherical collapse ($\lambda_1 = \lambda_2 = \lambda_3$) is rather improbable.

It is easy to see that the first invariant of the deformation tensor is the density contrast δ. Defining the ellipticity and prolateness of the deformation tensor as we did above for the density profile, an exact result for the joint conditional probability density of e and p given δ can be found (Sheth, Mo, and Tormen 2001):

$$f(e,p|\nu) = \frac{15^3}{3\sqrt{10\pi}} e(e^2 - p^2)\nu^5 \exp\left[-\frac{5}{2}\nu^2(3e^2 + p^2)\right]$$

($\delta = \nu\sigma_0$). The maximum of that distribution is at

$$p_{\max} = 0, \qquad e_{\max} = \frac{1}{\sqrt{5}\nu},$$

close to that for the density asymmetry distribution above (7.41). This formula is valid for a general point in the flow; adding the peak constraints would make it more complex, so it serves here as an illustration.

7.6.3 Clustering of peaks

The motivation for developing the peak statistics is that we can hope to identify the peaks with observed objects and thus to infer the properties of the random field itself. Apart from the number densities of peaks, the next simplest observational statistics is the correlation of the peaks. To model that we have to find the covariance function of the peak density field.

In order to find the covariance function of maxima we have to calculate the joint probability density of having maxima at two points separated by r. If the one-point density studied in the previous section had a 10×10 almost diagonal correlation matrix (7.28), the correlation matrix for the present distribution is, naturally, a 20×20 matrix. This matrix, however, has many nonzero, nondiagonal components, because correlations between the field and its derivatives at different points do not vanish. Bardeen et al. (1986) give recipes for a numerical solution of this problem. We shall use the approximation proposed by Kaiser (1984), calculating the density correlations of regions where the field is higher than some fixed level. When that level is small, these regions do not tell much about the maxima of the field, but when the level rises, it delineates the maxima better and better. If we suppose that maxima populate overdense regions roughly with a uniform number density, we can approximate the correlation function of maxima by that of the density correlation of overdense regions.

The overdensity correlation function is much easier to calculate because that depends only on the values of the density field at two points and does not involve the first and second derivatives.

In order to develop the joint distribution of the density contrast field (a homogeneous and isotropic zero-mean Gaussian field) $\delta_i = \delta(\mathbf{x}_i), i = 1, 2$ at two points, we shall follow Peebles' approach (1993) and introduce the auxiliary variables $\delta_+ = \delta_1 + \delta_2$ and $\delta_- = \delta_1 - \delta_2$. These variables are also Gaussian and they are independent, as

$$E\{\delta_+\delta_-\} = E\{\delta_1^2\} - E\{\delta_2^2\} = 0.$$

Their variance can be found as

$$E\{\delta_+^2\} = 2E\{\delta^2\} + 2E\{\delta_1\delta_2\} = 2\left[\xi(0) + \xi(r)\right],$$

(recall that the covariance function of the density contrast field coincides with the density correlation function $\xi(r)$). Similarly, the variance $E\{\delta_-^2\} = 2\left[\xi(0) - \xi(r)\right]$, $(r = |\mathbf{x}_1 - \mathbf{x}_2|)$. As δ_+ and δ_- are independent, their joint probability density is the product of two Gaussian densities

$$p(\delta_+, \delta_-) = \frac{1}{2\pi\sqrt{\xi^2(0) - \xi^2(r)}} \exp\left[-\frac{\delta_+^2}{4(\xi(0) + \xi(r))} - \frac{\delta_-^2}{4(\xi(0) - \xi(r))}\right].$$

Substituting $\delta_{+,-}$ by their definitions via $\delta_{1,2}$ and taking into account that the determinant of the Jacobian of the transformation $(\delta_{1,2} \to \delta_{+,-})$ equals one, we find the joint probability density of $\delta_{1,2}$:

$$p(\delta_1, \delta_2) = \frac{1}{2\pi\sqrt{\xi^2(0) - \xi^2(r)}} \exp\left[-\frac{\xi(0)\delta_1^2 + \xi(0)\delta_2^2 - 2\xi(r)\delta_1\delta_2}{2(\xi^2(0) - \xi^2(r))}\right].$$

If we choose now a lower density contrast limit $\delta_{\text{lim}} = \nu\sigma$, where $\sigma^2 = \xi(0)$ is the variance of the field, the probability that the density contrast at both points will be higher than δ_{lim} is

$$P_2(\nu) = \int_{\nu\sigma}^{\infty} \int_{\nu\sigma}^{\infty} p(\delta_1, \delta_2)\, d\delta_1 d\delta_2. \tag{7.42}$$

Considering such high-density regions as objects, the correlation function of the corresponding point process $\xi(r; \nu)$ is given by

$$1 + \xi(r; \nu) = P_2(r; \nu)/P_1^2(\nu),$$

where $P_1(\nu)$ is the one-point probability for the density to exceed the level $\nu\sigma$. For an exact procedure to find the integral in (7.42) and the exact correlation function we refer readers to the paper by Jensen and Szalay (1986). They represent the function as an infinite series that has to be summed numerically. We shall derive below an asymptotic expression, following Padmanabhan (1993).

7.6.4 High-peak asymptotics

For high peaks ($\nu \gg 1$) the integral in (7.42) simplifies. After a change of variables $y_i = \delta_i/\sigma$ the integral (7.42) can be written as

$$P_2(r;\nu) = \frac{1}{2\pi\sqrt{1-\psi^2(r)}}\int_\nu^\infty\int_\nu^\infty \exp\left[-\frac{y_1^2+y_2^2-2\psi(r)y_1y_2}{2\left(1-\psi^2(r)\right)}\right] dy_1 dy_2,$$

where $\psi(r) = \xi(r)/\xi(0)$ and we have used the fact that $\xi(0) = \sigma^2$.

Supposing that $\psi(r) \ll 1$ and ignoring the terms, which are quadratic in $\psi(r)$, we can write the integrand above as

$$\exp\left(-y_2^2/2\right)\exp\left[-\frac{\left(y_1-\psi(r)y_2\right)^2}{2}\right].$$

Integrating this over y_1 and supposing $\nu \gg 1$ we get

$$\exp\left(-y_2^2\right)\exp\left[-\frac{\left(\nu-\psi(r)y_2\right)^2}{2}\right],$$

which, after discarding quadratic terms in ψ, gives a similar integral for y_2. Integrating that, we get

$$P_2(\nu) = \frac{1}{2\pi}e^{-\nu^2}e^{\psi(r)\nu^2},$$

and as in the same approximation

$$P_1(\nu) = \frac{1}{\sqrt{2\pi}}e^{-\nu^2/2},$$

we get, finally, the correlation function

$$\xi(r;\nu) = \exp\left[\nu^2\frac{\xi(r)}{\xi(0)}\right] - 1. \tag{7.43}$$

This result was obtained by Politzer and Wise (1984).

Usually the argument of the exponent is small enough to justify the expansion of the exponent in series, giving in the first order

$$\xi(r;\nu) = \nu^2\frac{\xi(r)}{\xi(0)} = \frac{\nu^2}{\sigma^2}\xi(r). \tag{7.44}$$

This expression was obtained first by Kaiser (1984). Although the formulae (7.43, 7.44) are approximate, the comparison made by Jensen and Szalay (1986) shows that they do not differ much from the exact results for the correlation of high-density regions.

Another approximation made above was to substitute the full density field for the peak density field. Nevertheless, formula (7.44) has the same form as the asymptotic formula for the correlation of peaks, obtained by the exact procedure.

The only modification is the replacement of the density level ν by an effective density level (Bardeen et al. 1986):

$$\widetilde{\nu} = \frac{\nu - \gamma x}{1 - \gamma^2}, \tag{7.45}$$

where x is defined above (7.36). The asymptotic peak correlation function is

$$\xi(r; \nu) = \frac{\langle \widetilde{\nu} \rangle^2}{\sigma^2} \xi(r). \tag{7.46}$$

Here $\langle \widetilde{\nu} \rangle$ is the mean value of the effective density level (see (7.31, 7.32)) over the peak number density distribution,

$$\langle \widetilde{\nu} \rangle = \frac{1}{(2\pi)^2 R_\star^3 N_p(\nu)} \int_\nu^\infty e^{-\nu'^2/2} \int_0^\infty \frac{\nu' - \gamma x}{(1 - \gamma^2) F(x; \gamma, \gamma\nu')} \, d\nu' \, dx,$$

where $F(x)$ is given by (7.33) above. This can be integrated over ν' to get

$$\langle \widetilde{\nu} \rangle = \frac{\sqrt{2}}{\sqrt{\pi(1 - \gamma^2)} I(\nu)} \int_0^\infty f(x) e^{-x^2/2} \exp\left(-\frac{\nu - \gamma x)}{2(1 - \gamma^2)} \right) dx,$$

where $I(\nu)$ is given by (7.38) above (Mann, Heavens, and Peacock 1993).

The main result of this section is thus that high-density regions (peaks) are more correlated than the field in general, the amplification factor growing as the square of the density level. At first glance this result is rather surprising and needs clarification.

7.6.5 *Peak-background split*

Let us follow Kaiser (1984) and suppose that we can represent the density field as a sum of two independent random fields, a rather smooth background density $\rho_b(\mathbf{x})$ that describes the long-wavelength components of the density field, and a rapidly oscillating component $\rho_p(\mathbf{x})$ that is responsible for the peaks (the peak–background split). If we consider the correlation scale of the background field to be much larger than that of the peak field, we can apply the results obtained in the previous section locally.

The background field will change the effective threshold for peaks: a peak threshold δ in the full field will translate into $\delta - \delta_b(\mathbf{x})$ for the peak field, lower for the regions of maxima in the background. The peak level parameter ν_p is

$$\nu_p(\mathbf{x}) = \nu - \delta_b(\mathbf{x})/\sigma_0.$$

The background field will modulate the peak number densities:

$$n_p(\nu_p; \mathbf{x}) = n_p(\nu - \delta_b(\mathbf{x})/\sigma) \approx n_p(\nu) \left[1 - \frac{1}{n_p(\nu)} \frac{dn_p(\nu)}{d\nu} \frac{\delta_b(\mathbf{x})}{\sigma} \right].$$

This means that we can introduce the peak overdensity field, similar to the background overdensity, by

$$n_p(\nu_p;\mathbf{x}) = n_p(\nu_p)\left[1 + \delta n_p(\nu_p;\mathbf{x})\right].$$

Using the formula for $n_p(\nu)$ (7.31) and omitting the arguments ν_p and $\mathbf{x}$, we find

$$\delta n_p = \left(\nu - \frac{g}{\nu}\right)\frac{\delta_b}{\sigma_0}, \tag{7.47}$$

where the slowly changing function g is defined by

$$g(\gamma,\gamma\nu) = \frac{d\ln G(\gamma,\gamma\nu)}{d\ln\nu}. \tag{7.48}$$

The form (7.47) is chosen to capture the asymptotic dependence of δn_p on ν — the asymptotic formula for peak density (7.39) shows that for large ν, g is a constant ($g = 3$). Using the definition $\nu = \nu_p = \delta_p/\sigma_0$, the formula (7.47) can be rewritten as

$$\delta^L n_p = b^L(\nu_p,\sigma_0)\delta_b = \frac{\nu_p^2 - g}{\delta_p}\delta_b, \tag{7.49}$$

where we have introduced the peak overdensity bias $b^L(\nu_p)$. The superscript L means that the above formula describes the situation in the initial (Lagrangian) space (we shall change over to real space below).

As we see, the local peak overdensity is enhanced by the factor $b^L(\nu_p,\sigma_0)$ compared to the background overdensity δ_b. For high peaks, $\nu_p^2 \gg g$, this translates to the factor ν_p^2/σ_0^2 for the correlation function, reproducing the amplification found above.

To find the peak overdensity in real (Eulerian) space we have to account for the movement of the peaks. We can do that by considering the peaks as markers in the background field. In this case their density grows together with the density of the background,

$$n_p^E = n_p^L \rho_b,$$

where the superscript E denotes Eulerian quantities. For overdensities,

$$\delta n_p^E = (1 + \delta n_p^L)(1 + \delta_b) - 1 = \delta n_p^L + \delta_b.$$

We have assumed that the background overdensity $\delta_b \ll 1$; because it describes large scales, this assumption is justified. We have also assumed that its Eulerian value coincides with the Lagrangian one. This is a good approximation; an exact treatment can be found in Catelan et al. (1998). The peak overdensity in real space is now

$$\delta^E n_p = b^E(\nu_p,\sigma_0)\delta_b = \left(1 + \frac{\nu_p^2 - g}{\delta_p}\right)\delta_b, \tag{7.50}$$

where $b^E(\nu_p,\sigma_0)$ is the Eulerian peak bias.

As correlation functions scale as the square of the density contrast, we get the peak correlation function

$$\xi_p(r;\nu) = \left[b^E(\nu_p,\sigma_0)\right]^2 \xi(r). \tag{7.51}$$

This formula can be used for all peak heights, if we calculate the exact g (7.48).

The behavior of the peak bias can be analyzed, defining a typical peak amplitude $\delta_\star$ by

$$\delta_\star^2/\sigma^2 = g.$$

The expression for bias (7.50) tells us that the peak height $\delta_\star$ separates the different biasing regimes. Higher peaks are biased (their overdensity is higher than the background, mass density), but lower peaks are anti-biased (their overdensity is smaller than that of the background).

The discussion above is not rigorous, because it is difficult to separate the background and peak fields. A natural way would be to generate the background field by filtering the initial field and to get the peak field as the difference between the initial field and the background. In that case, however, the two fields will not be statistically independent. A perfect case would be if the power spectrum of the initial field had a wide gap, a region of wavenumbers where it is zero. The spectra that we have seen in cosmology, however, are not even remotely like that. An additional requirement is that there should be enough large-scale power to form a "background." For typical cosmological density power spectra this requirement is true for a rather wide wavenumber interval, from high wavenumbers down to about $k \approx 0.1h\text{Mpc}^{-1}$ ($L \approx 60h^{-1}$Mpc) (see Fig. 7.4).

Thus, the peak–background split remains a heuristic explanation. But it is useful to know that the higher the large-wavelength "plateaus" of a random field, the higher the peak number density in these regions.

7.6.6 Peak theory and cluster correlations

Galaxy clusters are high peaks in the present density field that is represented by galaxies. Thus it is natural to assume that they have originated from the high peaks of the initial density field (filtered on an appropriate mass scale). The initial density field was a homogeneous and isotropic Gaussian field with the variance σ much less than the mean density ρ. It makes the initial density contrast field δ a perfect zero-mean Gaussian field with practically no limitations on its amplitude.

However, the application of the peak theory to the observed galaxy distribution is not straightforward. First, we have to worry about one-to-one correspondence between the objects and the peaks, as smaller objects merge into larger ones during the evolution of the structure. This means that we have to select rather rare (high) peaks that have not had the possibility to move enough to merge, certainly not galaxies. Cen (1998) shows that the mean separation of galaxy clusters is larger than their possible total displacement, so galaxy clusters retain their identity.

The next problem is to find the connection between the basic arguments of the peak theory, the radius of the smoothing filter R_F and the height of a peak $\nu = \delta/\sigma_0$ (δ denoting here the smoothed density contrast field) with the observed parameters of galaxy clusters. There are several methods for that; we shall follow the approach of Cen (1998).

As the initial perturbations are small, the mass of a cluster is determined mainly by the smoothing radius of the density field. The relationship between these quantities can be derived, comparing the mass of an object inside the filter profile

$$M_F = (2\pi)^{3/2} R_F^3 \Omega_M \rho_c \tag{7.52}$$

(ρ_c is the critical cosmological average density) with a mass of a virialized spherical cluster

$$M_{vir} = \frac{4\pi}{3} R_{vir}^3 \bar{\rho}_{vir}, \tag{7.53}$$

where the mean density of a virialized cluster is

$$\bar{\rho}_{vir} \approx 178 \Omega_M^{-0.7} \Omega_M \rho_c \tag{7.54}$$

(see Chapter 6). In this picture all peaks of a field smoothed by a filter of radius R_F are assigned the same masses. As the mass range of real galaxy clusters is wide, we have to describe the cluster population by a set of smoothed fields for different filter radii.

The filter radius for a particular value of the virial mass can be selected, obviously, by demanding $M_F = M_{vir}$, which leads to the relation between R_F and r_{vir}:

$$R_F^3 = \frac{178}{3} \sqrt{\frac{2}{\pi}} \Omega_M^{-0.7} R_{vir}^3. \tag{7.55}$$

The masses of Abell clusters M_A are defined as the mass inside the Abell cluster radius $R_A = 1.5h^{-1}$ Mpc. This is a fixed radius at the present epoch; the filter radius (7.55) derived above describes the region in the initial density field that has collapsed to form the cluster. The larger this radius, the larger the mass of the galaxy cluster. To relate the integrated mass of an Abell cluster to the virial mass (7.53) we have to assume the cluster density profile. Cen (1998) chose a power-law density profile $\rho(r) \sim r^{-\alpha}$, which gives

$$M_A = M_{vir} \left(\frac{R_A}{R_{vir}} \right)^{3-\alpha}. \tag{7.56}$$

Substituting here (7.53) we find the relation between R_{vir} and M_A and using (7.55), we finally get a relation between R_F and M_A. This expression is rather cumbersome; we refer the reader to the paper by Cen (1998). Apart from the parameters of the cosmological model, the only adjustable parameter in this relation is α, which Cen found to be 2.3, comparing the cluster mass functions from numerical simulations with the peak number density relations.

The minimum peak height for a peak to be virialized by the present time and to be counted as a galaxy cluster is $\delta_c = 1.68$ for spherical collapse (Chapter 6). Cen argues that as real collapse could differ from that (generic collapse is anisotropic), it should also be treated as a free parameter. He finds by comparing the simulation cluster mass functions with peak theory results that it also has practically a constant value $\delta_c = 1.40$ for all cosmological models.

If the two parameters are fixed, there is no more additional freedom in the theory and it can be applied directly to observations. This is basically the way filter radii are connected with cluster masses in all other studies, including those based on the Press–Schechter formalism that we describe below.

Analyzing the behavior of the cluster mass function in the peak theory, Cen found that it depends mainly on the normalization of the spectrum σ_8, on the density parameter Ω_M, on the shape of the power spectrum, and slightly on Ω_Λ. The peak theory normalization

$$\sigma_8 \approx 0.5\Omega^{-0.43},$$

is slightly lower than that obtained from the application of the Press–Schechter theory (see Section 7.3).

It is much easier to restrict the range of cosmological models that are allowed by observations using the analytical peak theory than by extensive and costly numerical simulations. Cen (1998) carried out the first analysis of the parameter space and found six CDM models that possibly bracket all observational results. He also compared the correlation function of simulated clusters with the peak theory formula

$$\xi_p \approx \left(\frac{\langle\widetilde{\nu}\rangle^2}{\sigma_0^2} + 1 \right)^2 \xi(r),$$

which is the formula (7.46) derived above, modified for gravitational motions in Eulerian space, and found a good agreement. A general agreement of the amplification of correlations predicted by the peak theory was found earlier.

Another prediction of the peak theory is the dependence of the correlation function and the peak number densities on the density level ν that defines the peaks. Peaks with larger ν have higher amplitudes, they collapse earlier and turn into more massive (richer) galaxy clusters, because there is more time to accumulate galaxies by accretion. The richer the clusters, the smaller their number density (this can be described by the mean distance between clusters, $d_c = N_p^{-1/3}$). So we can compare the predicted amplification versus density level relation with observations. As the amplitude of the correlation function is described by the correlation length r_0, this test is presented as r_0 versus d_c dependence for different cluster samples.

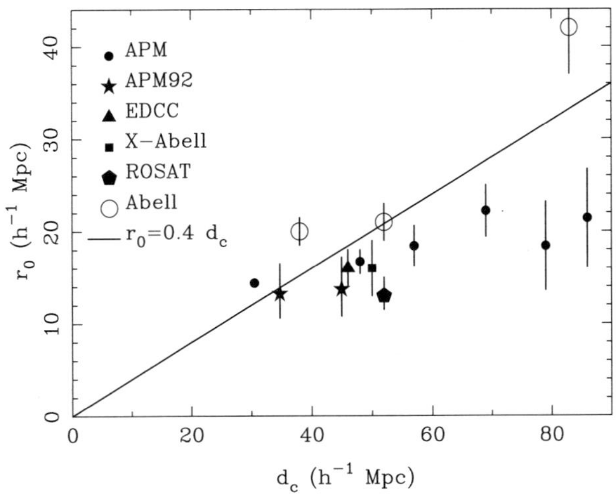

Figure 7.9 *The correlation radius — mean cluster distance dependence for cluster samples of different richness. The line shows the linear relation* $r_0 = 0.4d_c$ *of Bahcall and West (1992). The data do not confirm this trend. (Reproduced, with permission, from Croft et al. 1997, Mon. Not. R. Astr. Soc., 291, 305–313. Blackwell Science Ltd.)*

This test has an interesting history. In 1986, Bahcall and Burgett found on the basis of a few (four) samples a linear relation

$$r_0 \approx 0.5d_c \tag{7.57}$$

between the cluster correlation function and the mean distance between the clusters in a sample. Attempts to reproduce this relation on the basis of the peak theory failed. As shown by Mann, Heavens, and Peacock (1993) and by Croft and Efstathiou (1994), the peak theory gives substantially smaller amplification factors (smaller r_0 for a given d_c) than the relation quoted above. This can be easily seen — for a power-law galaxy correlation function $\xi(r) \approx (r/r_0)^{-2}$ the cluster correlation length will be amplified by ν/σ, but the mean cluster distance by $(n_p)^{-1/3} \approx \nu^{-1}\exp(\nu^2/6)$, so the d_c versus r_0 relation should level off with the growth of ν.

This contradiction was thought first to be a deficiency of the peak theory. However, attempts to explain the relation (7.57), using detailed numerical models of the formation of the clusters (Croft and Efstathiou 1994 used both methods), led, basically, to the same result — the r_0 versus d_c dependence leveled off for larger d_c. If the dependence (7.57) was right, the sites of formation of rich clusters should be more correlated than possible in a Gaussian field, and non-Gaussian distributions had to be invoked.

However, in recent years the number of independently selected cluster samples has grown, and observations now start to confirm the theory. We illustrate the situation in Fig. 7.9 from Croft et al. (1997). Here the data show the same trend as predicted by the Gaussian peaks theory. This tells us again that galaxy clusters could be born from high peaks of a Gaussian density field.

7.6.7 Peak–patch theory

The study of cluster mass functions and correlations has remained the only observational application of the statistics of Gaussian peaks. The reasons for that are several. First, it is not easy to identify peaks with observed objects. Due to the hierarchical nature of the structure, objects may include several peaks that merge together during their evolution. This restricts the application of the theory to clusters.

Another related problem is that the mass of a peak is not defined unambiguously. First, the method we used above for that is based on rather restrictive hypotheses of spherical collapse and universal power-law density profiles. Second, we saw above that one has to consider a set of smoothed fields with different radii to describe the observed mass range of clusters of galaxies. One peak may, however, be seen in different smoothed fields and thus is counted multiple times. This is called the "cloud-in-cloud" problem and it is also present in other statistical models for galaxy clusters.

Another factor has been the popularity of the Press–Schechter theory, which gives the needed predictions much more easily, although its theoretical foundations are less justified. This theory is presently one of the main tools for building statistical models; we describe it below.

The above difficulties can be resolved, using the peak theory for a detailed description of the evolution of structure. As the peak theory is based on geometrical properties of the peaks, it is well suited for that purpose. This approach is called the "peak–patch" method (Bond and Myers 1996a).

This approach was proposed to substitute numerical simulations with a semi-analytic treatment based on the peak theory. Given a realization of the initial density and velocity fields, the peak–patch method builds an object (cluster) catalog that models observations. There are several important steps in this procedure.

1. First, the peak list is constructed. This is done using the method proposed by Bond et al. (1991) — the density field is smoothed by a series of filters of diminishing radii and the peaks for any radius are found, excluding those that have already been listed (for larger radii). This is a step toward avoiding the "cloud-in-cloud" problem (we assign the largest possible volume to a peak).

   Peak patches are then defined around the peaks as regions that will collapse by the epoch of the simulated catalog (either using the mean overdensity for the spherical collapse picture or the mean strain tensor for ellipsoidal collapse).

Then the peak patches that are totally inside larger patches are eliminated, and the mass divided between partially overlapping peaks (the final solution of the "cloud-in-cloud" problem). The Lagrangian space is divided in this way into nonoverlapping patches of different mass.

2. The Lagrangian patches are moved into their final positions, using either the Zeldovich or higher-order approximations and the initial velocity fields at the patch, and patches are merged, if necessary.

3. The final density profiles for patches are calculated, based on the ellipsoidal collapse of a patch. The ellipsoidal collapse is solved here and above for self-consistent external tidal fields.

The procedure can be extended to include large-scale background power and to form objects at different redshifts, to simulate observations along a light cone. Although it is rather complex and not easy to program, it runs much faster than the usual N-body programs and can simulate larger volumes. The peak patches are groups of objects (galaxies). The comparison of the method with a N-body simulation, using identical initial data (Bond and Myers 1996b), shows that it reproduces the simulated groups very well. This could mean that in the future, when we return from the study of large volumes to that of individual objects, we have to use the language of peak patches; peak statistics is waiting for its time.

7.7 Press–Schechter method

The Press–Schechter method was developed to predict the mass distribution of hierarchically formed objects on the basis of initial power spectrum (Press and Schechter 1974). At first glance, this does not seem to be a problem of spatial statistics. However, this method is based on a specific picture of the spatial distribution of galaxies, has been used to construct statistical models for that distribution, and has been extended to predict correlations of objects of different mass. It is also one of the most used tools in cosmology.

The Press–Schechter approach starts searching for objects of mass M by smoothing the initial overdensity field by a filter of radius $R = R(M)$. This filter is usually taken as Gaussian, in which case, conventionally, $M(R) = \bar{\rho} V_G(R) = (2\pi)^{3/2} \bar{\rho} R^3$. Another assumption is that regions of overdensity $\delta \geq \delta_C$ have collapsed into observable objects. Such an overdensity limit follows from the spherical collapse solution described in Chapter 6. In order to describe mass distributions we have to consider fields with all smoothing radii at a given moment (at present) and assume that all objects have the overdensity δ_C. The overdensity field is usually normalized to its value at the present epoch (the value it would have if the evolution were linear). The typical numerical value of the collapse limit is then the well-known $\delta_C = 1.686$.

The usual assumption of Gaussianity of the initial overdensity field gives the probability that a given point belongs to a collapsed object (a massive halo)

$$P_G(\delta \geq \delta_C|R) = \frac{1}{2}\left[1 - \mathrm{erf}\left(\frac{\delta_C}{\sqrt{2}\sigma(R)}\right)\right], \tag{7.58}$$

where $\sigma^2(R)$ is the variance of the filtered field. A concern is that the maximum probability given by this formula is 1/2 and not unity (although it is not clear at all why all the mass in the universe should be concentrated in objects). Press and Schechter introduced a "fudge factor" and wrote for the fraction of mass in halos with mass greater than M (the integral mass distribution)

$$F(M) = 1 - \mathrm{erf}(\nu/\sqrt{2}),$$

where $\nu = \delta_C/\sigma(M)$. The mass function $n(M)$ is defined as a comoving number density of halos of mass M, so

$$Mn(M) = \bar{\rho}\left|\frac{dF}{dM}\right|,$$

where $\bar{\rho}$ is the mean comoving density. Using the expression for $F(M)$, we get

$$n(M) = \sqrt{\frac{2}{\pi}}\frac{\bar{\rho}}{M^2}\nu\left|\frac{d\ln\sigma(M)}{d\ln M}\right|e^{-\nu^2/2}. \tag{7.59}$$

This mass function can be immediately compared with observations. It depends on the cosmological model only through the mean density $\bar{\rho} = 3H_0^2\Omega_M/(8\pi G)$, and on the assumed power spectrum through the density variance

$$\sigma^2(M) = 4\pi\int_0^\infty |\widetilde{W}(k;M)|^2 P(k)\frac{k^2\,dk}{(2\pi)^3},$$

where $\widetilde{W}(k;M)$ is the Fourier transform of the real-space filter $W(x;M)$. If we are interested in the mass function at time t, we simply replace the collapse overdensity level by

$$\delta_C(t) = \frac{D_1(0)}{D_1(t)}\delta_C.$$

Early comparison with numerical experiments showed that the formula (7.59) describes simulations well, especially if we treated the critical overdensity level δ_C as a free parameter. The main problem then was the explanation of the "fudge factor." The problem was resolved after Peacock and Heavens (1990) and Bond et al. (1991) showed that this factor could be derived, if one considered a four-dimensional overdensity field $\delta(\mathbf{x}, R)$, where R is the filtering radius. For a set of widely spaced filtering radii the filtered fields could be considered essentially independent Gaussian random fields. For a fixed $\mathbf{x}$ the R-dependence of the amplitudes can be used to determine the mass associated with that point. The procedure is to start from the maximum R possible and to assign to each point $\mathbf{x}$ the filtering radius R when the smoothed overdensity at that point (the trajectory $\delta(R;\mathbf{x})$)

first upcrosses the collapse level δ_C. This filtering radius gives the largest mass around that point that has collapsed by the present time, hopefully also destroying any substructure during collapse. This eliminates the possibility of multiple counts of a mass point (the "cloud-in-cloud" problem).

The probability for a given point to belong to a halo of filtering radius R or larger can be written as (Peacock and Heavens 1990):

$$P(R) = P_G(\delta \geq \delta_C|R) = \int_{-1}^{\delta_C} \frac{dP_G}{d\delta} P_{up}(\delta, \delta_C)\, d\delta, \qquad (7.60)$$

where P_G is given by (7.58) and P_{up} is the probability that $\delta(R)$ could exceed the threshold δ_C at some filtering radius larger than R (the upcrossing probability). Thus, the statistics of the excursion sets of the trajectories of the smoothed overdensity $\delta(R; \mathbf{x})$ can be used to derive the mass functions.

The nature of these trajectories depends on the filters used. Peacock and Heavens (1990) derived approximate expressions for $P(R)$ for different filters. In all cases the result differs considerably from the simple Press–Schechter assumption, usually predicting more small-mass halos. Bond et al. (1991) demonstrated that the Press–Schechter formula, together with the "fudge factor" 2 can be obtained only in the case of a sharp k-space filter. In this case the trajectories execute Brownian walks — the increments of δ are formed by a sum of independent Fourier amplitudes and are independent. They recovered the Press–Schechter mass function with an ideal normalization (unity) by calculating the fraction of trajectories that have been absorbed by an overdensity barrier δ_C by the "time" $\sigma^2(R)$.

The independence of the overdensity increments in this picture shows that it is the mathematical expression of the peak–background split approximation. While for realistic filters the filtered density fields for different filter widths are correlated, the sharp k-space filter eliminates this correlation by definition. Thus, in addition to other assumptions, the Press–Schechter picture also includes the assumption of the exact peak–background split.

Bond et al. (1991) also derived the merger probability for masses, collapsed and virialized by an earlier moment t_1, to be included now in a larger mass. It is described by a similar Brownian walk with two absorbing barriers. The fraction of the mass of a large region of mass M_0 with variance σ_0^2 and overdensity limit δ_0 that has been brought by smaller halos of mass M_1 with the present linearly extrapolated overdensity limit δ_1 and variance σ_1, is

$$f(\sigma_1, \delta_1|\sigma_0, \delta_0)\frac{d\sigma_1^2}{dM_1}dM_1 = \frac{1}{\sqrt{2\pi}}\frac{\delta_1 - \delta_0}{(\sigma_1^2 - \sigma_0^2)^{3/2}} \exp\left[-\frac{(\delta_1 - \delta_0)^2}{2(\sigma_1^2 - \sigma_0^2)}\right]\frac{d\sigma_1^2}{dM_1}dM_1.$$

Mo and White (1996) used that relation to predict the halo bias and the enhancement of spatial correlations in the Press–Schechter picture. The average number of M_1 mass halos per a region of mass M_0 can be written as

$$n_h(1|0)dM_1 = \frac{M_0}{M_1} f(\sigma_1, \delta_1 | \sigma_0, \delta_0) \frac{d\sigma_1^2}{dM_1} dM_1.$$

This gives the overdensity of M_1 halos in M_0:

$$\delta_h^L(1|0) = \frac{n_h(1|0)}{n(M_1)V_0} - 1,$$

where $n(M)$ is the standard Press–Schechter mass function by (7.59) for the critical level $\delta_C(t_1)$ and V_0 is the volume occupied by the mass M_0 (a sphere of radius R_0). For a given power spectrum and filter the variance $\sigma^2(M)$ and the halo overdensity can be calculated exactly.

The asymptotic formula for $R_0 \gg R_1$ ($\sigma_0^2 \ll \sigma_1^2$) and $|\delta_0| \ll \delta_1$ is

$$\delta_h^L = b_h^L \delta_0 = \frac{\nu_1^2 - 1}{\delta_1} \delta_0,$$

where $\nu_1 = \delta_1/\sigma_1$ and b_h^L is the (Lagrangian) halo bias factor. The Eulerian bias factor can be derived in the same way as that for the peaks (7.50) and is

$$b_h^E = 1 + \frac{\nu_1^2 - 1}{\delta_1}. \tag{7.61}$$

We see that it does not depend on the characteristics of the larger region.

As these bias factors are mean values over regions, the Press–Schechter approach can predict only average correlations, defined as

$$\bar{\xi}(R_0) = \frac{3}{R_0^3} \int_0^{R_0} \xi(y) y^2 \, dy.$$

The average halo correlations are thus amplified as

$$\bar{\xi}_h(r) = \left(b_h^L(M_1, \delta_1)\right)^2 \bar{\xi}(r).$$

For scales sufficiently larger than R_1 the same relation applies to the correlation functions themselves.

In order to define the Press–Schechter biasing and anti-biasing regimes, we shall define a mass $M_\star$ by $\delta_1 = \sigma(M_\star)$. Using this approach, we vary masses of halos. For larger masses the variance $\sigma^2(M)$ is smaller, so by (7.61) more massive halos than $M_\star$ are (positively) biased and smaller halos are anti-biased.

Comparing the bias factors from the Press–Schechter approach with those obtained in the peak–background split approximation of the peak theory (7.49, 7.50) we see that they are very similar, the only difference being in the "constant" g that is unity for the Press–Schechter picture and can reach the value 3 in the high-peak approximation. Similar differences remain for higher-order correlations, derived by Mo, Ying, and White (1997). They also check both approximations by numerical simulations and observations and find that both work equally well — the status of observational checks does not yet allow us to choose between the two theories.

The Press–Schechter theory is meant to provide extensive distributions (different mass functions). It has also been used to describe merger processes in a hierarchical growth of the observed structure. It ignores many processes that are important for the growth of the structure (see Monaco 1999), but describes the mass functions rather well. The main difference with a better founded peak picture is that the Press–Schechter theory does not include the notion of a location of the halo. This has led to an implicit mixed theoretical paradigm of isolated virialized dark matter halos (peaks), which have a Press–Schechter type mass function, and are populated by galaxies.

In recent years, numerical simulations have shown that the standard Press–Schechter function over-estimates the abundance of high mass halos and under-estimates those of low mass (Sheth and Tormen 1996), as predicted by Peacock and Heavens (1990). We shall write the Sheth–Tormen mass function by noting first that a mass function can be written as

$$\nu f(\nu) = M^2 \frac{n(M)}{\bar{\rho}} \left| \frac{d\ln M}{d\ln \nu} \right|$$

for any variable $\nu = \nu(M)$. The original Press–Schechter theory predicts that the left-hand side of this relation depends only on ν and for $\nu = \delta_C/\sigma(M)$ it is:

$$\nu f(\nu) = \sqrt{\frac{2}{\pi}} \nu e^{-\nu^2/2}.$$

Later papers have started to use

$$\nu = \delta_C^2/\sigma^2(M), \tag{7.62}$$

and in this case

$$\nu f(\nu) = \frac{1}{\sqrt{2\pi}} \nu^{1/2} e^{-\nu/2}.$$

For the variable (7.62) Sheth and Tormen found from numerical simulations a fit:

$$\nu n(\nu) = A \left(1 + (\nu')^{-p}\right) (\nu')^{1/2} e^{-(\nu')/2},$$

where $\nu' = a\nu$. Their best fit has $a = \sqrt{2}$, $p = 0.3$, and A is defined by normalization, $\int n(\nu)\, d\nu = 1$. The standard PS function has the same form, if $a = 1$, $p = 0$, $A = 1/(2\sqrt{2\pi})$. A later work (Sheth, Mo, and Tormen 2001) showed that this mass function can be derived by the excursion set approach for elliptical collapse, where the barrier height is not constant, but depends on the ellipticity distribution.

The excursion set approach gives also bias functions, using the same methods as above. The (Eulerian) bias function derived by Sheth and Tormen (1996) is

$$b^E(\nu_1, \delta_1) = 1 + \frac{\nu_1' - 1}{\delta_1} + \frac{2p}{\delta_1(1 + (\nu_1')^p)}.$$

7.8 Halo model of galaxy clustering

The halo model of galaxy clustering has its roots in the Neyman and Scott model. It seeks the same goal that the Neyman–Scott model did, explanation of the spatial distribution of galaxies.

The motivation for that model stems from the failure of the dark-matter models to explain the observed power spectra and correlation functions of galaxies. The model spectra are too steep at small scales and have features at intermediate scales that conflict with the observed galaxy power spectrum that follows an almost perfect power law. This lead to postulating scale-dependent biasing of galaxies, but that biasing had to be explained.

The halo model was developed by Scherrer and Bertschinger (1991) and has become very popular again recently (see, e.g., Seljak 2000, Peacock and Smith 2000). The model consists of several assumptions that are rather well founded and can be checked independently. The components of the halo model are as follows:

1. The first component of the model is a population of massive virialized dark matter halos that form an inhomogeneous Poisson process, sampling the large-scale linear density field. This density field is described by its power spectrum $P_{lin}(k)$.
2. The halos are supposed to have a universal density profile. This assumption is supported by the latest N-body simulations, and the density profiles proposed are the NFW profile

$$\rho(r) = \frac{\rho_0}{y(1+y)^2}, \quad y \equiv r/r_c, \tag{7.63}$$

   and the M99 profile (see Peacock and Smith 2000 for more details):

$$\rho(r) = \frac{\rho_0}{y^{3/2}(1+y^{3/2})}, \quad y \equiv r/r_c, \tag{7.64}$$

   where r_c is the core radius. The mass of the halo is $M = (4\pi/3)\delta_v \bar{\rho} r_v^3$, where r_v is its outer radius and the virialization density contrast is taken as $\delta_v = 200$ (a popular convention instead of $\delta = 178$ given by spherical collapse, Chapter 6). Each profile has its advantages — the NFW profile has an analytical Fourier transform (White 2001) and the M99 profile has an exact mass-radius relation (Peacock and Smith 2000).
3. Although the next component is related to the halo profile, it is an independent and essential assumption. If we define the concentration parameter of a profile as $c = r_v/r_c$, this assumption says that c is a function of halo mass,

$$c(M) = C\left(\frac{M}{M_\star}\right)^{-\alpha}. \tag{7.65}$$

   Here $M_\star$ is the nonbiased halo mass, familiar from the Press–Schechter theory ($\sigma(M_\star) = \delta_C$, see Section 7.7). The parameters of the relation (7.65) are suggested by numerical simulations; a typical choice is $C \approx 10$, $\alpha \approx 0.2$ (more massive halos are less concentrated).

4. The fourth component is the halo mass function. The usual choice here is the recent modification of the standard PS function by Sheth and Tormen (1996). We shall follow their convention and redefine the density contrast parameter as $\nu = \delta_C^2/\sigma^2(M)$, where $\delta_C = 1.69$ and $\sigma(M)$ are defined as in the standard Press–Schechter theory. This mass function $n(\nu)$ is written as

$$\nu n(\nu) = A\big(1 + (a\nu)^{-p}\big)\sqrt{a\nu}e^{-a\nu/2},$$

with the standard choice $a = \sqrt{2}, p = 0.3$ and A coming from the normalization condition $\int \nu \, d\nu = 1$.

5. We know that the spatial distribution of massive halos is biased, so one of the ingredients of the model has to be the bias function. This function is derived on the basis of the mass function above (Sheth and Tormen 1996) and reads

$$b(\nu) = 1 + \frac{a\nu - 1}{\delta_c} + \frac{2p}{\delta_c\big(1 + (a\nu)^p\big)}.$$

These are all the assumptions needed to describe the clustering of dark matter. The power spectrum of dark matter $P_{dm}(k)$ is represented as a sum of two parts, one describing the halo–halo correlations and the other the short-range correlations within halos. To get the first part we have to take into account that halos are extended, so we have to convolve the initial density with halo profiles. In Fourier space this reduces to multiplication with an averaged filter:

$$P_{dm}^{hh}(k) = P_{lin}(k)\left[\int n(\nu)b(\nu)y(k,\nu)\, d\nu\right]^2. \tag{7.66}$$

The only new function in the above formula is the normalized Fourier transform of the halo density profile

$$y(k,\nu) = \frac{\widetilde{\rho}(k)}{M}, \qquad y(0) = 1$$

(the function $y(k,\nu)$ depends on ν via the halo mass M). This formula also introduces a constraint on the bias function (Seljak 2000). Because for large scales ($k \to 0$) the halo sizes do not matter, the power spectrum has to coincide there with the linear P_{lin}. This gives

$$\int n(\nu)b(\nu)\, d\nu = 1,$$

which in combination with the normalization condition for $n(\nu)$ says that when halos are biased ($b > 1$) in some mass range, there must be also a mass range where they are anti-biased ($b < 1$).

The second term of the power spectrum comes from correlation of mass within halos:

$$P^h_{dm} = \frac{1}{(2\pi)^3} \int n(\nu) \frac{M(\nu)}{\bar{\rho}} |y(k,\nu)|^2 \, d\nu \tag{7.67}$$

and the total power spectrum for dark matter is given by

$$P_{dm}(k) = P^{hh}_{dm}(k) + P^h_{dm}(k).$$

White (2001) extended this model to include redshift space distortions, describing these by the direction-averaged distortion factors. The amplification of the linear spectrum in the plane-parallel limit is

$$A_1(\mu) = (1 + f\mu^2)^2,$$

where μ is the cosine of the angle between the line-of-sight and the wave direction and $f \approx \Omega^{0.6}$ is the dimensionless growth factor; these distortions are explained in more detail in Chapter 8. The velocity distortions caused by virialized motions inside a halo suppress the spectrum by

$$A_2(k, \mu) = e^{-(k\sigma\mu)^2},$$

where σ is the one-dimensional variance of dynamical velocities in a halo, given by the virial theorem:

$$\sigma^2 = \frac{GM}{2r_v}.$$

The total effect of the redshift distortions can be written as the product of the two factors, and the total average redshift distortion is given by an integral

$$\mathcal{R}(k, \sigma) = \frac{1}{2} \int_{-1}^{1} A_1(\mu) A_2(k, \mu) \, d\mu.$$

Denoting $k\sigma = x$, the integral is (White 2001)

$$\mathcal{R}(x) = \frac{\sqrt{\pi}}{8} \frac{\text{erf}(x)}{x^5} [3f^2 + 4fx^2 + 4x^4] - \frac{e^{-x^2}}{4x^4} [3f^2 + 2f^2y^2 + 4fy^2].$$

Using these distortions, the formulae for the redshift space spectrum can be obtained, including the function $R(k\sigma)$ in the integrands in (7.66) and (7.67). This analytical model is compared with results of numerical simulation in Fig. 7.10. The power spectra in the figure are given in the power-per-log-interval presentation ($\Delta^2(k) = k^3 P(k)/(2\pi^2)$). Both the real space power spectrum and the velocity space power spectrum, predicted by the model, describe the simulation rather well. As the latter is the result of complex nonlinear dynamical evolution, it suggests that the power spectrum is insensitive to the details of this evolution and reflects only its initial and final stages.

The halo model would not be very useful, if it could predict only dark matter distributions. These are certainly extremely important, but in order to check theory

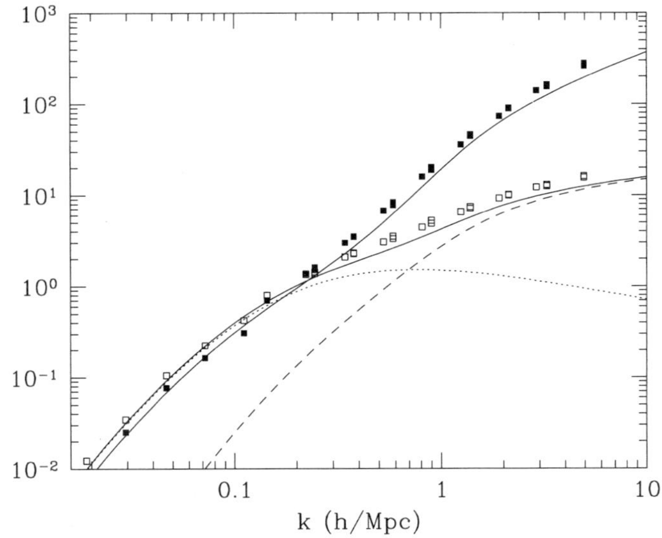

Figure 7.10 *Halo model predictions for real and redshift space power spectra (solid lines, the line close to the solid squares gives the real space power spectrum). The results of a N-body simulation with the same initial spectrum are shown by squares. Solid squares indicate real space power and open squares indicate redshift space power. The dashed and dotted lines show the halo–halo and in-halo contributions to the redshift space power spectrum. (Reproduced, with permission, from White 2001, Mon. Not. R. Astr. Soc. 321, 1–3. Blackwell Science Ltd.)*

with observations a model must describe galaxy distributions. The halo model can do that, if we add three more components to the model.

6. To get from halos to galaxies, an essential quantity is the average number of galaxies per halo, $N(M)$. This number should be a function of halo mass; a power-law slope $N(M)/M \sim M^{\psi}$ with $\psi \approx -0.2$ is suggested by N-body simulations.

7. Assuming that galaxies follow mass inside halos, the simplest function to describe the difference between the dark matter and galaxy correlations is the mean number of galaxy pairs per halo, $N_P(M) = \langle N(N-1) \rangle$. This also can be approximated by a power law; simulations suggest $\sqrt{N_P(M)}/M \sim M^{\psi}$ with the same value of $\psi \approx -0.2$ as above.

8. We assume that galaxies follow mass in halos. An important (but natural) assumption of the halo theory is that there is always the central galaxy. If the number of galaxies in a halo is large, this does not matter; the galaxy and dark matter correlations are similar, given by the convolution of the density profile.

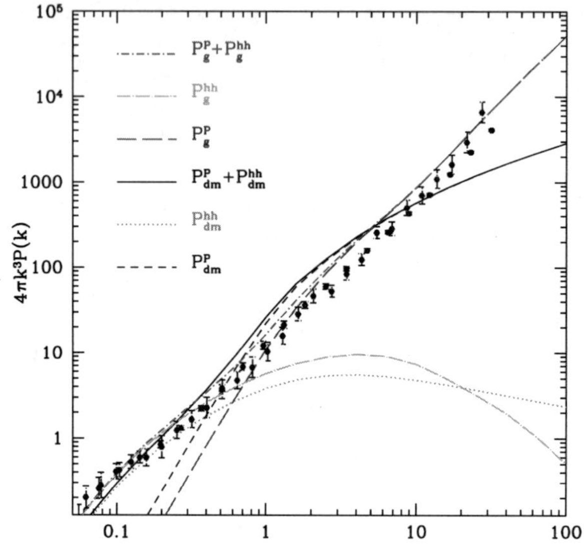

Figure 7.11 *Galaxy power spectrum predicted by a halo model (the line marked $P_g^P + P_g^{hh}$) compared to the APM galaxy power spectrum (dots with error bars). Different contributions to the galaxy and dark matter power spectra are shown; the notations in the legends differ only by using the superscript P for the single halo part of the spectra. (Reproduced, with permission, from Seljak 2000, Mon. Not. R. Astr. Soc., 318, 203–213. Blackwell Science Ltd.)*

If there are only two galaxies in a halo and one of them is in the center, the correlation function coincides with the density profile itself.

These assumptions modify the expressions for the power spectrum. The halo–halo part is now

$$P_g^{hh}(k) = P_{lin}(k)\left[\frac{\bar{\rho}}{\bar{n}}\int n(\nu)\frac{N(M)}{M}b(\nu)y(k,\nu)\,d\nu\right]^2, \tag{7.68}$$

where $\bar{n}$ is the mean number density of galaxies (in a sample). The factor $\bar{\rho}/\bar{n}$ comes from the normalization condition

$$\int \frac{N(M)}{M}n(\nu)\,d\nu = \frac{\bar{n}}{\bar{\rho}}.$$

The small-scale power from inside halos is

$$P_g^h = \frac{1}{(2\pi)^3}\left(\frac{\bar{\rho}}{\bar{n}}\right)^2\int n(\nu)\frac{M(\nu)}{\bar{\rho}}\frac{N_P(M)}{M^2}|y(k,\nu)|^p\,d\nu, \tag{7.69}$$

where the term $|y(k,\nu)|^p$ accounts for the different correlation functions for sparsely populated halos. If $N_P(M) > 1$, $p = 2$, if $N_P(M) \leq 1$, the correlation function is given by the halo profile and $p = 1$. The total power spectrum is

$$P_g(k) = P_g^{hh}(k) + P_g^{h}(k). \tag{7.70}$$

Fig. 7.11 shows an example of the galaxy power spectrum calculated by formulae (7.68–7.70). It shows both the dark matter and galaxy power spectrum term by term and compares it with the observationally determined (APM) power spectrum. The convention for the power spectrum is proportional to the Δ^2 measure used in the previous figure. We see that the halo model reproduces well the positive bias for smaller scales, due to the switch-over in the form of the correlation function, and is close to the observations for a wide wavenumber interval.

The halo model consists of several components that can be studied and improved separately. For example, Peacock and Smith (2000) calibrated the $N(M)$ dependence by using data on the luminosity function of galaxy groups and applied the halo model to halos found by numerical simulation.

In addition to the description above, they stress that the halo model also describes isolated galaxies with $N_P = 0$ in small halos. If we take into account that galaxies can orbit inside their parent halos, the small-scale biasing has to be nonlocal. They also show that the halo model can explain the low amplitude of the pairwise velocity variance of galaxies.

Predictions for galaxy power spectra can also include velocity distortions, as done above for the dark matter. Seljak (2001) has predicted the redshift space bias for several types of galaxies, showing that it is also nonmonotonic, as is the real space bias.

The halo model as delineated above predicts the power spectrum of the galaxy distribution. Scoccimarro and Sheth (2001) used the halo approach to generate model galaxy distributions. They replaced the time-consuming N-body modeling by approximate second-order Lagrangian perturbation theory predictions, used the halo merger tree algorithm of Sheth and Lemson (1999) to select virialized dark matter halos, and generated galaxies inside these halos according to the halo model rules. Their method is similar to the peak–patch method, but faster, and reproduces well the features of N-body halos. This method could be especially useful for maximum-likelihood calculations, when models have to be generated for a large parameter space.

7.9 Stochastic and nonlinear biasing

We have seen in previous sections that peak and halo fields are biased with respect to the matter density field. This is true for the spatial distribution of most objects, including galaxies. A simple proof for that is the observation that galaxies of different morphology are clustered differently, so they all cannot faithfully represent the matter density distribution. The usual assumption connects the galaxy density field $g(\mathbf{x})$ with that of matter overdensity $\delta(\mathbf{x})$ by

$$g(\mathbf{x}) = b_1 \delta(\mathbf{x}), \tag{7.71}$$

where b_1 is a constant bias factor. Dekel and Lahav (1999) have generalized this simple model. They note first that the common result $b > 1$ together with (7.71) cannot be applied for voids where it could predict negative galaxy densities. Second, they stress that there are certainly many other variables influencing galaxy formation other than the local overdensity alone. This leads to a stochastic biasing model where both δ and g are random fields and the biasing relation is determined by the biasing conditional distribution $P(g|\delta)$ for a given δ. The biasing relation (7.71) will be replaced then by the conditional mean

$$b(\delta)\delta = \langle g|\delta\rangle = \int P(g|\delta) g \, dg. \tag{7.72}$$

Dekel and Lahav define the weighted moments of $b(\delta)$ as

$$\bar{b} = \langle b(\delta)\delta^2\rangle / \sigma^2,$$

where σ^2 is the overdensity variance, and

$$\widetilde{b}^2 = \langle b^2(\delta)\delta^2\rangle / \sigma^2.$$

In the case of linear biasing they coincide with b_1.

As the galaxy density g is a random field, it is useful to define a zero-mean random biasing field ϵ:

$$\epsilon = g - \langle g|\delta\rangle.$$

The local variance of ϵ defines the biasing scatter

$$\sigma_b^2(\delta) = \langle \epsilon^2|\delta\rangle / \sigma^2.$$

Averaging this over δ and using the relation

$$\langle p(g)q(\delta)\rangle = \langle \langle p(g)|\delta\rangle_{g|\delta} q(\delta)\rangle_\delta, \tag{7.73}$$

where $p(g)$ and $q(\delta)$ are any functions of g and δ, we get the average biasing scatter

$$\sigma_b^2 = \langle \epsilon^2\rangle / \sigma^2.$$

Thus we have three natural biasing parameters instead of one — the mean biasing $\bar{b}$, the (second order) nonlinearity $\widetilde{b}/\bar{b}$, and the scatter $\sigma_b/\bar{b}$. These parameters are illustrated in Fig. 7.12, which is the result of a semi-analytic model of galaxy formation. In this model both the matter density and galaxy density fields are generated, the latter based on rather complicated empirical prescriptions for galaxy formation. We see that the model biasing relation is clearly nonlinear, has a pretty good scatter, and is different for different epochs.

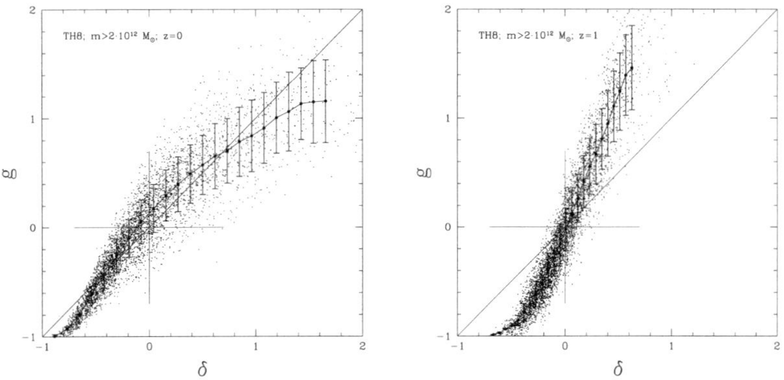

Figure 7.12 *Biasing of the galactic halo field versus the overdensity field in a N-body simulation. The left panel corresponds to the present moment ($z = 0$), the right panel to an earlier epoch ($z = 1$). The mean biasing relation is given by a solid curve and its scatter by error bars. (Reproduced, with permission, from Dekel and Lahav 1999, Astrophys. J., 520, 24–34. AAS.)*

Because different methods for estimating bias use different regressions, they measure different combinations of bias parameters. The most often used bias value is the ratio of variances

$$b_{var}^2 = \frac{\sigma_g^2}{\sigma^2} = \widetilde{b}^2 + \sigma_b^2,$$

which can be obtained by using the averaging rule (7.73). We also can write that relation as

$$b_{var} = \bar{b}\left(\frac{\widetilde{b}^2}{\bar{b}^2} + \frac{\sigma_b^2}{\bar{b}^2}\right).$$

This shows that b_{var} is biased compared to the mean bias $\bar{b}$. The galaxy–density covariance can be calculated to be

$$\langle g\delta\rangle = \bar{b}\sigma^2$$

and it does not depend on bias nonlinearity and scatter.

In the limiting case of linear and stochastic biasing

$$\bar{b} = \widetilde{b} = b_1, \qquad b_{var} = b_1\left(1 + \frac{\sigma_b^2}{b_1^2}\right),$$

and for nonlinear and deterministic biasing (as that derived for massive halos)

$$\widetilde{b} \neq \bar{b}, \qquad \sigma_b = 0, \qquad b_{var} = \widetilde{b}.$$

Dekel and Lahav (1999) analyze different bias estimates and show that they determine different combinations of bias parameters. They also extend their formalism to correlation functions and power spectra and to redshift distortions. The natural assumptions of nonlinearity and stochasticity of bias lead to more complicated relations between the descriptors of the galaxy distributions and descriptors of the total matter.

CHAPTER 8

Fourier analysis of clustering

8.1 Introduction

The current statistical model for the main cosmological fields (density, velocity, gravitational potential) is the Gaussian random field. This field is determined either by its correlation function or by its spectral density, and one of the main goals of spatial statistics in cosmology is to estimate these two functions.

While the correlation functions were the first to be studied, in recent years the power spectrum has become the main subject of interest. There are at least two reasons for that. The power spectrum is more intuitive physically, separating processes on different scales, and the Fourier amplitudes for different wavenumbers are statistically orthogonal,

$$E\left\{\widetilde{\delta}(\mathbf{k})\widetilde{\delta}^{\star}(\mathbf{k}')\right\} = (2\pi)^3\delta_D(\mathbf{k}-\mathbf{k}')P(\mathbf{k}).$$

Here $\widetilde{\delta}(\mathbf{k})$ is the Fourier amplitude of the density contrast field at a wavenumber $\mathbf{k}$, a star denotes complex conjugation, and $E\{\}$ denotes expectation values over realizations of the random field. The power spectrum $P(\mathbf{k})$ is the Fourier transform of the correlation function $\xi(\mathbf{r})$ of the field, $\delta_D(\mathbf{x})$ is the three-dimensional Dirac delta function, and the factor $(2\pi)^3$ comes from our Fourier transform convention:

$$\widetilde{f}(\mathbf{k}) = \int f(\mathbf{x})e^{i\mathbf{k}\cdot\mathbf{x}}\,d^3x,$$

$$f(\mathbf{x}) = \frac{1}{(2\pi)^3}\int \widetilde{f}(\mathbf{k})e^{-i\mathbf{k}\cdot\mathbf{x}}\,d^3k$$

(we can consider the volume element in the Fourier space $d^3k/(2\pi)^3$). Although we can see different Fourier transform conventions in cosmological literature, this is the most common one.

Estimation of power spectra from observations is a rather difficult task. Up to now the problem has been the scarcity of data; in the near future we will have the opposite problem of managing huge data sets. The development of statistical techniques here has been motivated largely by the analysis of CMB power spectra, where better data were obtained first, and has been parallel to that recently. We shall see that one of the possibilities to compress the information for large data sets is to calculate the correlation function as an intermediate step on the way to the power spectrum.

8.2 Estimation of power spectra

The first methods developed to estimate the power spectra were direct methods — a suitable statistic was chosen and determined from observations. We shall study these methods first; the more recent Bayesian approach will be described later.

8.2.1 Direct methods

Several approaches have been proposed to estimate the power spectrum for galaxy catalogs. Although they seem different at first glance, the real differences are slight, limited mainly to the choice of the weighting functions. The most detailed early presentations are given by Peebles in a series of papers beginning with Peebles (1973) and his book (1980). We shall follow here a more recent derivation by Tegmark (1995) and Tegmark et al. (1998).

Although our main model is the Gaussian density field, our observed samples consist of discrete objects. They can be modeled by an inhomogeneous point process (a Gaussian Cox process) of number density $n(\mathbf{x})$:

$$n(\mathbf{x}) = \sum_i \delta_D(\mathbf{x} - \mathbf{x}_i), \tag{8.1}$$

where $\delta_D(\mathbf{x})$ is the Dirac delta-function. The intensity of that process $\lambda(\mathbf{x}) = \langle n(\mathbf{x}) \rangle$ (angle brackets denote expectations over the point process) itself is a realization of a Gaussian density field. In cosmology we use instead of a density field a normalized zero-mean field of density contrast $\delta(\mathbf{x})$. Usually the density contrast is defined, comparing the density at $\mathbf{x}$ with the (constant) mean density of the universe (the density field, and, therefore, also the contrast field are supposed to be homogeneous and isotropic Gaussian random fields).

The first complication is that observed samples frequently have systematic density trends caused by selection effects, so the density contrast has to be defined here as

$$\delta(\mathbf{x}) = \frac{\rho(\mathbf{x}) - \bar{\rho}(\mathbf{x})}{\bar{\rho}(\mathbf{x})}, \tag{8.2}$$

where $\bar{\rho}(\mathbf{x})$ is the position-dependent mean density (selection function). Using the expression for the density (8.1), we can write the estimator of the density contrast in our sample as

$$D(\mathbf{x}) = \sum_i \frac{\delta_D(\mathbf{x} - \mathbf{x}_i)}{\bar{n}(\mathbf{x}_i)} - 1, \tag{8.3}$$

where $\bar{n}(\mathbf{x}) \sim \bar{\rho}(\mathbf{x})$ is the selection function expressed in the number density of objects.

The Fourier amplitudes of the field (8.2) are

$$\widetilde{\delta}(\mathbf{k}) = \int_V \psi(\mathbf{x})\delta(\mathbf{x})e^{i\mathbf{k}\cdot\mathbf{x}}\, d^3x - \widetilde{\psi}(\mathbf{k}). \tag{8.4}$$

Here V denotes the sample volume and $\psi(\mathbf{x})$ is a weight function that can be chosen to optimize the result. The term

$$\widetilde{\psi}(\mathbf{k}) = \int_V \psi(\mathbf{x}) e^{i\mathbf{k}\cdot\mathbf{x}}\, d^3x$$

depends only on the sample geometry and on the weight function.

Using (8.4, 8.3) we get the estimator for a Fourier amplitude

$$F(\mathbf{k}_i) = \sum_j \frac{\psi(\mathbf{x}_j)}{\bar{n}(\mathbf{x}_j)} e^{i\mathbf{k}_i\cdot\mathbf{x}} - \widetilde{\psi}(\mathbf{k}_i) \tag{8.5}$$

(for a finite set of frequencies $\mathbf{k}_i$).

The raw estimator for the spectrum is

$$P_R(\mathbf{k}_i) = F(\mathbf{k}_i) F^\star(\mathbf{k}_i). \tag{8.6}$$

The expected value for that is

$$\begin{aligned} E\left\{\langle F(\mathbf{k}_i) F^\star(\mathbf{k}_i)\rangle\right\} = & \int_V \int_V \frac{\psi(\mathbf{x})\psi(\mathbf{x}')}{\bar{n}(\mathbf{x})\bar{n}(\mathbf{x}')} E\left\{\langle n(\mathbf{x}) n(\mathbf{x}')\rangle\right\} e^{i\mathbf{k}_i(\mathbf{x}-\mathbf{x}')}\, d^3x\, d^3x' \\ & -|\widetilde{\psi}(\mathbf{k}_i)|^2. \end{aligned} \tag{8.7}$$

Here $E\{\}$ denotes expectations over realizations of the density contrast field, and angle brackets denote expectations over the point process.

The last term in (8.7) is obtained by substituting the zero-mean condition for the density contrast

$$E\left\{\widetilde{\delta}(\mathbf{k}_i)\right\} = 0,$$

into (8.4). This leads to

$$E\left\{\int_V \psi(\mathbf{x}) n(\mathbf{x}) e^{i\mathbf{k}_i\cdot\mathbf{x}}\, d^3x\right\} = \widetilde{\psi}(\mathbf{k}_i).$$

We have shown in (3.9) that the second-order intensity of an inhomogeneous Poisson process is

$$\langle n(\mathbf{x}) n(\mathbf{x}')\rangle = \lambda(\mathbf{x})\lambda(\mathbf{x}') + \delta_D(\mathbf{x}-\mathbf{x}')\lambda(\mathbf{x}). \tag{8.8}$$

The first term comes from the independence of the process in separate points, and the second term accounts for self-pairs — there is always a point of the process at zero distance. When calculating correlation functions, we discard self-pairs. Here they are usually included, as we sum over all points first to get the estimators of Fourier amplitudes and take a square later.

The definition of the density contrast (8.2) allows us to write

$$\lambda(\mathbf{x}) = \bar{n}(\mathbf{x})\left[1 + \delta(\mathbf{x})\right],$$

where $E\{\lambda(\mathbf{x})\} = \bar{n}(\mathbf{x})$. Let us expand now the expectation value in the integral (8.7):

$$E\{\langle n(\mathbf{x})n(\mathbf{x}')\rangle\} = \bar{n}(\mathbf{x})\bar{n}(\mathbf{x}')\left[1 + E\{\delta(\mathbf{x})\delta(\mathbf{x}')\}\right] + \delta_D(\mathbf{x}-\mathbf{x}')\bar{n}(\mathbf{x}). \quad (8.9)$$

Integrating the first term in the square brackets, we get a product of two integrals that gives $|\widetilde{\psi}(\mathbf{k}_i)|^2$, canceling the last term in (8.7). Integrating the third term in (8.9), we get the Poisson (shot) noise term

$$S = \int_V \frac{\psi^2(\mathbf{x})}{\bar{n}(\mathbf{x})}\, d^3x. \quad (8.10)$$

The second term in the square brackets in (8.9) gives

$$\int_V \int_V \psi(\mathbf{x})\psi(\mathbf{x}')E\{\delta(\mathbf{x})\delta(\mathbf{x}')\}\, e^{i\mathbf{k}_i(\mathbf{x}-\mathbf{x}')}\, d^3x\, d^3x'.$$

Using the convolution theorem, we get

$$\int \psi(\mathbf{x})\delta(\mathbf{x})e^{i\mathbf{k}_i\cdot\mathbf{x}}\, d^3x = \int \widetilde{\psi}^\star(\mathbf{k}_i - \mathbf{k}')\widetilde{\delta}(\mathbf{k}')\, \frac{d^3k'}{(2\pi)^3}.$$

Substituting this, we get

$$\int\int \widetilde{\psi}(\mathbf{k}_i - \mathbf{k}'')\widetilde{\psi}^\star(\mathbf{k}_i - \mathbf{k}')E\left\{\widetilde{\delta}(\mathbf{k}')\widetilde{\delta}^\star(\mathbf{k}'')\right\}\frac{d^3k'\, d^3k''}{(2\pi)^6}$$
$$= \int G(\mathbf{k}_i - \mathbf{k}')P(\mathbf{k}')\, \frac{d^3k'}{(2\pi)^3},$$

where the window function $G(\mathbf{k})$ is the square of the Fourier transform of the weight function:

$$G(\mathbf{k}) = |\widetilde{\psi}(\mathbf{k})|^2. \quad (8.11)$$

It is easy to see that this function is different from zero roughly in the region $k^\alpha < 1/L^\alpha$, where the index α denotes spatial components of a vector. For larger $\mathbf{k}$ the oscillating factor in the Fourier transform will tend to zero.

Thus, finally,

$$E\{\langle|F(\mathbf{k}_i)|^2\rangle\} = \int G(\mathbf{k}_i - \mathbf{k}')P(\mathbf{k}')\, \frac{d^3k'}{(2\pi)^3} + \int_V \frac{\psi^2(\mathbf{x})}{\bar{n}(\mathbf{x})}\, d^3x, \quad (8.12)$$

or, symbolically, we can obtain the estimate of the power spectra $\widehat{P}$ by inverting the integral equation

$$G \otimes \widehat{P} = P_R - S, \quad (8.13)$$

where $\otimes$ denotes convolution, P_R is the raw estimate of power (8.6), and S is the (constant) shot noise term. In order to get an unbiased estimate, the window function $G(\mathbf{k})$ has to satisfy the condition

$$\int G(\mathbf{k})\, \frac{d^3k}{(2\pi)^3} = 1,$$

that transforms to the normalization condition for the weight function

$$\int_V \psi^2(\mathbf{x})\, d^3x = 1 \tag{8.14}$$

by Parseval's theorem.

For small scales ($k \gg 1/L$) and in the case where the power spectrum can be expected to be fairly smooth, we can consider $P(\mathbf{k})$ constant throughout the width of the window $G(\mathbf{k})$. As the window is normalized, the estimate of the spectrum in that case is

$$\widehat{P} = P_R - S.$$

In general, we have to deconvolve the noise-corrected raw power to get the estimate of the power spectrum. This introduces correlations in the estimated amplitudes, so they are not statistically orthogonal any more. A sample of a characteristic spatial size L creates a window function of width of $\Delta k \approx 1/L$, correlating estimates of spectra at that wavenumber interval.

Fig. 8.1 shows an example of two window functions and a typical density power spectrum. Both window functions are rather sharp and describe deep samples. The window on the left is a typical window for a volume-limited sample. Such samples have sharp boundaries in real space, which cause wide sidelobes in Fourier space. The window on the right is typical for a pencil-beam survey. A pencil-beam survey is a deep survey in a very small region of the sky, and it cuts out a narrow beam in space. Such surveys have an anisotropic geometry that gives anisotropic window functions in Fourier space, where the largest extent of the window is determined by the smallest linear size of the sample. A deep survey in a thin slice of the sky has a similar Fourier space window function.

As we see, a wide window function will mix up the signals from different scales, severely limiting the resolution of the spectra we obtain. This problem is especially serious for small wavenumbers that are comparable with the width of the window, and also for regions where the amplitude of the spectrum is small. There the leakage of high-amplitude power can seriously distort the estimated signal.

8.2.2 Selection of weights

Because the raw estimates of the power spectra are weighted means of the true spectrum over the width of the window function, the deconvolved spectra has a limited resolution. The choice of the weight function affects both the resolution and the variance of the estimates of the spectrum.

The sample covariance of the power spectrum estimate is

$$E\{\Delta\widehat{P}(\mathbf{k})\Delta\widehat{P}(\mathbf{k}')\} = E\{\widehat{P}(\mathbf{k})\widehat{P}(\mathbf{k}')\} - P(\mathbf{k})P(\mathbf{k}'),$$

where $\Delta\widehat{P}(\mathbf{k}) = \widehat{P}(\mathbf{k}) - P(\mathbf{k})$ is the difference between the estimate $\widehat{P}(\mathbf{k})$ and the (unknown) true power spectrum $P(\mathbf{k})$. Assuming that the Fourier amplitudes

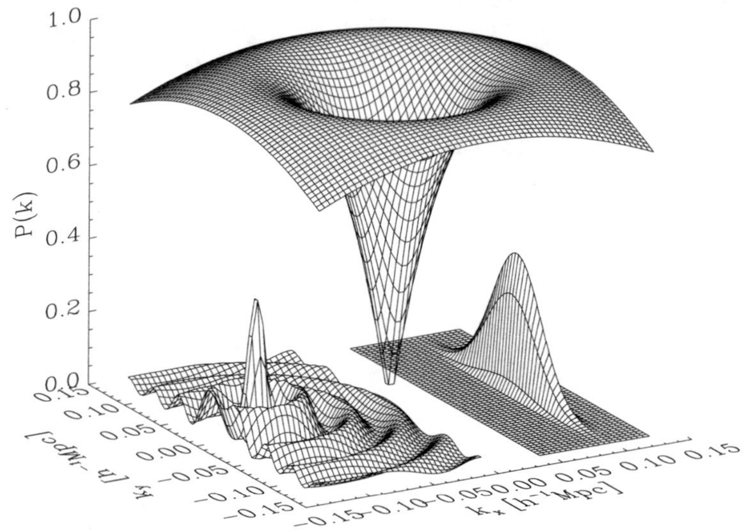

Figure 8.1 *A standard (CDM) density power spectrum (center) and two window functions. The window function on the left is typical for volume-limited samples, and that on the right is typical for low-dimensional surveys (e.g., for a pencil-beam survey). Both the power spectrum and the window functions are three dimensional; one dimension is suppressed in the figure. (Reproduced, with permission, from Tegmark 1995, Astrophys. J., 455, 429–438. AAS.)*

$F(\mathbf{k})$ are Gaussian variables, Feldman, Kaiser, and Peacock (1994) found

$$E\{\Delta\widehat{P}(\mathbf{k})\Delta\widehat{P}(\mathbf{k}')\} = |E\{F(\mathbf{k})F^\star(\mathbf{k}')\}|^2. \tag{8.15}$$

The assumption of Gaussianity is certainly justified for scales that follow linear dynamics.

The covariance of the raw estimates in (8.15) can be found in the same way that brought us to their variance in (8.12). This gives

$$\begin{aligned} E\{\langle F(\mathbf{k}_i)F^\star(\mathbf{k}_j)\rangle\} &= \int G(\mathbf{k}_i-\mathbf{k})G^\star(\mathbf{k}_j-\mathbf{k})P(\mathbf{k})\frac{d^3k}{(2\pi)^3} \\ &\quad + \int_V \frac{\psi^2(\mathbf{x})}{\bar{n}(\mathbf{x})}e^{i(\mathbf{k}_j-\mathbf{k}_i)\cdot\mathbf{x}}\,d^3x. \end{aligned} \tag{8.16}$$

For large $k \gg 1/L$ and a smooth power spectrum we can consider $P(\mathbf{k})$ approximately constant within the width of the window function. Using the convolution theorem, we can write

$$E\{\langle F(\mathbf{k}_i)F^\star(\mathbf{k}_j)\rangle\} \approx \int \psi^2(\mathbf{x})e^{i(\mathbf{k}_j-\mathbf{k}_i)\cdot\mathbf{x}}\left[P+\frac{1}{\bar{n}(\mathbf{x})}\right]d^3x, \tag{8.17}$$

where $P = P\big((\mathbf{k}_i+\mathbf{k}_j)/2\big)$. Because of the oscillating factor in this integral, raw estimates of spectrum are uncorrelated for $|\mathbf{k}_i-\mathbf{k}_j| \gg 1/L$ and well correlated for $|\mathbf{k}_i-\mathbf{k}_j| \ll 1/L$. This leads to the notion of coherence cell in the wavenumber space; estimates of raw power spectra are independent only for different coherence cells.

The formula for the covariance of the power spectrum (8.15) gives for the variance

$$\sigma_P^2(\mathbf{k}_i) = |P_R(\mathbf{k}_i)|^2.$$

Thus the rms error of a single power spectrum estimate (or the estimate for a single coherence cell) is equal to the raw spectrum; this estimate is extremely noisy. The only way to reduce the noise is to average over several coherence cells. As the cosmological spectra are usually assumed to be isotropic, the standard method to estimate the spectrum involves an additional step of averaging the estimates $\widehat{P}(\mathbf{k})$ over a spherical shell $k \in [k_i, k_{i+1}]$ of thickness $k_{i+1}-k_i > \Delta k = 1/L$ in the wavenumber space. The variance of this estimate is

$$\sigma_P^2(k) = \frac{2}{V_S^2}\int_{V_S}\int_{V_S}\left|E\{\langle F(\mathbf{k}_i)F^\star(\mathbf{k}_j)\rangle\}\right|^2 \frac{d^3k_i}{(2\pi)^3}\frac{d^3k_j}{(2\pi)^3}$$

(the factor 2 is due to the fact that the density field is real and only half of the coherence cells are independent). If the width of the shell is large compared to the coherence length, one of the integrals gives the volume of the shell. Using the result (8.17) and Parseval's theorem, we get

$$\sigma_P^2(k) = \frac{2P^2(k)}{V_S}\int_{V_S}\psi^2(\mathbf{x})\left[1+\frac{1}{\bar{n}(\mathbf{x})P(k)}\right]^2 d^3r. \tag{8.18}$$

The optimal weight function will give the minimum variance for the power spectrum estimates. Minimizing (8.18) together with the normalization condition (8.14) gives

$$\psi(\mathbf{x}) \sim \frac{\bar{n}(\mathbf{x})}{1+\bar{n}(\mathbf{x})P(k)}. \tag{8.19}$$

This weight function was derived first by Feldman, Kaiser, and Peacock (1994). Another derivation of this result is given by Tegmark et al. (1998); they also show that this weight function is optimal only for $k \gg 1/L$.

This weight function is, in principle, different for every wavenumber and demands a priori knowledge of the amplitudes of the power spectrum. However, if the weight function is normalized according to (8.14), the estimates will remain unbiased. An error in the value of P will only increase the variance of the estimates. Hence in applications a few constant values of P are usually used.

The weight function (8.19) gives the variance of the spectrum

$$\frac{\sigma_P^2(k)}{\widehat{P}^2(k)} \approx \frac{2}{V_s(k)V_e(k)},$$

where $V_s(k)$ is the volume of the shell in the wavenumber space and V_e is defined as:

$$V_e(k) = \int_V \left[\frac{\bar{n}(\mathbf{x})P(k)}{1+\bar{n}(\mathbf{x})P(k)}\right]^2 d^3x.$$

As the integrand is about one in regions where the signal dominates the noise, $P \gg 1/\bar{n}$, and zero where the noise dominates, this volume is called the effective volume. For a volume-limited survey ($\bar{n}$ is constant) we get

$$V_e(k) = V\left[1+\frac{1}{\bar{n}P(k)}\right]^{-2},$$

giving a better estimate for the coherence volume $V_c = 1/V_e$ than the simple result $V_c(k) \approx (\Delta k)^3 \approx 1/L^3 \approx 1/V$ (the coherence volume is defined in wavenumber space and thus it is measured in units of the inverse real space volume). So we can write

$$\frac{\sigma_P^2(k)}{\widehat{P}^2(k)} \approx \frac{2}{\mathcal{N}},$$

where $\mathcal{N}$ is the number of coherence volumes in the wavenumber space shell. The number of independent volumes is twice as small (the density field is real). We recall that the above results were obtained assuming that the probability distribution of spectral amplitudes is Gaussian.

It is instructive to compare the behavior of the estimates of the correlation function and the power spectrum, if we vary either the number of galaxies in the sample (the mean density $\bar{n}$) or the size of the sample L. If we observe more galaxies within the same volume, the variance of the correlation function decreases. Alternatively, we can increase the resolution of the correlation function, keeping the same variance level. We also can estimate the correlation function at smaller separations.

The main effect of a larger sample density on the power spectrum estimate is to reduce the shot noise level; the variance of the estimate practically does not change (it decreases only slightly due to the reduced contribution by shot noise). Thus, measuring positions of more galaxies in the same volume, we do not improve the estimate of the power spectrum. A sample of higher density allows us to extend the wavenumber space (the maximum wavenumber, for which we can expect to have reliable power spectrum estimates, is given by $k \approx 1/d$, where d is the average separation between galaxies).

If we increase the sample depth L, the variance of the correlation function estimate decreases; there are more pairs for a particular bin. As above, we can increase the resolution of the correlation function, but we cannot decrease the lower limit of pair separations. Naturally, we can estimate the correlation function for larger pair separations.

The variance of the estimate of the power spectrum for a coherence cell of width $\Delta k \approx 1/L$ does not depend on the depth L. Increasing the depth of the survey will shrink the coherence cells and increase the resolution of the power spectrum. The variance of the power spectrum will decrease only if we decide to keep the same final resolution; the final estimates will then be averaged over more coherence cells. A larger depth of a sample will also allow us to extend the spectral range to smaller wavenumbers.

The FKP weight function (8.19) has been the standard almost from the time of its derivation by Feldman, Kaiser, and Peacock (1994). Other methods that have been used include pure volume weighting (the weighting function is constant inside the survey volume) and a function $\psi(\mathbf{x}) = \bar{n}(\mathbf{x})$ that weights all objects equally. The FKP weight function uses volume weighting for regions where the mean density is large enough and the density is well sampled. For smaller densities (near the boundaries of a sample) it switches over to give equal weight to individual objects.

These weight functions are appropriate for estimating power spectra at large wavenumbers, $k \gg 1/L$. For smaller wavenumbers (large scales), where the window width is comparable to the wavenumber, the careful choice of proper weighting will be crucial. Tegmark (1995) has shown that the optimal selection function $\psi(\mathbf{x})$ is the eigenfunction of the equation

$$\left[\nabla^2 - \frac{\gamma}{\bar{n}(\mathbf{x})}\right] \psi(\mathbf{x}) = E\psi(\mathbf{x})$$

for the minimum eigenvalue E. The constant γ is a free parameter that gives the ratio of the rms error of the power spectrum estimates to the resolution in the wavenumbers (the ratio of "vertical error bars" to "horizontal error bars"). The selection function $\bar{n}(\mathbf{x})$ describes also the geometry of a sample, as $\bar{n}(\mathbf{x}) = 0$ outside sample boundaries.

8.2.3 Integral constraint

There is another effect that can distort the estimated power spectrum in the small wavenumber region. This error arises from the fact that we determine the mean number density of galaxies from the survey itself. The difference between this number and the real mean density characterizes the amplitude of the power spectrum on scales larger than the extent of the survey. Thus using $\bar{n}$ from the survey, we force the fluctuations on the scale of the survey and the power spectrum to zero.

To handle this let us suppose that the selection function $\bar{n}_0(\mathbf{x})$ we have used differs by a factor a from the true function:

$$\bar{n}(\mathbf{x}) = a\bar{n}_0(\mathbf{x})$$

(the density profile is easier to determine than the mean density). This would change the expectation of a Fourier amplitude estimate from zero to

$$E\{F(\mathbf{k}_i)\} = (a-1)\widetilde{\psi}(\mathbf{k}_i)$$

and the spectrum would be shifted by the square of this amount. Because we do not know the true mean density and a, the solution is to modify the weight function to make its Fourier transform vanish at $\mathbf{k}_i$.

The integral constraint

$$\int_V \left[\frac{n(\mathbf{x})}{a\bar{n}_0(\mathbf{x})} - 1\right] \psi(\mathbf{x})\, d^3x = 0$$

gives the estimate A for the correction factor a:

$$A = \frac{1}{\widetilde{\psi}(\mathbf{0})} \int_V \frac{n(\mathbf{x})}{\bar{n}_0(\mathbf{x})} \psi(\mathbf{x})\, d^3x = \frac{1}{\widetilde{\psi}(\mathbf{0})} \sum_j \frac{\psi(\mathbf{x}_j)}{\bar{n}_0(\mathbf{x}_j)}.$$

Substituting that into (8.4) we can write, after a few intermediary steps,

$$\widetilde{\delta}(\mathbf{k}_i) = \frac{a}{A} \int \frac{n(\mathbf{x})}{\bar{n}(\mathbf{x})} \psi_i(\mathbf{x})\, d^3x, \tag{8.20}$$

where

$$\psi_i(\mathbf{x}) = \left[e^{i\mathbf{k}_i\cdot\mathbf{x}} - \frac{\widetilde{\psi}(\mathbf{k}_i)}{\widetilde{\psi}(\mathbf{0})}\right] \psi(\mathbf{x}) \tag{8.21}$$

is the new weight function (we can safely approximate $a/A \approx 1$ in (8.20)). Its Fourier transform is

$$\widetilde{\psi}_i(\mathbf{k}) = \widetilde{\psi}(\mathbf{k}_i + \mathbf{k}) - \frac{\widetilde{\psi}(\mathbf{k}_i)}{\widetilde{\psi}(\mathbf{0})} \widetilde{\psi}(\mathbf{k}),$$

and it is easy to see that $\widetilde{\psi}_i(\mathbf{0}) = 0$.

We do not need to apply the new weights explicitly, but can follow the same procedure that we used to derive them. If we correct the selection function by A and use the previous estimator (8.5), we get the same Fourier amplitudes as by using the weights (8.21). Thus, the integral constraint can be considered as the second normalization condition (for the selection function; the first (8.14) was for the weight function).

The new weight functions change, however, the shot noise term and the window amplitude. The window integrates now to

$$W_i = \int \left| e^{i\mathbf{k}_i \cdot \mathbf{x}} - \frac{\widetilde{\psi}(\mathbf{k}_i)}{\widetilde{\psi}(\mathbf{0})} \right|^2 d^3r.$$

The new shot noise term will also depend on $\mathbf{k}_i$:

$$N_i = \int \left| e^{i\mathbf{k}_i \cdot \mathbf{x}} - \frac{\widetilde{\psi}(\mathbf{k}_i)}{\widetilde{\psi}(\mathbf{0})} \right|^2 \frac{\psi^2(\mathbf{x})}{\bar{n}(\mathbf{x})} d^3x.$$

So, the corrected estimate for the power spectrum will be

$$\left[G \otimes \widehat{P}\right]_i = \frac{P_R(\mathbf{k}_i) - N_i}{W_i}.$$

Different approximations have been used to calculate the integral constraint correction in the past. Tegmark et al. (1998) show that these usually give larger corrections than needed, leading to lower power spectrum values at small k.

As an example of the application of the direct method we show the power spectrum obtained by Miller and Batuski (2001) for a carefully selected deep sample of rich Abell/ACO clusters of galaxies. The total number of clusters in the sample is 545, and the sample geometry is typical — a double cone due to obscuration by dust near the plane of the Galaxy. In this sample the clusters with galactic latitude $b \in [-30°, 30°]$ were excluded. Also, observational projects needed to carry out full-sky surveys have to use several different instruments, and so frequently the survey depth is different in different directions in the sky. For this survey the northern galactic cone has a depth of 420 h^{-1} Mpc, and the southern cone has a depth of 300 h^{-1} Mpc. The spectrum is shown in Fig. 8.2 together with several other spectra obtained for smaller cluster samples. We see that the spectrum is rather close to a power law with a probable plateau (or maximum) around $k \approx 0.02\, h\,\mathrm{Mpc}^{-1}$. The reality of that maximum is not clear, however. The hatched region in the figure shows where the width of the window function is comparable with the wavenumber; thus the estimates in this region are inaccurate. But these estimates indicate that the spectrum is yet rising toward the wavenumber $k = 0.01h\,\mathrm{Mpc}^{-1}$. The spectra for smaller samples shown in the figure tend to decrease at the maximum scales of a sample; this could be the result of insufficient correction for the integral constraint in these studies.

The deepest galaxy sample obtained thus far is the ongoing 2dF redshift survey, described in Chapter 1. The power spectrum for this survey was estimated by a careful application of the direct method by Percival et al. (2001). The survey contained then about 160,000 galaxies and had a depth of 750 h^{-1} Mpc, deeper than the cluster survey described above. Fig. 8.3 shows the power spectrum of that survey, normalized to a model (CDM) power spectrum. This spectrum is a convolution of the true power spectrum with the survey window function; convolution

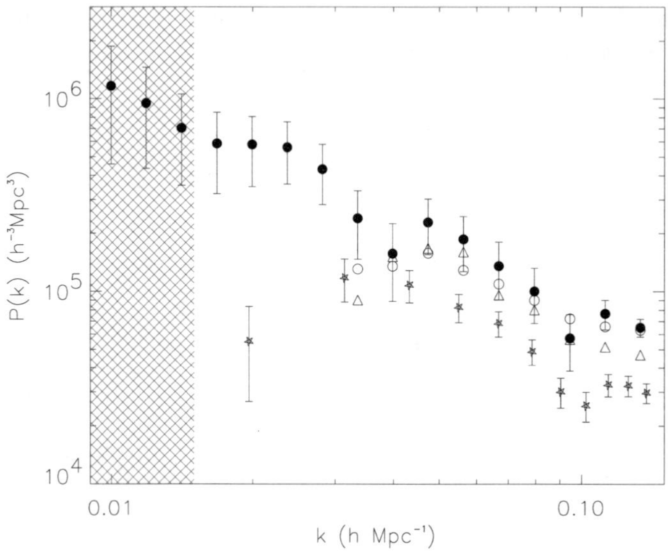

Figure 8.2 *Power spectrum of Abell/ACO clusters (solid circles) compared with those found for smaller samples. Open circles show the spectrum for very rich clusters in the same catalog, triangles show the spectrum for a smaller Abell/ACO sample, and stars show the spectrum for the APM cluster sample. Estimates of the power spectrum in the hatched region are inaccurate. (Reproduced, with permission, from Miller and Batuski 2001, Astrophys. J., 551, 635–642. AAS.)*

may distort the results at wavenumbers $k < 0.02h\,\text{Mpc}^{-1}$. The distortions of the power spectrum due to peculiar velocities and nonlinear processes were estimated using large-scale N-body simulations; these distortions compensate one another for $k < 0.15h\,\text{Mpc}^{-1}$. The region where the estimates of the power spectrum are reliable is constrained by vertical dotted lines. The error bars were determined from realizations of a Gaussian Cox process with a similar power spectrum in a similar survey volume.

The most interesting features of the 2dF survey power spectrum are its details (wiggles), which have been interpreted as traces of acoustic oscillations in the post-recombination power spectrum. Similar features were also found recently by Miller, Nichol, and Batuski (2001) on the basis of the Abell cluster sample, and by Silberman et al. (2001) on the basis of peculiar velocity data. Percival et al. (2001) found that the parameters of the best-fit cosmological model are $\Omega_M h = 0.21$, $\Omega_b/\Omega_M = 0.15$, which gives a rather high baryon fraction Ω_b. This concurs with recent results obtained from the CMB data.

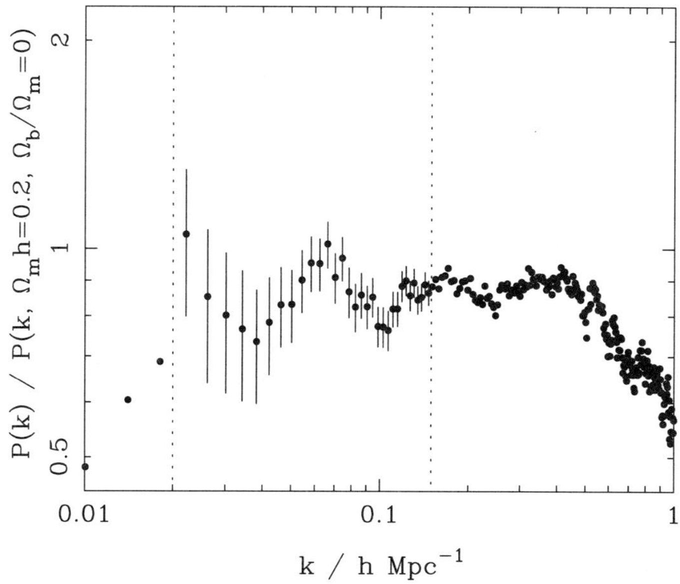

Figure 8.3 *Power spectrum of the 2dF redshift survey, divided by a linear-theory CDM power spectrum. The spectrum is not deconvolved. Error bars are determined from Gaussian realizations; the dotted lines show the wavenumber region that is free of the influence of the window function and of the radial velocity distortions and nonlinear effects. (Reproduced, with permission, from Percival et al., Mon. Not. R. Astr. Soc., in press. Blackwell Science Ltd.)*

8.2.4 Bayesian and maximum likelihood methods

The weight functions are usually chosen to minimize the variance of the estimate of the power spectrum. Thus, the FKP weight function (8.19) is derived using this condition for small scales ($k \gg 1/L$).

In principle, a better way to estimate the power spectrum is to use the minimum variance condition directly; this is called the maximum likelihood method. Any statistical model proposed to describe data has to supply the probability distribution function $L(\mathbf{d}; \mathbf{p})$, where $\mathbf{d} = \{d_i, i = 1 \ldots n\}$ are the data values and $\mathbf{p} = \{p_j, j = 1 \ldots m\}$ are the parameters of the model. For the purpose of the present section we can ignore the spatial arrangement of different d_i and consider both $\mathbf{d}$ and $\mathbf{p}$ as vectors.

The maximum likelihood method finds the estimates of the model parameters by finding the maximum of the distribution function L for an observed data set $\mathbf{d}$. As most model distribution functions are multivariate Gaussians, the method is usually applied by finding the minimum of the log-likelihood function $\mathcal{L}$:

$$\mathcal{L} = -\ln L(\mathbf{d};\mathbf{p}).$$

It can be shown that the maximum likelihood estimate is asymptotically unbiased and has the minimum variance (for large data sets). That makes it one of the most popular methods in statistics.

The usual frequentist viewpoint considers $L(\mathbf{d};\mathbf{p})$ as the distribution of all possible data samples that are described by the model (the sampling distribution). From an alternative, Bayesian viewpoint, the likelihood function is needed to describe the probability distribution of the model parameters for a given data set. The power spectrum analysis in cosmology usually takes the Bayesian approach, writing the Bayesian formula for the posterior probability densities of the model

$$P(\mathbf{p}|\mathbf{d},I) = P(\mathbf{p}|I)\frac{P(\mathbf{d}|\mathbf{p},I)}{P(\mathbf{d}|I)}, \tag{8.22}$$

where I denotes the prior information. The first factor on the right-hand side of (8.22) is the Bayesian prior that includes our prior knowledge about the values of the parameters $\mathbf{p}$. The nominator is the likelihood function (note that it can also depend on the prior information, in principle) and the denominator is the normalization constant (global likelihood). Confidence values in parameter space are defined by integrals of the likelihood function over subvolumes of that space. The normalization factor (global likelihood) is frequently difficult to estimate; in this case the ratio of integrals of the likelihood over subvolumes is used to find the "odds" that one subvolume of parameter space is more likely than the other in the light of the given data set. If we are interested only in a subset of model parameter, confidence values can be found by integrating over the remaining parameters (this is called "marginalization").

The problems usually associated with the Bayesian approach, the need to define the prior distribution $(\mathbf{p}|I)$ and difficulties in calculating the global likelihood, really stress the need to define exactly the statistical model we compare with observations. The real meaning of "more general" frequentist statistics is usually revealed only after extensive studies. For an introduction to the Bayesian approach in astronomy we recommend the review by Loredo (1990), which can be found together with more material at the Web site (*BIPS*).

In a Gaussian case the log-likelihood function

$$\mathcal{L} = (\mathbf{d}-\mathbf{m})\mathbf{C}^{-1}(\mathbf{d}-\mathbf{m})^{\dagger} + \frac{1}{2}\log(\det\mathbf{C}) + \text{const}, \tag{8.23}$$

where $\mathbf{m}$ and $\mathbf{C}$ are the model-predicted mean of data and the data covariance matrix. The notation $\mathbf{A}^{\dagger} = (\mathbf{A}^{T})^{\star}$ means the Hermitian conjugate of the matrix $\mathbf{A}$ (the complex conjugate of its transpose).

It would be difficult to build a model that would predict the coordinates of the observed objects. The power spectrum describes a continuous field $\delta(\mathbf{x})$; thus a natural extension of the formalism is to consider as observed data density values on a properly selected grid in space. This will reduce the data size in the case of large samples, but it is also useful if we concentrate on the large-scale features of power spectra. This procedure is called "pixelization." We define a new data set

$$d_i = \int \left[\frac{n(\mathbf{x})}{\bar{n}(\mathbf{x})} - 1 \right] \psi_i(\mathbf{x})\, d^3x, \tag{8.24}$$

where the functions $\psi_i(\mathbf{x})$ are well localized in space (i.e., have a compact support). These functions are close to the weight functions discussed earlier, so we have retained the same notation. The simplest $\psi_i(\mathbf{x})$ are those that divide the sample space into a collection of cells, then d_i are normalized counts of points in cells. In order to suppress the high frequency noise that such discrete cells produce, the kernels $\psi_i(\mathbf{x})$ can be taken as Gaussian (or the Epanechikov kernel could be used).

A pixelization can also be chosen to represent other coordinate systems (spherical coordinates) and the kernel functions can transform the data set into a collection of Fourier modes, if we choose $\psi_i(\mathbf{x}) = \psi_0(\mathbf{x}) \exp(i\mathbf{k}_i \cdot \mathbf{x})$. A popular choice is the pixelization introduced by Heavens and Taylor (1995), where the density field is expanded in terms of spherical harmonics and a discrete set of spherical Bessel functions. This helps isolate the radial velocity distortions (see Section 8.3).

The maximum likelihood approach is more general than the direct methods for estimating the power spectrum, but it is also more computer-intensive. The main work when minimizing the likelihood function is calculating the $n \times n$ covariance matrix $\mathbf{C}$ that is different for every point of the m-dimensional parameter space and inverting it. Due to the spatial correlations between data pixels the covariance matrix is dense and the number of operations to invert it is $O(n^3)$. This could be prohibitively large.

Although the likelihood function is linear in data, the covariance matrix depends, in general, nonlinearly on the model parameters. In order to estimate confidence regions for the parameters we have to integrate the likelihood function over different regions of parameter space.

Another way to estimate errors of the parameters is by using the Fisher information matrix used in the frequentist maximum likelihood method. This is equivalent to the Bayesian approach in the case of a flat prior (independent of parameters) and the assumption of a Gaussian likelihood function. The latter is also an assumption about the normalization, fixing the behavior of the posterior probability density in the wings (outer regions of the parameter space). The Fisher matrix $\mathbf{F}$ is defined as

$$F_{ij} = -E \left\{ \frac{\partial^2 \mathcal{L}}{\partial p_i \partial p_j} \right\}.$$

A thorough analysis and examples of the Fisher matrix are given by Tegmark, Taylor, and Heavens (1997). The inverse of the Fisher matrix $\mathbf{F}^{-1}$ can be thought of as the covariance matrix of parameter errors. It can be strictly proved that, first, the minimum variance obtainable for an estimate of a single parameter p_i (all others are known) is $\sigma_i^2 = F_{ii}^{-1}$ and, second, if all parameters together are estimated, this minimum is $\sigma_i^2 = (F^{-1})_{ii}$ (Kendall and Stuart 1967).

Intuitively, it is easy to see how the Fisher matrix describes the distribution of the estimated parameters. The log-likelihood function $\mathcal{L}$ in the vicinity of its maximum can be expanded as a quadratic function of parameters, because its first derivatives are zero. Because the likelihood function $L = \exp(-\mathcal{L})$, it will be Gaussian in this region. Neglecting higher-order terms in parameters, we may suppose L to be Gaussian everywhere. The covariances between parameters T_{ij} can be written then as the inverse matrix of the coefficients of the quadratic form at the maximum:

$$(T^{-1})_{ij} = \frac{\partial^2 \mathcal{L}}{\partial p_i \partial p_j}.$$

The Fisher matrix is the expectation value of these coefficients, and its inverse $\mathbf{F}^{-1}$ can be used as the covariance matrix of the parameter estimates.

In order to find the confidence regions for the parameters, we should integrate over the parameter space; to find error estimates for single parameters, we should marginalize over the parameter distribution, which involves more integrals. These multidimensional integrals are usually difficult to calculate. Using the Fisher matrix, we do not have to worry about integrals. Nevertheless, we have to keep in mind that this is essentially an approximate method (that works well most of the time).

As the Fisher matrix is an expectation value over realizations of a random field, it depends only on the statistical model. For a real multivariate Gaussian distribution, for example, the Fisher matrix can be calculated as

$$F_{ij} = \frac{1}{2}\mathrm{Tr}\left[\mathbf{C}^{-1}\mathbf{C}_{,i}\,\mathbf{C}^{-1}\mathbf{C}_{,j} + \mathbf{C}^{-1}(\mathbf{m}_{,i}\,\mathbf{m}_{,j}^T - \mathbf{m}_{,j}\,\mathbf{m}_{,i}^T)\right], \qquad (8.25)$$

where the commas denote derivatives with respect to model parameters, $(C_{,i})_{kl} = \partial C_{kl}/\partial p_i$. So the Fisher matrix can be calculated beforehand to define the observational strategy and to evaluate the model.

There are two popular choices of parameters to describe the density power spectrum in cosmology. One is a set of about 7 to 8 parameters describing the initial simple power spectrum (its amplitude and exponent), the number of neutrino flavors, the main cosmological parameters ($\Omega_M, \Omega_\Lambda, \Omega_{\text{baryon}}, H_0$), etc. Another choice is to describe the power spectrum by band-powers, its values at certain wavelengths, approximating the power spectrum by a step function. In the latter case the elements of the Fisher matrix give directly the covariance of the estimates of the power spectrum.

8.2.5 *Karhunen–Loèwe transform*

The "pixelized" data vector has strong correlations between different individual data values. This makes the covariance matrix $\mathbf{C}$ difficult to invert. Fortunately, there exist methods to transform the data vector so that the individual data values become uncorrelated.

The method used for that is the Karhunen–Loèwe transform. Its use in estimation of the cosmological power spectra began with the analysis of the cosmological microwave background and was developed further in parallel with the applications to galaxy clustering. Vogeley and Szalay (1996) were the first to show how to use that transform for analysis of galaxy samples. We shall follow here their presentation and that of Peacock (2000).

We rearranged our data above into a vector $\mathbf{d}$ with n components d_i. These components are the coefficients of expansion of $\mathbf{d}$:

$$\mathbf{d} = \sum_i d_i \mathbf{e}_i,$$

where $\mathbf{e}_i$ form an orthonormal basis in the space of data vectors. If we rotate the basis vectors to a new system $\mathbf{e}'_i$, the new coordinates of the data vector in this basis will be $d'_i = \mathbf{d} \cdot (\mathbf{e}'_i)^\star = (\mathbf{e}'_i)^\star \cdot \mathbf{d}$ (the new basis is also orthonormal, $(\mathbf{e}'_i)^\star \cdot \mathbf{e}'_j = \delta^K_{ij}$, where δ^K_{ij} is the Kronecker delta).

As the noise is usually statistically independent of the signal, the total data covariance matrix $\mathbf{C}$ can be written as a sum of the signal $\mathbf{S}$ and noise $\mathbf{N}$ covariance matrices,

$$\mathbf{C} = \mathbf{S} + \mathbf{N}. \tag{8.26}$$

We shall suppose first that we can neglect the noise. The (model) data covariance matrix

$$S_{ij} = E\left\{d_i d_j^\star\right\}$$

will be transformed to

$$S'_{ij} = (\mathbf{e}'_i)^\dagger E\left\{d_i d_j^\star\right\} \mathbf{e}'_j.$$

Demanding that the new covariance matrix $\mathbf{S}'$ is diagonal,

$$S'_{ij} = \lambda_i \delta_{ij}, \tag{8.27}$$

we get

$$\mathbf{S}\mathbf{e}'_i = \lambda_i \mathbf{e}'_i.$$

This means that the new basis vectors are the eigenvectors of the covariance matrix with eigenvalues $\lambda_i = \mathbf{S}'_{ii}$. This coordinate transformation is called the discrete Karhunen–Loèwe (K-L) transformation. Note that the covariance matrix $\mathbf{S}$ is the model covariance matrix that depends only on the proposed spectrum (it is also called the fiducial spectrum) and pixelization.

The condition (8.27) tells us that the new data values are statistically orthogonal. If the data has zero mean, then they are also uncorrelated, and if the data distribution is Gaussian, the new data values are statistically independent. All of these conditions are satisfied most of the time in cosmological problems.

Vogeley and Szalay (1996) show that the K-L transform is unique and that this basis is optimal to represent the data. If we arrange the modes by decreasing λ and select the first $m < n$ components of the new data vector, then the K-L basis gives the minimum rms truncation error.

8.2.6 Signal-to-noise eigenmodes

If the noise is constant per original data point (pixel), $\mathbf{N} = \sigma^2\mathbf{I}$, where $\mathbf{I}$ is the unit matrix, then the K-L transform, a coordinate rotation in the data space, leaves the noise matrix diagonal and the signal and noise remain uncorrelated.

If the noise per pixel is not constant, the noise matrix is diagonal, but not proportional to the unit matrix, and diagonalization of the signal covariance matrix will generate a nondiagonal noise covariance matrix. The remedy for that is to rescale the coordinates first to transform the noise covariance matrix to the unit matrix,

$$\mathbf{N}'' = \mathbf{W}^\dagger \mathbf{N} \mathbf{W} = \mathbf{I}.$$

We have written the formula above to be usable for more general transformations than simple rescaling. In the case of rescaling, if we write $\mathbf{N} = \mathrm{diag}(N_1, \ldots, N_n)$, the scaling matrix $\mathbf{W} = \mathrm{diag}(1/\sqrt{N_1}, \ldots, 1/\sqrt{N_n})$. Vogeley and Szalay (1996) call the coordinate transformation by $\mathbf{W}$ "prewhitening," as the scaled data vector has white noise. The covariance matrix for the prewhitened data is

$$\mathbf{C}'' = \mathbf{W}^\dagger \mathbf{C} \mathbf{W} = \mathbf{W}^\dagger \mathbf{S} \mathbf{W} + \mathbf{I}.$$

This gives the equation for the new K-L eigenfunctions

$$(\mathbf{W}^\dagger \mathbf{S} \mathbf{W} + \mathbf{I})\mathbf{e}'_i = \lambda_i \mathbf{e}'_i.$$

Using $\mathbf{W}\mathbf{W}^\dagger = \mathbf{N}^{-1}$ we get

$$\mathbf{N}^{-1}\mathbf{S}\mathbf{e}'_i = (\lambda_i - 1)\mathbf{e}'_i. \tag{8.28}$$

As the K-L eigenvectors of prewhitened data are the eigenvectors of $\mathbf{N}^{-1}\mathbf{S}$, they also are frequently called "the signal-to-noise eigenmodes." We also can write (8.28) as

$$\mathbf{S}\mathbf{e}'_i = \lambda_i \mathbf{N}\mathbf{e}'_i, \tag{8.29}$$

redefining the eigenvalues λ. Thus, the K-L eigenvectors can also be found as the solutions of the generalized eigenvalue problem (8.29). If the noise for different K-L modes is correlated, the prewhitening transform $\mathbf{W}$ can be more general, including both rescaling and rotation of the coordinates. The derivation above remains in force for that case, too.

These modes are optimized with respect to the signal-to-noise ratio, and those are the modes commonly used to estimate the power spectra. We also can find other K-L modes, optimized with respect to different parameters of the covariance matrix. As shown by Tegmark, Taylor, and Heavens (1997), the eigenvectors for this case can be obtained from

$$\frac{\partial \mathbf{C}}{\partial p_i}\mathbf{e}_i' = \lambda_i \mathbf{C}\mathbf{e}_i'. \tag{8.30}$$

The formula (8.29) above is a special case of (8.30) for the parameterization $\mathbf{C} = p_1\mathbf{S} + \mathbf{N}$ with $p_1 = 1$ for the fiducial spectrum. Tegmark, Taylor, and Heavens (1997) also show how to combine the representations of data, optimized for different parameters.

The total number of the K-L eigenmodes is the same as the length of the data vector (n) because they represent the data vector in the same space. As the rms amplitude of the noise is constant (unity for prewhitened data), the eigenvalues λ can be thought of as the signal-to-noise ratios. Usually many of the K-L modes have eigenvalues much less than one and we can truncate the total data vector by retaining only the modes with high eigenvalues. As the size of the covariance matrices is n^2, this compression frequently makes the likelihood problem solvable.

As an application of the K-L transform we show you the eigenfunctions found for the famous "CfA2 slice" of galaxies by Vogeley and Szalay (1996), who give a detailed pedagogical treatment of the K-L process in this paper.

The "CfA2 slice" was the first sufficiently deep sample of galaxies where the peculiarities of the galaxy distribution were clearly seen (de Lapparent, Geller, and Huchra 1986). The maximum extent of the sample on the sky is 117°, in the perpendicular direction 6° (from 8^h to 17^h in RA and from 26.5° to 32.5° in declination), and its depth is 150 h^{-1}Mpc, so it is really a wedge-like slice in space. Because it is so thin, we shall take it as twodimensional for the present example. Fig. 8.4 shows the positions of galaxies in this sample.

Vogeley and Szalay binned this galaxy distribution, prewhitened the data vector, and computed the K-L eigenmodes, using the known selection function and the power spectrum. The K-L transform also can be thought of as the selection of a new pixelization with the weight functions

$$\psi_i'(\mathbf{x}) = \sum_j (\mathbf{e}_i')_j^\star \psi_j(\mathbf{x}), \tag{8.31}$$

the continuous K-L eigenmodes. While it seemed natural to choose localized weight functions to represent data, we see now that optimal weight functions include contributions over the whole sample volume.

The most significant 12 eigenmodes for the "CfA2 slice" are shown in Fig. 8.5.

Fig. 8.6 shows the binned data (1,225 bins), its representation by the 500 most significant K-L modes (with the signal-to-noise ratio larger than unity), and the error of that truncation, represented by the remaining 725 modes. The data compression ratio is rather modest here because of the two-dimensional nature of the data.

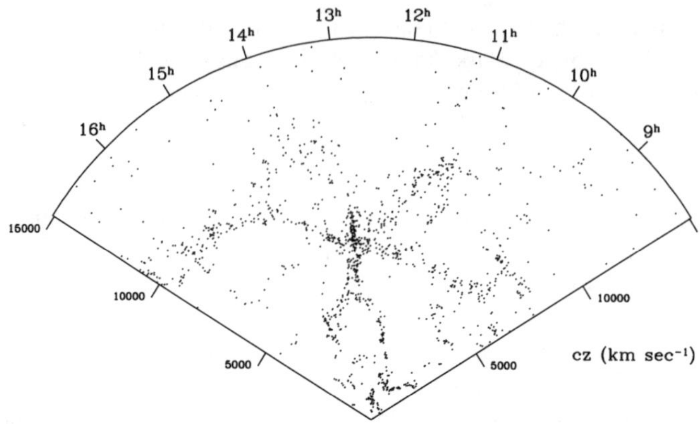

Figure 8.4 *Two-dimensional projection of galaxy positions in the CfA2 north sample, "the CfA2 slice" (see Fig. 1.11). (Reproduced, with permission, from de Lapparent, Geller, and Huchra 1986, Astrophys. J. Lett., 302, L1–L5. AAS.)*

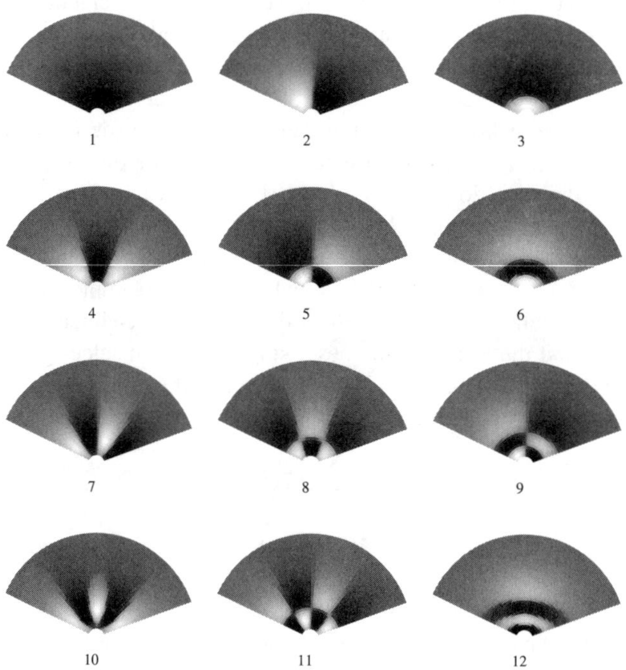

Figure 8.5 *The most significant 12 Karhunen–Loèwe eigenmodes for the CfA2 north slice. The gray density is proportional to the mode amplitude. (Reproduced, with permission, from Vogeley and Szalay 1996, Astrophys. J., 465, 34–53. AAS.)*

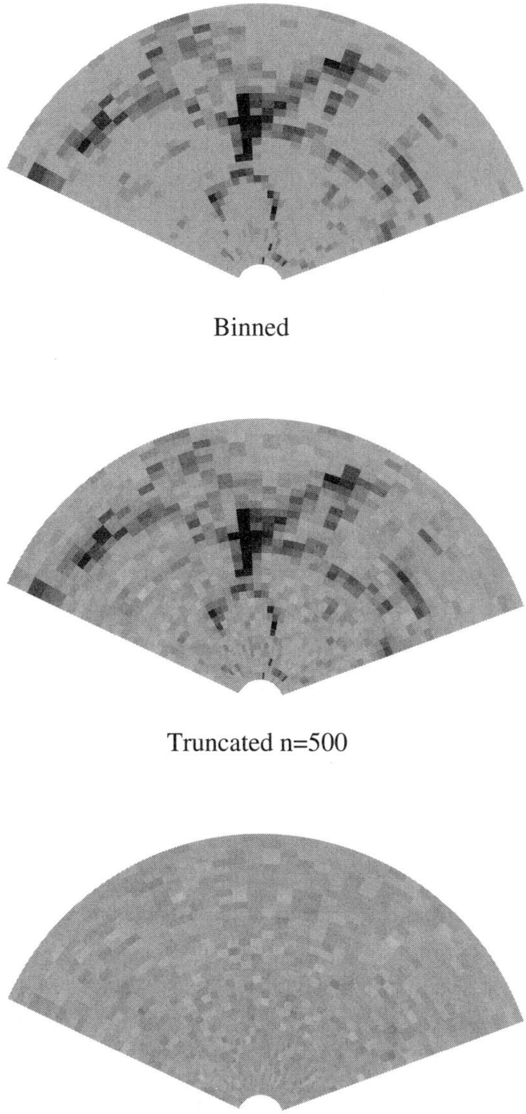

Figure 8.6 *Demonstration of the K-L transform. The top panel shows the binned density distribution for the "CfA2 slice," the middle panel shows its reconstruction by the 500 most significant K-L modes, and the bottom panel shows the truncation error. The gray density is proportional to the number density in bins. (Reproduced, with permission, from Vogeley and Szalay 1996, Astrophys. J., 465, 34–53. AAS.)*

We can choose the original weight functions to satisfy the integral constraint:

$$E\{d_i\} = 0. \tag{8.32}$$

This is easy to do for pixelized data, as we shall show below. By the derivation similar to that for the expectation of power amplitudes above we can get the expression for the signal covariance matrix $\mathbf{S}$:

$$S_{ij} = \int \widetilde{\psi}_i(\mathbf{k})\widetilde{\psi}_j^\star(\mathbf{k})P(\mathbf{k})\,\frac{d^3k}{(2\pi)^3}. \tag{8.33}$$

For the diagonal elements we can write

$$\lambda_i = S_{ii} = \int G_i(\mathbf{k})P(\mathbf{k})\,\frac{d^3k}{(2\pi)^3}, \tag{8.34}$$

where the window function

$$G_i(\mathbf{k}) = \left|\widetilde{\psi}_i(\mathbf{k})\right|^2.$$

The noise covariance matrix is a unit matrix now, so

$$C_{ii} = \lambda_i = w_i P_i + 1,$$

where P_i is the mean power at the wavenumber interval determined by the window function $G_i(\mathbf{k})$ (the band-power). Prewhitening means that this window is not normalized to unity, but to $w_i = (WW^\dagger)_{ii}$. For simple rescaling and a constant initial weight for all pixels in (8.24) the window normalization is $w_i = \bar{n}$ and the K-L band-power estimator is

$$P_i = \frac{\lambda_i - 1}{\bar{n}}.$$

The K-L band-powers are not meant to directly estimate the power spectrum, because they are not optimized for resolution in the wavenumber space. The spectral windows for the most significant modes tend to be rather wide, and the overlap between the modes is substantial (see the examples in Vogeley and Szalay 1996).

The calculation of the model covariance matrix is easy for the parameters that we used to define the K-L transform; it is diagonal in the basis formed by the K-L eigenmodes. When we minimize the log-likelihood function, the covariance matrix does not remain exactly diagonal any more, but it will be sparse and easier to handle than a dense covariance matrix.

The K-L modes depend both on the properties of the sample (geometry and the selection function) and on the fiducial power spectrum, chosen a priori. As these modes constitute an invertible transformation of the original data pixels, the choice of a fiducial spectrum does not bias the final estimates of the power spectrum. As tested in practice, a seriously incorrect choice of the fiducial spectrum can only slightly enlarge the errors of the estimates.

8.2.7 Quadratic compression

Because the number of significant K-L modes could be rather large for large data samples, another data representation can be useful. This is called quadratic compression and was first applied for galaxy samples by Padmanabhan, Tegmark, and Hamilton (2000). This paper gives a careful and detailed analysis of the method and examples of its use.

The method is based on representing the power spectrum as a piecewise function of m values $p_i, i = 0, \dots, m$. The parameters $p_i = P(k), k_i < k < k_{i+1}$ are called band-powers. (We have used here the fact that all cosmological model spectra known thus far are isotropic.)

This model for the power spectrum gives the covariance matrix

$$\mathbf{C} = \sum_{i=0}^{m} p_i \mathbf{C}_{,i} \,,$$

where the derivative $\mathbf{C}_{,i} = \partial \mathbf{C} / \partial p_i$ of the covariance matrix with respect to the band-power can be computed as

$$(C_{,i})_{ab} = \int_{k_i \le |\mathbf{k}| < k_{i+1}} \widetilde{\psi}_a(\mathbf{k}) \widetilde{\psi}_b^\star(\mathbf{k}) \, \frac{d^3 k}{(2\pi)^3} .$$

The noise term is included in this sum by defining $\mathbf{C}_{,0} = \mathbf{N}$ and choosing the coefficient $p_0 = 1$. If we apply the quadratic compression before the K-L transform, the weight functions here describe the original pixelization. This compression also can be applied after K-L transformation, with continuous K-L modes (8.31) as the weight functions.

Now we compress the data vector $\mathbf{d}$ into a new data set $\mathbf{q}$ of size $m + 1$ by

$$q_i = \frac{1}{2} \mathbf{d}^\dagger \mathbf{C}^{-1} \mathbf{C}_{,i} \, \mathbf{C}^{-1} d .$$

The Fisher information matrix for a Gaussian $\mathbf{d}$ and $\mathbf{m} = \mathbf{0}$ is (see 8.25)

$$\mathbf{F}_{ij} = \frac{1}{2} \left[\mathbf{C}^{-1} \mathbf{C}_{,i} \, \mathbf{C}^{-1} \mathbf{C}_{,j} \right] .$$

Tegmark (1997) showed that the band-powers can be found directly, without using a costly maximum-likelihood procedure. The mean and covariance of the compressed data set $\mathbf{q}$ are determined by the Fisher matrix:

$$E\{\mathbf{q}\} = \mathbf{F}\mathbf{p}, \tag{8.35}$$

$$E\left\{\mathbf{q}\mathbf{q}^T\right\} - E\{\mathbf{q}\} E\{\mathbf{q}\}^T = \mathbf{F}. \tag{8.36}$$

Let us separate now the shot noise term. Formula (8.35) gives

$$E\{q_i\} = F_{i0} + \sum_{j=1}^{m} F_{ij} p_j$$

(the parameter $p_0 = 1$). So we can restrict the data and parameter vectors and the Fisher matrix to the index range $i, j = 1, \ldots, m$ and write

$$E\{\mathbf{q}\} = \mathbf{F}\mathbf{p} + \mathbf{f},$$

where the shot noise contribution f is the first column of the original Fisher matrix, $f_i = F_{i0}$. Generalizing the simple unbiased estimator (see 8.35)

$$\widehat{\mathbf{p}} = \mathbf{F}^{-1}\mathbf{q} \tag{8.37}$$

(a hat denotes estimator) we can write

$$\widehat{\mathbf{p}} = \mathbf{M}(\mathbf{q} - \mathbf{f}). \tag{8.38}$$

Equations (8.35, 8.36) now give

$$E\{\widehat{\mathbf{p}}\} = \mathbf{W}\mathbf{p}, \tag{8.39}$$

$$E\left\{\widehat{\mathbf{p}}\widehat{\mathbf{p}}^T\right\} - E\{\widehat{\mathbf{p}}\}E\{\widehat{\mathbf{p}}\}^T = \mathbf{M}\mathbf{F}\mathbf{M}^T, \tag{8.40}$$

where $\mathbf{W} = \mathbf{M}\mathbf{F}$ is the window matrix. Formula (8.39) shows that any band-power estimate is a weighted sum of true band-powers with a row of that matrix, so these rows are the (discrete) window functions.

One can choose the matrix $\mathbf{M}$ in different ways. The choice $\mathbf{M} = \mathbf{F}^{-1}$ gives the raw estimate of band-powers. The covariance matrix of that estimate is $\mathbf{F}^{-1}$, which usually gives rather large rms errors, anti-correlated between neighboring bands.

Another choice is to use

$$M_{ij} = \left(\sum_{k=1}^{m} F_{ik}\right)^{-1} \delta_{ij} = \frac{1}{a_i}\delta_{ij},$$

(Tegmark 1997), which gives

$$\widehat{p}_i = \frac{1}{a_i}\mathbf{F}\mathbf{p}_i = \frac{1}{a_i}E\{\mathbf{q}\}_i,$$

the window-scaled quadratic amplitudes. This estimate is guaranteed to be positive and has minimum variance, although the errors of the amplitudes are correlated.

If we wish to use the band-powers for further maximum likelihood analysis, correlated errors make our work difficult. Fortunately, there exists a possibility to get band-power estimates with uncorrelated errors. This is the choice $\mathbf{M} = \mathbf{F}^{-1/2}$ (Hamilton and Tegmark 2000) that makes the covariance matrix of the parameter estimates (8.40) diagonal. An exact expression for $\mathbf{M}$ in this case is

$$M_{ij} = \left[\sum_{k=1}^{m} \left(F^{1/2}\right)_{ik}\right]^{-1} \left(F^{-1/2}\right)_{ij}, \tag{8.41}$$

where the normalization in (8.41) is caused by the condition that the rows of the

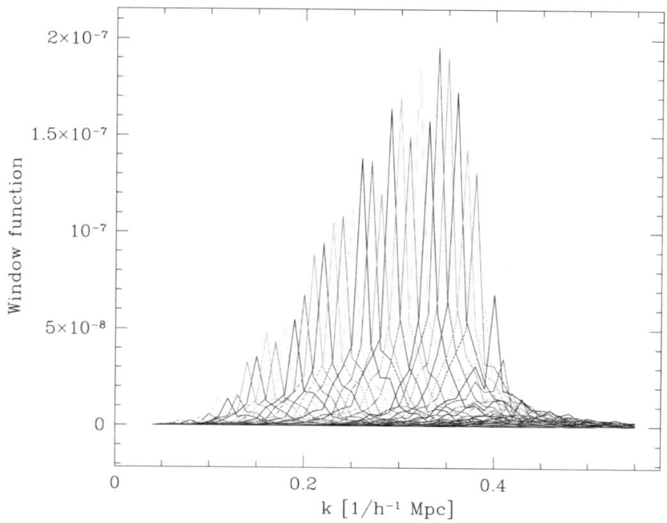

Figure 8.7 *The window functions for the 130* h^{-1} *Mpc depth volume-limited CfA/SSRS sample using the decorrelation method. The windows have been rescaled to the inverse variance of the band-power estimate. (Reproduced, with permission, from Padmanabhan, Tegmark, and Hamilton 2000, Astrophys. J., 550, 52–64. AAS.)*

window matrix $\mathbf{W} = \mathbf{MF}$ have to sum to unity. The window functions obtained by Padmanabhan, Tegmark, and Hamilton (2000) for this case are illustrated in Fig. 8.7. They are highly peaked at the k_i where the band-powers were chosen and overlap little. The real window functions have to be normalized to a unit integral; in this figure they have been normalized to $1/\sigma^2(p_i)$ to show the relative variance of the band-power estimates.

Compression to band-powers is lossless; they contain all the information about the power spectrum (Tegmark 1977). Thus the parameters of the model power spectrum can be found by maximum-likelihood (or Bayesian) methods, using band-powers as the data vector. As the number of band-powers is much smaller than the size of the original data vector, the gain in speed is appreciable.

8.2.8 Pixelized integral constraint

For the pixelized case the (probably biased) mean of data $\mathbf{m}$ can be written as

$$m_i = \int \psi_i(\mathbf{x}) d^3x.$$

Because we do not know its right normalization, we can choose coordinates that do not depend on it. In a vector space these coordinates span the subspace perpendicular to $\mathbf{m}$. Expressing the data vector in these coordinates is equivalent to projecting the data vector into that subspace. The simplest projection operator for that is

$$\mathbf{\Pi} = \mathbf{I} - \mathbf{m}(\mathbf{m}^\dagger\mathbf{m})^{-1}\mathbf{m}^\dagger.$$

The projected data vector is $\mathbf{d}' = \mathbf{\Pi d}$, $E\{\mathbf{d}'\}$ will be identical to zero, and the new covariance matrix will be $\mathbf{C}' = \mathbf{\Pi C \Pi}^\dagger$. The new data vector is isolated from possible uncertainties in the mean density. This, however, comes at a price; the new covariance matrix is singular. If we use the maximum likelihood method, we need to invert that. Tegmark et al. (1988) recommend replacing $\mathbf{C}^{-1}$ by the pseudo-inverse $\mathbf{\Pi}\left[\mathbf{C} + \gamma\mathbf{m}\mathbf{m}^\dagger\right]^{-1}\mathbf{\Pi}$. The result does not depend on γ.

If we estimate the power spectrum using quadratic compression, there exists a more convenient choice of the projection (Tegmark et al. 1998). The optimal (minimum-variance) projection operator is

$$\mathbf{\Pi} = \mathbf{I} - \mathbf{m}\left(\mathbf{m}^\dagger\mathbf{C}^{-1}\mathbf{m}\right)^{-1}\mathbf{m}^\dagger\mathbf{C}^{-1}.$$

For this operator only the compressed data have to be changed by $\mathbf{d}' = \mathbf{\Pi d}$; no other change is necessary. Because the old covariance matrix remains in use, there will be no problems with matrix inversion.

8.3 Redshift distortions

The treatment of the samples in the previous sections assumed that we knew the spatial coordinates of objects (galaxies). Regrettably, this is not the case with the real samples. The 3-D samples are based on three coordinates per galaxy — its two spherical coordinates in the sky, and its redshift z (or the radial velocity $v = cz$). The last coordinate can be used to obtain the radial (comoving) distance; for nearby objects this is given by $r = cz/H_0$, where H_0 is the present Hubble's constant. These coordinates describe the spatial arrangement of objects in redshift space. Thus, the methods to estimate power spectra that we described above give power spectra for redshift space.

The observed radial velocities are, in fact, a sum of the Hubble recession velocity cz and the peculiar velocity caused by gravitational dynamics (collapse velocities and velocities in virialized systems). In a gravitating environment no object can stay in one place; all equilibria involve motion. Typical distortions caused by peculiar velocities are illustrated in Fig. 8.8, from the review by Hamilton (1998). This figure shows the density profiles in redshift space caused by a spherical overdensity in real space. There are two main effects — squashing of the density profiles at large scales, due to infall of the outer regions, and the finger-of-God effect at small scales, due to the gravitational equilibrium of the central region. The left panel illustrates the situation when the object is far from the observer. The right panel shows a sphere that is close to the observer; the density profiles are more complex.

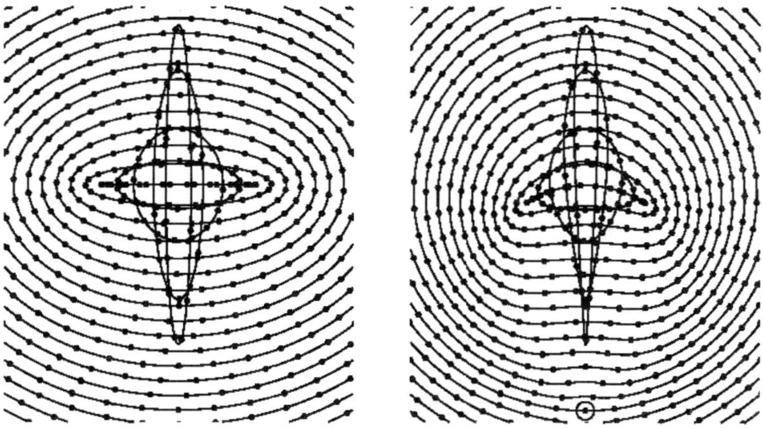

Figure 8.8 *The redshift space view of a evolving spherical overdensity in real space. The panels show the redshift space density contours in an equatorial plane of the sphere. The left panel shows a sphere that is far from the observer; the right panel shows a nearby sphere. (Reproduced, with permission, from Hamilton 1998, in The Evolving Universe, Kluwer Academic Publishers.)*

8.3.1 General case

This effect was first treated analytically by Kaiser (1987). We follow here the more general derivation by Hamilton (1998). The redshift distance of a luminous object (galaxy) s can be written as

$$\mathbf{s} = \mathbf{r} + \widehat{\mathbf{r}} \cdot \mathbf{v}/H_0,$$

where r is the distance of the galaxy in real space, $v = \widehat{\mathbf{r}} \cdot \mathbf{v}$ is the peculiar velocity of the galaxy (the projection of the galaxy's velocity $\mathbf{v}$ on the direction from the observer to the galaxy $\widehat{\mathbf{r}}$), and H_0 is the present Hubble's constant. We shall consider nearby samples where the geometry of the space is Euclidean and the change of Hubble's constant with redshift can be ignored. We have also assumed that the observer is at rest. We can define that with respect to the CMB rest frame, subtracting our (well-known) velocity with respect to that frame from the radial velocity of the galaxy.

The distance distortion cannot change the number of galaxies, so

$$n^s(\mathbf{s})d^3s = n(\mathbf{x})d^3x,$$

where n^s and n are the number densities of galaxies in redshift space and in real space, respectively. Introducing the density contrast, the above relation can be written as

$$\bar{n}(\mathbf{s})\left[1 + \delta^s(\mathbf{s})\right] s^2 ds\, d\Omega = \bar{n}(\mathbf{r})\left[1 + \delta(\mathbf{r})\right] r^2 dr\, d\Omega,$$

where δ^s are δ are the density contrasts in redshift space and in real space, respectively, and $d\Omega$ is the appropriate region on the sky. The average densities $\bar{n}^s(\mathbf{s})$ and $\bar{n}(\mathbf{r})$ are the selection functions of the sample and are the same to the first order. More accurate treatment of the selection functions can be found in Hamilton (1998). Let us discard the nearby regions of the sample. The peculiar velocities then satisfy the condition $v/(H_0 r) \ll 1$. We also assume that $\delta \ll 1$ (the linear approximation that describes large-scale dynamics rather well). Then we get

$$1 + \delta^s(\mathbf{s}) = \frac{\bar{n}(\mathbf{r})}{\bar{n}(\mathbf{r} + \hat{\mathbf{r}}v/H_0)} \left(1 + \frac{v}{H_0 r}\right)^{-2} \left(1 + \frac{1}{H_0}\frac{\partial v}{\partial r}\right)^{-1} [1 + \delta(\mathbf{r})] .$$

This gives

$$\delta^s(\mathbf{s}) = \delta(\mathbf{r}) - \frac{1}{H_0}\frac{\partial v}{\partial r} - \left[2 + \frac{r}{\bar{n}(\mathbf{r})}\frac{d\bar{n}(\mathbf{r})}{dr}\right]\frac{v}{H_0 r}.$$

Introducing the dimensionless derivative of the selection function $\alpha(\mathbf{r})$ by

$$\alpha(\mathbf{r}) = 2 + \frac{d\ln(r^2\bar{n}(\mathbf{r}))}{d\ln(r)}, \tag{8.42}$$

we can write

$$\delta^s(\mathbf{r}) = \delta(\mathbf{r}) - \left(\frac{\partial v}{\partial r} + \frac{\alpha(\mathbf{r})}{r}\right)\frac{v}{H_0}.$$

We have used here the fact that in the first order $\delta^s(\mathbf{s}) = \delta^s(\mathbf{r})$.

The velocity $\mathbf{v}$ is connected with the density contrast δ by the equation of continuity. In the linear approximation this connection is given by the formula (see Chapter 6):

$$H(a)f(a)\delta_M = \nabla \cdot \mathbf{v}, \tag{8.43}$$

where $H(a)$ is the Hubble function and $f(a)$ is the dimensionless growth rate of the growing mode:

$$f(a) = \frac{d\ln\left(D_1(a)\right)}{d\ln(a)}.$$

We have written δ_M in (8.43) to stress that it describes the total density. The galaxy density contrast $\delta = \delta_G$ is, in the first approximation, proportional to that:

$$\delta_G = b\delta_M,$$

where b is the biasing factor. Recalling the fact that the growing mode of the velocity in the linear approximation is irrotational, we can write

$$\mathbf{v} = \nabla\Psi,$$

where Ψ is the velocity potential. This potential can be found from the equation of continuity (8.43):

$$\Delta\Psi = \nabla^2\Psi = -H_0\frac{f_0}{b}\delta, \tag{8.44}$$

where we have used the present values of H and f. Thus, we can write

$$\mathbf{v} = -H_0\beta\nabla(\nabla^{-2}\delta),$$

where $\beta = f_0/b$ and ∇^{-2} is the integral operator that inverts the equation (8.44). The radial velocity distortions can now be written as

$$\delta^s(\mathbf{r}) = \mathbf{S}\delta(\mathbf{r}), \tag{8.45}$$

where the radial distortion operator $\mathbf{S}$ is

$$\mathbf{S} = 1 + \beta\left(\frac{\partial^2}{\partial r^2} + \frac{\alpha(\mathbf{r})}{r}\frac{\partial}{\partial r}\right)\nabla^{-2}. \tag{8.46}$$

When applying this formula, we should remember that $\partial/\partial r = \widehat{\mathbf{r}} \cdot \nabla$. The correlation function in redshift space can be written as

$$\xi^s(\mathbf{x}_1, \mathbf{x}_2) = \xi^s(r_{12}, r_1, r_2) = \mathbf{S}_1\mathbf{S}_2\xi(r_{12}). \tag{8.47}$$

We see that an isotropic real space correlation function transforms to an anisotropic redshift space correlation function. The first equality above is caused by the fact that the correlation does not depend on the orientation of the triangle $\{\mathbf{0}, \mathbf{x}_1, \mathbf{x}_2\}$ in space. Calculating the correlation functions by (8.47) is difficult, as the operator $\mathbf{S}$ is an integro-differential operator in real space. The usual approach is to use the Fourier transform.

It is also possible to invert formula (8.45), writing

$$\delta(\mathbf{s}) = \mathbf{S}^{-1}\delta^s(\mathbf{s}),$$

where $\mathbf{S}^{-1}$ is the inverse radial distortion operator (Taylor and Valentine 1999). This is given by

$$\mathbf{S}^{-1} = 1 - \frac{\beta}{1+\beta}\left(\frac{\partial^2}{\partial s^2} + \frac{\alpha'(s)}{s}\frac{\partial}{\partial s}\right)\nabla^{-2},$$

where

$$\alpha'(s) = \frac{d\ln\phi^{-1}(s)}{d\ln s} + 2,$$

and

$$\alpha(r) + \alpha'(r) = 4.$$

Taylor and Valentine (1999) give also a clear review of different velocity distortion effects and of their treatment.

8.3.2 Far-field approximation

The first studies of radial velocity distortions were done in the far-field approximation, where we can suppose that the density profile is a plane wave oriented at an angle to the line-of-sight direction. Looking along the z-direction, we get

$$\mathbf{S}^P = 1 + \beta\frac{\partial^2}{\partial z^2}\nabla^{-2}.$$

This expression is very simple in the Fourier space, where $\partial^2/\partial z^2 = k_z^2/k^2 = \mu^2$, where μ is the cosine between the line of sight and the wavevector $\mathbf{k}$ (the normal to the plane). So the distortion operator reduces to

$$\mathbf{S}^P = 1 + \beta\mu^2,$$

meaning that a redshift space Fourier mode is proportional to a real space Fourier mode:

$$\widetilde{\delta}^s(\mathbf{k}) = \left(1 + \beta\mu^2(\mathbf{k})\right)\widetilde{\delta}(\mathbf{k}).$$

This gives a very simple result for the power spectra

$$P^s(\mathbf{k}) = \left(1 + \beta\mu^2(\mathbf{k})\right)^2 P(k).$$

The constant β is usually approximated by

$$\beta = \frac{\Omega_M^{0.6}}{b}$$

(see Chapter 6 for the justification of the approximation) and its value is about 0.5 to 1.0. Thus, depending on the orientation, the redshift space spectral amplitudes could be up to four times higher than real space spectrum; the velocity distortion effect is substantial. Expanding the square, we can write

$$P^s(\mathbf{k})/P(k) = \left(1 + \frac{2}{3}\beta + \frac{1}{5}\beta^2\right) + \left(\frac{4}{3}\beta + \frac{4}{7}\beta^2\right)P_2(\mu) + \frac{8}{35}\beta^2 P_4(\mu),$$

where $P_2(\mu)$ and $P_4(\mu)$ are the Legendre polynomials that describe the quadrupole and hexadepole harmonics, respectively.

Hamilton (1992) showed that in the plane-parallel approximation we can get similar expressions for the redshift space correlation function:

$$\xi^s(\mathbf{r}) = \xi_0(r) + \xi_2(r)P_2(\mu) + \xi_4(r)P_4(\mu),$$

where

$$\begin{aligned}
\xi_0(r) &= \left(1 + \frac{2}{3}\beta + \frac{1}{5}\beta^2\right)\xi(r), \\
\xi_2(r) &= \left(\frac{4}{3}\beta + \frac{4}{7}\beta^2\right)\left[\xi(r) - \bar{\xi}(r)\right], \\
\xi_4(r) &= \frac{8}{35}\beta^2\left[\xi(r) + \frac{5}{2}\bar{\xi}(r) - \frac{7}{2}\bar{\bar{\xi}}(r)\right]
\end{aligned}$$

and

$$\begin{aligned}
\bar{\xi}(r) &= \frac{3}{r^3}\int_0^r \xi(x)x^2\,dx, \\
\bar{\bar{\xi}}(r) &= \frac{5}{r^5}\int_0^r \xi(x)x^4\,dx.
\end{aligned}$$

If we ignore redshift distortions, we get from observations the angle-averaged redshift correlation function

$$\langle \xi^s(r) \rangle = \xi_0(r)$$

from above. For $\beta = 1$ its amplitude is twice that of the real space correlation function.

If a galaxy pair shares the same plane density wave, the vector $\mathbf{r}$ will be parallel to the normal to the density wave. Thus, the quantity μ from the Fourier picture corresponds to the cosine of the angle θ between the line-of-sight and the vector of the galaxy pair $\mathbf{r} = \mathbf{x}_2 - \mathbf{x}_1$. In the plane-parallel approximation the arguments r and μ are frequently replaced by the transversal distance $\sigma = r_\perp = r \sin\theta$ and by the radial distance $\pi = r_\| = r\cos\theta = r\mu$.

Observationally, these distances are determined as

$$\pi = s_\| = |s_1 - s_2|$$

and the transversal distance

$$\sigma = s_\perp = \sqrt{s^2 - s_\|^2},$$

where s_i are the redshift distances for the two galaxies and s is the redshift space separation of the pair.

Thus, the observed redshift space correlation function really depends on two arguments, $\xi(\mathbf{r}) = \xi(\pi, \sigma)$ (in the far-field approximation).

The above derivation used the linear approximation to dynamics and thus describes only large-scale effects. There is as yet no theory to describe the effects of nonlinear dynamics and this is approximated phenomenologically, supposing that virial motions simply erase small-scale velocity correlations. This leads to the convolution of the linear redshift correlation function by the one-dimensional random pairwise velocity distribution of galaxies $f(v)$. In Fourier space it reduces to the multiplication of the corresponding Fourier transforms.

Two model distributions have been used, the Gaussian velocity distribution

$$f(v) = \frac{1}{\sqrt{2\pi}\sigma} e^{-v^2/2\sigma^2}$$

with the Fourier transform

$$\widetilde{f}(\mathbf{k}) = e^{-[\sigma k \mu(\mathbf{k})]^2/2}$$

(note that it depends only on the line-of-sight component $k_\|$ of the wavevector) and the exponential distribution

$$f(v) = \frac{1}{\sqrt{2}\sigma} e^{-\sqrt{2}|v|/\sigma}$$

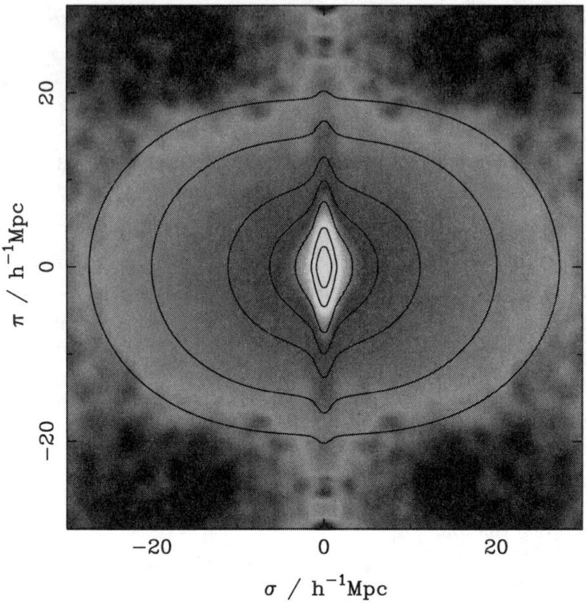

Figure 8.9 *The redshift space correlation function of the 2dF survey, shown as the gray density map, and a model correlation function, shown with contour lines. The y-axis shows the radial separation π and the x-axis shows the transversal separation σ. (Reproduced, with permission, from Peacock et al. 2001, Nature, 410, 169–173. Macmillan Publishers Ltd.)*

with the Fourier transform

$$\widetilde{f}(\mathbf{k}) = \frac{1}{1 + \frac{1}{2}\left[\sigma k \mu(\mathbf{k})\right]^2}.$$

The velocity variance σ^2 here is the pairwise velocity variance that is two times larger than the single-point velocity variance.

A model of the redshift space correlation function together with the results from the 2dF redshift survey (Peacock et al. 2001), is shown in Fig. 8.9. The model uses the estimated power spectrum from the APM survey that we shall discuss below, and assumes a best-fit $\beta = 0.4$ and the exponential pairwise small-scale distortion with $\sigma = 400$ km s^{-1}. Both the large-scale squashing of the correlation function contours and the finger-of-God effect are predicted rather well by the model.

8.4 Velocity distortions in power spectrum

For estimation of power spectra from observed samples the plane-parallel approximation is not exact enough and we have to use the full radial distortion operator.

The main effect of this operator is to introduce a strong coupling of redshift space Fourier modes. Tegmark et al. (1998) write the covariance function of the Fourier components as

$$E\left\{\widetilde{\delta^s}(\mathbf{k})(\widetilde{\delta^s})^\star(\mathbf{k}')\right\} = (2\pi)^3 \int g(\mathbf{k},\mathbf{k}',\mathbf{k}'')P(\mathbf{k}'')\,d^3k'', \tag{8.48}$$

where

$$\begin{aligned} g(\mathbf{k},\mathbf{k}',\mathbf{k}'') &= \delta_D(\mathbf{k}-\mathbf{k}'')\delta_D(\mathbf{k}'-\mathbf{k}'') \\ &\quad +\beta\left[\delta_D(\mathbf{k}-\mathbf{k}'')f(\mathbf{k}',\mathbf{k}) + \delta_D(\mathbf{k}'-\mathbf{k}'')f^\star(\mathbf{k},\mathbf{k}')\right] \\ &\quad +\beta^2 f^\star(\mathbf{k},\mathbf{k}'')f(\mathbf{k}',\mathbf{k}'') \end{aligned} \tag{8.49}$$

and

$$f(\mathbf{k},\mathbf{k}') = \int \left[\mu^2(\mathbf{r},\mathbf{k}') - \frac{i\alpha(\mathbf{r})}{k'r}\mu(\mathbf{r},\mathbf{k}')\right] e^{i(\mathbf{k}'-\mathbf{k})\cdot\mathbf{r}} d^3r. \tag{8.50}$$

8.4.1 Fourier–Bessel expansion

Direct use of the above formulae leads to complicated expressions for the covariance matrix $\mathbf{C}$ that is needed for the maximum-likelihood method. Two approaches have been used to overcome this. The first is to use a spherical coordinate system that is more natural for the problem. An example is the analysis of the PSCz redshift survey by Tadros et al. (1999) who use the pixelization introduced by Heavens and Taylor (1995). The basis functions for that are spherical harmonics (for angular modes) and spherical Bessel functions for radial modes. The pixelization is

$$\widetilde{n}_{lmn} = c_{ln} \int n(\mathbf{s})w(s)j_l(k_{ln}s)Y^\star_{lm}(\theta,\phi)\,d^3s, \tag{8.51}$$

where c_{ln} are normalization constants, j_l are spherical Bessel functions, k_{ln} are discrete wavenumbers, and Y_{lm} are spherical harmonics. We shall describe the Fourier–Bessel expansion in more detail in Chapter 9.

The pixelization (8.51) above gives pixels in the wavenumber space of spherical harmonics. Expanding the velocity, the data vector D_{lmn} is defined as the normalized difference between the amplitudes (8.51) and similar harmonics for the mean density (selection function). This leads to the relation

$$\mathbf{D} = (\mathbf{\Phi} + \beta\mathbf{V})\widetilde{\mathbf{d}},$$

where $\widetilde{\mathbf{d}}$ are the Fourier modes in real space and the matrices $\mathbf{\Phi}$ and $\mathbf{V}$ describe the mixing of modes due to the geometry of the survey (incomplete sky coverage, for example) and the radial distortion, respectively. The choice of spherical coordinates allows us to separate the matrices $\mathbf{\Phi}$ and $\mathbf{V}$ into the angular and radial parts, writing

$$\mathbf{D} = \mathbf{SW}\,(\mathbf{\Phi} + \beta\mathbf{V})\,\widetilde{\mathbf{d}}.$$

We have redefined the matrices $\mathbf{\Phi}$ and $\mathbf{V}$ above; now they describe the radial parts of previous $\mathbf{\Phi}$ and $\mathbf{V}$. The matrix $\mathbf{W}$ describes mixing of angular modes (due to angular selection or incomplete sky coverage) and the "scattering" matrix $\mathbf{S}$ describes the influence of dynamical velocities (convolution with the one-dimensional random pairwise velocity distribution). The expression for the covariance matrix is rather complicated, but simpler than the general formulae (8.48–8.50) above:

$$\mathbf{C} = \mathbf{SW}\left(\mathbf{\Phi} + \beta\mathbf{V}\right)\mathbf{P}\left(\mathbf{\Phi} + \beta\mathbf{V}\right)^T \mathbf{W}^\dagger\mathbf{S}^T + \mathbf{RW}, \tag{8.52}$$

where $\mathbf{P}$ is the diagonal prior power spectrum matrix and $\mathbf{R}$ is the radial part of the noise matrix. The detailed expressions for these matrices are rather complex; we refer the reader to the paper by Tadros et al. (1999).

In addition to the Fourier–Bessel expansion there exist other possibilities. Hamilton and Culhane (1996) note that when the selection term $\alpha(\mathbf{r})$ is constant, the radial distortion operator is scale-free and commutes with the operator $\partial/\partial \ln r$. Then one can choose a complete set of commuting Hermitian operators:

$$i\left(\frac{\partial}{\partial \ln r} + \frac{3}{2}\right) = -i\left(\frac{\partial}{\partial \ln k} + \frac{3}{2}\right), L^2, L_z.$$

The last two (rotation) operators give the usual spherical harmonics and the first operator — logarithmic waves either in real space or in Fourier space. The orthonormal eigenfunctions for this set are

$$Z_{\omega lm}(\mathbf{r}) = \frac{1}{\sqrt{2\pi}} e^{-(3/2+i\omega)\ln r}\, Y_{lm}(\widehat{\mathbf{r}}),$$

where Y_{lm} are spherical harmonics. When the observed number density is expanded in this basis, the redshift space Fourier amplitudes will be proportional to those for real space. It means that the matrix $\mathbf{V}$ in (8.52) will be diagonal, simplifying the calculation of the covariance matrix. Because the scales where we measure power spectra usually extend over several decades, logarithmic pixelization in scale is better in this respect, too.

As an example of the application of the machinery described above we show the real space spectrum obtained for the PSCz survey by Hamilton, Tegmark, and Padmanabhan (2000). They expanded the galaxy density in logarithmic spherical waves, applied the Karhunen–Loèwe signal-to-noise transform, and used the quadratic compression. This spectrum has also been corrected for linear redshift distortions by allowing a scale-dependent β. The resulting power spectrum for large scales is shown in Fig. 8.10. As we see, the errors of the spectrum are especially large at large scales (a bit smaller for the correlated power spectrum). There are interesting details in the spectrum at large scales but deeper surveys are clearly needed to establish their reality. The model power spectrum shown describes observations rather well, except for the large wavenumber region, where nonlinear dynamics may change the results.

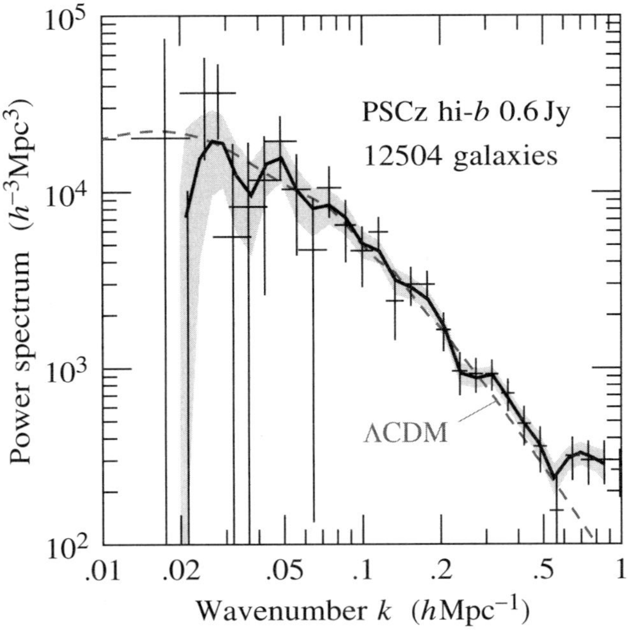

Figure 8.10 *The real space power spectrum for galaxies from the PSCz redshift survey. The solid line is the correlated power spectrum and the shaded region its rms error. Points with rms error bars show the decorrelated power spectrum and the dashed line gives a model power spectrum. (Reproduced, with permission, from Hamilton, Tegmark, and Padmanabhan 2000, Mon. Not. R. Astr. Soc., 317, L23–L27. Blackwell Science Ltd.)*

8.4.2 Modeling the correlation function

Another possibility is to calculate the model correlation function (data covariance matrix) directly in data space. Although it seems difficult at first glance, Szalay, Matsubara, and Landy (1998) have found a simple exact formula for that. As we mentioned above, the anisotropic two-point correlation function $\xi(\mathbf{x}_1, \mathbf{x}_2)$ is the function of the triangle $\{\mathbf{0}, \mathbf{x}_1, \mathbf{x}_2\}$, but not its orientation in space. They parameterize the size of a triangle by $r = |\mathbf{x}_1 - \mathbf{x}_2|$ and its shape Δ by two angles. The first angle is θ, the half-angle between the directions to the two galaxies:

$$\cos 2\theta = \frac{\mathbf{x}_1 \cdot \mathbf{x}_2}{r_1 r_2}.$$

The second angle is γ, the angle between the ray bisecting the angle between $\mathbf{x}_1$ and $\mathbf{x}_2$, and the direction of the far-away side $\mathbf{x}_1 - \mathbf{x}_2$ of the triangle. This angle is not an independent parameter; it can be found from

$$\cos\gamma = \cos\left(\frac{\theta(r_1 - r_2)}{r}\right).$$

The redshift space correlation function is then

$$\xi^s(r,\Delta) = c_{00}\xi_0^{(0)} + c_{02}\xi_2^{(0)} + c_{04}\xi_4^{(0)} + c_{11}\xi_1^{(1)} + c_{13}\xi_3^{(1)} + c_{20}\xi_0^{(2)} + c_{22}\xi_2^{(2)},$$

where

$$\xi_l^{(n)}(r) = \frac{1}{2\pi^2}\int k^{2-n}P(k)j_l(kr)\,dk$$

give the scale dependence and the coefficients $c_{ij}(\Delta)$ describe the dependence on the shape of the triangle. The functions $j_l(x)$ are the spherical Bessel functions.

The largest shape coefficients are the following:

$$\begin{aligned}
c_{00} &= 1 + \frac{2}{3}\beta + \frac{1}{5}\beta^2\cos^2\theta\sin^2\theta, \\
c_{02} &= -\left(\frac{4}{3}\beta + \frac{4}{7}\beta^2\right)\cos 2\theta P_2(\cos\gamma) - \frac{2}{3}\left(\beta - \frac{1}{7}\beta^2 + \frac{4}{7}\beta^2\sin^2\theta\right)\sin^2\theta, \\
c_{04} &= \frac{8}{35}\beta^2 P_4(\cos\gamma) - \frac{4}{21}\beta^2\sin^2\theta P_2(\cos\gamma) - \frac{1}{5}\beta^2\left(\frac{4}{21} - \frac{3}{7}\sin^2\theta\right)\sin^2\theta.
\end{aligned}$$

The functions $P_n(x)$ above are Lagrange polynomials.

The remaining shape coefficients are substantially smaller, in general. These coefficients also depend on the selection function via the constants

$$\alpha_i = r_i\alpha(\mathbf{x}_i);$$

see (8.42) for the definition of $\alpha(\mathbf{x})$. These coefficients are:

$$\begin{aligned}
c_{11} &= (\alpha_1 - \alpha_2)\left(\beta + \frac{3}{5}\beta^2 - \frac{4}{5}\beta^2\sin^2\theta\right)\cos\theta P_1(\cos\gamma) \\
&\quad + (\alpha_1 + \alpha_2)\left(\beta + \frac{3}{5}\beta^2 - \frac{4}{5}\beta^2\cos^2\theta\right)\sin\theta P_1(\sin\gamma), \\
c_{13} &= \frac{1}{5}(\alpha_1 - \alpha_2)\beta^2\cos\theta\left[sin^2\theta P_1(\cos\gamma) - 2P_3(\cos\gamma)\right] \\
&\quad + \frac{1}{5}(\alpha_1 + \alpha_2)\beta^2\sin\theta\left[cos^2\theta P_1(\sin\gamma) - 2P_3(\sin\gamma)\right], \\
c_{20} &= \frac{1}{9}\alpha_1\alpha_2\beta^2(4\cos\theta - 1), \\
c_{22} &= \frac{1}{3}\alpha_1\alpha_2\beta^2\sin^2\theta - \frac{2}{3}\alpha_1\alpha_2\beta^2 P_2(\cos\gamma).
\end{aligned}$$

Matsubara, Szalay, and Landy (2000) used these formulae to find the redshift space covariance matrix for the Las Campanas redshift survey, which has a rather complex geometry. They neglected the last four shape coefficients, integrated the correlation function numerically for nearby pixels,

$$S_{ij} = \frac{1}{V_i V_j} \int_{V_i} \int_{V_j} \xi^s(\mathbf{x}_1, \mathbf{x}_2)\, d^3x_1 d^3x_2,$$

and used the Taylor approximation for distant pixel pairs.

8.5 Methods for estimating power spectra

The number of different methods to estimate power spectra seems bewildering at first glance, but they all have their own areas of application. The pros and cons of different methods have been compared in the review paper by Tegmark et al. (1998). Their recommendations and later practice can be summarized in the following steps:

1. The power spectrum on small scales (about $k \leq L/10$) should be found by a direct method from raw data.
2. Second, the data should be pixelized into pixels substantially smaller than $L/10$.
3. The pixelized data should be compressed using the Karhunen–Loèwe transform. Although this compression is not too strong, it conserves phase information and the compressed data can be used both to estimate velocity distortions and to analyze the assumptions made in the analysis; these show up clearly in the error distribution.
4. Quadratic compression should be used to get the band-powers; the final maximum likelihood analysis is most efficiently done using the band-powers as the final data vector.

As an example of such a combined approach we demonstrate in Fig. 8.11 the latest estimate of the real space power spectrum of the PSCz redshift survey over the whole accessible wavenumber range by Hamilton and Tegmark (2000). The wavenumber range for this power spectrum extends from $k = 0.0137h\,\mathrm{Mpc}^{-1}$ to $k = 316h\,\mathrm{Mpc}^{-1}$, almost 4.5 decades.

The linear part of the spectrum has been obtained by the machinery described above and is the same as that shown in Fig. 8.10. As we noted above, it is not particularly usable for small scales; the density distribution is not Gaussian any more and the number of pixels will be prohibitively large. Here the nonlinear part has been obtained using a version of the direct method. Because the redshift space spectrum is anisotropic, it was expanded into spherical harmonics. The harmonics of the band-powers were defined as

$$P_l^s(k,\mu) = \int W(k,k')(2l+1)\mathcal{P}_l(\mu)P^s(\mathbf{k}')\,\frac{d^3k'}{(2\pi)^3}, \tag{8.53}$$

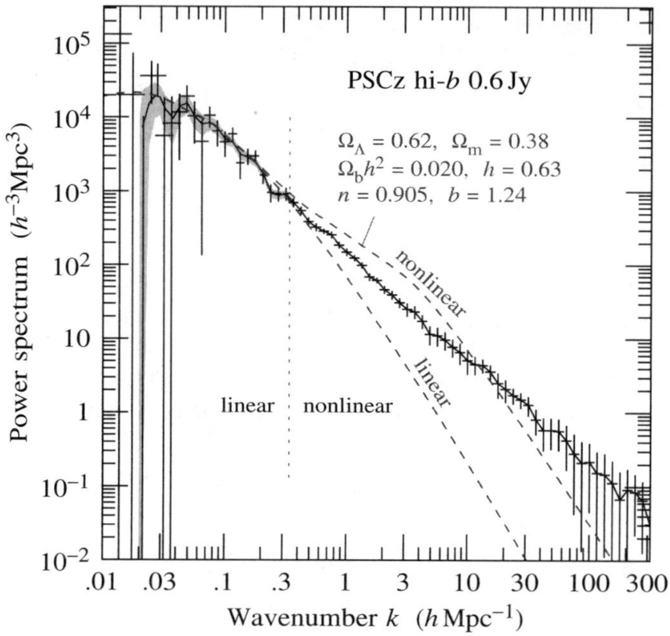

Figure 8.11 *The real space power spectrum for galaxies from the PSCz redshift survey. The spectrum is a combination of the linear calculations and the direct estimator in the nonlinear regime. The solid line shows the correlated power spectrum and the shaded region shows its rms error. Points with rms error bars show the decorrelated power spectrum in the linear wavenumber region and the correlated power spectrum in the nonlinear region. The dashed line gives the power spectrum for a CDM model with the parameters shown in the figure. (Reproduced, with permission, from Hamilton and Tegmark, Mon. Not. R. Astr. Soc., in press. Blackwell Science Ltd.)*

where $W(k, k')$ is the window function, $\mathcal{P}_l(x)$ are the Legendre polynomials of index l, and $P^s(\mathbf{k})$ is the redshift space power spectrum. The integral in (8.53) can be transformed to an integral over real space

$$P_l^s(k, \mu) = \int W_l(k, r)(2l + 1)\mathcal{P}_l(\mu)\xi^s(\mathbf{x})\, d^3x,$$

where $W_l(k, r)$ is a spherical Bessel transform of $W(k, k')$:

$$W_l(k, r) = 4\pi i^l \int_0^\infty j_l(kr)W(k, k')\frac{(k')^2 dk'}{(2\pi)^3}.$$

The redshift space covariance function $\xi^s(r,\mu)$ is computed in the plane-parallel approximation by using the Hamilton (1993) estimator

$$1+\xi^s(r,\mu)=\frac{\langle DD\rangle\langle RR\rangle}{\langle DR\rangle^2},$$

where the quantities in angle brackets are data and random reference pair counts, the latter done here as integrals over the catalog volume. For a direct method the band-power windows can be chosen by hand; Hamilton and Tegmark chose

$$W(k,k')\sim\left(\frac{k'}{k}\right)^n e^{-(k'/k)^2}.$$

These windows eliminate the mean mode problem ($W(0,k)=0$) and they are narrow for large n. The exponent chosen was $n=72$, giving the width of the window at half maximum $\Delta\ln k\approx 1/12$. As we see, quadratic data compression is done here by calculating the correlation function.

The redshift distortion problem was solved by analyzing the redshift space power spectrum. As the real space power spectrum is supposed to be isotropic, it has to coincide with the redshift space power spectrum in the transversal direction:

$$P(k)=P^s(k_{\|}=0,k_{\perp}=k), \tag{8.54}$$

where $k_{\|}$ and $k_{\perp}$ are the radial and transverse wave vectors. Hamilton and Tegmark argue that the estimate (8.54), although unbiased, is rather noisy and the use of the redshift information is equivalent to improving the spectral resolution by $R/(H_0\sigma)$ times, where R is the depth of the survey and σ^2 is the typical one-dimensional velocity variance. For the PSCz survey this ratio is about 16.

Redshift distortions can be quantified by defining the correction function

$$f(\mathbf{k})=\frac{P^s(\mathbf{k})}{P(k)}.$$

As their analysis showed that the standard pairwise distribution functions did not describe this ratio well enough, Tegmark and Hamilton (2000) approximated the correction function by a finite sum over even harmonics

$$f(\mathbf{k})=\sum_{l=0}^{l_{\max}(k)} f_l(k)P_l(\mu).$$

The real space band-powers can be obtained then as a sum of the redshift space band-power harmonics, weighted by the correction function. The maximum harmonics was chosen as the nearest even integer to $l_{\max}(k)=16k^{1/2}$ to resolve radial scales corresponding to the standard one-dimensional rms velocity error $\sigma/H_0=3h^{-1}$ Mpc.

Because the density distribution at small wavelengths is not Gaussian, we cannot use the Fisher matrix to estimate the variances of the band-power estimates.

Hamilton and Tegmark (2000) found the variances using the fluctuations of the estimates in the survey. The catalog was divided into a number of subvolumes, and the fluctuation of a band-power estimate for a subvolume i was defined as

$$\Delta\widehat{P}_i = \frac{1}{2} w_i \frac{\partial \widehat{P}}{\partial w_i},$$

where w_i is a relative weight assigned to the subvolume i. The variance of $\widehat{P}$ can now be written as a sum over pairs of subvolumes:

$$\langle \Delta\widehat{P}^2 \rangle = \sum_{ij} \Delta\widehat{P}_i \Delta\widehat{P}_j. \tag{8.55}$$

The problem with this method is that the sum of fluctuations over all subvolumes vanishes, leading to a zero variance, and we have to restrict the number of pairs in the sum (8.55). This is done by including into the sum (8.55) pairs of subvolumes with growing separation and canceling summation when the sum has attained its maximum value (Hamilton 1993b).

The resulting power spectrum is pretty close to a power law over a large wavenumber region. As a check of the methods used, the linear and nonlinear parts of the spectrum match well. The full line shows the correlated estimate of the power spectrum and the shaded region shows its rms error for linear scales. The error bars show the decorrelated spectrum for linear scales; because there is no direct analog of that for nonlinear scales, the error bars there describe the correlated spectrum. The model spectrum shown in the figure by a dashed line does not fit the observations well; the easiest way out is to postulate scale-dependent biasing. The power spectrum has two regions where errors are large. The high wavenumber region describes scales that are too close to the sizes of galaxies; a meaningful limit to the galaxy power spectra would be around wavenumbers k =30–100 $h\,\mathrm{Mpc}^{-1}$ ($\lambda \approx$30–100 h^{-1} kpc). The large errors in the small wavenumber limit are due to insufficient depth of the catalog and also show that the Fisher matrix approximation for rms errors does not work well in that region. The solution is to use there a full-blown maximum likelihood method that gets confidence regions by integrating over the parameter (band-power) space. And, as before, we see that there is not yet enough evidence for a turnover in the power spectrum. It must be rather close, because the CMB spectrum measurements show the mean power spectrum of opposite slope $P(k) \sim k$ at scales near $1,000h^{-1}$ Mpc ($k \approx 0.003h\,\mathrm{Mpc}^{-1}$).

8.6 Bispectrum

The bispectrum is identically zero for Gaussian fields. Thus a nonzero amplitude of the observed bispectrum will be a clear indication of deviations from Gaussianity. However, because many different mechanisms may generate such deviations, it is rather difficult to entangle different contributions to the bispectrum.

8.6.1 Models of the bispectrum

Although the bispectrum of a Gaussian density field is zero, gravitational dynamics, velocity distortion, and biasing all distort the field and contribute to the bispectrum. These contributions can be predicted, but only for small (linear) effects. This has led to introduction of purely phenomenological clustering models. The most popular of these models is the hierarchical model (see, e.g., Scoccimarro and Frieman 1999, where other models are also proposed). This model starts by assuming that all higher-order irreducible correlation functions can be represented by a polynomial of the two-point correlation functions. The expressions for higher-order functions are rather complicated, but for the three-point connected correlation function the clustering model predicts

$$\zeta_{123} = Q(\xi_{12}\xi_{23} + \xi_{13}\xi_{12} + \xi_{13}\xi_{23}), \tag{8.56}$$

where Q is a constant and we have used the notation $\xi_{12} \equiv \xi(\mathbf{x}_1, \mathbf{x}_2)$. This gives a model for the bispectrum:

$$B(\mathbf{k}_1, \mathbf{k}_2, \mathbf{k}_3) = Q\big[P(\mathbf{k}_1)P(\mathbf{k}_2) + P(\mathbf{k}_1)P(\mathbf{k}_3) + P(\mathbf{k}_2)P(\mathbf{k}_3)\big]. \tag{8.57}$$

Perturbation theory for gravitational dynamics (Fry 1984) predicts a similar expression in the second-order approximation (in first order the bispectrum remains zero). It gives, however, different coefficients for every term in (8.57):

$$B_{123} = Q_{12}P_1P_2 + Q_{13}P_1P_3 + Q_{23}P_2P_3,$$

where we have used a notation similar to that used for correlation functions, $P_1 \equiv P(\mathbf{k}_1)$, etc. The coefficients Q_{ij} are determined by the geometry of the triangle $\{123\}$:

$$Q_{ij} = (1+\kappa) + \cos\,\theta_{ij}\left(\frac{k_i}{k_j} + \frac{k_j}{k_i}\right) + (1-\kappa)\cos^2\theta_{ij},$$

where θ_{ij} is the angle between the two sides of the triangle with lengths k_i and k_j. The coefficient κ describes the cosmological model and varies only slightly around $\kappa = 3/7$. Velocity distortions and bias also retain the structure of this expression, changing only the coefficients Q_{ij} in a complicated way (Verde et al. 1998, Scoccimarro 2000). This leads to the definition of the reduced bispectrum

$$Q_{123} = \frac{B_{123}}{P_1P_2 + P_2P_3 + P_1P_3}. \tag{8.58}$$

While the power spectrum and the bispectrum vary strongly with a typical wavenumber, for a power-law spectrum the reduced bispectrum depends mostly on the shape on the triangle formed by the vectors $\mathbf{k}_1, \mathbf{k}_2, \mathbf{k}_3$ and its amplitude does not vary much. We show the expected bispectrum for different physical models in Fig. 8.12. It is higher for elongated triangles, depends on the assumed spectral index, depends little on the cosmological model, and velocity distortions do not change it much.

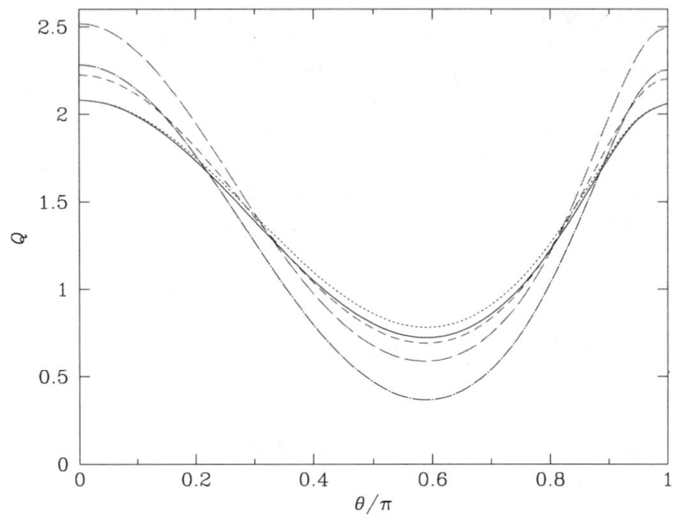

Figure 8.12 *The dependence of Q on the shape of the triangle. The power spectrum $P(k) \sim k^{-1.4}$. The triangles are described by the side ratio $k_1/k_2 = 2$ and by the angle θ between these sides (the horizontal axis). The solid curve shows the result in real space, the short-dashed curve in redshift space for $\Omega_M = 0.3$, and the long-dashed curve in redshift space for $\Omega_M = 1$. The dotted curve stands for the case $\Omega_M = 0$. The dot-dashed curve shows the effect of a (small) local bias. (Reproduced, with permission, from Scoccimarro et al. 2001, Astrophys. J., 546, 652–664. AAS.)*

The maximum effect is caused by local bias. As dynamical effects have to be calculated at least to second order, in bispectrum studies it also is customary to use a second-order bias law

$$\delta_{gal} = b_1 \delta + \frac{b_2}{2} \delta^2 .$$

The correction for bias transforms the coefficients Q to

$$Q_{gal} = \frac{Q}{b_1} + \frac{b_2}{b_1^2} .$$

The main effect is the change of the overall amplitude of the bispectrum. The dot-dashed curve in Fig. 8.12 describes the case of a rather mild (anti)bias, $b_1 = 0.8, b_2 = -0.3$.

As the galaxy bias b is usually difficult to estimate (velocity distortions give the combination $\beta = f_0/b$), determination of the bispectrum can give independent evidence on b.

8.6.2 Estimation of the bispectrum

A thorough review on the estimation of the bispectrum from observations and on the influence of different observational effects is given by Scoccimarro (2000). In principle, estimation of the bispectrum proceeds along the same lines as described above for the power spectrum.

The estimator for the Fourier amplitudes of the density field $F(\mathbf{k}_i)$ is given by (8.5). When calculating their three-point covariances we have to use the formula for the irreducible three-point density correlation function for a point process (see Bertschinger 1992):

$$E\{\delta_1\delta_2\delta_3\} = \zeta_{123} + \bar{n}^{-1}\left[\delta^D_{12}\xi_{23} + \delta^D_{23}\xi_{31} + \delta^D_{31}\xi_{12}\right] + \bar{n}^{-2}\delta^D_{12}\delta^D_{23},$$

where we have used the notation $\delta^D_{12} = \delta^D(\mathbf{x}_1 - \mathbf{x}_2)$ for brevity. This is similar to the formula for the two-point correlation (8.9) above; here there are more discreteness correction terms.

We can use a similar derivation as in Section 8.2 to get the expectation value $E\{F_1F_2F_3\}$. This, however, leads to complicated expressions. Matarrese, Verde, and Heavens (1997) derived a simpler formula, using the approximation of narrow window functions, that allows us to get the power spectra out of the integrals. In our notation of Section 8.2 it gives

$$\begin{aligned} E\{F_1F_2F_3\} &= B_{123}\int_V \psi^3(\mathbf{x})\,d^3x + (P_1 + P_2 + P_3)\int_V \frac{\psi^3(\mathbf{x})}{\bar{n}(\mathbf{x})}\,d^3x + \\ &\quad + \int_V \frac{\psi^3(\mathbf{x})}{\bar{n}^2(\mathbf{x})}\,d^3x, \end{aligned} \tag{8.59}$$

where $\psi(\mathbf{x})$ is the weight function and $\bar{n}(\mathbf{x})$ is the local mean density of the survey. The estimator for the power spectrum is in the same approximation

$$E\{|F_1|^2\} = P_1\int_V \psi^2(\mathbf{x})\,d^3x + \int_V \frac{\psi^2(\mathbf{x})}{\bar{n}(\mathbf{x})}\,d^3x. \tag{8.60}$$

Scoccimarro (2000) has shown that the optimal weight function $\psi(\mathbf{x})$ for the bispectrum is the same FKP function (8.19) that was found to be optimal for estimation of the power spectrum. Normalizing the weights for the power spectrum by

$$\int_V \psi^2(\mathbf{x})\,d^3x = 1$$

as before, we get the estimators

$$\widehat{P} = \langle |F|^2 \rangle - I_{21},$$

where the average is over a shell in the wavenumber space, and

$$\widehat{B}_{123} = \frac{\langle F_1F_2F_3 \rangle}{I_{30}} - (\widehat{P}_1 + \widehat{P}_2 + \widehat{P}_3)\frac{I_{31}}{I_{30}} - \frac{I_{32}}{I_{30}}.$$

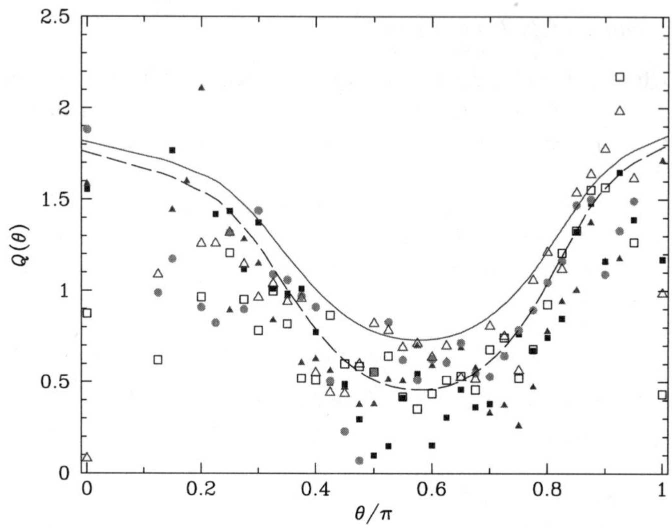

Figure 8.13 *PSCz survey bispectrum for triangles of the side ratio* $k_2/k_1 = 0.4 - 0.6$, *parameterized by the angle* θ *(horizontal axis). The solid curve is the result for all matter in a numerical* Λ*CDM model, and the dashed curve shows the model result for galaxies, obtained using the best-fit bias parameters (see text). Symbols show the observational results for triangles of different size* k_1*: filled triangles for 0.20–0.24* $h\,Mpc^{-1}$*; filled squares for 0.24–0.28* $h\,Mpc^{-1}$*; filled circles for 0.28–0.32* $h\,Mpc^{-1}$*; open circles for 0.32–0.36* $h\,Mpc^{-1}$*; and open squares for 0.36–0.42* $h\,Mpc^{-1}$*. (Reproduced, with permission, from Feldman et al. 2001, Phys. Rev. Let., 86, 1434–1437. Copyright 2001 by the American Physical Society.)*

This result was obtained first by Matarrese, Verde, and Heavens (1997). Here the assumption that the random density field is isotropic allows us to average over all orientations of the triangles of the same size and shape as $\{123\}$ in the wavenumber space. We have used the notation

$$I_{NM} = \int_V \frac{\psi^N(\mathbf{x})}{\bar{n}^M(\mathbf{x})} d^3x$$

for the normalization integrals above. The reduced bispectrum estimator is then

$$\widehat{Q}_{123} = \frac{\widehat{B}_{123}}{\widehat{P_1}\widehat{P_2} + \widehat{P_2}\widehat{P_3} + \widehat{P_1}\widehat{P_3}}. \tag{8.61}$$

While the estimators for the power spectrum and for the bispectrum are unbiased, at least in the approximations used above, the estimator for the reduced bispectrum is biased. Moreover, its one-point distribution is skewed, as found by Scoccimarro (2000), so care is needed when fitting observed estimators to models.

The bispectrum estimates are rather noisy. The reason for that is simple — as explained at the beginning of this chapter, there is a finite number of coherence cells in the volume of the wavenumber space that we can study. The estimates of the power spectra are improved by assuming isotropic spectra and averaging individual estimates over k-space shells. The parameter space for the bispectrum is much larger, we can average only over identical triangles of different orientation, and, consequently, the number of coherence cells that can be used to get a single estimate of the bispectrum is small.

In conclusion, we show in Fig. 8.13 the bispectrum of the PSCz survey, obtained by Feldman et al. (2001). Different symbols show different triangle configurations and the lines show the model results (solid line means unbiased, dashed line means the best bias fit). We see that, although the scatter is large, the reduced bispectrum is definitely larger than zero and follows approximately theoretical predictions. The estimate for the linear bias $1/b \approx 1.2$ is close to the PCSz survey bias $b \approx 0.84$ found from velocity distortions, and the fit also gives a nonlinear bias term $b_2/b_1^2 \approx -0.5$.

8.7 Low-dimensional samples

The methods described above used 3-D data samples to estimate the properties of the galaxy distributions. Galaxy surveys may frequently have lower dimensionality, either lacking the radial distances (redshifts) or restricted by the angular limits of the survey. Such surveys were frequent about 20 to 30 years ago, and they still provide important information. They fall into two different classes.

The first class is photometric surveys, where no attempt has been made to obtain redshift data. Because redshifts are much harder to measure, have rather large measurement errors, and are subject to distortions by unknown dynamical velocities, photometric surveys are more exact and usually contain a much larger number of galaxies. An example of a photometric survey that has been used extensively in recent years is the APM galaxy survey. These surveys give the angular coordinates of galaxies and their apparent magnitudes.

The second class of low-dimensional surveys includes redshift surveys of a special geometry. As the measurements of redshifts are extremely time-consuming, to get a deep survey we have to restrict its angular extent. Typical surveys of this kind are "pencil-beams," where the redshifts of all galaxies visible in a small spot on the sky are measured, and "slices," where galaxies are observed in a thin strip on the sky.

Both types of surveys can be used to estimate properties of the full three-dimensional galaxy distribution. The methods, however, differ. We shall start with photometric surveys, where we observe the projection of the spatial galaxy distribution on the sky.

8.7.1 Limber's equation

The projected density contrast $\delta(\widehat{\mathbf{r}})$ ($\widehat{\mathbf{r}}$ is the direction on the sky) is

$$\delta(\widehat{\mathbf{r}}) = \int_o^\infty \delta(\mathbf{y}) y^2 \phi(y)\, dy \tag{8.62}$$

where $\delta(\mathbf{y})$ is the real density contrast, $y = |\mathbf{y}|$ is the radial distance, and $\phi(y)$ is the selection function, normalized by $\int \phi(y) y^2 dy = 1$.

The angular correlation function w is defined similarly to the spatial two-point correlation function:

$$E\left\{\delta(\widehat{\mathbf{r}}_1)\delta(\widehat{\mathbf{r}}_2)\right\} = w(\theta),$$

depending by isotropy only on the angle θ between the two directions. Using the expression (8.62), we get

$$w(\theta) = \int\int \xi(\mathbf{y}_1, \mathbf{y}_2) y_1^2 y_2^2 \phi(y_1)\phi(y_2)\, dy_1 dy_2. \tag{8.63}$$

Two-dimensional surveys are usually deep, with the depth L much larger than the correlation radius r_0. This means that we can use the small-angle approximation:

$$r^2 = |\mathbf{y}_1 - \mathbf{y}_2|^2 = x^2 + y^2\theta^2,$$

where $x = y_1 - y_2$ and $y = (y_1 + y_2)/2$. Changing to the variables x and y in the integral (8.63) we get

$$w(\theta) = \int_0^\infty y^4 \phi^2(y)\, dy \int_0^\infty \xi\left(\sqrt{x^2 + y^2\theta^2}\right)\, dx. \tag{8.64}$$

This formula is known as Limber's equation. The integration limits are formal, reflecting the assumption that $\phi(y)$ is different from zero only in a limited interval.

This formula predicts that the angular correlation function scales with the sample depth D. Let us define this depth by $\phi(y > D) \approx 0$. The normalization condition tells us that the overall amplitude of $\phi \sim D^{-3}$. If the distribution of galaxies is homogeneous on average, the main part of the integral comes from the regions $y \approx D$ and the correlation function can be written as

$$w(\theta) \approx \frac{1}{D} W(\theta D).$$

This scaling relation has been checked in practice (see a review in Peebles 1980, § 57). It confirms the theory and also provides one of the arguments against the fractal nature of galaxy distribution at large scales. As an example, Fig. 8.14 shows the scaling of the angular correlation function of the APM galaxy survey (Baugh 1996). The different correlation functions were obtained by selecting galaxies by their apparent magnitude — the larger the apparent magnitude, the larger the mean depth of the survey. As expected, the shape of the angular correlation function remains the same, but it is shifted (by the same amount) both in the log-amplitude and in the log-argument.

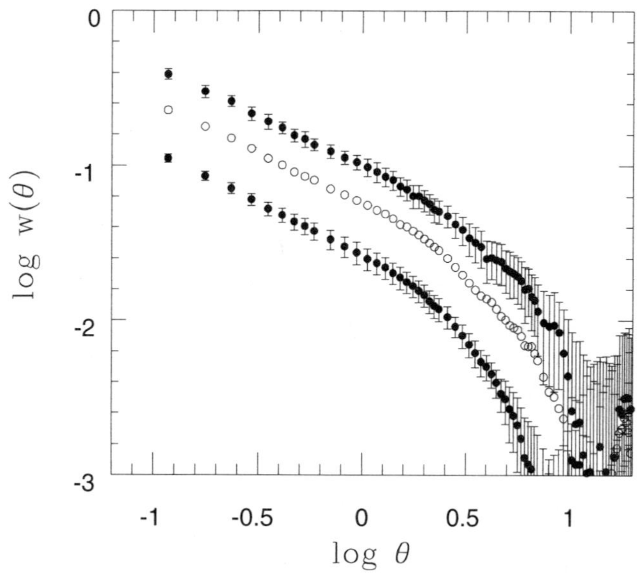

Figure 8.14 *Scaling of the angular correlation function of the APM galaxy survey. The different functions correspond to the magnitude intervals of* $17 \leq b_J \leq 18$, $17 \leq b_J \leq 19$, *and* $17 \leq b_J \leq 20$, *counting from above. For clarity, the errors of the middle curve have been omitted. (Reproduced, with permission, from Baugh 1996, Mon. Not. R. Astr. Soc., 280, 267–275. Blackwell Science Ltd.)*

As photometric surveys are usually deep enough to feel the curvature effects, the relativistic version of Limber's equation is commonly used. Let us recall the definition of the volume element from Chapter 2

$$dV = R_0^3 S_k^2(\omega)\, d\omega\, d\Omega,$$

where ω is the dimensionless comoving distance, $S_k(\omega)$ depends on the curvature of the universe, and $d\Omega$ is the angular surface element on the sky. It gives us instead of (8.63)

$$w(\omega_1, \omega_2, \theta) = R_0^6 \int\int \phi(\omega_1)\phi(\omega_2) S_k^2(\omega_1) S_k^2(\omega_2) \xi(r)\, d\omega_1 d\omega_2,$$

where r is the three-dimensional separation. In the small-angle approximation

$$r^2 = R_0^2 \left(d\omega^2 + S_k^2(\omega) d\theta^2 \right).$$

Writing the radial comoving separation as $x = R_0 \Delta\omega$ and introducing

$$y = R_0 S_k(\omega) \tag{8.65}$$

(the Mattig comoving distance), we can transform the relativistic Limber formula to a form close to the nonrelativistic one:

$$w(\theta) = \int \phi^2(y) \frac{y^4}{C_k(y)}\, dy \int \xi\left(\sqrt{x^2 + y^2\theta^2}\right) dx. \qquad (8.66)$$

All information on the cosmological model is contained in $C_k(y)$ — it is the function $dS_k(\omega)/d\omega$, where ω is replaced by y:

$$C_k(y) = \left[1 - k\frac{y}{R_0}\right]^{1/2}.$$

The integration limits depend on the curvature k of the space. For $k = 0$ and $k = -1$ the limits can be chosen as in the classical Limber's equation, but for closed models the integration limits are finite. The normalization condition for the selection function now becomes

$$\int \phi(y) \frac{y^2}{C_k(y)}\, dy = 1.$$

The formula (8.66) describes the situation when the three-dimensional separation is measured in comoving coordinates (the units of $R_0\omega$ are h^{-1} Mpc). Many papers use a slightly different form of Limber's equation, based on a different form for the space-time metric and using physical separations:

$$w(\theta) = \int \phi^2(y) a^6(y) \frac{y^4}{F^2(y)}\, dy \int \xi\left(\sqrt{x^2/F^2(y) + y^2\theta^2}\right) dx,$$

where $\phi(y)$ is normalized by

$$\int \phi(y) a^3(y) \frac{y^2}{F(y)}\, dy = 1$$

and $F(y)$ describes the curvature effects (see Peebles 1980, § 50, for exact expressions).

8.7.2 *Evolution of correlations*

Curvature effects are easy to include if we know the cosmological parameters. A realistic model should also include evolution effects. As the structure in the universe evolves from small perturbations to its present state, the spatial correlation function depends on the lookback time, and thus on the distance argument y. This dependence is known only approximately, so it is usually included in a simple parametric form.

There are two limiting cases that we can use. During the linear stage of evolution of structure, the density contrast evolves according to $\delta \sim D_1(z)$, where $D_1(z)$ is the growing mode of linear perturbations. Because the correlation function is quadratic in the contrast, we can write

$$\xi(r, z)_{\text{lin}} \sim D_1^2(z) \approx (1 + z)^{-2},$$

where the last equality is true for the case $\Omega_M = 1, \Omega_\Lambda = 0$. Another limiting case is the final stage of the evolution of structure, when most galaxies are collected into virialized clusters. The form of the correlation function is then defined by the density profile of the clusters that is constant in physical coordinates. Approximating the correlation function by its present form $\xi(r) \sim (ar)^{-\gamma}$ (r is the comoving distance), its dependence on redshift becomes

$$\xi(r, z) \sim (1 + z)^{\gamma - 3}.$$

The factor -3 in the exponent arises from the normalization of the density contrast by the mean density $\bar{\rho} \sim (1 + z)^3$. This case is called "stable clustering" and that assumption has been used frequently to describe the evolution of correlations.

Elaborate machinery has been developed to approximate the evolution of the correlation function (and the power spectrum) in the intermediate regime between the two asymptotes. This is known as the HKLM procedure after its authors (Hamilton et al. 1991). The idea of this procedure is to represent the nonlinear correlation function as a universal function of the linear correlation amplitudes for rescaled arguments. Peacock and Dodds (1996) extended this procedure to predict the nonlinear power spectrum on the basis of the linear power spectrum. As the HKLM procedure is analytical, it is used rather frequently to predict nonlinear power spectra for large parameter spaces of cosmological models, where direct N-body modeling would not be feasible. The HKLM procedure is explained in detail in Peacock's book (1999).

The selection function $\phi(y)$ in Limber's equation (8.66) is usually estimated by noting that

$$\phi(y) \frac{y^2}{C_k(y)} \, dy = \frac{dN}{dz} \, dz, \tag{8.67}$$

where dN/dz is the observed redshift distribution of galaxies (normalized to unity). This distribution can be estimated, e.g., by pencil-beam surveys of the same depth.

Limber's formula can be inverted using Mellin's transform (see Peebles 1980, § 53). Later work has used Lucy's algorithm (Lucy 1974; we describe it below) to numerically invert Limber's formula. This was the approach used by Baugh (1996) to find the spatial correlation function of the APM galaxy survey. He calculated the spatial correlation function for the scales up to $50h^{-1}$ Mpc, and showed also that the correlation function does not differ appreciably from the present one. Since the mean redshift of the APM survey is about 0.15, depending on the magnitude limit, we cannot expect strong evolution effects.

Another possibility to apply Limber's formula is to approximate the spatial correlation function by a power law

$$\xi(r,z) = \left(\frac{r}{r_0}\right)^{-\gamma} f(z),$$

as we did above when discussing evolution effects. Then the angular correlation function is

$$w(\theta) = A\theta^{1-\gamma}, \tag{8.68}$$

where the coefficient A depends on the galaxy redshift distribution, on the cosmological parameters, and on the assumption about the evolution of the spatial correlation function. Finding the amplitude and the exponent of the observed angular correlation function, we can solve (8.68) to find γ and r_0.

As the exponent of the correlation function does not depend in this approximation on any unknown parameters, it is a useful probe of the shape of the spatial correlation function in the past (at large redshifts). Giavalisco et al. (1998) applied this technique to a sample of distant Lyman-break galaxies at $z \approx 3$ and found that their spatial correlation function has about the same slope as that of the present-time galaxies. As their sample spans a relatively small redshift interval, they also could determine the correlation length without making any assumptions about the evolution of the correlation function. Their results show that the correlation function does not change appreciably with redshift.

8.7.3 Power spectra

Using Limber's formula, we also can derive the relation between the angular correlation function and the power spectrum of density fluctuations. Substituting the correlation function in (8.64) by

$$\xi(r) = 4\pi \int_0^\infty P_3(k) \frac{\sin(kr)}{kr} \frac{k^2\,dk}{(2\pi)^3},$$

we can integrate over x and get

$$w(\theta) = 2\pi \int_0^\infty \phi^2(y) y^4 dy \int_0^\infty P_3(k) J_0(ky\theta) \frac{k\,dk}{(2\pi)^2}, \tag{8.69}$$

where $J_0(x)$ is a Bessel function. We shall denote the power spectrum of the three-dimensional galaxy distribution in this section as $P_3(k)$ to distinguish it from the power spectrum of the projected distribution $P_2(K)$. Comparing this formula with the relation between the angular correlation function and the two-dimensional power spectrum

$$\omega(\theta) = 2\pi \int_0^\infty P_2(K) J_0(K\theta) \frac{K\,dK}{(2\pi)^2},$$

where K is the two-dimensional wavenumber, we get

$$P_2(K) = \int_0^\infty P_3\left(\frac{K}{y}\right)\phi^2(y)y^2dy. \tag{8.70}$$

The spectrum $P_2(K)$ describes the amplitudes of a plane-wave decomposition of the density distribution, so the above formulae are valid for the small-angle approximation only. The corresponding relativistic formulae include the factor $C_k^{-1}(y)$ as above. The formula (8.70) can also be used to derive the scaling law for the two-dimensional spectrum. Proceeding as we did in case of the angular correlation function above, we find

$$P_2(K) \approx \frac{1}{D^3}p(K/D),$$

where $p(x)$ is a universal function.

Baugh and Efstathiou (1993, 1994) have used the above formulae to estimate the power spectrum of the APM galaxy survey from the angular correlation function. Using the selection function from (8.67), we can write

$$w(\theta) = \int_0^\infty G_w(\theta, k)P_3(k)k\,dk, \tag{8.71}$$

where the kernel $G_w(\theta, k)$ is given by

$$G_w(\theta, k) = \frac{1}{2\pi R_0}\int_0^\infty \left(\frac{dN}{dz}\right)^2 (1+z)^{-\alpha} J_0\big(ky(z)\theta\big)\left(\frac{d\omega}{dz}\right)^{-1} dz.$$

The second factor in the integrand describes the evolution of the spatial correlation function. The functions $\omega(z)$, $y(z) = R_0 S_k\left(\omega(z)\right)$ and the curvature radius R_0 depend on the parameters of the cosmological model. The factor $C_k(y)$ is canceled by a similar factor from the derivative dy/dz. The power spectrum of galaxy distribution can be found now from (8.71) by deconvolving the angular correlation function. The kernel $G_w(\theta, k)$ is rather wide and oscillating, so deconvolution is not easy.

The formula for the two-dimensional spectrum reduces in this approach to

$$KP_2(K) = \int_0^\infty G_P(K/k)P_3(k)\,dk, \tag{8.72}$$

where the kernel $G_P(y)$ is

$$G_P(y) = \left(\frac{dN}{dz}\right)^2 (1+z)^{-\alpha} C_k(y)\left(\frac{dz}{dy}\right)^2$$

and $y(z)$ is given by (8.65) above.

Baugh and Efstathiou (1994) found the two-dimensional power spectrum of the APM galaxy survey by a direct method. The three-dimensional power spectrum was obtained by deconvolution. The kernel $G_P(y)$ is also rather wide, but positive, which makes deconvolution easier.

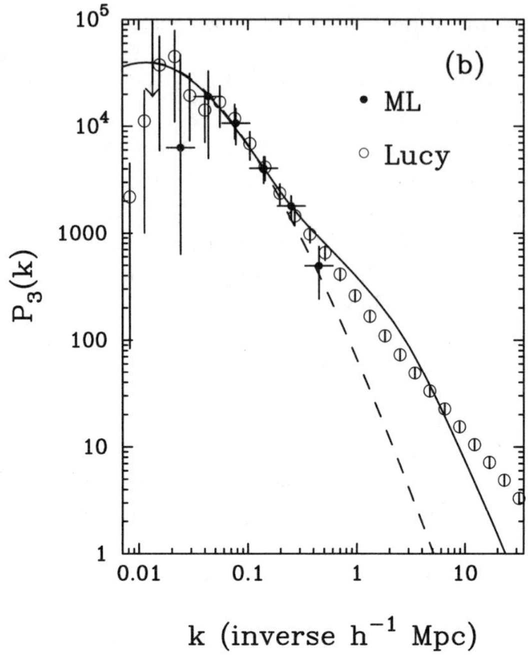

Figure 8.15 *Power spectrum from the APM galaxy survey. Filled circles show the maximum likelihood estimates, and open circles show the spectrum obtained by deconvolution of the two-dimensional power spectrum. Lines show model predictions. (Reproduced, with permission, from Efstathiou and Moody 2001, Mon. Not. R. Astr. Soc., 325, 1603–1615. Blackwell Science Ltd.)*

Baugh and Efstathiou (1993, 1994) determined the scatter of the power spectrum estimates by breaking the survey into four distinct regions. Both methods gave a similar result that was long considered the most reliable power spectrum estimate. The size of the APM survey was much larger than that of any redshift surveys available, and this estimate does not depend on the treatment of redshift distortions. In a recent paper, however, Efstathiou and Moody (2001) re-estimated the APM power spectrum by applying the maximum likelihood method to the APM pixel map. The results are compared in Fig. 8.15. The maximum-likelihood results are shown by the filled circles and the results of the previous deconvolution are shown by the open circles. While the results agree rather well, the clear maximum around $k \approx 0.02h\,\mathrm{Mpc}^{-1}$ that is seen in the sequence of open circles is not confirmed by the ML method (the ML point at $k \approx 0.013h\,\mathrm{Mpc}^{-1}$ is outside the graph). This shows once more that direct methods do not give reliable results for scales close to the size of the survey volume.

8.7.4 Lucy deconvolution

Baugh and Efstathiou chose the Lucy deconvolution method (Lucy 1974) to get the power spectrum. Because this method has a probabilistic motivation and can be useful in further applications, we shall describe it here.

If we have to convert the integral equation

$$f(x) = \int K(x, \chi) g(\chi)\, d\chi, \tag{8.73}$$

where $f(x)$ and $K(x, \chi)$ are known, we can rescale all these functions first to have a unit integral:

$$\int f(x)\, dx = \int g(\chi)\, d\chi = \int K(x, \chi)\, dx = 1.$$

Now we can think of the functions $f(x)$ and $g(\chi)$ as probability densities for x and χ, and, similarly, suppose that $K(x, \chi) = K(x|\chi)$, the conditional probability density for x, given χ. Then we can write

$$g(\chi) = \int L(\chi|x) f(x)\, dx, \tag{8.74}$$

where $L(\chi|x)$ is the conditional probability density for χ, given x. This density can be obtained from Bayes theorem:

$$L(\chi|x) = \frac{g(\chi) K(x|\chi)}{f(x)}. \tag{8.75}$$

Lucy (1974) showed that we can now proceed by iterations, choosing $g(\chi)$ first to be a reasonable smooth function (Baugh and Efstathiou chose a power law for $P_3(k)$). Denoting the first choice $g_1(\chi)$, we find $f_1(x)$ from (8.73) and $L_1(\chi|x)$ from (8.75). Using that in (8.74) we get the next approximation $g_2(\chi)$. This process leads to the rule

$$g_{i+1}(\chi) = g_i(\chi) \int \frac{f(x)}{f_i(x)} K(x|\chi)\, dx,$$

where $f_i(x)$ and $g_i(\chi)$ are connected by (8.73). Lucy (1974) analyzes the stopping criterion and shows that the best results are usually obtained after a few iterations; further iterations amplify small-scale features in $g(\chi)$ that are caused by (frequently random) fluctuations in $f(x)$.

8.7.5 Wide-angle surveys

The above formulae were all written in the small-angle approximation (for flat regions in the sky). The angular extent of the APM survey is about 120°, so this approximation needs improvement. We shall give a brief outline of the exact formulae here and refer for a full treatment to Peebles (1973). Please note that in this paper Peebles uses different Fourier conventions and also treats point source samples.

For a wide-angle survey we can expand the density contrast as

$$\delta(\widehat{\mathbf{r}}) = \sum_{l=0}^{\infty} \sum_{m=-l}^{l} a_l^m Y_{lm}(\widehat{\mathbf{r}}),$$

where $\widehat{\mathbf{r}}$ is the unit vector of the direction in space and $Y_{lm}(\theta, \phi)$ are spherical harmonics. Spherical harmonics (see Abramowicz and Stegun 1965, Press et al. 1992) are defined as

$$Y_{lm}(\theta, \phi) = \sqrt{\frac{2l+1}{4\pi} \frac{(l-m)!}{(l+m)!}} P_l^m(\cos\theta) e^{im\phi}, \tag{8.76}$$

where $P_l^m(x)$ are the associated Legendre polynomials. As spherical harmonics form an orthonormal basis, the coefficients a_l^m can be found by

$$a_l^m = \int_0^{2\pi} d\phi \int_0^{\pi} \sin\theta \, d\theta \delta(\theta, \phi) Y_{lm}^{\star}(\theta, \phi)$$

(θ is the zenith distance, $z = \cos\theta$). As we see, the coefficients a_l^m are the amplitudes of the power spectrum. The quadratic statistics (correlation function and power spectrum) are related by

$$w(\theta) = \frac{1}{4\pi} \sum_{l=0}^{\infty} \sum_{m=-l}^{l} E\left\{|a_l^m|^2\right\} P_l(\cos\theta), \tag{8.77}$$

$$E\left\{|a_l^m|^2\right\} = 2\pi \int_{-1}^{1} w(\theta) P_l(\cos\theta) \, d\cos\theta. \tag{8.78}$$

The functions $P_l(x)$ are the Legendre polynomials; note that $|a_l^m|^2$ does not depend on m.

The exact Limber's equation does not need spherical harmonics; the nonrelativistic version reads

$$w(\theta) = \int_0^{\infty} \int_0^{\infty} \phi(s)\phi(t) s^2 t^2 \xi(r) \, ds \, dt,$$

and r has to be found from the exact formula

$$r^2 = s^2 + t^2 - 2st\cos\theta.$$

The selection function is normalized as above, and relativistic generalization proceeds in the same fashion.

The relation between the power spectra can be written as (Peacock 1999):

$$E\left\{|a_l^m|^2\right\} = \int_0^{\infty} G_l^2(k) P(k) \frac{k^2 \, dk}{(2\pi)^3},$$

where the (square root of) the weight function $G_l(k)$ is

$$G_l(k) = 4\pi \int_0^\infty j_l(ky)\phi(y)y^2 dy,$$

and $j_l(x)$ is the Bessel function.

This formalism was used by Hauser and Peebles (1973) for the first power spectrum estimation in cosmology; they applied it to Abell clusters of galaxies.

8.7.6 Pencil-beams and slices

One of the most intriguing results on the large-scale distribution of galaxies was obtained by a pencil-beam survey. Broadhurst et al. (1990) combined the results of deep pencil-beam surveys, one toward the north galactic pole, another toward the south galactic pole. The northern beam consisted of several subbeams of width $5'$ inside a region of width $40'$; the southern data came from a single beam of width $20'$. The depth of both beams was about 1,200 h^{-1} Mpc. The binned radial distribution of galaxies along the combined beam and its one-dimensional power spectrum are shown in Fig. 8.16. The distribution exhibits strong periodicity with a period of $\sim 130h^{-1}$ Mpc and the power spectrum has a corresponding strong maximum.

This result provoked a strong interest in the problem of how well such low-dimensional surveys describe the full spatial galaxy distribution. We saw above that projected surveys retain most of the information; is that also true for "cutouts"?

It is not difficult to derive the relationship between the power spectra if we suppose that the survey is strictly onedimensional. Let us align the pencil-beam with the z-axis; it measures then the density contrast $\delta_1(z) = \delta(0,0,z)$. The power spectrum of this one-dimensional field is

$$P_1(k) = \int_{-\infty}^{\infty} dk_x \int_{-\infty}^{\infty} dk_y P_3(k_x, k_y, k).$$

Changing to polar coordinates $(dk_x, dk_y \to dr, d\phi)$ we get

$$P_1(k) = \frac{1}{(2\pi)^2} \int_0^{2\pi} d\phi \int_0^\infty P_3(\sqrt{k^2+r^2})r\,dr = \frac{2}{\pi}\int_k^\infty P_3(y)y\,dy, \quad (8.79)$$

where the last integral comes from the change of integration variable to $y^2 = k^2 + r^2$.

We see that the one-dimensional power spectrum at a certain wavenumber k collects contributions from all higher wavenumbers (smaller scales). We also see that the one-dimensional power spectrum is a monotonic function of the wavenumber, as

$$\frac{dP_1(k)}{dk} = -\frac{2}{\pi}kP_3(k).$$

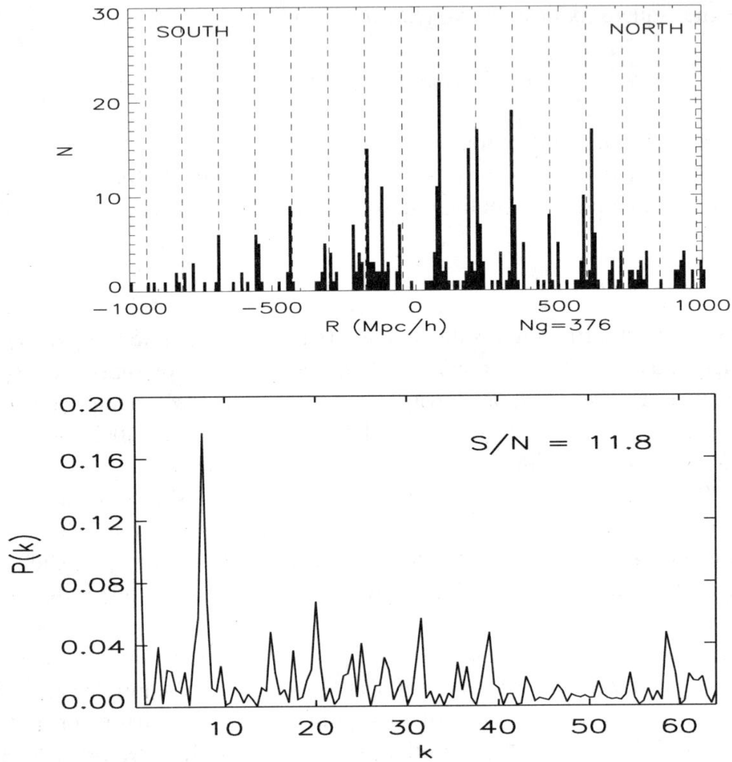

Figure 8.16 *Distribution of galaxies in the Broadhurst et al. (1990) pencil-beam survey. The upper panel shows the binned distribution along the beam; dashed lines mark the* 130 h^{-1} *Mpc periodicity. The lower panel shows the one-dimensional power spectrum of the survey. (Reproduced, with permission, from Yoshida et al. 2001, Mon. Not. R. Astr. Soc., 325, 803–816. Blackwell Science Ltd.)*

When the usual spatial power spectrum $P(k) \to 0$ when $k \to 0$, the one-dimensional power spectrum for large scales approaches a constant value. This has been interpreted as describing a Poisson distribution of small-scale clumps (Kaiser and Peacock 1991).

A real pencil-beam has a finite size on the sky and finite length. To account for that, the formula (8.79) has to be modified:

$$P_1(k) = \int |\widetilde{W}_z|^2(k_z - k)|\widetilde{W}|_r^2(k_x, k_y)P_3(k)\,\frac{d^3k}{(2\pi)^3},$$

where $\widetilde{W}_z(k)$ and $\widetilde{W}(k_x, k_y)$ are the Fourier transforms of the selection functions along the beam and for the transverse directions, respectively. If we approximate the beam by a cylinder of radius R and length $L \ll R$, the formula becomes

$$P_1(k) = \frac{1}{2\pi} \int_k^\infty |\widetilde{W}|^2(\sqrt{y^2 - k^2}) P_3(y) y \, dy,$$

where $\widetilde{W}(k) = 2J_1(kR)/(kR)$ ($J_1(x)$ is the Bessel function of order 1). The spectral window is a disk with the width $\Delta k_r \approx 1/R$ The thickness of the window function along the beam, which was ignored in the approximation above, is $\Delta k_z \approx 1/L$. For the Broadhurst survey the ratio of these dimensions is $\Delta k_r / \Delta k_z \approx 100$.

A pencil-beam is really a cone; Kaiser and Peacock (1991) derive the exact formulae for that case, but the results do not differ much.

Thus, the spectral resolution of a pencil-beam survey is determined by its transverse dimensions and not by its depth. Based on that, most of the papers devoted to the analysis of the Broadhurst et al. (1990) data discarded the result as a statistical fluke, caused by the high-frequency modes. The case is open, however, because the periodogram analysis of the distribution of distances between distinct clumps of galaxies showed that it is too regular to be explained by random models (Dekel et al. 1992). Extensive modeling of pencil-beams by numerical simulations also has been unable to reproduce the regularity at that level (see Yoshida et al. 2000).

The case of the two-dimensional slices is similar to that above. The power spectrum for an infinitely thin slice is

$$P_2(k) = \frac{1}{2\pi} \int_{-\infty}^\infty P(\sqrt{k^2 + k_z^2}) \, dk_z = \frac{1}{\pi} \int_k^\infty P(y) \frac{y}{\sqrt{y^2 - k^2}} \, dy.$$

The finite thickness of a slice introduces a window $|\widetilde{W}(k)|^2$ in the direction perpendicular to the slice; for a constant density slab with a thickness $2D$, the transform of the selection function is $\widetilde{W}(k) = \sin(kL)/(kL)$.

The three-dimensional windows for a slab (slice) resemble long needles and the spectral resolution is again limited by the smallest dimension of the survey region. Because of that, the power spectra of low-dimensional samples are severely distorted by the leakage from the high-wavenumber regions. The correlation functions of pencil-beams and slices differ from the spatial correlation functions only by greater large-scale noise. This illustrates clearly the basic difference between the power spectrum and the correlation function — while the power spectrum separates different scales (as well as possible), correlation functions mix contributions over all scales.

However, if we relax the assumptions about isotropy, then the pencil-beams and slices can help tremendously in measuring the anisotropic features of galaxy distributions.

CHAPTER 9

Cosmography

9.1 Introduction

The basic statistical model used for the description of the spatial distribution of galaxies is the inhomogeneous Poisson point process, where the intensity function is determined by a realization of a Gaussian random field. The goal of cosmography (cosmic cartography) is to study this realization. The motivation is, first, to see and understand the universe around us. Second, the better we map our local neighborhood, the better we are able to answer the questions about its statistical properties and its past.

As usual, we start with frequentist methods. The simplest methods of density estimation are to use counts-in-cells or kernel estimators on the observed survey data. As the density of galaxies varies greatly, we could use adaptive kernels (Silverman 1986) or wavelets. Adaptive density estimation methods are usually used for discriminating structures (groups and clusters of galaxies) in redshift space (see, e.g., Pisani 1996).

If we wish to find the local intensity in real space, we have to worry about velocity distortions, as the survey data is usually obtained for redshift space. We have seen above that the differences could be of the same order as the densities itself, or even much more — recall the finger-of-God effects. While this distortion is better to quantify analytically when we determine power spectra, for map-making the first goal would be to get rid of these fingers. In our immediate neighborhood, for example, rich clusters of galaxies distort severely the galaxy maps (see Fig. 8.4, the "Slice of the Universe," where the the stick-man really does not have a body; it is the finger-of-God of the Coma cluster).

This subject has not raised much interest in recent years. The first attempt to correct for these fingers was made by Gramann, Cen, and Gott (1994). They found the cluster fingers by a clustering algorithm (usually called "friends-of-friends" in cosmology), where the neighbor distance was scaled differently in the radial and transversal directions. This scaling can be determined from the observed density profiles of clusters of galaxies in redshift space; the scaling ratios usually are about 6–10. After identifying the cluster fingers, they collapsed those to disks at the median redshift of the cluster. They were not interested in cartography; a better way was to compress them to spherical distributions, as was done later by Monaco and Efstathiou (1999) and Monaco et al. (2000) for the PSCz reconstruction. Of course, because masses and mean densities of galaxy clusters vary, the scaling ratios vary, too. The compression of cluster fingers really needs an application of adaptive matching filters, as proposed for automatic identification of galaxy clusters from photometric and redshift surveys (Kepner et al. 1999). The idea is to fit parametric cluster models over the whole survey volume and analyze the peaks of the resulting likelihood map. This demands serious computer resources and is only now becoming feasible.

Another problem with direct density estimation is that of bias — we cannot be sure that the galaxy distribution traces the total distribution of matter. Fortunately, as gravitational forces are universal for all types of objects, there are ways to map the total density.

9.2 Potent method

Although the main distance measurement method by far is by galaxy redshifts, there exist independent methods of distance determination. Such methods are based on the "distance ladder" described in Chapter 1, and they are very labor-intensive and rather inexact, the best "instrumental errors" being about 20%. Thus, these distance estimators cannot be used directly to estimate densities of large-scale structures. However, the existence of independent distance estimators allows us to delineate the peculiar velocity fields. The first attempt was made by Lynden-Bell et al. (1988). They estimated the distances and peculiar velocities for 400 elliptical galaxies and discovered that there exists a mean flow toward a region behind the galactic plane that they called "the Great Attractor." This region had not been detected before and the velocity fields were useful in describing the local density distribution.

This result motivated Bertschinger and Dekel (1989) to propose a method to recover the full velocity field. They used the fact that in the linear approximation of structure evolution the velocity field is irrotational,

$$\mathbf{v}(\mathbf{x}) = \nabla\Psi(\mathbf{x}), \tag{9.1}$$

where $\Psi(\mathbf{x})$ is the velocity potential. Thus, the observed radial velocity $v_r(\mathbf{x})$ contains the same amount of information as the velocity potential, and the potential can be recovered by integrating the radial velocity along radial rays from the observer,

$$\Psi(r,\theta,\phi) = \int_0^r v_r(r',\theta,\phi)\,dr'. \tag{9.2}$$

Once we have the potential, we can find the full velocity field by (9.1). In the linear approximation, the continuity equation gives

$$\delta = -\frac{1}{f}\nabla\cdot\mathbf{v} \tag{9.3}$$

($f \approx \Omega_M^{0.6}$ is the nondimensional density growth rate), so we can find the local density distribution. The important point here is that δ in (9.3) describes the full density distribution, dark matter plus galaxies, and thus it is free of assumptions about biasing.

The practical realization of the method, called "POTENT," was proposed by Dekel and Bertschinger (1990). Although the method is simple in principle, several sources of possible error were found during its application in later years. The second version of POTENT (Dekel et al. 1999) has been designed to eliminate these errors. The steps of POTENT are as follows:

1. Correction of observed peculiar velocities. As the independent distance determinations reduce to the estimation of the absolute luminosities of galaxies, these distances suffer from the Malmquist bias, and must be corrected. The types of Malmquist bias depend on the distance determination method. A thorough review of different types of Malmquist bias for peculiar velocity catalogs and correction methods is given in Strauss and Willick (1995). Correction of distances implies correction of peculiar velocities.

2. Smoothing of the observed radial velocity field. This has to be done for two reasons — first, the formulae written above are based on the linear approximation of the evolution of structure that is applicable only for comparatively large scales, at about a few tens of Mpc and larger. Second, the peculiar velocity surveys are sparse. The largest survey done thus far is the MARK III survey (Willick et al. 1997), which includes about 3,300 galaxies in a volume of a radius of about 90 h^{-1} Mpc. So, in order to be able to integrate along radial rays we have to interpolate the radial velocity field first over the survey volume. The smoothing kernel chosen is a spherical Gaussian, and due to sparsity of data the width of the window R_G is typically chosen at about 10–12 h^{-1} Mpc.

   Because the radial directions in neighboring regions differ, the interpolation has to use tensor weights and has to assume a velocity model. Typical models used are a constant flow near a point $\mathbf{v}(\mathbf{x}) = \mathbf{B}$, $|\mathbf{x} - \mathbf{x}_i| \leq R_G$ or a linear shear field $\mathbf{v}(\mathbf{x}) = \mathbf{B} + \mathbf{T}(\mathbf{x} - \mathbf{x}_i)$, where $\mathbf{T}$ has to be a symmetric matrix to ensure that the flow is irrotational. The combination of tensor weights with a spherical window introduces a bias called "the tensor window bias." Using higher-order models can reduce that bias.

   As the mean radial velocity changes through the window, nonuniform sampling gives rise to a bias that is called "the sampling-gradient bias." It can be reduced by using volume-weighting recipes for interpolation instead of the straightforward galaxy weighting.

   The velocities have also to be weighted to take account of the random distance measurement errors.

3. Integrating the interpolated velocity field to recover the velocity potential and the three-dimensional velocity field.

4. Estimating the density field. The formula (9.3) is only true for small values of density contrast δ and we can improve on that.

The only assumption we have used in recovering the velocity field thus far is the existence of the velocity potential. Recall that in the Zeldovich approximation the flow of matter stays irrotational until the first crossings, up to infinite density. Thus, in this step methods are used that also work in mildly nonlinear regimes to improve the dynamic range of the prediction.

There are several higher-order approximations that describe the structure evolution. The POTENT choice is the Eulerian version of the Zeldovich approximation (Nusser et al. 1991). The Zeldovich approximation predicts present-time Eulerian positions of mass elements (objects) as a function of their Lagrangian coordinates:

$$\mathbf{x}(\mathbf{q}) = \mathbf{q} + \frac{1}{f(\Omega)}\mathbf{v}(\mathbf{q}), \tag{9.4}$$

where the function f is the dimensionless growth rate of perturbations. Assuming that there have been no flow crossings, the mapping (9.4) is unique and can be inverted:

$$\mathbf{q}(\mathbf{x}) = \mathbf{x} - f^{-1}\mathbf{v}\left[\mathbf{q}(\mathbf{x})\right]. \tag{9.5}$$

The mass conservation law

$$\rho_x d^3x = \rho_q d^3q = \bar{\rho} d^3q$$

gives the expression for the Eulerian density contrast

$$\delta(\mathbf{x}) = \left| I - f^{-1}\frac{\partial \mathbf{v}}{\partial \mathbf{x}} \right|. \tag{9.6}$$

This (rather crude) approximation works better than the higher-order expressions. For POTENT use, the Jacobian in (9.6) is modified by changing slightly the coefficients of powers of f (Dekel et al. 1999).

5. As the errors of this complex analysis are difficult to estimate, the last step includes evaluation of the errors, using the POTENT method on mock N-body catalogs to find the influence of observational effects.

The results of the POTENT analysis provide an independent measure of the total mass density and thus help to clarify the nature of the galaxy bias. Similarly, POTENT gives also a direct way to estimate Ω_M (recall that the dependence of f on Ω_Λ is small; see Chapter 6).

The density and velocity maps of our local neighborhood obtained by POTENT are shown in Fig. 9.1. Due to the wide filter (the rms width is $R_G = 12\,h^{-1}$ Mpc) the map is rather smooth. The peculiar velocity field shows a smooth flow over the region converging at the Great Attractor. The relief map in the lower panel shows the main features of the local density distribution — the Great Attractor (GA); the Great Wall (GW) that was first seen in the "Slice of the Universe"; the Perseus–Pisces supercluster (PP) and its extension, the Southern Wall (SW); the Local Group of galaxies (LG) in the center and the Sculptor Void (Void). The Local Supercluster itself is the low shoulder to the left of the Local Group.

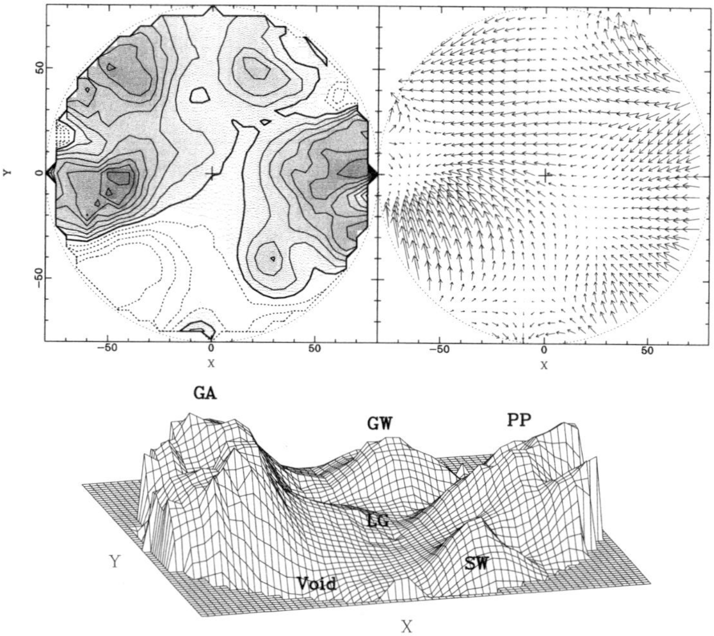

Figure 9.1 *POTENT maps of the supergalactic plane from the Mark III peculiar velocity catalog. The upper left panel shows the density contrast: the heavy contour corresponds to* $\delta = 0$*, darker areas have larger densities, and the contour spacing is* $\Delta\delta = 0.2$*. The upper right panel shows the projections of the peculiar velocities on the supergalactic plane. Distances are in units of* h^{-1} *Mpc. The lower panel shows the relief of the density contrast. (Reproduced, with permission, from Dekel et al. 1999, Astrophys. J., 522, 1–38. AAS.)*

This distribution is similar to the reconstruction of the local neighborhood obtained by different methods, showing that gravitation is really the main force behind the evolution of structure. Comparison of reconstructions on density–density or velocity–velocity basis helps clarify further our picture of that evolution.

There are several reconstruction and velocity comparison methods similar to POTENT; a review of these methods is given by Dekel (1997). These methods can be roughly divided into two classes. The first class is represented by POTENT, which attempts to reconstruct the velocity and density fields in real space. Comparison of the reconstructed density field with density fields found from redshift surveys allows us to determine the parameter $\beta = bf(\Omega_M)$ (b is the bias and $f(\Omega_M)$ is the dimensionless density growth rate). These methods are called the "density–density" (d-d) comparison methods.

The second class contains the methods that attempt to reconstruct the observed velocity fields in redshift space and to compare them with the velocity fields found from redshift surveys. The best known of these methods is the VELMOD maximum likelihood method (Willick et al. 1997), which reconstructs velocity fields by maximizing the probability of observing the peculiar velocity indicators (apparent magnitude and velocity width) of the sample galaxies, given their redshifts. Because in this case β is estimated by comparing velocity fields, these methods are called the "velocity–velocity" (v-v) comparison methods. Thus far, the value of β found by (d-d) methods is about twice the value found by the (v-v) methods; the reason for that is not yet known. The latest velocity–velocity analysis by Silberman et al. (2001) gives the estimates of Ω_M that are close to the estimates found from the analysis of the CMB data and of the deep redshift surveys.

9.3 Wiener filtering

The main effort in the application of POTENT is correction for various systematic errors. These errors are caused, first, by large intrinsic errors of the distance determination methods, but they also are due to a very sparse sampling of the survey volume. An alternative method that treats the errors rigorously is Wiener filtering. A general review of this subject is given by Rybicki and Press (1992); the description of its application to galaxy surveys is provided by Zaroubi et al. (1995).

Let us describe our survey data as a vector $\mathbf{d}$ of length N and the signal we wish to estimate as a vector s of length M, and consider the case where the mapping between data and signal is linear:

$$\mathbf{d} = \mathbf{R}\mathbf{s} + \mathbf{n}, \tag{9.7}$$

where $\mathbf{R}$ is the response function (a $N \times M$ matrix) and $\mathbf{n}$ is the noise vector. This formula covers both popular estimates — estimation of local intensities from galaxy positions and from peculiar velocity surveys. Direct inversion of relation (9.7) is either impossible, when $N < M$, or noisy, so we proceed to define a linear estimator for s:

$$\widehat{\mathbf{s}} = \mathbf{F}\mathbf{d}$$

and demand that it be optimal in the sense of minimizing the variance of the discrepancy

$$\mathbf{r} = \mathbf{s} - \widehat{\mathbf{s}},$$

where s is the true value of the signal vector. The variance we have to minimize is

$$\langle \mathbf{r}\mathbf{r}^\dagger \rangle = \left\langle (\mathbf{s} - \mathbf{F}\mathbf{d})(\mathbf{s}^\dagger - \mathbf{d}^\dagger\mathbf{F}^\dagger) \right\rangle,$$

where the angular brackets denote averages over realizations. As the data and signal vectors also may describe the Fourier amplitudes, we have to use Hermitian conjugation, denoted by a dagger (in the case of real matrices that will reduce to transposition). Adding to both sides of the above expression the term $\langle \mathbf{sd}^\dagger \rangle \langle \mathbf{dd}^\dagger \rangle^{-1} \langle \mathbf{ds}^\dagger \rangle$ (completing the square) we get

$$\begin{aligned} \langle \mathbf{rr}^\dagger \rangle &= \left(\mathbf{F} - \langle \mathbf{sd}^\dagger \rangle \langle \mathbf{dd}^\dagger \rangle^{-1}\right) \langle \mathbf{dd}^\dagger \rangle \left(\mathbf{F} - \langle \mathbf{sd}^\dagger \rangle \langle \mathbf{dd}^\dagger \rangle^{-1}\right)^\dagger \\ &\quad - \langle \mathbf{sd}^\dagger \rangle \langle \mathbf{dd}^\dagger \rangle^{-1} \langle \mathbf{ds}^\dagger \rangle + \langle \mathbf{ss}^\dagger \rangle. \end{aligned}$$

Minimization of this expression with respect to $\mathbf{F}$ gives

$$\mathbf{F} = \langle \mathbf{sd}^\dagger \rangle \langle \mathbf{dd}^\dagger \rangle^{-1}. \tag{9.8}$$

The second term above can be rewritten as

$$\langle \mathbf{dd}^\dagger \rangle = \mathbf{D} = \mathbf{R} \langle \mathbf{ss}^\dagger \rangle \mathbf{R}^\dagger + \langle \mathbf{nn}^\dagger \rangle = \mathbf{RSR}^\dagger + \mathbf{N},$$

where $\mathbf{D}$, $\mathbf{S}$, and $\mathbf{N}$ are the data, signal, and noise covariance matrices, respectively. This derivation assumes that data and noise are not correlated. The first term in (9.8) reduces to

$$\langle \mathbf{sd}^\dagger \rangle = \langle \mathbf{ss}^\dagger \rangle \mathbf{R}^\dagger = \mathbf{SR}^\dagger.$$

The minimum-variance estimator for the signal is, then,

$$\widehat{\mathbf{s}} = \mathbf{SR}^\dagger (\mathbf{RSR}^\dagger + \mathbf{N})^{-1} \mathbf{d}. \tag{9.9}$$

This expression is easier to understand if we introduce a new noise vector $\mathbf{n}'$ by $\mathbf{n} = \mathbf{Rn}'$. This type of noise is typical when estimating densities from redshift surveys, where the shot noise from discrete sampling is proportional to density. The above formula reduces then to

$$\widehat{\mathbf{s}} = \mathbf{S}(\mathbf{S} + \mathbf{N}')^{-1} \mathbf{R}^{-1} \mathbf{d} \tag{9.10}$$

(where $\mathbf{N}'$ is the new noise covariance matrix) and states that the estimator is obtained by inverting the initial expression (9.7) and weighting the obtained result by the ratio of signal to signal-plus-noise. The mean square residual can be transformed to

$$\langle \mathbf{rr}^\dagger \rangle = \mathbf{SR}^\dagger (\mathbf{RSR}^\dagger + \mathbf{N})^{-1} \mathbf{N} (\mathbf{R}^\dagger)^{-1}, \tag{9.11}$$

or, in the shot noise case, to

$$\langle \mathbf{rr}^\dagger \rangle = \mathbf{S}(\mathbf{SN}')^{-1} \mathbf{N}'.$$

In cosmological applications the matrices $\mathbf{S}$ and $\mathbf{R}$ depend frequently only on the distances between data points. This means that the Wiener estimators also can be used for predicting the signal values for new points; the only extra argument we need is the location of that point.

The above formulae are very similar to those we used for the estimation of power spectrum; the two approaches are closely connected. Tegmark et al. (1998) show that Wiener filtering becomes diagonal in the Karhunen–Loève basis and is thus easy to implement. Seljak (1998) has carried out a detailed comparison of Wiener filtering and power spectrum estimation. He gave the formulae for the Wiener filter in the signal-to-noise eigenmode basis and showed how to compute minimum variance power spectrum estimates from the Wiener-filtered modes.

9.3.1 Filtering in spherical basis

The Wiener filter method can be applied both to the redshift surveys and to the peculiar velocity surveys. The usual approach in the case of redshift surveys is to work in spherical coordinates and to use the Fourier–Bessel expansion to describe the density contrast. Detailed descriptions of that procedure can be found in the first application of that method to the 1.2 Jy IRAS redshift survey by Fisher et al. (1994) and in the Wiener reconstruction of the PSCz survey by Schmoldt et al. (1999). The Fourier–Bessel expansion for the density contrast is

$$\delta(r,\widehat{\mathbf{r}}) = \sum_{lmn} Y_{lm}(\widehat{\mathbf{r}}) j_l(k_{ln}r)\delta_{lmn}, \tag{9.12}$$

where $\widehat{\mathbf{r}}$ is the direction in the space (angular coordinates), $Y_{lm}(\widehat{\mathbf{r}})$ are spherical harmonics, $j_l(x)$ are spherical Bessel functions, and the discrete set of radial wavenumbers k_{ln} is determined by the boundary conditions. There exist different choices for that (see Fisher et al. 1994). The most popular choice is to ignore possible density fluctuations after the boundary of the survey $\left(\delta(\mathbf{r}) = 0, r > R\right)$ and to demand that the logarithmic derivative of the gravitational potential should be continuous at the boundary. In this case, the wavenumbers k_{ln} are the zeros of $j_{l-1}(kR)$.

The inverse transform is

$$\delta_{lmn} = C_{ln} \int_V \delta(r,\widehat{\mathbf{r}}) j_l(k_{ln}r) Y^\star_{lm}(\widehat{\mathbf{r}}), \tag{9.13}$$

where C_{ln} are the normalization constants, defined as:

$$\int_0^R j_l(k_{ln}r) j_l(k_{ln'}r) r^2\, dr = \delta^K_{nn'} C^{-1}_{ln}$$

(δ^K_{ij} is the Kronecker delta). For the boundary condition used here

$$C^{-1}_{ln} = \frac{R^3}{2}\left(j_l(k_{ln}R)\right)^2.$$

There exist different conventions for the Fourier–Bessel transform. Fisher et al. (1994) include the coefficients C_{ln} in the sum for the direct transform (9.12) and omit them in (9.13); we follow the convention adopted in Schmoldt et al. (1999).

The amplitudes δ_{lmn} now describe the real density field. Limiting the index range in the sum amounts to smoothing, as higher-order angular and radial modes are usually noisy. The usual method is to choose l to limit the angular resolution and to limit n by choosing

$$n_{\max}(l) + l/2 = R/r_{\min},$$

where $r_{\min}$ is the smallest scale resolved.

The next step is to correct for the radial velocity distortions. Summation over galaxy coordinates gives the Fourier–Bessel redshift space density coefficients:

$$\rho^s_{lmn} = C_{ln} \sum_i w(s_i) Y^\star_{lm}(\widehat{\mathbf{r}}_i) j_l(k_{ln} s_i),$$

where the summation is over all the objects i, s_i and $\widehat{\mathbf{r}}_i$ are their redshift distance and the position on the sky, respectively, and $w(s)$ is the weight function. The radial velocity correction assumes the role of the response function $\mathbf{R}$ above, as the relation between the contrast amplitudes in real and redshift space can be written as

$$\delta^s_{lmn} = \sum_{n'} Z_{lnn'} \delta_{lmn'}, \tag{9.14}$$

where the redshift space density contrast is computed as

$$\delta^s_{lmn} = \frac{1}{\bar{\rho}} \rho^s_{lmn} - \delta^K_{l0} \delta^K_{m0} O_n.$$

The term O_n is the monopole term (the transform of the constant 1):

$$O_n = \sqrt{4\pi} R^3 \left(\frac{j_1(k_{0n} R)}{k_{0n} R} \right).$$

Using the radial velocity distortion operator (8.46), the response matrix can be written as

$$\begin{aligned} Z_{lnn'} = \delta^K_{nn'} \quad - \quad & \beta \int_0^R j_l(k_{ln} r) \left[\left(\frac{l(l+1)}{k^2_{ln'} r^2} - 1 \right) j_l(k_{ln'} r) + \right. \\ & \left. + \frac{j'_l(k_{ln'} r)}{k_{ln'} r} \frac{d \ln \phi(r)}{d \ln r} \right] r^2 \, dr, \end{aligned} \tag{9.15}$$

where we have used the inverse weighting, following Fisher et al. (1994), $w(r) = 1/\phi(r)$, $\phi(r)$ is the radial selection function of the survey. Generalization of these formulae for a general weighting function can be found in Schmoldt et al. (1999).

9.3.2 Density interpolation

We saw in the section on the estimation of power spectra that radial and angular modes are mixed due to the limited angular window of the survey. The reconstruction ideology uses a different approach — extrapolating the observed density distribution over the full sky. Predicting the structures that we cannot yet observe is logically also one of the goals of reconstruction.

The first studies manually extrapolated the density into the masked-out regions. Later approaches use the minimum-variance method to do that. Schmoldt et al. represent the observed redshift space density distribution by an ansatz

$$\rho^s(s, \widehat{\mathbf{r}}) = \sum_{lmn} \rho^s_{lmn} Y_{lm}(\widehat{\mathbf{r}}) j_0(k_{0n} s),$$

that separates radial and angular modes at the expense of postulating the radial dependence. The expansion coefficients are estimated by minimizing the integrated variance

$$\Sigma^2 = \int \phi(\widehat{\mathbf{r}}) \left| \rho^s(s, \widehat{\mathbf{r}}) - \sum_{lmn} \rho^s_{lmn} Y_{lm}(\widehat{\mathbf{r}}) j_0(k_{0n} s) \right|^2 r^2 \, d\Omega \, dr,$$

where $\phi(\widehat{\mathbf{r}})$ is the angular selection function and $d\Omega$ is the solid angle element. This gives the system of equations for ρ^s_{lmn}:

$$\int \phi(\widehat{\mathbf{r}}) Y^\star_{lm}(\widehat{\mathbf{r}}) j_0(k_{0n} s) s^2 \, d\Omega ds = C^{-1}_{0n} \sum_{l'm'} \rho^s_{l'm'n} \int \phi(\widehat{\mathbf{r}}) Y_{l'm'}(\widehat{\mathbf{r}}) Y^\star_{lm}(\widehat{\mathbf{r}}) \, d\Omega. \tag{9.16}$$

Inverting (9.16), we find the redshift space density amplitudes ρ^s_{lmn} and supplement the catalog by filling in the masked regions (generating galaxies by an inhomogeneous Poisson process). Fig. 9.2 shows an example (from Schmoldt et al. 1999) that compares the galactic x-z coordinate plane of the PSCz redshift survey before and after filling the masked zones. It can be seen that the procedure interpolates the density distribution, taking into account the structure of the survey data.

Saunders and Ballinger (2000) generalized this approach for three-dimensional interpolation of the density field into masked regions. They assumed that the density field is log-normally distributed and the observational survey is randomly sampled for that field. The density logarithm can be expanded in the Fourier–Bessel basis,

$$\rho^s(\mathbf{x}) = \exp\left[\sum_n a_n f_n(\mathbf{x})\right],$$

where n is the collective index (lmn). Neglecting correlations, they write the log-likelihood $\mathcal{L}$ as

$$\mathcal{L} = \ln L = \sum_m \sum_n a_n f_n(\mathbf{x}_m)$$

(the index m denotes summation over galaxies) and maximize it together with the constraint that the total number of galaxies should be that observed. This gives the system of equations

$$\frac{\partial \mathcal{L}}{\partial a_n} = \sum_m f_n(\mathbf{x}_m) - \int_V \psi(\mathbf{x}) f_n(\mathbf{x}) \exp\left[\sum_n a_n f_n(\mathbf{x}_m)\right] d^3x,$$

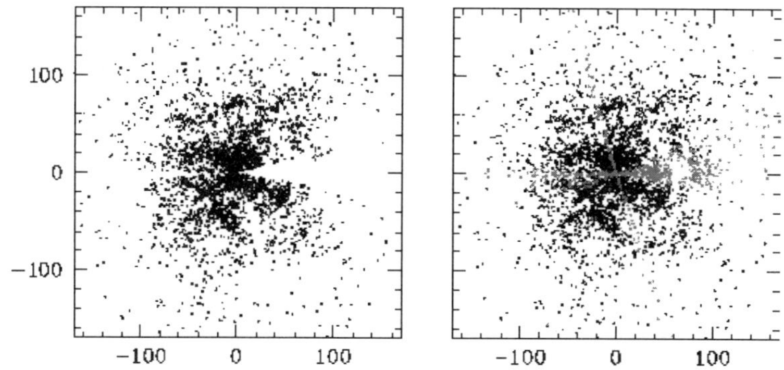

Figure 9.2 *Projections of a slice of the PCSz survey galaxies to the galactic x-z plane (the y range is $(-20, 20)h^{-1}$ Mpc). The x-axis (directed toward $l = 0°, b = 0°$) is horizontal, the z-axis (toward $b = 90°$) is vertical, units are in h^{-1} Mpc. The left panel shows the raw data, the right panel shows the structure after filling in the masked zones, the horizontal wedge of the Zone of Avoidance, and the high-altitude mask strip. The added galaxies are shown by light gray. (Reproduced, with permission, from Schmoldt et al. 1999, Astronom. J., 118, 1146–1160. AAS.)*

which can be solved for the full set of (redshift space) density amplitudes. They test the procedure on simulated catalogs and find that it works well.

9.3.3 Wiener reconstruction

Equation (9.14) shows that for a full-sky survey the response matrix (**Z** in our case) can be solved separately for all angular modes. For that we need the expressions for the noise covariance matrix

$$N_{lnn'} = C_{ln}C_{ln'} \int_0^R w(r) j_l(k_{ln}r) j_l(k_{ln'}r) r^2\, dr$$

(recall that we have chosen here $w(r) = 1/\phi(r)$) and the signal covariance matrix

$$\begin{aligned} S_{lmnn'} &= C_{ln}C_{ln'} \int_0^R \int_0^R j_l(k_{ln}r_1) j_l(k_{ln'}r_2) Y_{lm}^{\star}(\widehat{\mathbf{r}}_1) Y_{lm}(\widehat{\mathbf{r}}_2) \times \\ &\qquad \times \xi(\mathbf{r}_1 - \mathbf{r}_2) r_1^2 r_2^2\, dr_1 dr_2, \end{aligned}$$

where $\xi(r)$ is the correlation function of the density field. Writing that as a Fourier transform of the power spectrum

$$\xi(\mathbf{r}_1 - \mathbf{r}_2) = \int P(k) e^{-i\mathbf{k}\cdot(\mathbf{r}_1 - \mathbf{r}_2)} \frac{k^2\, dk}{(2\pi)^3}$$

and using the expansion of plane waves in spherical harmonics,

$$e^{i\mathbf{k}\cdot\mathbf{r}} = 4\pi \sum_{lm} i^l j_l(kr) Y^\star_{lm}(\widehat{\mathbf{r}}) Y_{lm}(\widehat{\mathbf{k}})$$

we get

$$\begin{aligned} S_{lnn'} &= \frac{2}{\pi} C_{ln} C_{ln'} \int_0^\infty P(k) k^2\, dk \int_0^R j_l(k_{ln} r_1) j_l(k r_1) r_1^2\, dr_1 \times \\ &\quad \times \int_0^R j_l(k_{ln} r_2) j_l(k r_2) r_2^2\, dr_2. \end{aligned}$$

A frequently used approximation is to extend the integration in one of the integrals over distance to infinity, getting the closure relation for spherical Bessel functions:

$$\int_0^\infty j_l(k_{ln} r) j_l(kr) r^2\, dr = \frac{\pi}{2k^2} \delta_D(k - k_{ln}).$$

Another integral over r gives the normalization constant, and finally we get

$$S_{lnn'} = \delta^K_{nn'} C_{ln} P(k_{ln}).$$

This shows that for very narrow windows the signal covariance matrix is diagonal with elements proportional to the values of the power spectrum at discrete wavenumbers. None of the matrices $\mathbf{Z}, \mathbf{N}, \mathbf{S}$ depends on m.

In order to calculate the covariance matrix $\mathbf{S}$ we have to choose the prior power spectrum. This spectrum also can be obtained by a maximum likelihood fit of the same data, but that is not strictly necessary.

Applying now the Wiener filter (9.10) with the response matrix $\mathbf{R} = \mathbf{Z}$, we reconstruct the density field in real space:

$$\widehat{\mathbf{s}} = \mathbf{S}(\mathbf{S} + \mathbf{N}')^{-1} \mathbf{Z}^{-1} \mathbf{d}.$$

This requires that we invert the matrices $\mathbf{S}$ and $\mathbf{Z}$ for every l.

Knowing the density field, we can also reconstruct the gravitational potential Φ by

$$\Delta\Phi = \frac{3}{2} \Omega_M H_0^2 \frac{\delta}{b}$$

(the δ we observe is the galaxy overdensity, connected with the mass overdensity δ_M by $\delta = b\delta_M$, where b is the bias factor). Expanding Φ in the Fourier–Bessel series and using the fact that $j_l(kr) Y_{lm}(\widehat{\mathbf{r}})$ is an eigenfunction of the Laplacian,

$$\Delta \left[j_l(kr) Y_{lm}(\widehat{\mathbf{r}}) \right] = -k^2 j_l(kr) Y_{lm}(\widehat{\mathbf{r}}),$$

we find that the Fourier–Bessel amplitudes of the gravitational potential are simply

$$\Phi_{lmn} = -\frac{3\Omega_M H_0^2}{2b} \frac{\delta_{lmn}}{k_{ln}^2}. \tag{9.17}$$

The linear approximation velocity field is

$$\mathbf{v} = -\frac{2}{3}\frac{f(\Omega_M)}{\Omega_M H_0}\Delta\Phi.$$

That gives for the radial velocity field

$$u(r,\widehat{\mathbf{r}}) = H_0\beta \sum_{lmn} \frac{j_l'(k_{ln}r)}{k_{ln}} Y_{lm}(\widehat{\mathbf{r}})\delta_{lmn}.$$

The formulae for the transverse components are more complex; they can be found in Fisher et al. (1994).

9.3.4 Maps

We illustrate the Wiener reconstruction procedure in Fig. 9.3. It shows the steps in density reconstruction of the IRAS 1.2 Jy survey by Fisher et al. (1994). The upper left panel shows the raw redshift space density field, smoothed by a Gaussian kernel with the width growing toward larger radii. The upper right panel shows the density obtained by the Fourier–Bessel expansion of redshift data. Here the smoothing is smaller and the density field is noisier. The lower left panel shows the result of the Wiener filtering in redshift space. We see that the Wiener filter has suppressed high-frequency noise, with stronger smoothing in the outskirts of the survey, where the shot noise is stronger. The lower right panel shows the final result, the application of the Wiener filter with correction for radial velocity distortions. The main differences between the last two density distributions are higher density amplitudes and more anisotropic structures in redshift space. The overall reconstructed density distribution is similar to that obtained by the POTENT method and illustrated above, but more detailed, due to the much larger size of the survey.

Fig. 9.4 shows the other reconstructed fields along with the density field (upper left panel). The gravitational potential in the upper right panel is much smoother than the density. Its main features are the large positive region, corresponding to the Sculptor void, and two deep minima at the regions of the Great Attractor and the Perseus–Pisces superclusters. The radial velocity field is shown in the lower left panel and the full peculiar velocity field is shown in the lower right panel. The main features of the velocity field are the infall patterns toward the two main density maxima.

Comparison of this picture with the POTENT solution (Fig. 9.1) shows that while there are slight differences in the density distributions, the velocity field predicted on the basis of the density distribution does not describe the divergence-free component of the flow through the region revealed by peculiar velocity measurements. It can be caused by an insufficient depth of the 1.2 Jy IRAS survey — the velocity field feels gravitational (tidal) forces from larger distances. This assumption is proved by the reconstruction of the velocity field on the basis of

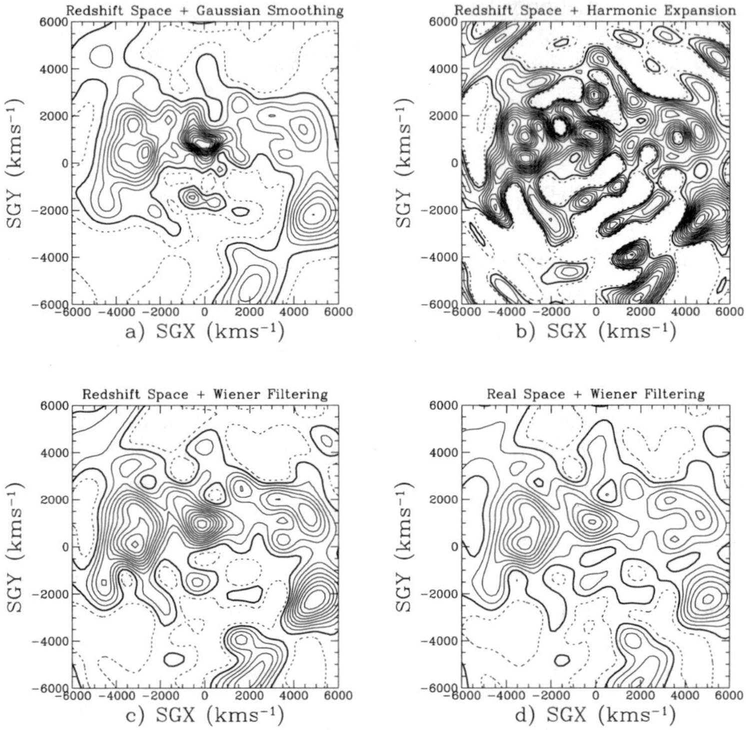

Figure 9.3 *Reconstruction of the 1.2 Jy IRAS density field in the supergalactic plane. The density contour interval is* $\Delta\delta = 0.5$, *the heavy contour denotes* $\delta = 0$, *and dashed lines show negative contours. Panel a) shows the raw redshift space density distribution, smoothed with a variable width Gaussian filter. Panel b) shows the redshift space density in the Fourier–Bessel expansion with* $l_{\max} = 15$ *and* $k_n R < 100$. *Panel c) gives the redshift space density after smoothing with the Wiener filter, and panel d) shows the Wiener reconstructed real space density field. (Reproduced, with permission, from Fisher et al. 1994, Mon. Not. R. Astr. Soc., 272, 885–910. Blackwell Science Ltd.)*

the PSCz redshift survey (Schmoldt et al. 1999). This survey has about two times more galaxies than the 1.2 Jy survey and is about twice as deep. This field is shown in Fig. 9.5.

We see that the velocity pattern in the central region is already more similar to the POTENT solution. The backfall into the Great Attractor region (the Centaurus and Pavo–Indus superclusters) is clearly seen (although the authors do not yet consider it firmly established) and the Shapley supercluster has emerged in the velocity map.

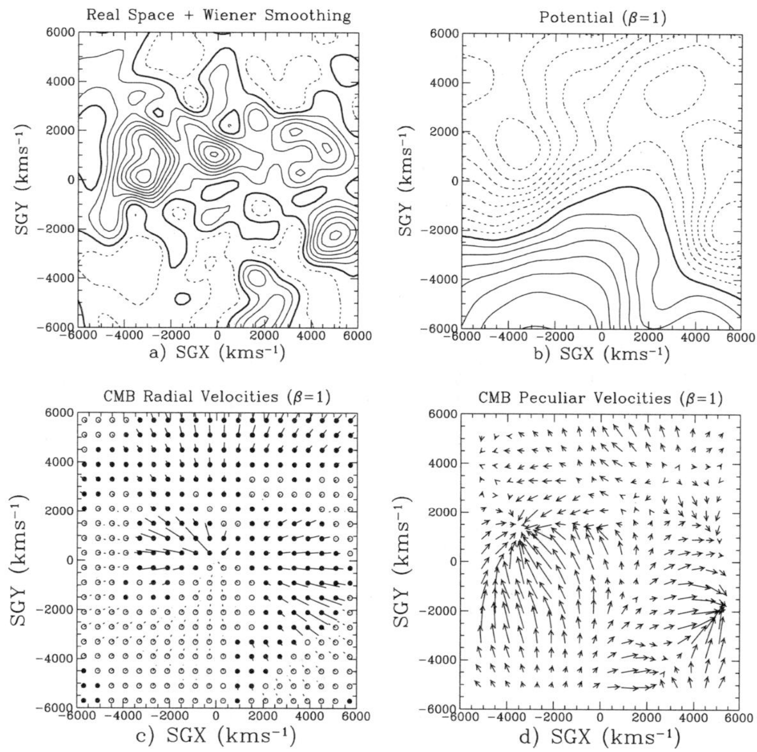

Figure 9.4 *Reconstruction of the 1.2 Jy IRAS density, potential, and velocity fields for the supergalactic plane. Panel a) shows the reconstructed density field, and panel b) shows the gravitational potential, assuming* $\beta = 1$. *Panel c) shows the radial velocity field, with open dots showing negative velocities and closed dots indicating positive velocities. Panel d) shows the projection of the full peculiar velocity field on the plane. (Reproduced, with permission, from Fisher et al. 1994, Mon. Not. R. Astr. Soc., 272, 885–910. Blackwell Science Ltd.)*

The reconstruction generates three-dimensional maps of the density, potential, and velocity fields in our neighborhood. Because these maps are difficult to represent, the usual maps are those for the supergalactic plane. To break with this tradition, we show in Fig. 9.6 the three-dimensional reconstruction of the real space galaxy density of the PSCz survey. The figure also shows known and tentatively identified superclusters. A movie of a reconstruction of the three-dimensional density fields, based on sky projections of shells of different distance, can be seen in the PSCz survey Web pages (*PSCz*).

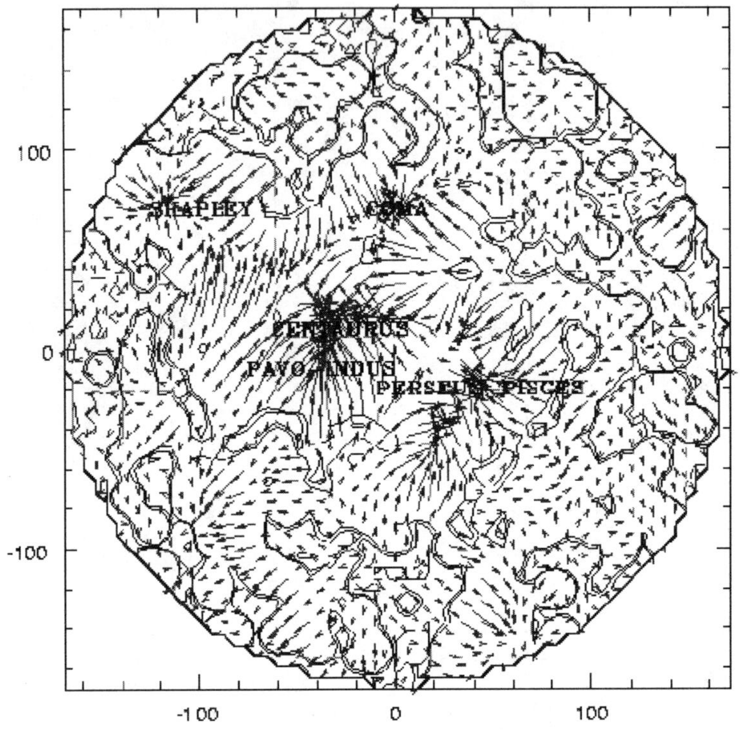

Figure 9.5 *Reconstruction of the peculiar velocity field for the PSCz redshift survey. The figure shows the projection of the velocities on the supergalactic $SGX - SGZ$ plane (the horizontal axis is SGX, the vertical axis is SGZ). Coordinate units are h^{-1} Mpc. The locations of the main superclusters are marked. Double contours show the zero density contrast level. (Reproduced, with permission, from Schmoldt et al. 1999, Astronom. J., 118, 1146–1160. AAS.)*

9.3.5 Velocity reconstruction

The Wiener filter reconstruction can also be applied to the peculiar velocity data. The data–signal formula describes in this case the relation between the full peculiar velocity and its radial component, with an additional noise term:

$$\mathbf{u} = \widehat{\mathbf{r}} \cdot \mathbf{v} + \mathbf{n}.$$

Here $\mathbf{u}$ is the collection of the observed radial peculiar velocities, $\mathbf{v}$ are the true peculiar velocities, $\widehat{\mathbf{r}}$ is the direction of the radius-vector, and $\mathbf{n}$ are the "instrumental" errors of measurement of radial velocities. In components, this relation

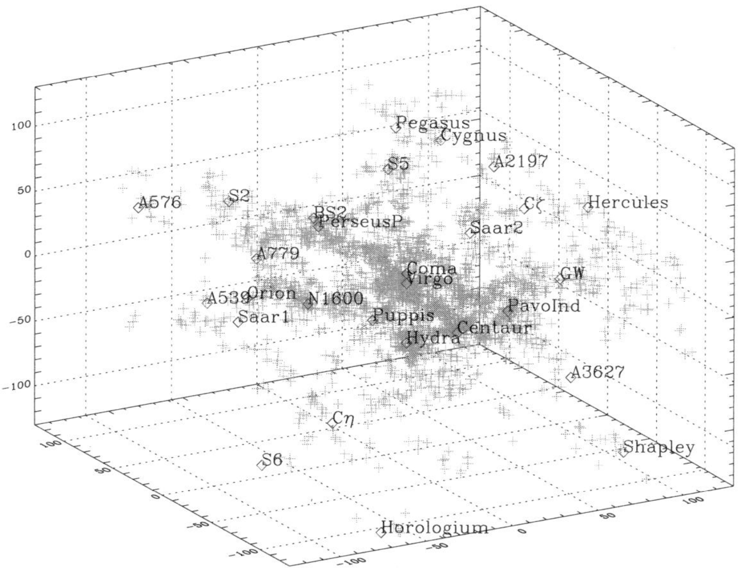

Figure 9.6 *Reconstruction of the real space density for the PSCz redshift survey. The figure shows the three-dimensional density distribution in supergalactic coordinates; the coordinate units are* h^{-1} *Mpc. The locations of the known superclusters are marked. (Reproduced, with permission, from Schmoldt et al. 1999, Astronom. J., 118, 1146–1160. AAS.)*

is

$$u_i = \sum_{\alpha} \widehat{r}_i^{\alpha} v_i^{\alpha} + n_i,$$

where the lower (Latin) indices enumerate galaxies and the upper (Greek) indices that may have values from 1 to 3, the vector components. To avoid confusion, we shall write explicitly below all sums over indices.

The Wiener estimator for the velocity can now be written as

$$\widehat{v}_k^{\alpha} = \sum_{i,j} \langle v_k^{\alpha} u_i \rangle D_{ij}^{-1} u_j = \sum_{ij} V_{ki}^{\alpha} D_{ij}^{-1} u_j, \tag{9.18}$$

where the data covariance matrix is

$$D_{ij} = \langle u_i u_j \rangle = \sum_{\alpha,\beta} \widehat{r}_i^{\alpha} \langle v_i^{\alpha} v_j^{\beta} \rangle \widehat{r}_j^{\beta} + \delta_{ij}^K \sigma_i^2.$$

The linear approximation relation between velocity and density contrast (9.3) allows us to write a similar Wiener estimator for the density contrast:

$$\widehat{\delta}_k = \sum_{i,j} \langle \delta_k u_i \rangle D_{ij}^{-1} u_j = \sum_{ij} T_{ki} D_{ij}^{-1} u_j. \tag{9.19}$$

The velocity covariance tensor is the main object of study in turbulence (classical turbulence describes incompressible liquids, so velocity is the only random field there). The velocity covariance tensor can be written as (Monin and Yaglom 1975):

$$\langle v_i^\alpha v_j^\beta \rangle = S_{ij}^{\alpha\beta} == \Psi_\perp(r_{ij})\delta_{\alpha\beta}^K + \left[\Psi_\parallel(r_{ij}) - \Psi_\perp(r_{ij})\right]\widehat{r}_{ij}^\alpha \widehat{r}_{ij}^\beta, \tag{9.20}$$

where $\Psi_\parallel(r)$ and $\Psi_\perp(r)$ are the radial and transverse velocity correlation functions, $\mathbf{r}_{ij} = \mathbf{r}_i - \mathbf{r}_j$ and $r_{ij} = |\mathbf{r}_{ij}|$.

In cosmology velocity is tightly correlated with density perturbations, and the velocity correlation functions can be expressed via the density power spectrum (Gorski 1988):

$$\Psi_\parallel(r) = \frac{H_0^2 f^2(\Omega_M)}{2\pi^2}\int_0^\infty \frac{j_1(kr)}{kr}P(k)dk, \tag{9.21}$$

$$\Psi_\perp(r) = \frac{H_0^2 f^2(\Omega_M)}{2\pi^2}\int_0^\infty \left(j_0(kr) - 2\frac{j_1(kr)}{kr}\right)P(k)dk, \tag{9.22}$$

where $j_0(x)$ and $j_1(x)$ are the spherical Bessel functions. Using (9.20), the expression for the data covariance matrix simplifies to

$$D_{ij} = \Psi_\perp(r_{ij})\widehat{\mathbf{r}}_i \cdot \widehat{\mathbf{r}}_j + \left[\Psi_\parallel(r_{ij}) - \Psi_\perp(r_{ij})\right](\widehat{\mathbf{r}}_i \cdot \widehat{\mathbf{r}}_{ij})(\widehat{\mathbf{r}}_j \cdot \widehat{\mathbf{r}}_{ij}) + \delta_{ij}^K \sigma_i^2.$$

The velocity–radial velocity covariance matrix in (9.18) can be written as

$$V_{ij}^\alpha = \sum_\beta S_{ij}^{\alpha\beta}\widehat{r}_i^\beta = \Psi_\perp(r_{ij})\widehat{r}_j\alpha + \left[\Psi_\parallel(r_{ij}) - \Psi_\perp(r_{ij})\right]\widehat{r}_i^\alpha$$

and the density–velocity covariance matrix is

$$T_{ij} = \Psi_\delta(r_{ij})\widehat{\mathbf{r}}_{ij} \cdot \widehat{\mathbf{r}}_j,$$

with the density–velocity correlation function $\Psi_\delta(r)$ given by (Zaroubi, Hoffman, and Dekel 1999)

$$\Psi_\delta(r) = -\frac{H_0 f(\Omega_M)}{2\pi^2}\int_0^\infty P(k)j_1(kr)k\,dk. \tag{9.23}$$

Additional density and velocity smoothing can be done by multiplying the power spectrum $P(k)$ in formulae (9.21 and 9.23) by a squared Fourier transform of a smoothing filter, for Gaussian smoothing of width R, for example, by $\exp(-k^2R^2)$.

Zaroubi, Hoffman, and Dekel (1999) applied the Wiener filter to the Mark III peculiar velocity catalog. The comparison of the Wiener-filtered peculiar velocity field from Mark III and the fields predicted by the IRAS 1.2 Jy survey are compared in Fig. 9.7. The density and velocity fields (upper panels, the IRAS field on the left, the Mark III field on the right) are only roughly similar. The difference also might be due to the complex nonlinear reconstruction procedure of the IRAS

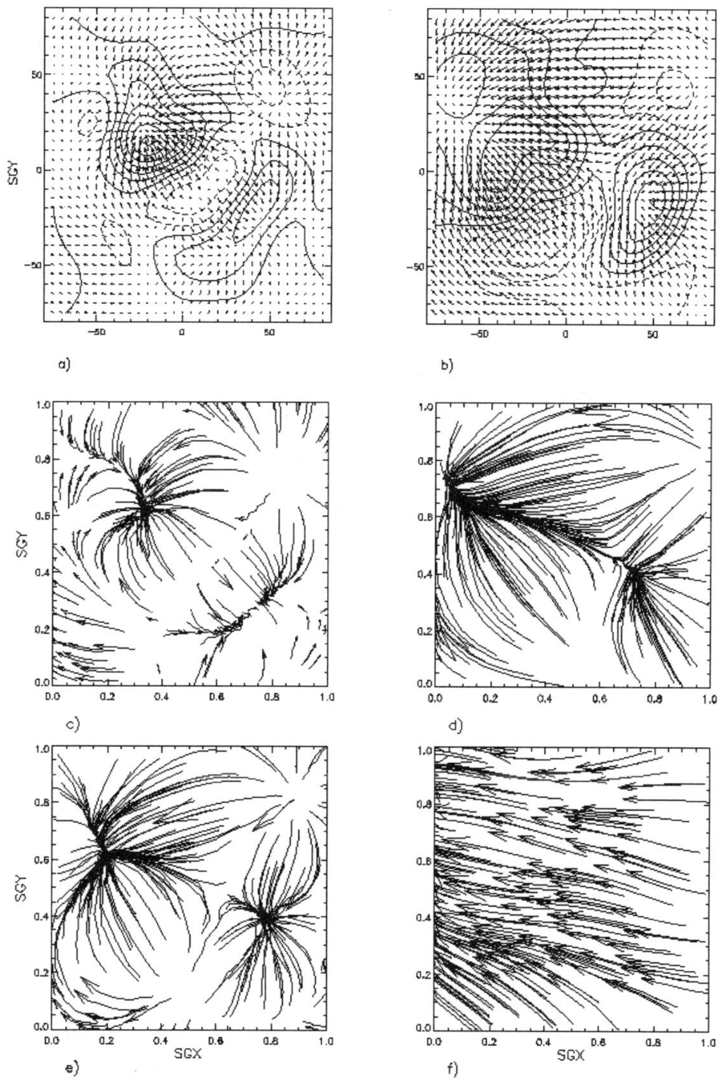

Figure 9.7 *Comparison of the density and velocity fields obtained by Wiener filtering of a redshift catalog (IRAS 1.2 Jy survey) and a peculiar velocity catalog (Mark III catalog). The upper panels show the density and velocity distributions in the supergalactic $SGX - SGY$ plane, the IRAS 1.2 Jy survey on the left, the Mark III on the right. The middle panels show the flow lines for these catalogs (the lengths of the lines are proportional to the velocity amplitude). The lower panels show the different components of the Mark III velocity field, the divergent component responsible for the local density structure on the left, the remaining tidal component on the right. (Reproduced, with permission, from Zaroubi, Hoffman, and Dekel 1999, Astrophys. J., 520, 413–425. AAS.)*

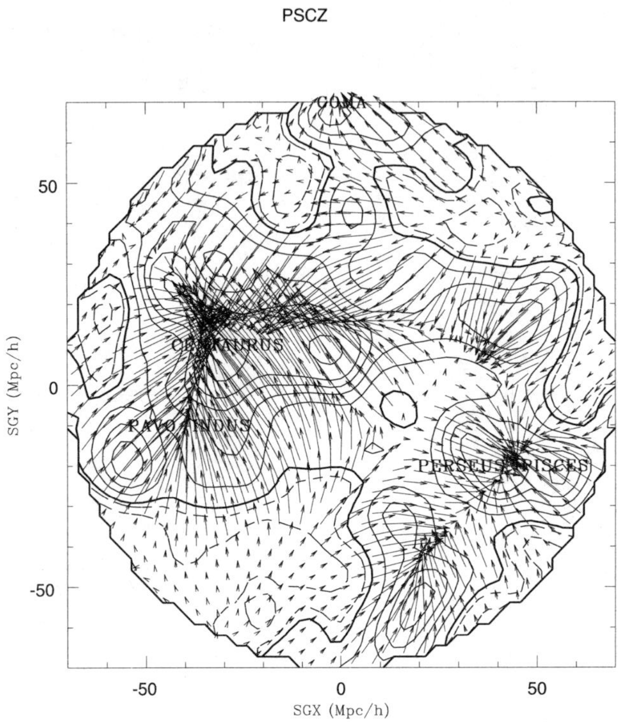

Figure 9.8 *Wiener reconstruction of the density and velocity distributions of the PSCz redshift survey in the nearby region of the supergalactic plane. (Reproduced, with permission, from Schmoldt et al. 1999, Astronom. J., 118, 1146–1160. AAS.)*

1.2 Jy density field. Comparison with the straight Wiener-filtered PSCz density map (Fig. 9.8) shows a better agreement.

More interesting is the comparison of the velocity fields in the middle panels, which look very different. The lower panels show a decomposition of the Mark III velocity field into its divergent and tidal components. As expected, the divergent fields are similar, but the redshift survey reconstruction has not recovered the tidal field. Even the large-scale reconstruction in Fig. 9.5 does not show that component, so it must be due to more distant regions of space.

Another problem where the effect of the large-scale velocity field is clearly seen is the monopole problem — prediction of the Local Group velocity. This can be estimated by different methods, and as the surveys get deeper, the direction of that velocity changes. We know the final answer; it should be the same as our velocity with respect to the CMB background, but thus far differences remain.

9.4 Constrained fields

The Wiener filter reconstruction is usually rather smooth, especially for noisy data. It can demonstrated by writing it in the Fourier domain (see Press et al. 1992):

$$\widetilde{\delta}_W(\mathbf{k}) = F_W(\mathbf{k})\widetilde{\delta}_O(\mathbf{k}),$$

where the Wiener filter $F_W(\mathbf{k})$ is

$$F_W(\mathbf{k}) = \frac{\langle|\widetilde{\delta}(\mathbf{k})|^2\rangle}{\langle|\widetilde{\delta}(\mathbf{k})|^2\rangle + \langle|\widetilde{n}|^2\rangle}.$$

Tildes denote Fourier amplitudes: $\widetilde{\delta}$ are the true Fourier amplitudes, $\widetilde{\delta}_O$ are the raw observational estimates, $\widetilde{\delta}_W$ are the Wiener filter estimates, and $\widetilde{n}$ are the noise amplitudes.

The variance of the Wiener filter estimate is

$$\begin{aligned}\langle F_W^2 \delta_O^2\rangle &= \left(\frac{\langle|\delta\delta|^2\rangle}{\langle|\delta\delta|^2\rangle + \langle|\delta n|^2\rangle}\right)^2 (\langle|\delta\delta|^2\rangle + \langle|\delta n|^2\rangle) \\ &= \left(\frac{\langle|\delta\delta|^2\rangle}{\langle|\delta\delta|^2\rangle + \langle|\delta n|^2\rangle}\right) \langle|\delta\delta|^2\rangle.\end{aligned}$$

This is always smaller than the variance of the true field. Also, in the outer regions of surveys, where shot noise may dominate, the Wiener-filtered signal will approach zero. This means that the reconstruction of the true field is not uniform over the survey volume, which makes comparison with other methods difficult.

Two approaches are used to change this property of Wiener filtering. One approach, the power-preserving filter, proposed by A. Yahil, is described and applied to galaxy surveys by Sigad et al. (1998). This filter is

$$F_Y(\mathbf{k}) = \sqrt{F_W(\mathbf{k})}.$$

Calculating the variance of the estimate as above we find that $\langle F_Y^2|\widetilde{\delta}_O|^2\rangle = \langle|\widetilde{\delta}|^2\rangle$, so the variance of the field is preserved. The filter is almost optimal, with the mean square residual only slightly larger than that for the Wiener filter. The mean field will also approach zero in the regions dominated by noise, but more slowly.

Another approach is that of restoring the small-scale features of the true field. In order to apply the filters, we had to choose the prior power spectrum, so we already have used knowledge about the true field. Statistical homogeneity of the reconstructed field is used to produce random realizations of the field that are consistent both with the prior and the data.

Let us develop the signal-data model

$$\mathbf{d} = \mathbf{R}\mathbf{s} + \mathbf{n} \tag{9.24}$$

further and suppose that the true signal $\mathbf{s}$ and the noise $\mathbf{n}$ are Gaussian homogeneous (zero-mean) processes of dimension M, the corresponding probability densities being multivariate Gaussians,

$$f_s(\mathbf{s}) = \frac{1}{(2\pi)^{m/2}\sqrt{|\mathbf{S}|}} \exp\left[-\frac{1}{2}\mathbf{s}\mathbf{S}^{-1}\mathbf{s}^\dagger\right], \tag{9.25}$$

and a similar formula for $f_n(\mathbf{n})$ with the noise covariance matrix $\mathbf{N}$. The probability density (9.25) describes a free field, with the statistical properties determined in our case by the prior power spectrum. But our realizations have to satisfy the M observational constraints (9.24). The conditional probability density for such a field is, by Bayes theorem,

$$f(\mathbf{s}|\mathbf{d}) = \frac{f_s(\mathbf{s})f(\mathbf{d}|\mathbf{s})}{f_d(\mathbf{d})}.$$

As the probability $f(\mathbf{d}|\mathbf{s})$ of data conditioned on signal is just the probability of noise $\mathbf{n} = \mathbf{d} - \mathbf{R}\mathbf{s}$, we can write

$$f(\mathbf{s}|\mathbf{d}) \propto \exp\left(-\frac{1}{2}\left[\mathbf{s}^\dagger\mathbf{S}^{-1}\mathbf{s} + (\mathbf{d}-\mathbf{R}\mathbf{s})^\dagger\mathbf{N}^{-1}(\mathbf{d}-\mathbf{R}\mathbf{s})\right]\right).$$

The expression in the exponent can be transformed by completing the square to

$$f(\mathbf{s}|\mathbf{d}) \propto \exp\left(-\frac{1}{2}\left[\mathbf{u}^\dagger\mathbf{Q}^{-1}\mathbf{u}\right]\right), \tag{9.26}$$

where

$$\mathbf{u} = \mathbf{s} - \mathbf{S}\mathbf{R}^\dagger(\mathbf{R}\mathbf{S}\mathbf{R}^\dagger + \mathbf{N})^{-1}\mathbf{d} \tag{9.27}$$

and

$$\mathbf{Q} = (\mathbf{S}^{-1} + \mathbf{R}^\dagger\mathbf{N}^{-1}\mathbf{R})^{-1}.$$

Although these expressions might seem a little complicated, they show that the constrained field is also a Gaussian field (but not homogeneous any more) with a mean value

$$\bar{\mathbf{u}} = \mathbf{S}\mathbf{R}^\dagger(\mathbf{R}\mathbf{S}\mathbf{R}^\dagger + \mathbf{N})^{-1}\mathbf{d} \tag{9.28}$$

and the covariance matrix $\mathbf{Q}$. Because (9.28) coincides with the Wiener filter estimate, it shows that the Wiener filter gives the mean value of the conditional signal distribution. For the Gaussian distribution this is also the most probable value for the signal vector.

We can also see from above that the distribution of the residual field $\mathbf{u}$ depends only on the covariance matrices $\mathbf{S}$ and $\mathbf{N}$ and on the form of the response operator $\mathbf{R}$, and not on the values of the measurements $\mathbf{d}$ themselves ($\mathbf{u}$ is a zero-mean Gaussian field). This leads us to the procedure of constructing constrained realizations of the field s, first proposed by Hoffman and Ribak (1991). We start with known $\mathbf{d}, \mathbf{S}, \mathbf{N}$, and the response operator $\mathbf{R}$. The next steps are as follows:

1. Generate a realization of the Gaussian random field $\mathbf{s}'$ with the prior power spectrum and find the values of the measurements $\mathbf{d}' = \mathbf{R}\mathbf{s}'$ for this particular realization. These values will certainly differ from the measured values $\mathbf{d}$ we have.

2. Find the residual field $\mathbf{u}$ using (9.27); this field will also be the residual field of the final constrained realization.
3. Find the mean value of the constrained realization using (9.28) and add to it the residual field.

This sums up to the formula

$$\mathbf{s} = \mathbf{s}' + \mathbf{SR}^{\dagger}(\mathbf{RSR}^{\dagger} + \mathbf{N})^{-1}(\mathbf{d} - \mathbf{d}').$$

Because the matrices $\mathbf{SR}^{\dagger}$ and $\mathbf{D}^{-1}$ had to be found for the Wiener estimate, anyway, generating constrained realizations does not require extra effort. We have described methods of generation of Gaussian random fields in Chapter 7.

An example of a Wiener-filtered mean field and its constrained realizations, given in Fig. 9.9, describes the density and velocity distribution of the Mark III catalog. The upper left panel shows the reconstruction (the density field is smoothed by an extra Gaussian of width 9 h^{-1} Mpc). The upper right panel shows the signal–noise ratio of the density field, defined as

$$S/N = \frac{|\delta_W(\mathbf{x})|}{\sigma_W(\mathbf{x})},$$

where δ_W is the predicted density contrast and σ_W is the rms residual from (9.11). The first contour of S/N starts at unity, and we see that clear structure is seen only in the two large supercluster regions and in the Sculptor void. The middle and lower panels show the constrained realizations of the same density and velocity fields. We see that as the main features remain approximately same, the Wiener filter estimate still leaves considerable freedom for small-scale details.

Nevertheless, the small-scale structure is affected by the constraints imposed by the large-scale environment. An example of an analysis of such an influence is the work of van de Weygaert and Hoffman (2000) on the peculiarities of the cosmic flow in our neighborhood. It is a long-known and uncomfortable fact that the flow velocity here is appreciably less than the predictions of N-body models. van de Weygaert and Hoffman show that it is possible to build constrained realizations of a local cold flow, conditioned by the strong tidal shear in our neighborhood.

9.4.1 *Constrained realizations for models*

The derivation above used the fact that the distribution of measurements conditioned by the field coincided with the noise distribution. However, the formalism of constrained realizations of Gaussian fields also can be applied to the case where there is no noise, and the operator $\mathbf{R}$ describes constraints we wish to apply to a field (e.g., for simulation). In fact, generation of constrained initial fields for numerical simulations was the motivation to develop the formalism (Bertschinger 1987). In that case we can write

$$f(\mathbf{d}|\mathbf{s}) \propto \exp\left[-\frac{1}{2}\mathbf{d}\mathbf{C}^{-1}\mathbf{d}\right],$$

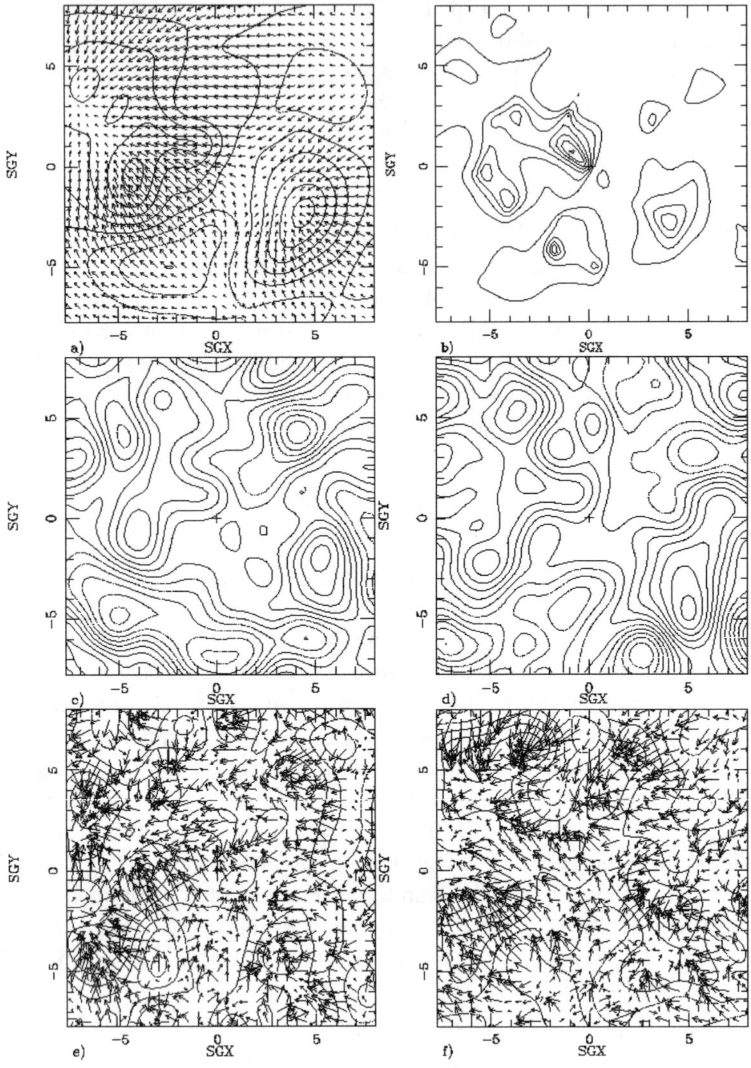

Figure 9.9 *Wiener reconstruction of the Mark III catalog. The upper left panel shows the reconstructed density and velocity fields; the upper right panel shows their signal–noise ratio (with the contour spacing one). The density has been smoothed by a Gaussian of width 9* h^{-1} *Mpc. The middle and lower panels show different constrained realizations of the data in the upper left panel: the middle panels show only the density, the lower panels show both the density and velocity fields. The spacing of density contours is 0.15 in the upper and middle panels and 0.3 in the lower panels. (Reproduced, with permission, from Zaroubi, Hoffman, and Dekel 1999, Astrophys. J., 520, 413–425. AAS.)*

because the constraints are linear, and the covariance matrix of constraints $\mathbf{C}$ given the field $\mathbf{d}$ is

$$\mathbf{C} = \mathbf{R}^\dagger \mathbf{S} \mathbf{R}.$$

Using that, the formula for $f(\mathbf{s}|\mathbf{d})$ will coincide with the formula (9.26) if we omit there the terms containing $\mathbf{N}$:

$$\mathbf{s} = \mathbf{s}' + \mathbf{S}\mathbf{R}^\dagger \mathbf{C}^{-1}(\mathbf{d} - \mathbf{d}').$$

For a true continuous representation of a random field linear constraints are linear functionals of the field, including values of the field and its derivatives at selected points. A detailed description of constructing such realizations and examples of their application for generating the initial fields for numerical simulations are given by van de Weygaert and Bertschinger (1996). Linear constraints can be imposed as convolutions of the field $s(\mathbf{x})$ with a kernel $H_i(\mathbf{x}; \mathbf{x}_i)$:

$$d_i = \int s(\mathbf{x}) H_i(\mathbf{x}; \mathbf{x}_i)\, d^3x.$$

In this case the constrained field can most easily be found from its Fourier transform $\widetilde{s}(\mathbf{k})$:

$$\widetilde{s}(\mathbf{k}) = \widetilde{s}'(\mathbf{k}) + P(k)\widetilde{H}_i(\mathbf{k}) C_{ij}^{-1}(d_j - d_j'),$$

where $\widetilde{s}'(\mathbf{k})$ are the Fourier amplitudes of the unconstrained realization, $P(k)$ is the power spectrum of the field, $\widetilde{H}_i(\mathbf{k})$ is the Fourier transform of the convolution kernel, d_j is the value of the j-th constraint, and d_j' is the value of the same constraint for the unconstrained realization. The covariance matrix of the constraints C_{ij} is given by the Fourier integral

$$C_{ij} = \int P(k)\widetilde{H}_i^\star(\mathbf{k})\widetilde{H}_j(\mathbf{k})\frac{d^3k}{(2\pi)^3}.$$

The main features of constrained realizations are illustrated by a simple example in Fig. 9.10. This figure shows the density profiles for slices of constant height z through a constrained density realization. The smooth density distribution (constraints) is represented by three ellipsoids in the upper left panel. While the constraints for the observational catalogs usually take the values of the field at selected points, an ellipsoidal density peak has to be determined by 10 constraints (its height, position, orientation, and shape; van de Weygaert and Bertschinger 1996). Specifying the velocity field will add another 10 constraints per peak. We have not done that; the constrained realizations in Fig. 9.10 satisfy 30 constraints.

The three remaining panels show different z-slices (denoted by the slice number) of the same constrained realization. The field was generated in a 32^3 cube, so these slices are all close to the center of the cube. The high density constraint in the central slice has effectively suppressed fluctuations at the locations of the two main peaks, but the smaller peak is severely distorted. Traces of regular structure disappear quickly with increasing distance from the center. In the three centers

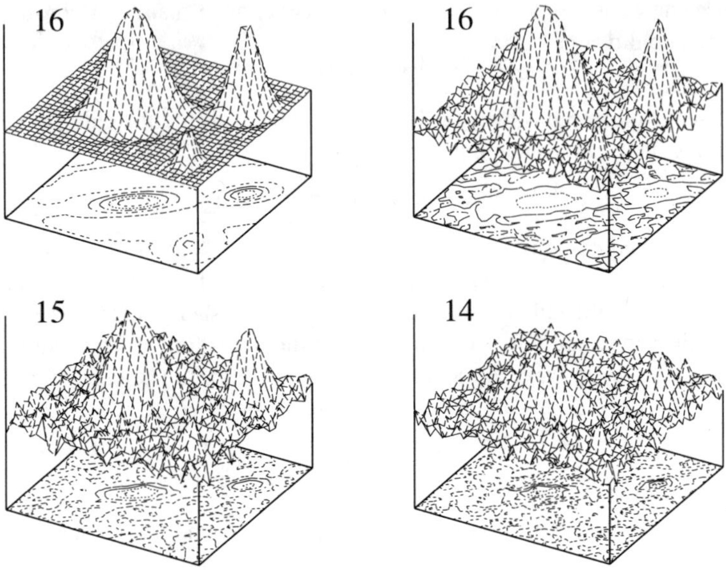

Figure 9.10 *Constrained realizations of a Gaussian density field (the height of the surface gives the density contrast). The constraints are specified as the three ellipsoids in the upper left panel. The remaining panels show the realizations in different z-slices of a 32^3 computational cube; the slice numbers are shown.*

the dispersion of the field is zero, but it grows fast outward and soon achieves its value for the free field. The outer regions of the cube in the upper left panel show a picture typical for a Wiener-filtered field, and the realizations give there a uniform random density field.

The realizations in Fig. 9.10 were computed using Bertschinger's COSMICS package of generation of initial conditions for cosmological simulations. It can be found at the (*COSMICS*) Web page.

A useful generalization of the above formalism was developed by Sheth (1995). He showed that we can use similar methods to construct constrained realizations of random fields that can be obtained by transformations of a Gaussian field. This class includes log-normal fields, χ_n^2 fields, and Rayleigh ($\sqrt{\chi_n^2}$) fields. Such fields are frequently used as statistical models in cosmology.

9.5 Time machines

Reconstruction of physical fields in our neighborhood can be developed further, asking how they looked in the past. That knowledge would allow us to check the basic assumptions of the structure formation theories and enable us simply to satisfy our curiosity about the past. We can never observe that, because our observations take us along the light cone; the past we see is the past of far-away regions in space.

There are also technical requirements for that. Creating Gaussian constrained realizations for present-day reconstructions is wrong, in principle, as the density distribution at small scales is no longer Gaussian. The right way to add small-scale structure is to take the present structure into the past where it should be Gaussian, to build constrained realizations there, and to evolve these dynamically back to the present time. The resulting fields are used to create reference catalogs to compare with the observed data and to estimate the errors of the analysis. This approach has been realized by Kolatt et al. (1996) and Bistolas and Hoffman (1998).

The main problem here is the choice of a good "time machine" that would calculate for us the density and velocity distribution at an early time, given the present-day data. Straight numerical N-body modeling will not work. Due to the noise in the present-day reconstruction, there is always a mixture of the decaying mode of perturbations, and N-body modeling would amplify it into the past, soon losing the growing mode we are looking for. Thus, the "time machines" are usually based on dynamical approximations that exclude the decaying mode. There are several such methods; a review and a comparison of these methods are given by Narayanan and Croft (1999) (the only method published later is that of Monaco and Efstathiou 1999).

Narayanan and Croft divide the methods into three classes. The first class starts with the linear solution, where the density is proportional to the growth rate of the growing mode $D_1(t)$ that is a known function of time, and its higher-order generalizations. A second approach assumes that the rank order of the density field does not change during evolution; the highest density peaks remain the highest and the lowest densities remain the lowest. The present density distribution is re-mapped to a Gaussian distribution and this is taken into the past with one of the time machines of the first class. This gives a better initial density distribution than the first methods, but cannot be used when we want to study the one-point probability distribution of density, because it is transformed into a Gaussian by force. The third class of methods is based on the least action principle, and, in our view, is the most interesting; we shall describe that approach below.

Peebles (1989) proposed using the least action principle to trace galaxy orbits back in time. He applied that method to nearby galaxies; Croft and Gaztañaga (1997) reformulated the method for cosmological reconstruction of large galaxy catalogs.

In the Hamiltonian formulation of dynamics the properties of a mechanical system are described by a functional called action:

$$\mathcal{S} = \int_0^t \mathcal{L}(\mathbf{x}, \dot{\mathbf{x}})\, dt,$$

where the Lagrangian $\mathcal{L}$ is the difference of the kinetic energy $\mathcal{K}$ and the potential energy $\mathcal{W}$,

$$\mathcal{L} = \mathcal{K} - \mathcal{W}.$$

The equations of motion of a system of particles can be found from the requirement that the action has to be stationary (minimal) when the particle trajectories are varied. Thus, the requirement of least action is equivalent to finding all true paths of particles.

For a system of points in an expanding background the energies are (Peebles 1989):

$$\mathcal{K} = \frac{1}{2}\sum_i m_i a^2 |\dot{\mathbf{x}}_i|^2, \tag{9.29}$$

$$\mathcal{W} = \frac{G}{a}\sum_{i\neq j} \frac{m_i m_j}{|\mathbf{x}_i - \mathbf{x}_j|} + \frac{2}{3}\pi G\bar{\rho}a^2 \sum_i m_i |\mathbf{x}_i|^2, \tag{9.30}$$

where $m_i, \mathbf{x}_i$, and $\dot{\mathbf{x}}_i$ are the masses, coordinates, and velocities of the particles, a is the dimensionless scale factor, G is the gravitational constant, and $\bar{\rho}$ is the mean density of the universe. The dependence of these energies on the scale factor is due to the expansion and the second term in the potential energy is the relativistic curvature term, the energy of the mean smooth distribution of matter. The kinetic and potential energies are connected by the cosmic energy conservation equation (Peebles 1980, § 24)

$$\frac{d}{dt}\left[a(\mathcal{K} + \mathcal{W})\right] = -\mathcal{K}\dot{a}.$$

We can write the Zeldovich approximation for particle paths and velocities as:

$$\mathbf{x}_i(t) = \mathbf{x}_i(0) + D(t)\mathbf{\Psi}_i, \tag{9.31}$$

$$\dot{\mathbf{x}}_i(t) = \dot{D}(t)\psi_i, \tag{9.32}$$

where we have supposed that the particle velocities in the past were small, $\dot{\mathbf{x}}_i(0) \approx 0$. Substituting these relations into the formula (9.29) we get

$$\mathcal{K} = \frac{1}{2}a^2\dot{D}^2 \sum_i m_i |\mathbf{\Psi}_i|^2.$$

The sum above defines the mean square particle displacement

$$|\mathbf{\Psi}|^2 = \sum_i m_i |\mathbf{\Psi}_i|^2.$$

The energy conservation equation gives

$$\mathcal{W} = -\mathcal{K} - \frac{|\mathbf{\Psi}|^2}{2a}\int_o^a a^2\dot{D}^2\, da$$

and the total action is

$$\mathcal{S} = |\mathbf{\Psi}|^2 \int_0^1 \left[\dot{D}^2 a^2 + \frac{1}{2a} \int_0^a \dot{D}^2 a^2 \, da \right] \frac{da}{\dot{a}}.$$

We see that the action is proportional to the mean square displacement $|\mathbf{\Psi}|^2$; the function given by the integral depends only on the background cosmological model and not on the particle distribution.

Croft and Gaztañaga proposed an ingenious method to find the minimum of $\mathcal{S}$. They start with a homogeneous distribution of the past locations of the particles and select at random particle pairs, a past particle with a present particle. Each pair gives a displacement $\mathbf{\Psi}_i$ and all the pairs together give the mean square displacement. Then two pairs are chosen at random and their end points, the present particles, are swapped. If this change makes the sum of the path lengths squared for these two pairs smaller, it is accepted; otherwise, it is rejected. This procedure can be repeated until the decrease of the mean square displacement slows down. Because the pair swaps are independent, the procedure is fast enough for a particle number around 10^6.

Having found the final pair arrangement, the initial density field can be constructed by scaling down the displacements and using them to move the initial particles from their uniform density grid, by (9.31). The initial velocities can be found either from the Zeldovich approximation or from solving the Poisson equation for the gravitational potential and using the linear theory connection between the initial velocity and acceleration (the "quiet start" procedure for N-body simulations). Croft and Gaztañaga (1997) called this method the "path interchange Zeldovich approximation" (PIZA).

The comparison of different methods by Narayan and Croft (1999) showed that this method was one of the best. One additional trick is to choose more than one initial "PIZA particle" per galaxy; this reduces considerably the discreteness noise.

This method also can be used to estimate peculiar velocities from redshift catalogs, as done by Valentine, Saunders, and Taylor (2000). The possibility is clear, because the final peculiar velocities are simply rescaled displacements. The main problem when applying PIZA to a redshift survey is dealing with the redshift distortions.

We can write the redshift space displacement δ^s as (Taylor and Valentine 1999)

$$\xi_i^s = \mathcal{P}_i j \xi_j$$

(ξ_i is the real space displacement), where the redshift space projection tensor is

$$\mathcal{P}_{ij} = \delta_{ij}^K + \beta \widehat{r}_i \widehat{r}_j.$$

The inverse relation is

$$\xi_i = \mathcal{P}_{ij}^{-1} \xi_j^s, \tag{9.33}$$

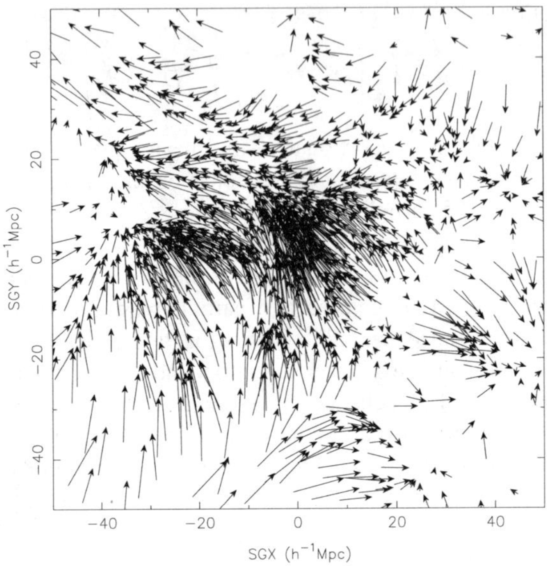

Figure 9.11 *Gen-PIZA-reconstructed PSCz velocity field in the supergalactic plane (a slice $20h^{-1}$ Mpc thick). The arrows show the galaxy displacements, starting at the initial position and ending in the PSCz real-space position. (Reproduced, with permission, from Valentine, Saunders, and Taylor 2000, Mon. Not. R. Astr. Soc., 319, L13–L17. Blackwell Science Ltd.)*

where the inverse redshift projection tensor is

$$\mathcal{P}_i^{-1}j = \delta_{ij}^K - \frac{\beta}{1+\beta}\widehat{s}_i\widehat{s}_j$$

(s is the radial vector in redshift space). The formula (9.33) gives an expression for the square of the real-space displacement

$$\xi_i^2 = (\xi_i^s)^2 - \left(1 - \frac{1}{(1+\beta)^2}\right)|\boldsymbol{\xi}_i^s \cdot \widehat{\mathbf{r}}|^2.$$

There are a few additional finesses. First, in the case of a magnitude-limited survey the masses assigned to galaxies (and PIZA particles) are inversely proportional to the selection function and low-mass pairs could get large displacements, so additional constraints must be used to eliminate large displacements for small-mass pairs. Second, as Taylor and Valentine (1999) have shown, the velocity reconstruction is less noisy when the Local Group velocity (the dipole term) is subtracted. The resulting reconstruction of the PSCz velocity field is shown in Fig. 9.11.

If we compare this velocity field with the Wiener-filtered direct velocity measurements of the Mark III catalog (Fig. 9.7) we see that the two velocity fields are rather close. The PIZA reconstruction has captured well the divergent component of the flow and, to some extent, the tidal component.

9.6 Gravitational lensing

Gravitational lensing is a powerful tool that allows us to measure directly the total matter content in the universe. This is a very broad and well-developed topic, with applications that are closer to image reconstruction than to three-dimensional spatial statistics. Although the theory of application of gravitational lensing for studying the large-scale structure of the universe was developed in the early 1990s, the first observational detections appeared only recently. Here we present only a brief introduction to the subject.

9.6.1 Physics of gravitational lensing

General reviews of gravitational lensing can be found in Schneider, Ehlers, and Falco (1992) and Narayan and Bartelmann (1999). Our attention will be focused on weak lensing and we follow below mainly the review by Bartelmann and Schneider (2001).

Gravitational lensing is based on the well-known fact that light is deflected by gravitating masses, predicted by general relativity. The deflection angle $\widehat{\alpha}$ for a light ray passing at a distance (impact parameter) ξ from a point mass (star) is

$$\widehat{\alpha} = \frac{4GM}{c^2\xi},$$

where M is the mass of the star and c is the speed of light. If the mass distribution is extended, the total deflection is given by

$$\widehat{\boldsymbol{\alpha}} = \frac{4G}{c^2}\int \frac{\boldsymbol{\xi}-\boldsymbol{\xi}'}{|\boldsymbol{\xi}-\boldsymbol{\xi}'|^2}\,dm = \frac{4G}{c^2}\int d^2\xi' \int \frac{\boldsymbol{\xi}-\boldsymbol{\xi}'}{|\boldsymbol{\xi}-\boldsymbol{\xi}'|^2}\rho(\boldsymbol{\xi}',z)\,dz,$$

where $\widehat{\boldsymbol{\alpha}}$ is the two-dimensional deflection angle on the sky, $\boldsymbol{\xi}$ is the corresponding impact parameter vector, and z is distance from the observer. This expression is obtained using the approximation of small deflection angles, which is almost always satisfied. Introducing the surface (projected) mass density

$$\Sigma(\boldsymbol{\xi}) = \int \rho(\boldsymbol{\xi},z)\,dz,$$

the deflection angle can be written as

$$\widehat{\boldsymbol{\alpha}} = \frac{4G}{c^2}\int \frac{\boldsymbol{\xi}-\boldsymbol{\xi}'}{|\boldsymbol{\xi}-\boldsymbol{\xi}'|^2}\Sigma(\xi')\,d^2\xi'.$$

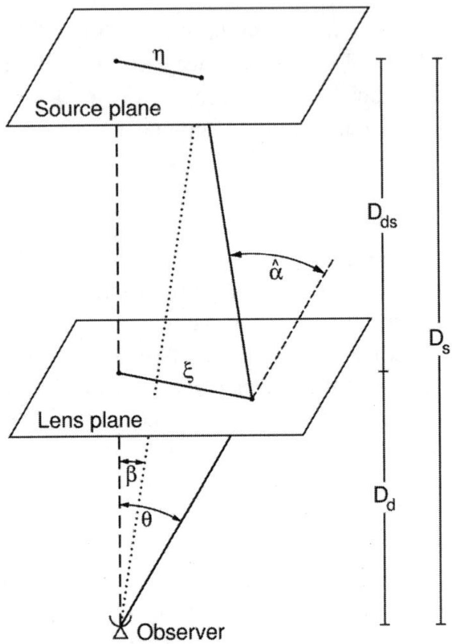

Figure 9.12 *Geometry of an extended gravitational lens system. (Reproduced, with permission, from Bartelmann and Schneider 2001, Phys. Rep., 340, 291–472. Elsevier Science.)*

Let us now write the lens equation that relates the true position of the source with its apparent position on the sky. Fig. 9.12 shows the geometry of a gravitational lens. The position of the source $\boldsymbol{\eta}$ in the source plane can be written as

$$\boldsymbol{\eta} = \frac{D_s}{D_d}\boldsymbol{\xi} - D_{ds}\widehat{\boldsymbol{\alpha}}(\boldsymbol{\xi}),$$

where the distances D have been defined in the figure. Replacing the position vectors $\boldsymbol{\xi}$ and $\boldsymbol{\eta}$ by the corresponding angles $\boldsymbol{\theta}$ and $\boldsymbol{\beta}$, we get

$$\boldsymbol{\beta} = \boldsymbol{\theta} - \frac{D_{ds}}{D_s}\widehat{\boldsymbol{\alpha}}(D_d\boldsymbol{\theta}) = \theta - \boldsymbol{\alpha}(\boldsymbol{\theta}),$$

introducing the scaled deflection angle $\boldsymbol{\alpha}$. As the lenses and sources are usually far away, and the distances above are used to describe angles, these distances are cosmological angular diameter distances (Chapter 2).

The dimensionless surface mass density (or convergence) κ is defined as

$$\kappa(\boldsymbol{\theta}) = \frac{\Sigma(D_d\boldsymbol{\theta})}{\Sigma_{cr}},$$

where

$$\Sigma_{cr} = \frac{c^2}{4\pi G}\frac{D_s}{D_d D_{ds}}$$

is the critical surface mass density, which divides the occurrences of strong (multiple images) and weak lensing. The scaled deflection angle is now

$$\boldsymbol{\alpha}(\boldsymbol{\theta}) = \frac{1}{\pi}\int \kappa(\boldsymbol{\theta}')\frac{\boldsymbol{\theta}-\boldsymbol{\theta}'}{|\boldsymbol{\theta}-\boldsymbol{\theta}'|^2}\, d^2\theta'.$$

The last equation shows that the deflection angle is a gradient of the deflection potential ψ, $\boldsymbol{\alpha} = \nabla\psi$, where the deflection potential is given by the two-dimensional Poisson equation

$$\Delta\psi(\boldsymbol{\theta}) = 2\kappa(\boldsymbol{\theta}).$$

Gravitational deflection will change apparent positions of light sources and will distort the images of extended sources (galaxies). Locally the source–image mapping can be described by

$$d\boldsymbol{\beta} = \mathbf{A}d\boldsymbol{\theta}, \tag{9.34}$$

where the Jacobian $\mathbf{A}$ is

$$\mathbf{A}(\boldsymbol{\theta}) = \left(\frac{\partial\beta_i}{\partial\theta_j}\right) = \left(\delta_{ij} - \frac{\partial^2\psi(\boldsymbol{\theta})}{\partial\theta_i\partial\theta_j}\right) = \begin{pmatrix} 1-\kappa-\gamma_1 & -\gamma_2 \\ -\gamma_2 & 1-k+\gamma_1 \end{pmatrix}$$

and the shear components γ_1 and γ_2 are given by

$$\gamma_1 = \frac{1}{2}(\psi,_{11} - \psi,_{22}), \qquad \gamma_2 = \psi,_{12}\,. \tag{9.35}$$

The shear is a trace-free part of the symmetric matrix and can be represented by a complex number

$$\gamma = \gamma_1 + i\gamma_2.$$

It is not a vector, because it transforms under a coordinate rotation by an angle ϕ as $\exp(2i\phi)$; other complex orientation descriptors that we define below are similar. This is related to the fact that an ellipse is invariant under rotations $\phi = n\pi$, where n is an integer.

Shear and convergence are related. Using their representation by the deflection potential (9.35) we obtain

$$\gamma(\boldsymbol{\theta}) = \frac{1}{\pi}\int D(\boldsymbol{\theta}-\boldsymbol{\theta}')\kappa(\boldsymbol{\theta}')\, d^2\theta', \tag{9.36}$$

where

$$D(\boldsymbol{\theta}) = \frac{\theta_2^2 - \theta_1^2 - 2i\theta_1\theta_2}{|\boldsymbol{\theta}|^4}$$

To clarify the meaning of the convergence κ and the shear γ, we write the linearized mapping (9.34) in the neighborhood of the image point $\boldsymbol{\theta}$, which corresponds to the source point $\boldsymbol{\beta}_0$, as

$$I(\boldsymbol{\theta}) = I_s\big(\boldsymbol{\beta}_0 + \mathbf{A}(\boldsymbol{\theta}-\boldsymbol{\theta}_0)\big).$$

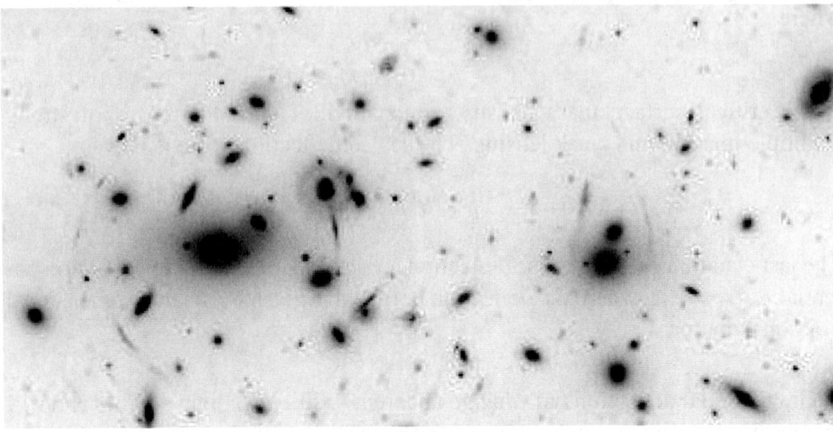

Figure 9.13 *The central part of the Abell 2218 galaxy cluster, with strong gravitational lensing effects. (Reproduced, with permission, from Kneib et al. 1996, Astrophys. J., 471, 643–656. AAS.)*

Here I represents the surface brightnesses of the image and the source, respectively. This mapping shows that a circular source of a unit radius will be transformed to an ellipse with semiaxes

$$a = \left(1 - \kappa - |\gamma|\right)^{-1}, \quad b = \left(1 - \kappa + |\gamma|\right)^{-1}.$$

The total fluxes from the image and the (unlensed) source are proportional to the areas of the image and the source, and their ratio is the magnification $\mu(\boldsymbol{\theta}_0)$:

$$\mu = \frac{1}{\det \mathbf{A}} = \frac{1}{(1-\kappa)^2 - |\gamma|^2}$$

The regions in the image plane, where the transformation is singular and the magnification is formally infinite, $\det \mathbf{A} = 0$, are called critical curves. Near critical curves the magnification and distortion of images are great.

Fig. 9.13 shows the Hubble Space Telescope image of the well-known lensing cluster Abell 2218, where many strongly deformed images of background galaxies (arclets) are seen, together with a number of less deformed images. Such a map can be used to restore the projected mass distribution of the far-away cluster (its distance is $525h^{-1}$ Mpc). Because lensing is strong, several critical curves and multiple images can be identified here (Kneib et al. 1996).

9.6.2 *Weak lensing*

Gravitational lensing effects usually are weak, and the shear in the image plane has to be determined by measuring ellipticities of background galaxies.

The shape of a galaxy image can be described by the tensor of brightness moments M_{ij}:

$$M_{ij} = \int W(\theta) I(\boldsymbol{\theta}) \theta_i \theta_j d^2\theta,$$

where the angles $\boldsymbol{\theta}$ are calculated in a coordinate system centered on the image and $W(\theta)$ is a normalized weight function. The complex ellipticity e is defined as

$$e = \frac{1}{M_{11} + M_{22}} (M_{11} - M_{22} + 2iM_{12})$$

The lens mapping (9.34) induces a relation between the source moments $\mathbf{M}^s$ and the image moments $\mathbf{M}$:

$$\mathbf{M}^s = \mathbf{A}\mathbf{M}\mathbf{A}^T,$$

which gives the relation between the complex ellipticities of the source e^s and image e, respectively:

$$e^s = \frac{e - 2g + g^2 e^\star}{1 + |g|^2 - 2\Re(g e^\star)}. \tag{9.37}$$

This relation depends only on the reduced shear

$$g(\boldsymbol{\theta}) = \frac{\gamma(\boldsymbol{\theta})}{1 - \kappa(\boldsymbol{\theta})}.$$

Taking the expectation value of (9.37), the reduced shear can be expressed via the measured average image ellipticities. For a weak lensing regime, when $\kappa \ll 1$, $|\gamma| \ll 1$, and $|g| \ll 1$, we get from (9.37) the relation

$$e \approx e^s + 2g = e^s + 2\gamma.$$

As the orientation of background galaxies is isotropic, the expectation value of this relation gives

$$\langle|\gamma|\rangle = \frac{1}{2}\langle e \rangle.$$

Another measure of complex ellipticity is frequently used:

$$\epsilon = \frac{e}{1 + (1 - |e|^2)^{1/2}},$$

for which

$$\langle|\gamma|\rangle = \langle \epsilon \rangle.$$

Thus, we can estimate the shear by locally averaging the complex ellipticities of galaxy images. Once we know the shear distribution in the image plane $\gamma(\boldsymbol{\theta})$, we can, in principle, reconstruct the projected surface density by inverting the convergence–shear relation (Kaiser and Squires 1993):

$$\kappa(\boldsymbol{\theta}) = \frac{1}{\pi} \int \Re\left[D^\star(\boldsymbol{\theta} - \boldsymbol{\theta}')\gamma(\boldsymbol{\theta}')\right] d^2\theta' + \kappa_0.$$

Another approach is to use maximum-likelihood and Bayesian methods for reconstruction. A list of references and application of a maximum entropy method can be found in Bridle et al. (1998).

9.6.3 Cosmic shear

Reconstruction of selected lensing mass distributions has been feasible for several years. Another application of gravitational lensing is measurement of cosmic shear, produced by the overall effect of the large-scale gravitational fields. This signal was first detected only in 2000.

Using a similar reasoning that leads us from the lens equation to the expression for the convergence (see, e.g., Schneider et al. 1998), the cosmic convergence can be written as

$$\kappa(\boldsymbol{\theta}, \omega) = \frac{3}{2}\frac{H_0^2 R_0^2}{c^2}\Omega_M \int_0^{\omega} \frac{S_k(\omega - \omega')S_k(\omega')}{S_k(\omega)} \frac{\delta[S_k(\omega')\boldsymbol{\theta}, \omega']}{a(\omega')}\, d\omega'. \qquad (9.38)$$

Here ω is the dimensionless comoving distance, δ is the three-dimensional density contrast, Ω_M is the mass density parameter, H_0 is Hubble's constant, R_0 is the radius of curvature, and a is the scale factor. The function $S_k(\omega)$ describes the geometry of the model and is defined in (2.2); the angular diameter distance is given by $d_a = R_0 S_k(\omega)$. The cosmic convergence field (9.38) depends on the source distance ω. Introducing the normalized source density distribution $p(\omega)$, the total convergence becomes

$$\kappa(\boldsymbol{\theta}) = \frac{3}{2}\frac{H_0^2 R_0^2}{c^2}\Omega_M \int_0^{\omega_H} g(\omega) S_k(\omega) \frac{\delta[S_k(\omega)\boldsymbol{\theta}, \omega]}{a(\omega)}\, d\omega, \qquad (9.39)$$

where ω_H is the comoving horizon distance and

$$g(\omega) = \int_{\omega}^{\omega_H} p(\omega') \frac{S_k(\omega' - \omega)}{S_k(\omega')}\, d\omega'$$

is the source-averaged distance ratio D_{ds}/D_s (see Fig. 9.12). Assuming, as usual, that the density contrast is a Gaussian random field, the convergence field is also a Gaussian random field, and it can be described by its power spectrum. Calculating the dependence between the convergence power spectrum and the three-dimensional density contrast power spectrum is similar to calculating the relation between the projected and spatial density power spectra (see Chapter 8). The only difference is in the weighting functions. A detailed derivation can be found in Schneider et al. (1998) and in Bartelmann and Schneider (2001) (please note that both these papers use the comoving coordinate $w = R_0\omega$, while we use the dimensionless comoving coordinate ω). The result is

$$P_\kappa(K) = \frac{9}{4}\frac{H_0^4 R_0^4}{c^4}\Omega_M^2 R_0^{-3} \int_0^{\omega_H} \frac{g^2(\omega)}{a^2(\omega)} P_\delta\left(\frac{K}{R_0 S_k(\omega)}; \omega\right) d\omega, \qquad (9.40)$$

where $P_\kappa(K)$ is the convergence power spectrum, K is the modulus of the two-dimensional wave vector, and P_δ is the density contrast power spectrum. Its dependence on ω stresses the fact that this spectrum evolves with time along the light cone.

Using the relation between the convergence and shear (9.36), we can find the variance of the modulus of the shear in top-hat windows of radius θ_c:

$$\langle|\gamma|^2\rangle = \frac{2}{\pi\theta_c^2}\int_0^\infty P_\kappa(K)J_1^2(K\theta_c)\frac{dK}{K}. \tag{9.41}$$

This is the simplest measure of cosmic shear. Another measure, the aperture mass, is defined as

$$M_{ap} = \int_{\theta<\theta_c} \kappa(\boldsymbol{\theta})\, U(\theta)\, d^2\theta, \tag{9.42}$$

where $U(\theta)$ is a zero-mean (compensated) filter. Schneider et al. (1998) showed that the aperture mass variance is given by the convergence power spectrum:

$$\langle M_{ap}^2\rangle = \frac{288}{\pi\theta_c^4}\int_0^\infty P_\kappa(K)J_4^2(K\theta_c)\frac{dK}{K}.$$

Here and above we have used the standard notation for the Bessel functions $J_n(x)$. It is useful to define for each galaxy the tangential and radial shears γ_t and γ_r with respect to the center of the aperture as

$$\gamma_t = -\gamma_1\cos(2\phi) - \gamma_2\sin(2\phi), \tag{9.43}$$
$$\gamma_r = -\gamma_2\cos(2\phi) + \gamma_1\sin(2\phi), \tag{9.44}$$

where ϕ is the angle between the x-axis and the line from the aperture center to the galaxy. Now we can express the aperture mass by the tangential shear:

$$M_{ap} = \int_{\theta<\theta_c} \gamma_t(\boldsymbol{\theta})\, Q(\theta)\, d^2\theta, \tag{9.45}$$

where the new filter $Q(\theta)$ is given by

$$Q(\theta) = \frac{2}{\theta^2}\int_0^\theta U(\theta')\theta' d\theta' - U(\theta).$$

An interesting fact is that when we replace γ_t by γ_r in (9.45), the integral vanishes. This can be used to check the origin of the signal; if it is due to intrinsic ellipticities of galaxies, the integral is not zero.

As an example of a recent measurement of cosmic shear, we refer the reader to the article by van Waerbeke et al. (2001). They measured ellipticities for about 1.2 million faint galaxies, and estimated the $\langle|\gamma|^2\rangle$ and $\langle M_{ap}^2\rangle$ statistics. Fig. 9.14 shows their results for the aperture mass statistics.

The upper panel shows the statistic itself, compared with predictions of different cosmological models, and the lower panel shows the radial shear check. The definition of the radial and tangential shear (9.43) shows that the tangential

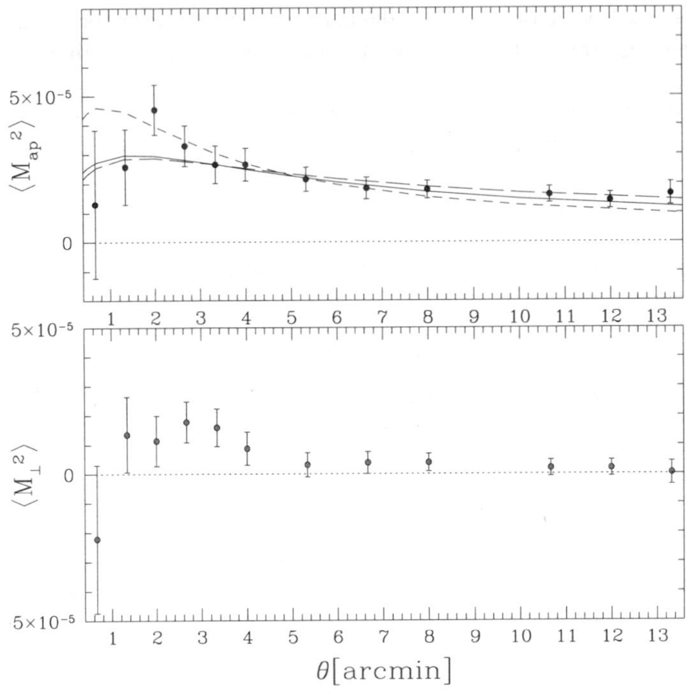

Figure 9.14 *Aperture mass statistics for different aperture sizes θ. The upper panel shows the M_{ap} statistics and different model predictions. The lower panel shows the R-mode, which has to be zero for pure gravitational lensing. (Reproduced, with permission, from van Waerbeke et al. 2001, Astr. Astrophys., 296, 374, 757–769).*

shear can be made radial by rotating a galaxy by 45°. van Waerbeke et al. (2001) call the mass aperture statistic for 45° rotated galaxies the R-mode; in regions where it differs from zero, the statistic is probably influenced by intrinsic galaxy ellipticities or by instrumental effects. As we see, this region is rather limited, and the mass aperture estimate in the upper panel truly represents the cosmic shear.

As this shear can be predicted, given the parameters of the cosmological model and the model density contrast power spectrum, cosmic shear measurements can be used as an independent check of these models.

Fig. 9.14 also shows that the amplitude of the signal is extremely small. Thus, in order to estimate this value with necessary confidence, a large number of faint galaxy images have to be processed. This requires considerable effort and also involves careful elimination of various systematic instrumental effects.

Another integrated gravitational lens effect is the magnification bias of galaxy number counts. If the surface density of galaxies with flux greater than S is $N(S)$, then gravitational lensing changes this density to $N'(S) = \mu^{-1}N(S\mu^{-1})$. The first factor μ^{-1} arises because solid angles are magnified by μ; thus, the number density is diluted. The second factor changes the detectable flux limit, and magnification allows us to see weaker sources. The magnification bias is, then,

$$q(\mu, S) = \frac{N'}{N} = \frac{N(S\mu^{-1})}{\mu N(S)}.$$

Number counts frequently have a power-law form, $N \sim S^{-\alpha}$, which gives the magnification bias $q = \mu^{\alpha-1}$. Thus for the case $\alpha = 1$ gravitational lensing does not change the source counts. The steeper the number counts relation $N(S)$, the stronger the magnification bias. For weak lensing the magnification is $\mu = 1+2\kappa$; thus, the expected change in counts is $\Delta N/N = 2\kappa|\alpha - 1|$.

CHAPTER 10

Structure statistics

10.1 Introduction

The picture that emerged from the description of the galaxy distribution in Chapter 1 is one of a network of galaxies concentrated in clusters, filaments, and walls surrounding large empty voids. While the statistical measures described in Chapters 3 and 8 do provide a wealth of information about the scale, amplitude, and even the nature of the deviations from a uniform distribution, they at best offer only suggestive statistical measures for these structural patterns.

In this chapter we present the tools that have been developed to highlight one or more of these structural features in the galaxy distribution. We start with a recapitulation of the observational situation with respect to the structural features, trying to define carefully which aspects should be described.

The first aspect of the density field that we want to describe is the topology of the galaxy distribution. A particularly good measure to judge whether we have a cellular or a sponge-like distribution of galaxies is the Euler characteristic of a density field, characterized by the genus. The theory behind this quantity is presented, together with a discussion of what is expected in a Gaussian density field. The results of measurements in observational catalogs are presented. We discuss the techniques for obtaining the smooth density field from the discrete galaxy distribution and their influence on determination of the genus.

While the topological parameter provides some information on the nature of the galaxy distribution, a real statistical description of the walls, filaments, and clusters starts to emerge only with the development of structure functions. We treat here the shape statistics introduced by Luo and Vishniac (1995), in addition to the mathematically clean description defined by the Minkowski functionals.

Two older statistical techniques that focused in particular on the detection of filaments in the galaxy distribution are percolation analysis and minimal spanning trees. In particular, the latter may have the potential to be useful not only as a descriptor in its own right, but also as a tool to estimate other statistical quantities (Barrow 1992).

We will discuss the techniques that have been developed to identify voids in the galaxy distribution. Wavelets have been introduced in cosmology as structure detectors. We will review what they are and which aspects are useful for the analysis of galaxy clustering. We present different algorithms, many of them based on the previous techniques, to find clusters within clustered point processes. Finally, the appropriate statistical tools for studying any possible periodicity of structures in the galaxy distribution are discussed.

10.2 Topological description

10.2.1 The theory of topological analysis: the genus

Topological analysis of the large-scale structure is how cosmologists refer to the art of measuring the degree of connectivity of the matter distribution in the universe once the redshift surveys have been smoothed with an appropriate filter function, as explained in Chapters 3 and 9. This is done by means of the topological genus (Gott, Dickinson, and Melott 1986; Mellot 1990; Coles 1992). The genus of a surface G is basically

$$G = \text{(number of holes)} - \text{(number of isolated regions)} + 1.$$

For example, a sphere has topological genus equal to 0, and a torus has genus $+1$, while N disjoint spheres have $G = -(N-1)$.

The Gauss–Bonnet theorem provides the relationship between the curvature of the surface and its topological genus. According to this theorem, the integral of the Gaussian curvature of a compact two-dimensional surface is

$$\int \mathcal{K} dS = 2\pi\chi = 4\pi(1-G)$$

where χ is the Euler–Poincaré characteristic, related to the fourth Minkowski functional (Stoyan, Kendall, and Mecke 1985).

Gott et al. (1986) proposed an algorithm for calculating the topological genus of an isodensity surface, the boundary surface between two regions with density above and below a given density threshold. An important property of a Gaussian field is an analytically calculable genus curve (Doroshkevich 1970; Adler 1981). The genus per unit volume $g \equiv (G-1)/V$

$$g(\nu) = N(1-\nu^2)e^{-\nu/2}, \tag{10.1}$$

where ν is the number of standard deviations from the mean density of the density threshold and N is the normalization constant that depends on the power spectrum of the smoothed density field

$$N = \frac{1}{4\pi^2}\left(\frac{\langle k^2\rangle}{3}\right)^{3/2},$$

where

$$\langle k^2\rangle = \frac{\int k^2 P(k) d^3k}{\int P(k) d^3k}.$$

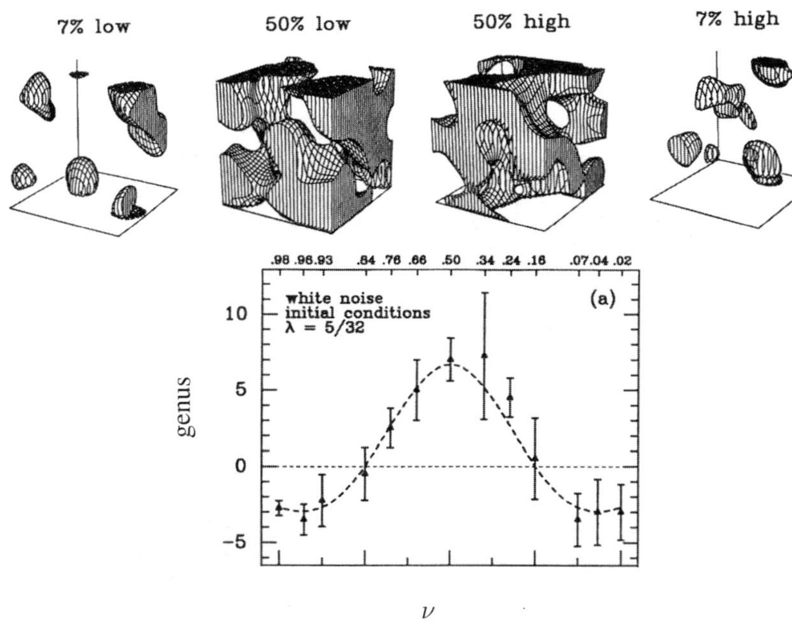

Figure 10.1 *Several isodensity contours corresponding to a Gaussian field are shown in the top panels. At the bottom, the genus curve calculated empirically (triangles and error bars) is displayed together with the expected analytical expression (10.1). The smoothing radius $\lambda = \sqrt{2}\omega$ is shown in units of the side length. (Reproduced, with permission, from Weinberg, Gott, and Melott 1987, Astrophys. J., 321, 2–27. AAS.)*

In fact, for a power-law spectrum, $P(k) \propto k^n$, the constant can easily be calculated

$$N = \frac{1}{(2\pi)^2 2^{3/2} \omega^3} \left(\frac{3+n}{3} \right)^{3/2}, \qquad n \leq -3,$$

where ω is the smoothing length of a Gaussian filter. The genus curve, corresponding to the relation (10.1), is symmetric about 0 in ν, or in other words it is symmetric around the mean density, with the overdense regions and the underdense regions statistically indistinguishable. This is the Gaussian or random-phase topology corresponding to a "sponge-like" surface. For $|\nu| < 1$, $g(\nu) > 0$ corresponding to a surface with many holes, multiply connected and negatively curved. For $|\nu| > 1$, $g(\nu) < 0$; therefore, there are isolated clusters and isolated voids.

Isodensity surfaces can be labeled by means of the fraction $f_{\rm vol}$ of the volume contained in regions with density exceeding a given threshold, or by the number,

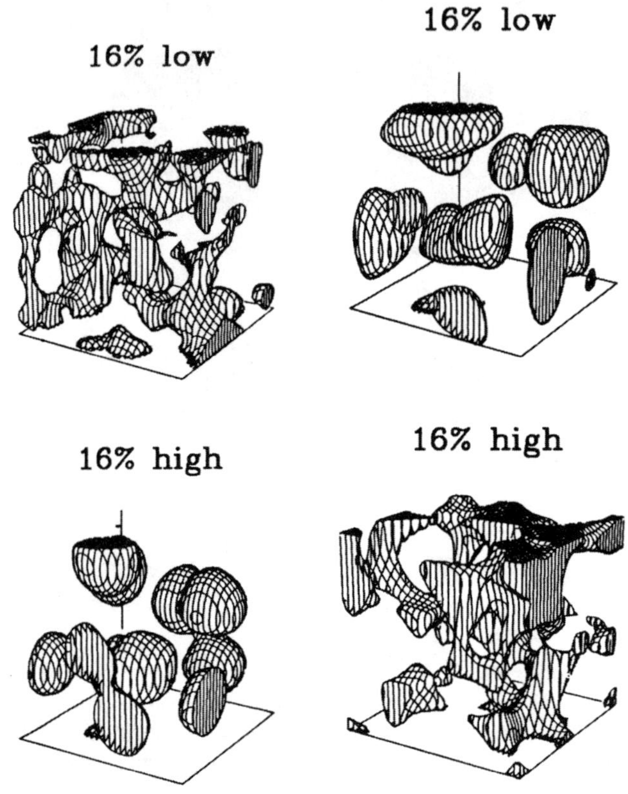

Figure 10.2 *The isodensity contours for two models. On the left, eight clusters have been superimposed over a uniform background. On the right, eight bubbles have been simulated. As we see the roles of the high-density regions and low-density regions have been interchanged in both models. (Reproduced, with permission, from Weinberg, Gott, and Melott 1987, Astrophys. J., 321, 2–27. AAS.)*

$\nu_{\rm vol}$, defined by

$$f_{\rm vol} = \frac{1}{\sqrt{2\pi}} \int_{\nu_{\rm vol}}^{\infty} e^{-t^2/2} dt.$$

Note that, for a Gaussian density field, $\nu_{\rm vol}$ is equivalent to the number of standard deviations ν that a given density threshold is above or below the average density, but, in general, this is not the case. The topological analysis consists of studying how the genus of an isodensity surface varies with $f_{\rm vol}$ or $\nu_{\rm vol}$.

Fig. 10.1 shows the isodensity contours for a Gaussian field in a box, where the notation $x\%$ high ($x\%$ low) means that the depicted contours are encompassing

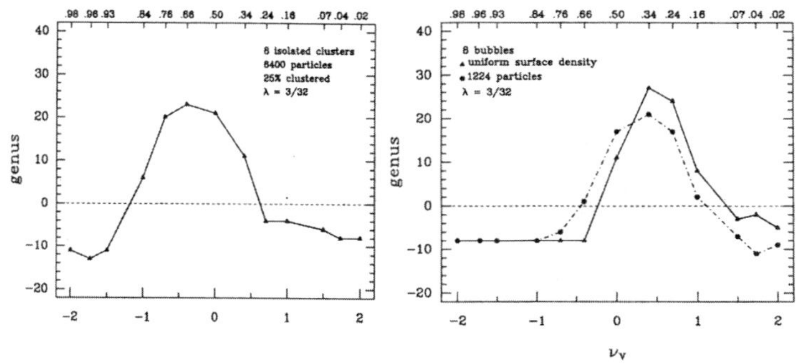

Figure 10.3 *The genus curve as a function of* $f_{\rm vol}$ *(top axis) and* $\nu_{\rm vol}$ *(bottom axis) calculated empirically for the models of Fig. 10.2. (Reproduced, with permission, from Weinberg, Gott, and Melott 1987, Astrophys. J., 321, 2–27. AAS.)*

the most dense (less dense) $x\%$ of the total volume of the box. The corresponding genus curve is shown in the same figure calculated over four realizations of this model; the triangles show the average and rms fluctuations are represented by error bars.

If the curves of the genus are biased to the left the topology corresponds to isolated clusters superimposed on a smooth background. It is referred to as "meatball" topology. Finally, if the curve is biased to the right, the topology is the "Swiss-cheese" kind, with empty voids surrounded by one connected high density region. Weinberg, Gott, and Melott (1987) and Melott (1990) analyzed several point distributions by means of this formalism. In a model dominated by clusters, 6,400 points were distributed in a 32^3 lattice, 75% randomly and 25% on eight clusters. A similar model but with eight bubbles was also generated. Fig. 10.2 shows some isocontours corresponding to these models and Fig. 10.3 shows the corresponding genus curves illustrating the behavior explained above.

10.2.2 Estimation of the topology, technicalities

The first step in applying this kind of analysis to real three-dimensional galaxy distribution is to smooth the point process to obtain a representation of the continuous density field. Some methods were presented in Section 3.3.1. A detailed study is presented in Melott and Dominik (1993). Once the discrete point distribution has been smoothed, the genus statistic is applied to the isodensity contours.

The shape of the genus curve is clearly affected by the election of the bandwidth λ. Larger values of λ tend to oversmooth the data, creating positive genus, while smaller values of λ produce undersmoothing and, thus, negative genus. This variability is also useful when applying the genus statistic to real galaxy catalogs:

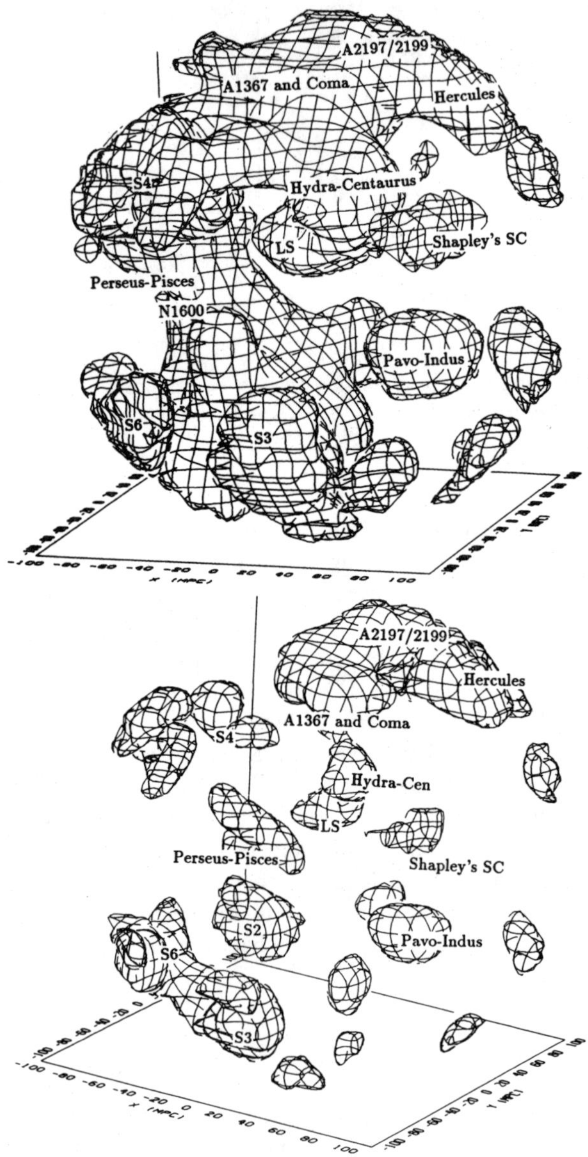

Figure 10.4 *Isodensity contours of the QDOT survey. The top panel shows the contours enclosing the regions corresponding to one third high-density volume and the bottom panel shows the contours enclosing the regions corresponding to one tenth high-density volume. (Reproduced, with permission, from Moore et al. 1992, Mon. Not. R. Astr. Soc., 256, 477–499. Blackwell Science Ltd.)*

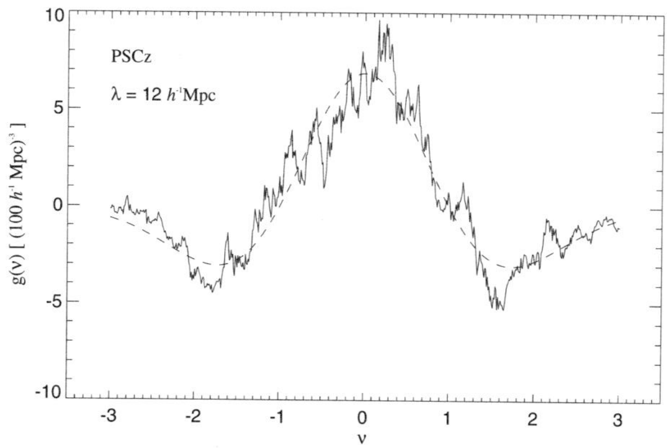

Figure 10.5 *The genus curve of the PSCz redshift survey for a smoothing radius* $\lambda = 12\,h^{-1}$ *Mpc (solid line). The dashed line is the best Gaussian fit to the empirical curve. (Reproduced, with permission, from Canavezes et al. 1998, Mon. Not. R. Astr. Soc., 297, 777–793. Blackwell Science Ltd.)*

It has become standard to show the genus curve applied to the same data set for different values of the smoothing radius.

When a given isodensity surface is approximated by a network of polygonal faces with the grid size much smaller than the smoothing radius, the genus of the surface G verifies

$$\sum_i D_i = 4\pi(1 - G), \tag{10.2}$$

where

$$D_i = 2\pi - \sum_i V_i$$

is the angle deficit around vertex i and V_i are the angles around the vertex. As an example, we can approximate a sphere by a cube. It has 8 vertices. The angle deficit at each vertex is $\pi/2 = 2\pi - 3\pi/2$, and therefore the expression (10.2) provides a value for the genus $G = 0$, as expected for the sphere.

10.2.3 Topological measurements: observations

Several galaxy redshift catalogs have been analyzed thus far using the genus curve, the QDOT catalog (Moore et al. 1992), the CfA2 catalog (Vogeley et al. 1994), and the PSCz catalog (Canavezes et al. 1998). Fig. 10.4 shows the isodensity contours of the QDOT survey encompassing roughly one third (left panel) and one tenth (right panel) of the total volume. Different large scale structures, clusters, and superclusters, mentioned in Chapter 1, are highlighted in these plots. Although analysis of old surveys seemed to support a slight departure of the sponge-like topology at small scales $\leq 10\,h^{-1}$ Mpc to a filamentary distribution, the analysis of Canavezes at al. (1998) has shown that the PSCz catalog supports the hypothesis of random phases for the initial conditions. No significant phase correlations are detected on scales from 10 to 32 h^{-1} Mpc. The expected universal w-shape genus curve fits well the observations, as can be appreciated in Fig. 10.5.

10.3 Structure functions

10.3.1 Three-dimensional shape statistics

As illustrated in Chapter 1, the galaxy distribution exhibits large-scale features such as clusters, filaments, walls, and voids. The sheet-like and filamentary structure is particularly evident in most of the redshift slices compiled thus far. Luo and Vishniac (1995) have introduced shape statistics to obtain a quantitative description of the number of filaments and pancakes (planar shapes) in the galaxy distribution. These statistics are based on the first and second moments of the point distribution within a window. For a ball of radius R centered at the galaxy at position $\mathbf{x}_0$, the moments are defined in the following way:

$$M_i = \frac{1}{N}\sum_k (x_k^i - x_0^i)$$

and

$$M_{ij} = \frac{1}{N}\sum_k (x_k^i - x_0^i)(x_k^j - x_0^j)$$

where the sums are taken over all galaxies in the window.

From these quantities Luo and Vishniac (1995) have defined the following cubic structure statistics as (repeated indices in a term means summation from 1 to 3)

$$\begin{aligned} S_l &= \frac{1}{(M^{ii})^3}\{-M^{ii}(M^{ij} - M^iM^j)^2 \\ &+\frac{1}{2}M^{ii}(M^{kk} - (M^j)^2) \\ &+3M^{ij}(M^{jk} - M^jM^k)(M^{ik} - M^iM^k) \\ &-\frac{3}{2}M^{ij}(M^{ii} - M^kM^k)(M^{ij} - M^iM^j)\} \end{aligned} \qquad (10.3)$$

$$\begin{aligned} S_p &= \frac{1}{(M^{ii})^3}\{+4M^{ii}(M^{ij} - M^iM^j)^2 \\ &-4M^{ii}(M^{kk} - (M^j)^2) \\ &-12M^{ij}(M^{jk} - M^jM^k)(M^{ik} - M^iM^k) \end{aligned}$$

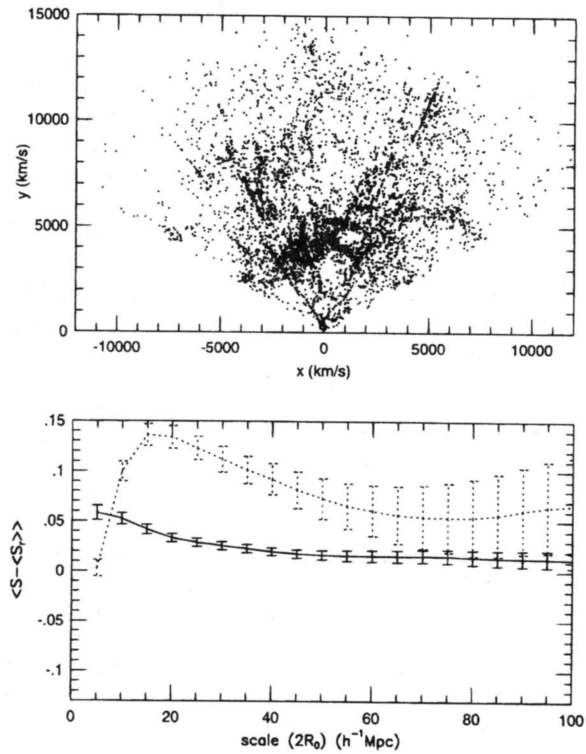

Figure 10.6 *The Perseus–Pisces catalog and the shape statistics,* $\langle S - \langle S_r \rangle \rangle$*, calculated on this sample. (Reproduced, with permission, from Luo, Vishniac, and Martel 1996, Astrophys. J., 468, 62–74. AAS.)*

$$+12M^{ij}(M^{ii} - M^k M^k)(M^{ij} - M^i M^j)\} \tag{10.4}$$

Luo, Vishniac, and Martel (1996) have applied these statistics to galaxy redshift surveys. They denote by $\langle S_r \rangle$ the mean values of the shape statistics calculated for ten realizations of a binomial process with the number of points equal to the number of galaxies in the galaxy survey and within the same volume. Fig. 10.6 shows the average for all the windows of a given radius R_0, $\langle S - \langle S_r \rangle \rangle$, as a function of the diameter of the window, $2R_0$, for the Perseus–Pisces sample. It can be appreciated that the plane shape statistic has a peak around 15 h^{-1} Mpc. This is probably a consequence of the alignment of fingers-of-God. The line shape statistic declines monotonically from 5 h^{-1} Mpc, indicating that the filamentary structures are more evident at smaller scales.

10.3.2 Minkowski functionals

Minkowski functionals are well-known measures in the field of stochastic geometry. These quantities, which serve to characterize the shape and connectivity of point patterns, were introduced in cosmology by Mecke, Buchert, and Wagner (1994).

The scalar functionals are applied to subsets of the Euclidean space $\mathbb{R}^d$ that are finite unions of convex bodies. The conditions that the functional must verify are invariance under translations and rotations, additivity (the functional of the union of two bodies is the sum of the functionals of each body minus the functional of the intersection), and the convex continuity (the functionals of approximations of a convex body converge, using the Hausdorff metric, to the functional of the convex body).

Hadwiger (1957) has shown that there are only $d+1$ independent functionals verifying these conditions. These $d+1$ measures provide us with a full characterization of the global morphology of the convex set. In $\mathbb{R}^3$, the four functionals have simple interpretation in terms of well-known geometric quantities: they are the volume V, the surface area A, the integral mean curvature H, and the Euler–Poincaré characteristic, χ. The four Minkowski functionals in their more common notations are

$$M_0 = V, \quad M_1 = \frac{A}{8}, \quad M_2 = \frac{H}{2\pi^2}, \quad M_3 = \frac{3\chi}{4\pi}, \tag{10.5}$$

where the quantities A, H, and χ of a compact convex body in $K \subset \mathbb{R}^3$ with smooth boundary ∂K and principal curvature radii R_1 and R_2 are defined in terms of the surface integrals

$$A = \int_{\partial K} dS,$$

$$H = \frac{1}{2}\int_{\partial K}\left(\frac{1}{R_1} + \frac{1}{R_2}\right) dS,$$

and

$$\chi = \frac{1}{4\pi}\int_{\partial K} \frac{1}{R_1 R_2} dS.$$

Note that in the last expression $\chi \equiv \chi(K) = \chi(\partial K)/2$ (Mecke, Buchert, and Wagner 1994). Other standard notation for the Minkowski functionals is

$$V_\nu = \frac{\omega_{3-\nu}}{\omega_3} M_\nu$$

where

$$\omega_\eta = \frac{\pi^{\eta/2}}{\Gamma(1+\eta/2)}$$

is the volume of the η-dimensional unit sphere. The characterization of point fields with Minkowski functionals is performed by associating to the point distribution

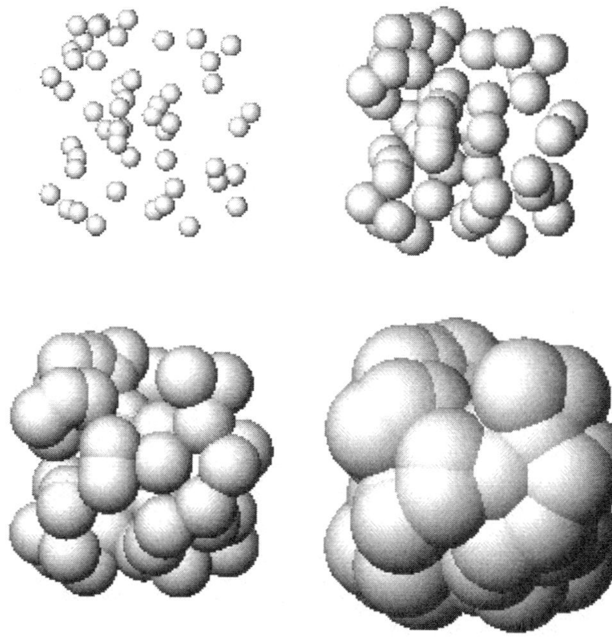

Figure 10.7 *An illustration of the Boolean grain model. The union set A_r is displayed for different values of the diagnostic parameter r for an underlying Poisson distribution. (Reproduced, with permission, from Schmalzing, Kerscher, and Buchert 1996, in Dark Matter in the Universe, IOS Press, Società Italiana di Fisica.)*

a union of convex sets in the following way. Each point of the process is decorated with a ball of radius r; the radius used acts as a parameter (Mecke, Buchert, and Wagner 1994). The functionals are then applied to the sets $A_r = \cup_{i=1}^{N} B_r(\mathbf{x}_i)$. If the underlying point process is a finite realization of a Poisson process, this model is a special case of a Boolean grain model. Fig. 10.7 shows how this model provides different sets A_r as the parameter r varies for an underlying random point process.

We can generalize the Boolean model by replacing the underlying Poisson point process by any general point process. This is known as the germ–grain model or coverage process (Hanisch 1981; Hall 1988; Stoyan, Kendall, and Mecke 1995). Depending on the clustering properties of the underlying process, the sets A_r will present different shapes, and thus different Minkowski functionals. In this manner, Minkowski functionals represent a rather complete tool to study the clustering properties of galaxy and galaxy cluster distributions (Kerscher et al. 1997; Kerscher et al. 1998).

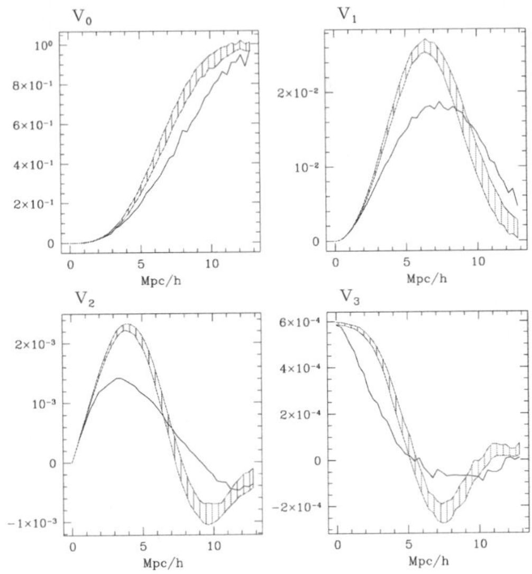

Figure 10.8 *The volume densities of the four Minkowski functionals, V_ν, $\nu = 0, 1, 2, 3$ for a volume-limited sample of the CfA1 galaxy catalog with 80h^{-1} Mpc depth. The shaded areas represent the rms error range of empirical calculations on 15 realizations of a Poisson distribution. (Reproduced, with permission, from Kerscher and Martínez 1998, Bull. Int. Statist. Inst., 57-2, 363–366.)*

For a Poisson process, the mean volume densities of the Minkowski functionals are analytically known (Mecke and Wagner 1991; Schmalzing, Kerscher, and Buchert 1996), and thus, they are usually used as a standard of reference. Fig. 10.8 shows the Minkowski functionals for a volume-limited sample of the CfA1 catalog, together with the Poisson values.

The edge correction for the estimation of the Minkowski functionals relies upon the principal kinematic formula (Santaló 1976) that provides the Minkowski functional of the intersection of convex bodies (Mecke and Wagner 1991).

Minkowski functionals also have been used to characterize the isodensity surfaces of random fields in the analysis of the cosmic background radiation (Schmalzing and Górski 1998). This approach unifies and extends several statistical quantities presented throughout this book (Kerscher 2000). For example, the volume density of the first Minkowski functional, V_0, is just the spherical contact distribution function, defined in Section 3.7 and the fourth Minkowski functional, V_3 is the Euler characteristic, related to the genus of isodensity surfaces of smoothed cosmological fields, as explained in Section 10.2.

Sahni, Sathyaprakash, and Shandarin (1998) and Kerscher (2000) show how particular ratios of the Minkowski functionals can be used to define shape finder quantities (planarity and filamentarity), similar to the quantities introduced in Section 10.3.1, and therefore provide a diagnostic about the presence of filaments, sheet-like structures, and ribbons in the galaxy distribution.

10.4 Cluster and percolation analysis

The Minkowski functionals described in the previous section are meant to describe properties of individual objects, formed by clustering from the original point set. These clusters are not always formal; they also could be physical objects. Cluster analysis was probably first used in cosmology by Turner and Gott (1976) to objectively define groups of galaxies. In later years it has been used to compile catalogs of superclusters (e.g., Einasto et al. 1997). Cluster analysis in cosmology is usually called the "friends-of-friends" (FOF) method, after Press and Davis (1982), although other techniques are also available (see Section 10.7).

The main problem in cosmological cluster analysis is the choice of a "right" neighborhood radius (the radius of the ball built on a sample point) R. This gives us the classification of points (galaxies) into clusters and allows us to find their geometrical properties by Minkowski functionals. This is not an easy choice in practice, as the observed clustering is hierarchical with a continuous range of scales.

We can turn this deficiency of the method into an advantage, considering the variation of the cluster properties with the neighborhood radius. Let us study the process of growth of clusters, substituting the neighborhood radius for time. For small radii the unions of the balls (ball clusters) are more or less isolated and their total volumes are small. As the process continues, clusters grow and join together, until at a certain radius the largest system will connect opposite boundaries of the sample volume. If our sample was a typical region of the universe, we could have a connected structure (cluster) flashing through the whole universe. This moment is called percolation and the corresponding neighborhood radius is called the percolation radius.

Percolation properties of a point sample describe the topology of the spatial distribution — the more filamentary it is, the easier it is to achieve percolation. The usual percolation characteristic for a point sample is the average number of points B_c in a sphere with the percolation radius R_c. Percolation does not depend on the mean density of the sample. This can be found by numerical simulations; for a Poisson distribution it is $B_c \approx 2.7$.

Use of the percolation technique in cosmology was advocated by Shandarin (1983) and Einasto et al. (1984) were the first to apply it to the study of galaxy distributions. There are a few serious problems, however, in the straightforward application of the percolation technique to the observed galaxy catalogs. The most obvious one is the complicated sample geometry; as the opposite sides of a sample

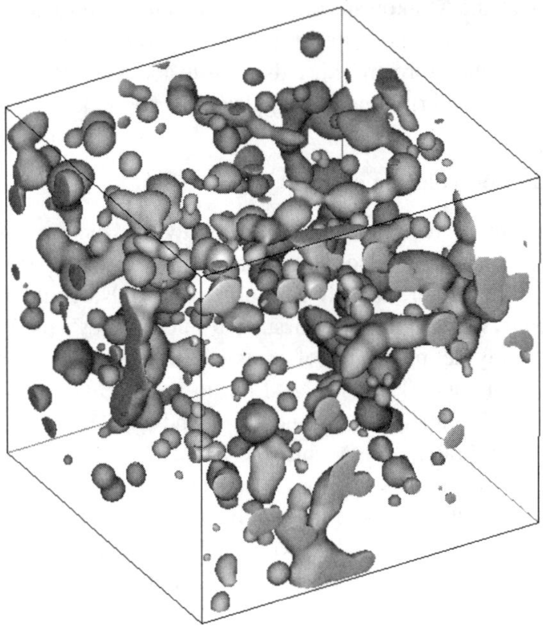

Figure 10.9 *A percolating cluster (for filled regions) for a $P \sim k^{-1}$ cosmological simulation. Its main features are compact clusters and filamentary bridges that connect the clusters together. (Reproduced, with permission, from Colombi, Pogosyan, and Souradeep 2000, Phys. Rev. Let. 26, 5515–5518. Copyright 2000 by the American Physical Society.)*

are not easy to define, percolation has to be established by following the growth of the diameter (maximum extent) or the volume (mass) of the largest cluster. The most serious problem, however, is the Poisson noise caused by the discreteness of the point set. As galaxy catalogs are usually magnitude limited, their mean density drops toward the sample boundaries and we should use a distance-dependent neighborhood radius to track structures through those regions. This can be done, but it also amplifies the Poisson noise.

Thus, in later years, percolation analysis has been used mostly to analyze results of cosmological numerical simulations. A thorough review is given by Klypin and Shandarin (1993); we shall follow their presentation below. As these simulations produce a density field on a lattice, we can carry over the results of percolation studies from condensed matter physics. Percolation techniques have a long history in this branch of science; a good review is given in Ziman (1979). For a density field the role of the neighborhood radius is taken over by the critical overdensity level δ_c; we mark all lattice points of overdensity $\delta \geq \delta_c$ as filled and other points as empty and form clusters of lattice points, considering a given number of nearest lattice points as neighbors (in three dimensions, usually six).

The structure of a typical percolating cluster for cosmological large-scale structure is rather complex, as illustrated by the three-dimensional view in Fig. 10.9. This describes the result of a scale-free simulation with $P(k) \sim k^{-1}$ (appropriate for scales from a few tens to a few hundreds of Mpc). We see that the network consists of more or less spherical clusters, connected by filamentary bridges. Colombi, Pogosyan, and Souradeep (2000) show that local features of the cluster can be classified on the basis of the three-dimensional curvature of the density field.

The overdensity level δ determines the fraction of filled points (cells) p. Every cell can be either empty with the probability $1-p$, can be a member of the percolating cluster with a probability μ_∞ (this cluster is called also the infinite cluster), or can be a member of a finite cluster with a probability $\sum_v vn(v)$. Here v is the size (number of cells) of a cluster and $n(v)$ is density of clusters of size v (the multiplicity function). These quantities describe all properties of clustering on the lattice; they have to satisfy the obvious relation

$$1 - p + \mu_\infty + \sum_v vn(v) = 1.$$

Two clustering characteristics are used in practice to check for percolation. The first is the normalized volume of the largest cluster

$$\mu_\infty = v_{\max}/N^3 \sim (p - p_c)^\beta, \tag{10.6}$$

where $v_{\max}$ is the number of cells in the largest cluster and N^3 is the total number of grid cells (we have assumed a cubic grid volume of N cells per side, for simplicity). The power-law relation near the percolation threshold p_c is true for large filling factors $p > p_c$ ($\mu_\infty = 0, p < p_c$) and is predicted by condensed matter physics. The value of p_c for a Poisson lattice (randomly assigned filled and empty cells) is $p_c \approx 0.313$. The exponent $\beta = 0.4$ for a Poisson lattice has been found not to differ much for cosmological simulations (Klypin and Shandarin 1993 get $\beta \approx 0.5$). At percolation the $\mu_\infty(p)$ curve ("the percolation curve") grows very fast, but the exact percolation threshold is not easy to determine from this curve alone. A better estimate of the percolation threshold p_c can be found by fitting (10.6) to the percolation curve.

The second characteristic is the weighted mean square size of all other (finite) clusters

$$\mu^2 = \frac{1}{N^2} \frac{\sum_v v^2 n(v)}{\sum_v n(v)} \sim |p - p_c|^{-\gamma},$$

where the factor $1/N^2$ is chosen for normalization and the power-law relation comes again from condensed matter theory; the exponent $\gamma \approx 1.7$ is both for the

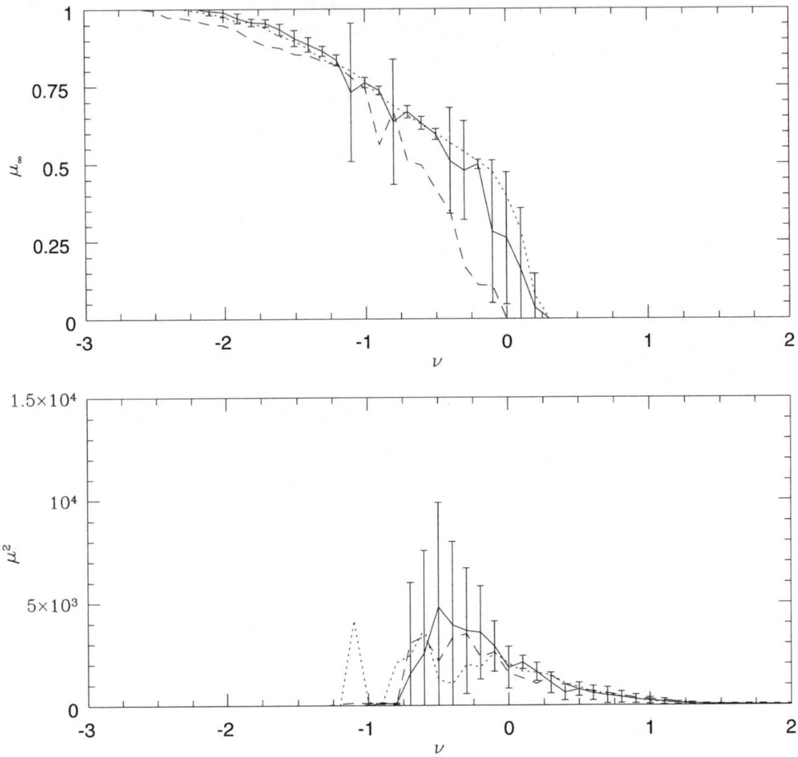

Figure 10.10 *Percolation statistics for various model distributions. The upper panel shows the μ_∞ versus overdensity level curve and the lower panel shows the μ^2 statistics. Different line types correspond to different N-body models. (Reproduced, with permission, from Coles et al. 1998, Mon. Not. R. Astr. Soc., 294, 245–258. Blackwell Science Ltd.)*

Poisson lattice and for cosmological models. As seen from the formula above, the $\mu^2(p)$ dependence peaks at $p = p_c$.

An example of the behavior of these two statistics is shown in Fig. 10.10. This figure shows the percolation curves for galaxy cluster density distribution in different cosmological N-body models. There are a few differences from the methods described above. First, the curves are shown as functions of the normalized overdensity level $\nu = \delta/\sigma$, where σ is the rms overdensity. Second, the μ_∞ statistics is determined on the basis of the volume of a cluster spanning the calculation volume, so it is zero until the start of the percolation. We see that both statistics are rather noisy (error bars are obtained by comparing several realizations). The dynamical evolution of the topology of the large-scale structure

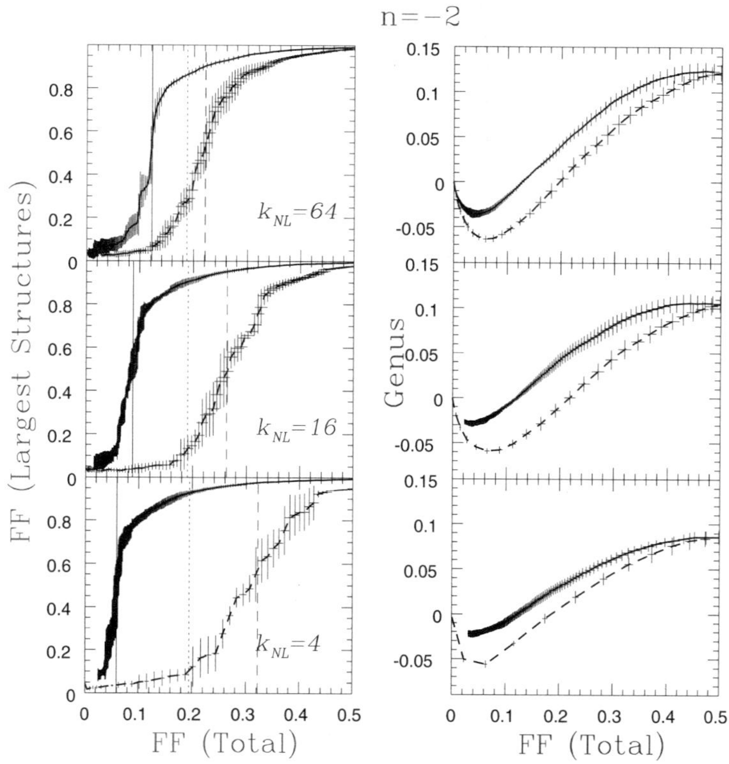

Figure 10.11 *Percolation and genus curves for a $P(k) \sim k^{-2}$ N-body model for various epochs. Thick solid lines describe filled regions; thick dashed lines represent voids. Thin lines indicate percolation levels, solid lines indicate filled regions, and dashed lines indicate voids. The dotted line shows the initial percolation level $p \approx 0.2$. Error bars are obtained by comparing several realizations and the three panels correspond to different epochs — the higher the panel, the earlier the epoch. (Reproduced, with permission, from Sahni, Sathyaprakash, and Shandarin 1997, Astrophys. J. Lett., 476, L1–L5. AAS.)*

is illustrated in Fig. 10.11, where the percolation curves $\mu_\infty(p)$ are calculated for three different moments for several realizations of a $P(k) \sim k^{-2}$ N-body simulation. The epochs are described by the wavenumber k_{NL} where the dynamics just becomes nonlinear (larger k_{NL} correspond to smaller scales and to earlier times).

This figure also shows another advantage of the lattice approach to percolation. While it is difficult to define voids in the case of point samples (see Section 10.8), voids and filled regions can be treated on a lattice in the same way; voids on a lattice are defined as clusters of empty cells. Because all cosmological simulations start from a realization of a Gaussian density field, the initial percolation properties of voids and filled regions are identical (the Gaussian distribution is symmetric). During the evolution the percolation curves start to differ — topology of filled regions becomes more filamentary during the evolution and that of empty regions becomes more spherical.

Fig. 10.11 also shows the genus curves for the same epochs, described earlier in this chapter. As we see, the $g(p)$ relation is very insensitive to dynamical evolution (but it is sensitive to the form of the power spectrum, as shown by Sahni, Sathyaprakash, and Shandarin 1997).

Thus far, the application of percolation techniques to observed samples has suffered from Poisson noise. With the expected rapid growth of the size of future galaxy catalogs the situation should improve; even the volume-limited samples necessary for percolation studies will have enough galaxies to repress the influence of noise.

10.5 Minimal spanning trees

Several statistics used to find filamentary shapes in the galaxy distribution were introduced in the previous sections. Here we show a particular technique to quantify the filamentary character of the galaxy clustering based on the minimal spanning tree.

The minimal spanning tree (MST for short) is a graph-theoretical construct that was introduced by Kruskal (1956) and Prim (1957). The MST of a set of N points is the *unique* network of $N-1$ edges (each linking two points) providing a route between any pair of points while minimizing the sum of the lengths of the edges. We have already used the MST in Chapter 4 to provide an estimator for the Hausdorff dimension. Here, we present this graph-theoretical construction in more detail. The MST can be formulated as follows.

The data set is a *graph G*, consisting of a *vertex set V* (the points) and *edge set E* (an edge is a straight line connecting two points, E is a subset of $V \times V$), each having a "length" or "weight" (in our case the Euclidean distance between the two vertices). A sequence of edges joining vertices is a *path*; a closed path is called a *circuit*. If there is a path between any pair of vertices the graph is called *connected*. A connected graph containing no circuits is called a *tree*. If the tree of a connected graph contains all the vertices of the vertex set then it is called a *spanning tree*. The length of a tree is defined to be the sum of the weights of the component edges. The MST is the spanning tree of minimal length.

The algorithm we used for calculating the MST of a given point set starts by choosing an arbitrary point of the set and finding its nearest neighbor. These two points and the corresponding edge form the subtree T_1. For each *isolated point* (point not yet in the subtree) the identity and distance to its nearest neighbor within the subtree is stored; by definition, this distance is called the distance of the isolated point to the subtree. These potential MST edges are called *links*. The Mth

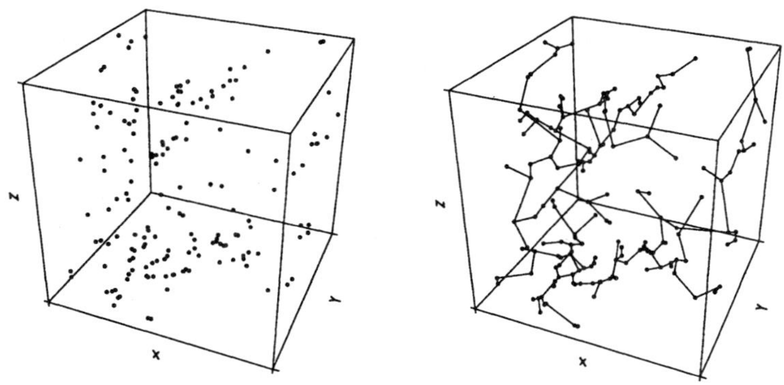

Figure 10.12 *A subsample of the CfA1 redshift survey enclosed in a cubic window. There are 153 galaxies; their MST is shown in the right panel. Several filamentary structures are traced by the graph. (Reproduced, with permission, from Martínez et al. 1990, Astrophys. J., 357, 50–61. AAS.)*

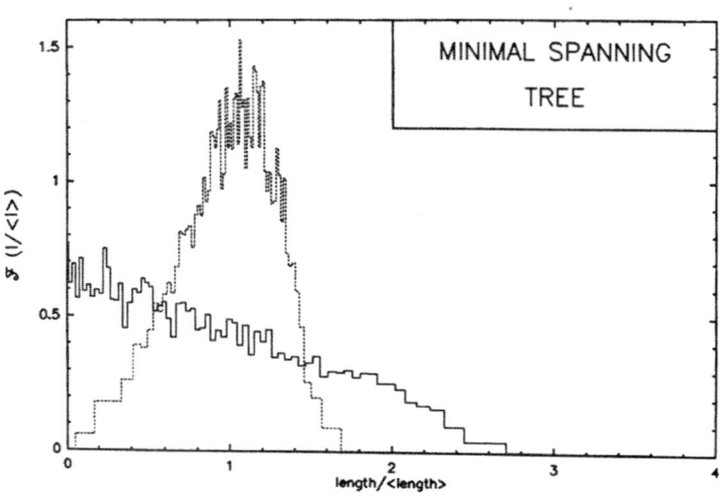

Figure 10.13 *The length distribution function, normalized to the mean edge length, of the MST corresponding to a binomial process (dashed line) and to a Voronoi model (solid line). The distributions reflect the clustering of the underlying point patterns. (Reproduced, with permission, from van de Weygaert 1991, Voids and the Geometry of the Large Scale Structure, Leiden University.)*

subtree, T_M, is formed by adding to T_{M-1} the *nearest neighbor of the subtree*, the isolated point whose distance to the subtree is minimal, with the corresponding link. After adding this new point to the subtree the new links are updated, i.e., the distance from each isolated point to the new subtree vertex is calculated to see whether it is smaller than the previous distance of the point to the subtree. The resulting T_{N-1} is the MST. If there are M points in the subtree the updating requires computation proportional to $N - M$, for M going from 1 to $N - 1$, so that the total computation time is proportional to $N(N - 1)/2$. Moreover, the computation time is just linearly proportional to d, the embedding dimension of the point set.

The use of the MST in cosmology was anticipated by Barrow, Sonoda, and Bhavsar (1985) as a statistic to find clusters and to detect filaments in the galaxy distribution (see Fig. 10.12). Bhavsar and Ling have demonstrated that the MST is a useful tool to show that the filaments are real and not chance alignments; in particular, they have shown that these linear structures appear to radiate from the central regions of rich clusters, an effect that confirms the early view of Einasto, Joeveer, and Saar (1980) that clusters were placed at the intersection of interwoven filaments.

Several quantitative distributions can be extracted from the MST, for example, the number of edges per vertex or the distribution of edge length within the MST (van de Weygaert 1991; Krzewina and Saslaw 1996). The length distribution function of the branches of the MST corresponding to different point patterns is a good clustering measure of the point process. Fig. 10.13 illustrates these differences by means of the histogram showing the edge length distribution functions of a binomial process with 5,000 points and a Voronoi model (6,700 vertices) generated from binomial distributed nuclei (1,000 points). The clustering of the Voronoi vertices is reflected in the plot. The curve corresponding to the Voronoi model reflects the presence of branches with length in a broader range, at small distances connecting the points within the cluster, and at large distances connecting the clusters themselves. Doroshkevich et al. (1999) have made extensive use of the frequency histograms of the MST edge lengths to analyze the galaxy distribution and to compare it with mock catalogs drawn from cosmological simulations.

10.6 Wavelets

10.6.1 Wavelet theory

Wavelet transforms are mathematical tools recently introduced in analysis of signals and images. This integral transform is suitable for getting information localized both in space and scale. A review on wavelets and applications (to turbulence) can be found in Farge (1992) and good mathematical introductions in Chui (1992), Burrus, Gopinath, and Guo (1998), Starck, Murtagh, and Bijaoui, (1998), and Walker (1998). Here, we summarize only some of the main aspects of the wavelet transforms. For simplicity, in this section we shall deal with one-dimensional wavelets. The decomposition of scale performed by the wavelet analysis is made by translation and dilation of a single *parent* function, which is usually called the analyzing wavelet. This is a square integrable function that has to verify the admissibility condition

$$\int_0^{+\infty} \frac{|\widehat{g}(k)|^2}{|k|} dk < \infty \tag{10.7}$$

where

$$\widehat{g}(k) = \int_{-\infty}^{+\infty} e^{-ixk} g(x) dx$$

is the Fourier transform. If $\widehat{g}$ is differentiable, the condition in (10.7) implies that g is of zero mean, or equivalently $\widehat{g}(0) = 0$.

In general wavelets are complex-valued functions. To illustrate how wavelets are really localized both in space and scale we show in Fig. 10.14 two well-known examples of analyzing wavelets together with their corresponding Fourier transforms. The family of wavelets used by Paul (1985) in the context of quantum mechanics is

$$g_p(x) = \Gamma(p+1) \frac{i^p}{(1 - ix)^{1+p}}.$$

The other example is the Marr wavelet, which is the second derivative of a Gaussian (this is a real-valued wavelet)

$$g(x) = (1 - x^2) e^{-x^2/2}.$$

The wavelet transform of a signal $f(x)$ with respect to the analyzing function g is

$$T_g(a, b) = \frac{1}{\sqrt{a}} \int_{-\infty}^{+\infty} f(x) g^* \left(\frac{x - b}{a} \right) dx,$$

where the star denotes the complex conjugate. The parameter $a > 0$ is the dilation of the length scale and b is the position. Therefore, the wavelet transform is just the inner product $\langle f, g_{ab} \rangle$, where

$$g_{ab}(x) = \frac{1}{\sqrt{a}} g \left(\frac{x - b}{a} \right).$$

To stress the main differences between the wavelet transform and the Fourier transform Farge (1992) has proposed considering the Fourier transform of a smooth function $f(x)$ with some singularities. Although the information about $f(x)$ remains in $\widehat{f}(k) = \langle f(x) | e^{ikx} \rangle$, the Fourier spectrum does not describe the spatial distribution at all, i.e., it is not possible to find the singularities in the Fourier space. On the contrary, the local character of the wavelet transform allows us to localize the singularities; in fact, the wavelet coefficients remain smooth while $f(x)$ is locally smooth, and in the neighborhood of a singularity, the amplitude of the coefficients take remarkably high values.

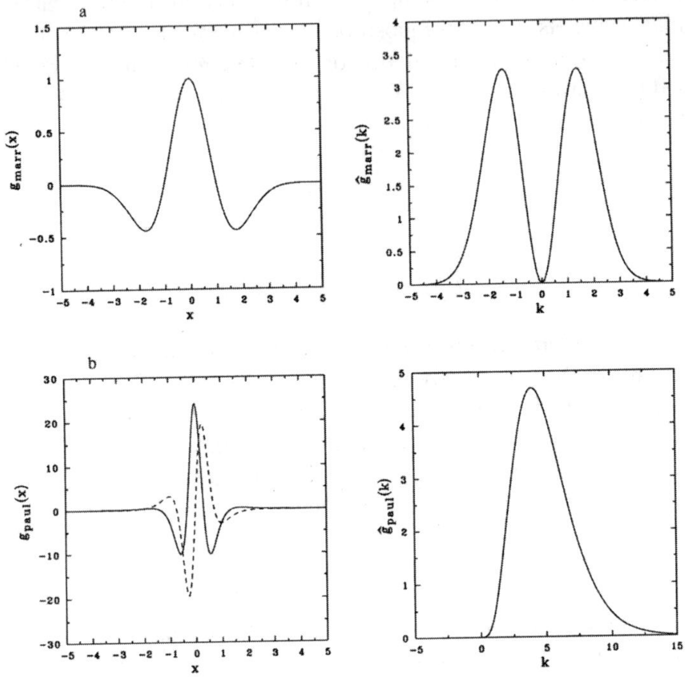

Figure 10.14 *Two examples of wavelet functions together with their Fourier transforms. (Reproduced, with permission, from Martínez, Paredes, and Saar 1993, Mon. Not. R. Astr. Soc., 260, 365–375. Blackwell Science Ltd.)*

One interesting property of the wavelet window function is the fact that it has zero mean

$$\int_{-\infty}^{\infty} g(x)dx = 0.$$

10.6.2 Wavelets and multifractals

The most important property that makes the wavelet transform useful in the context of fractals is the preservation of the scaling. In fact, Holschneider (1988) has shown that if f behaves as

$$|f(x_0 + \lambda x) - f(x_0)| \sim \lambda^{\alpha(x_0)} |f(x_0 + x) - f(x_0)|$$

then the wavelet transform verifies

$$T_g(\lambda a, x_0 + \lambda b) = \lambda^{\alpha(x_0)} T_g(a, x_0 + b).$$

We shall use this property in the calculation of the Hölder exponents from the scaling of the wavelet transforms.

For example, we may ask where the singularities of the fractal measure lie, or in other words, whether it is possible to know the spatial locations of the different structures interweaving in the multifractal object. This goal may be reached by means of the wavelet transform. If the fractal measure defined over $\mathbb{R}^n$ is represented by $d\mu(\mathbf{x}) = f(\mathbf{x})d^n x$, $f(\mathbf{x})$ being the density, the mass within a ball centered at the point $\mathbf{y}$, and radius r is just the integral of $d\mu(\mathbf{x})$ over that ball,

$$\mu(B(\mathbf{y}, r)) = \int_{B(\mathbf{y},r)} d\mu(\mathbf{x}).$$

In $\mathbb{R}^n$, the wavelet transform of the measure with respect to the analyzing function $g(\mathbf{x})$ is defined by (Argoul et al. 1989),

$$T_g(a, \mathbf{b}) = \frac{1}{a^m} \int g^*(a^{-1}r^{-1}(\mathbf{x} - \mathbf{b}))d\mu(\mathbf{x}) \quad a > 0, \quad \mathbf{b} \in \mathbb{R}^n, \tag{10.8}$$

where r represents the n-dimensional rotation operator. The exponent m is freely chosen to get the best visual representation.

One of the most frequently used wavelet functions is the Marr wavelet (Argoul et al. 1989; Slezak, Bijaoui, and Mars 1990):

$$g(\mathbf{z}) = (n - |\mathbf{z}|^2)e^{-|\mathbf{z}|^2/2}.$$

This function is also called the Mexican hat wavelet in 2-D for obvious reasons (see Fig. 10.15). Due to its radial symmetry, this analyzing wavelet works well only for isotropic processes. Therefore, the rotation operator in Eq. 10.8 can be dropped. It also should be possible to use anisotropic wavelets. The possibility of using nonisotropic wavelets could be very interesting in the analysis of the specific shapes introduced by the peculiar velocities in the redshift catalogs.

The wavelet transform may be considered as a mathematical microscope (Arnéodo, Grasseau, and Holschneider 1988), whose magnification and position are given by a^{-1} and $\mathbf{b}$, respectively. The optics of this microscope should be determined by the choice of g. This is illustrated in Fig. 10.16, where the wavelet transform of the the first CfA2 slice is shown for different values of a. In this illustration, the galaxy distribution has been considered two dimensional, ignoring the declination.

This variation can be used as a measure of the local scaling properties of the multifractal set due to the fact that wavelet transforms inherit the preservation of the scaling. As a consequence of the scaling invariance of the wavelet transform, if the measure has multifractal scaling behavior

$$\mu(B(\mathbf{y}, r)) \sim r^{\alpha(\mathbf{y})}$$

the wavelet transform reproduces the scale invariance in the following way (Arnéodo, Grasseau, and Holschneider 1988; Ghez and Vaienti 1989; Freysz et

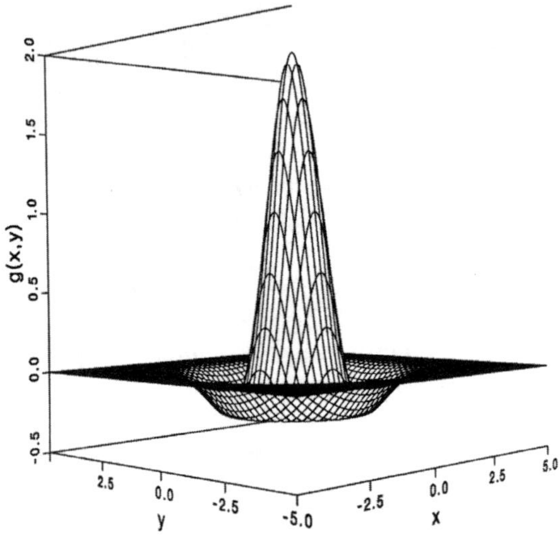

Figure 10.15 *The Mexican hat wavelet function. (Reproduced, with permission, from Martínez, Paredes, and Saar 1993, Mon. Not. R. Astr. Soc., 260, 365–375. Blackwell Science Ltd.)*

al. 1990)

$$|T_g(\lambda a, \mathbf{b})| \sim \lambda^{\widehat{\alpha}(\mathbf{b})} |T_g(a, \mathbf{b})| \qquad \text{with} \quad \widehat{\alpha}(\mathbf{b}) = \alpha(\mathbf{b}) - m.$$

The last equation implies that we can obtain the Hölder exponent $\alpha(\mathbf{b})$ associated with a singularity located at $\mathbf{b}$, just by taking the slope of the log–log plot of the absolute value of the wavelet transform versus the scale a (Martínez, Paredes, and Saar 1993).

Wavelets were introduced in astronomy by Slezak, Bijaoui, and Mars (1990). These authors used the wavelet transform as a structure identifier in cluster analysis performed on Schmidt plates. It also can be used as a powerful tool in the analysis of the subclustering in rich clusters of galaxies (Escalera and Mazure, 1992) and for detecting voids and high-density regions in the CfA slice (Slezak, de Lapparent, and Bijaoui 1993), as shown in Fig. 10.17.

10.7 Cluster-finding algorithms

The definition of a "cluster" in terms of a given distribution of galaxies seen in projection on the sky is a complex process, involving both the nature of the original survey and the selection criteria. Certainly, in the past, the cluster identification methods were subjective and difficult to quantify, yet even now with automatically

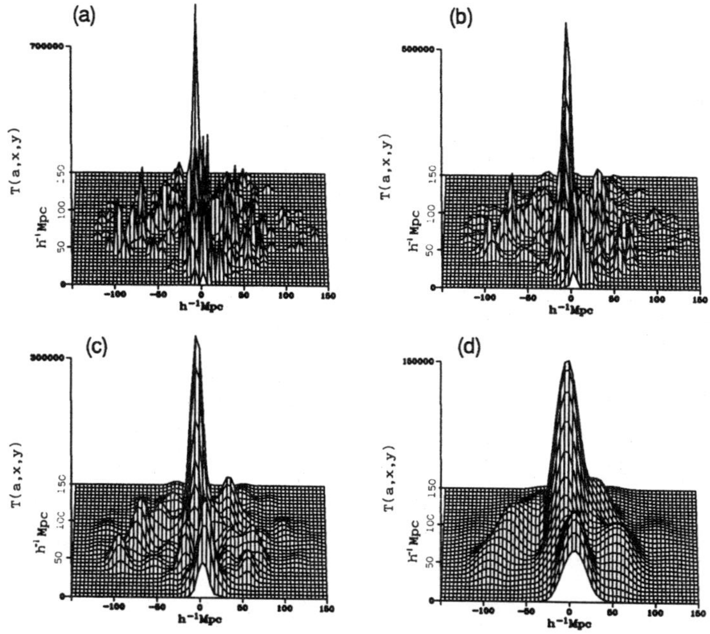

Figure 10.16 *The wavelet transform acting on the first CfA2 slice (see Fig. 8.4). We see how different values of the magnification parameter reveal different structures of the galaxy distribution. (Reproduced, with permission, from Martínez, Paredes, and Saar 1993, Mon. Not. R. Astr. Soc., 260, 365–375. Blackwell Science Ltd.)*

scanned survey plates and with computer algorithms being used to find and classify clusters, the situation does not appear to be that much simpler.

What is clear is that we can identify "isolated" galaxies simply by looking to see how many neighbors galaxies have within a specified radius. Successive removal of galaxies that are isolated according to this criterion from a sample leaves a set of galaxies that are clustered, and then we can select subsets of those that define clusters having specific properties. The clusters found by this method depend on the choice of a length scale and a density contrast.

An alternative and quite popular strategy is to identify clusters of points by a "friends-of-friends" algorithm. Two galaxies are within the same cluster if they are within a specified distance of each other. This is closely related to the identification of clusters by graphical methods such as the MST construct. The clusters found by this method depend on the choice of a length scale denoting how close a galaxy must be to another to be attached to the same cluster.

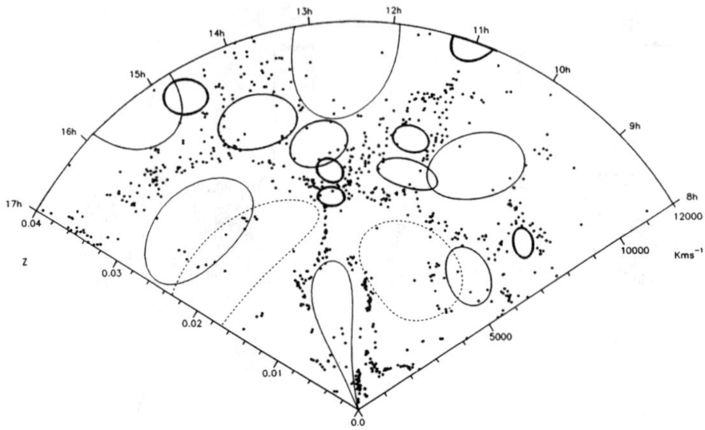

Figure 10.17 *The wavelet transform is used here to detect the voids in the galaxy distribution of the CfA2 slice, after removal of the Coma cluster. (Reproduced, with permission, from Slezak, de Lapparent, and Bijaoui 1993, Astrophys. J., 409, 517–529. AAS.)*

10.7.1 MST

The minimal spanning tree (MST) introduced in Section 10.5 may be used as a base to find clusters. Fig. 10.18 shows the MST associated with clustered point process within a box of $100h^{-1}$ Mpc sides. Here we describe how the MST might be used to select clusters. Given a length scale r_s, if we remove the edges with length greater than r_s, different disconnected subsets still remain linked with edges smaller than r_s. We can consider all these subsets as clusters associated with the length scale r_s. Obviously, as r_s decreases, the number of clusters increases, while their size gets smaller. This effect is clearly illustrated in Fig. 10.19, which shows the clusters selected by means of this pruning method for the value of $r_s = 2\,h^{-1}$ Mpc.

10.7.2 Modified friends-of-friends

In Section 10.4 we explained the friends-of-friends algorithm introduced by Press and Davis (1982). A modification of this technique was introduced by Croft and Efstathiou (1994) to select clusters from their simulation N-body simulations. Let us illustrate how this process works on our clustered point process. For each point of the process, we count how many neighbors lie within a distance r_s, and we consider all these balls (as many as points) as the first generation of clusters. We find the center of mass of each cluster and delete any cluster within r_s of a cluster of larger mass. The process continues by counting the number of particles within balls of radius r_s placed in the centers of mass of the surviving clusters. These

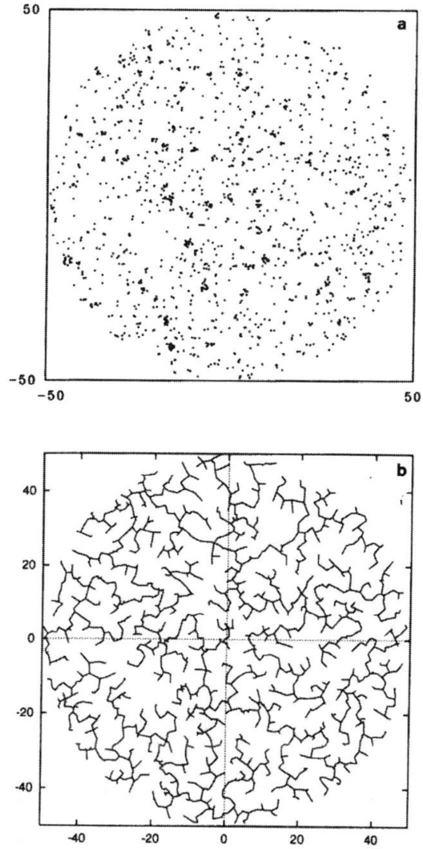

Figure 10.18 *A 2-D realization of clustered point process: (a) the point set; (b) the corresponding MST. (Reproduced, with permission, from Paredes, Jones, and Martínez 1995, Mon. Not. R. Astr. Soc., 276, 1116–1130. Blackwell Science Ltd.)*

are the clusters of the second generation. Again, we compute the new centers of mass and delete in the same way the overlap clusters (we consider that two spheres have enough overlap if the distance between the centers is smaller than their radius). The process is repeated several times. Croft and Efstathiou (1994) start the process from the percolation centers, instead of from all the points, as tested here.

Fig. 10.20 shows how the clusters are selected using this algorithm for values of $r_s = 1$ and $r_s = 2\,h^{-1}$ Mpc.

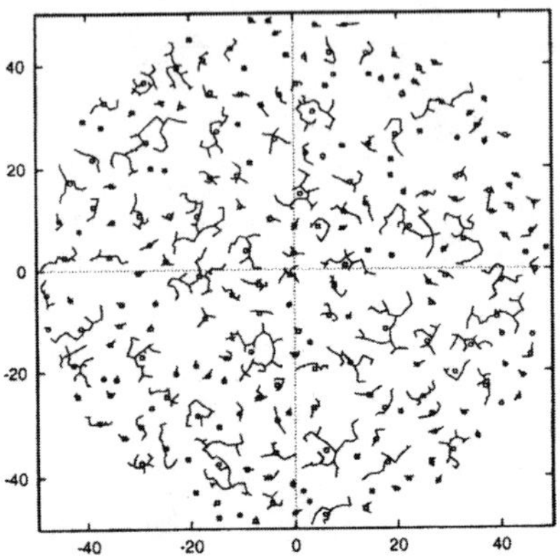

Figure 10.19 *Illustration of pruning the MST to select clusters from a clustered point process with length scale* $r_s = 2\,h^{-1}$ *Mpc. We can see the surviving connected branches of the tree and the centers of mass of the corresponding clusters. (Reproduced, with permission, from Paredes, Jones, and Martínez 1995, Mon. Not. R. Astr. Soc., 276, 1116–1130. Blackwell Science Ltd.)*

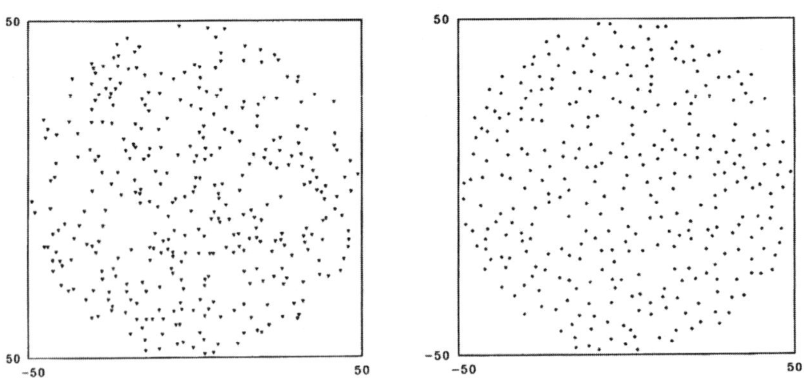

Figure 10.20 *Clusters selected using the modified friends-of-friends algorithm. In the left panel,* $r_s = 1\,h^{-1}$ *Mpc, and in the right panel* $r_s = 2\,h^{-1}$ *Mpc. (Reproduced, with permission, from Paredes, Jones, and Martínez 1995, Mon. Not. R. Astr. Soc., 276, 1116–1130. Blackwell Science Ltd.)*

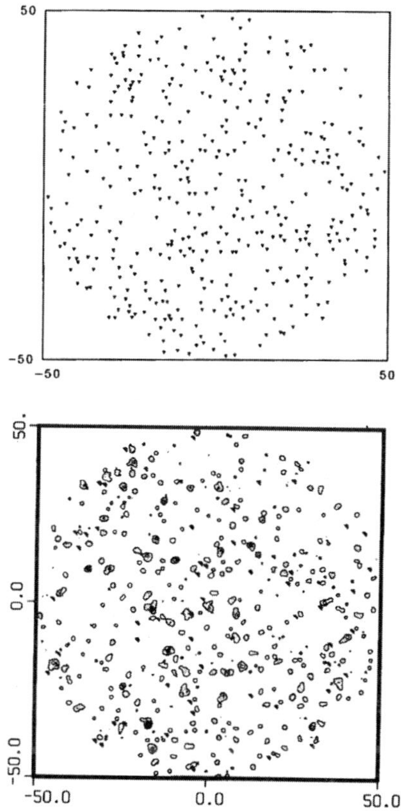

Figure 10.21 *Clusters selected using the algorithm based on the wavelet transform for the value of* $r_s = 1\,h^{-1}$ *Mpc (a = 0.93). The bottom plot shows the isocontours of the wavelet coefficients. (Reproduced, with permission, from Paredes, Jones, and Martínez 1995, Mon. Not. R. Astr. Soc., 276, 1116–1130. Blackwell Science Ltd.)*

10.7.3 Wavelets

Wavelet transforms can be used to select clusters from the point process. We have seen how the wavelet transform acts as a mathematical microscope in which $\mathbf{b}$ is the location and a^{-1} is the magnification. Therefore, changing the scale a, the wavelet transforms reveal different structures as a function of their size.

The bottom panel of Fig. 10.21 shows the isocontours of the wavelet transform applied to our clustered set for $a = 0.93$. If a is increased the small structures are erased. For a given value of a, we will consider, as the centers of the clusters, the local maxima of the function $T_g(a, \mathbf{b})$. After erasing overlapping clusters as in the previous method, we end up with a set of positions for the selected clusters. The top panel of Fig. 10.21 shows how this method selects clusters in the clustering model for $r_s = 1$.

A comparison of the efficiency of the previous methods is presented in Paredes, Jones, and Martínez (1993). For finding rich clusters the three methods provide identically good results: they are equally efficient. The differences appear when the methods are used to find small groups formed by pairs, triplets, and so on. The dependence on the scale length r_s is then crucial.

The MST method is very sensitive to the chosen value of the pruning length r_s. This is because, if r_s is large, different clusters percolate and are considered as a single structure. The wavelet method works well for finding clusters, although it does miss some that are found by the modified friends-of-friends algorithm. One of the most promising techniques that can be exploited for finding clusters from the homogeneous and isotropic matter distribution was recently introduced by Sanz, Herranz, and Martínez-Gónzalez (2001). These authors present a method with very good performance based on the selection of optimal pseudofilters once the density profile of the sources to be detected has been assumed. The method provides an unbiased and efficient estimator of the amplitude of the source.

The quantity of new cosmological data that will become available in the next years (2df, SDSS) necessitates the development of more efficient algorithms for finding clusters in multidimensional databases. This effort is being made as part of a huge "Computational AstroStatistics" collaboration (Nichol et al. 2000). The algorithms developed by this collaboration will be part of the statistical analysis tools of the "Virtual Observatory" (see *Virtual*).

10.8 Void statistics

Voids are easy to see in the galaxy distributions. To study their properties, we need an operational definition of a void. Surprisingly, this concept has been rather difficult to define. In the first study of voids Einasto, Einasto, and Gramann (1989) used the definition of a void as an empty sphere of a maximum radius. This can be made more precise, defining the distance field (Aikio and Mähönen 1998) by

$$D(\mathbf{x}) = \min_i \{|\mathbf{x} - \mathbf{x}_i|\}, \tag{10.9}$$

where $\mathbf{x}_i$ are the coordinates of sample points (galaxies). The local maxima of $D(\mathbf{x})$ are then the centers of spherical voids and the values of D at these points are the void sizes.

Of course, this void definition can be used to compare observational samples with numerical models, but it does not correspond to our intuitive notion of a void. To remedy this, a number of algorithms were later proposed. We describe two that give the closest result to the voids we detect visually.

The first algorithm is the "void finder" algorithm of El-Ad and Piran (1997). Its logic is probably very close to the human void recognition mechanism. We

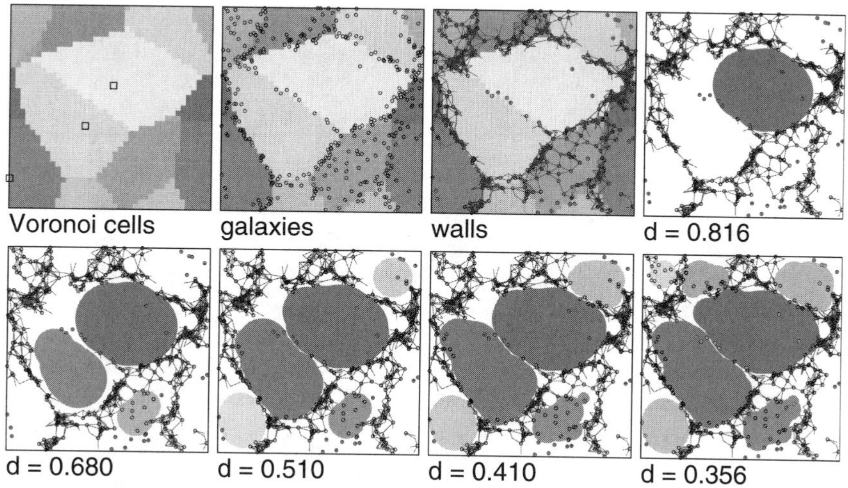

Figure 10.22 *The "void finder" applied to a Voronoi tessellation. The first two panels show the original tessellation and the galaxies generated at cell walls. The third panel shows the result of the work of the wall builder. Subsequent panels show successive stages of the void finder at work. (Reproduced, with permission, El-Ad and Piran 1997, Astrophys. J., 491, 421–435. AAS.)*

demonstrate how it works in Fig. 10.22. Basically, this algorithm determines voids as unions of empty spheres of different radii. The problem with such a natural definition is that it is difficult to prohibit percolation of voids. We know that both in numerical simulations and in observations the empty regions percolate, but we perceive neighboring voids as different entities. For this reason their algorithm starts with the "wall builder" that separates galaxies into two classes, the "wall galaxies" that delimit voids and "field galaxies" that do not. Wall galaxies are defined as galaxies that have at least n other wall galaxies at a distance l around it. The values of n and l are free parameters of the algorithm; El-Ad and Piran choose $n = 3$ and l, which is related to the density of the sample. The wall builder has done its work by panel 3 (walls) of Fig. 10.22.

After the walls have been defined, the void finder starts fitting empty spheres in the voids. This process works on a grid in sample space and proceeds in steps of different sphere sizes. First, unions of spheres of slightly different radii are found that fit into the voids. This produces the fourth panel ($d = 0.816$) in Fig. 10.22. The grid points that have been assigned to a void will not be reassigned in the future. If older voids exist, spheres of the same radii are used to add extra grid points to them. Then the radius is decremented and the procedure is repeated until

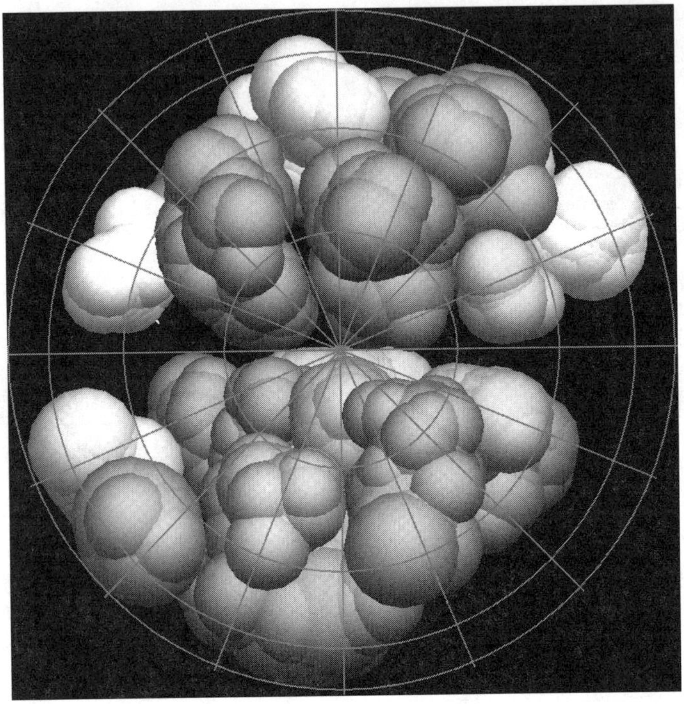

Figure 10.23 *Voids in the IRAS 1.2-Jy redshift survey. There are 24 separate voids. The zone of avoidance is seen in the center of the figure; the Great Attractor is shown as space devoid of voids at the left; and the Perseus-Pisces supercluster is shown at lower right. (Reproduced, with permission, from El-Ad, Piran, and da Costa 1997, Mon. Not. R. Astr. Soc., 287, 790–798. Blackwell Science Ltd.)*

a preassigned minimum sphere size (also a free parameter of the procedure) is reached.

The hierarchical nature of the algorithm ensures that two nearby voids will eat up all the space available to them and will not percolate remaining separate entities. This process is clearly seen following the gradual growth of voids in Fig. 10.22. The free parameters of the scheme allow variation of the smoothness of voids.

A three-dimensional picture of the voids found by this algorithm in the IRAS 1.2-Jy redshift survey is shown in Fig. 10.23. It is clearly seen that the voids are unions of different spheres and occupy most of the space. Our visual voids are certainly close to the "void finder" voids.

The "void finder" algorithm works well, but it is rather complex. Fortunately, another algorithm was proposed recently that gives similar results but is much simpler (Aikio and Mähönen 1998). This starts from the distance field $D(\mathbf{x})$ (10.9), defined on a grid. Starting now from a grid point $\mathbf{x}_i$ we set up a path toward the fastest growth of $D(\mathbf{x})$ until the path ends at a local maximum M_j. Then the point $\mathbf{x}_i$ and all points on the path are assigned to the subvoid v_j with the center at M_j. In this way, all points are assigned to subvoids in one pass of the algorithm (and if a path reaches an assigned point before a maximum, the path can be assigned at once).

Second, subvoids are collected together into voids. If the centers of two subvoids M_1 and M_2 are closer than $\max\{D(M_1), D(M_2)\}$, the subvoids are joined together. Using the same criterion, other subvoids can be joined to the large void in the "friends-of-friends" manner, forming a subvoid cluster. This completes the algorithm — all empty cells are assigned to voids.

The algorithm is fast, robust, and versatile. It has one free parameter — cells closer than a limiting distance $D_{\min}$ to a point can be left free. Cells belonging to border voids can be left unassigned, too; this is probably the best strategy for an observed sample. The algorithm can also be used for different density levels on a grid, defining filled and empty regions by the critical density level.

The voids that the algorithm produces are similar to our visual impression and to those produced by the "void finder."

10.9 Checking for periodicity

Periodic structures are sometimes seen in the distribution of galaxies and their clusters. The best known is the discovery of periodicity in the distribution of galaxies in a deep pencil-beam survey by Broadhurst et al. (1990). We describe these results and the statistical methods used to study them in detail in Chapter 8.

Traces of periodicity are also seen in the distribution of galaxy clusters (Einasto et al. 1997b). Fig. 10.24 shows the distribution of clusters of galaxies in rich superclusters (higher density regions) in the southern galactic hemisphere. This distribution resembles that of points (clusters) lying along the edges of a cubic grid. The spatial grid happens to be well aligned with the supergalactic coordinates, so the projection effects do not distort the picture.

Attempts to quantify this picture have used the details (oscillations) of the cluster correlation function (Einasto et al. 1997a). As this is an isotropic statistic, it smoothes the anisotropic signal and is not the best statistic to describe anisotropic structures. A natural method would be to use either anisotropic correlation functions or power spectra, but these are very noisy. The 3-D power spectrum has a variance that is equal to the estimate itself (see Chapter 8). The anisotropic correlation function dilutes the data pairs over a 3-D space of coordinate shifts, compared to the one-dimensional space of pair distances in the isotropic case, with a corresponding increase in variance.

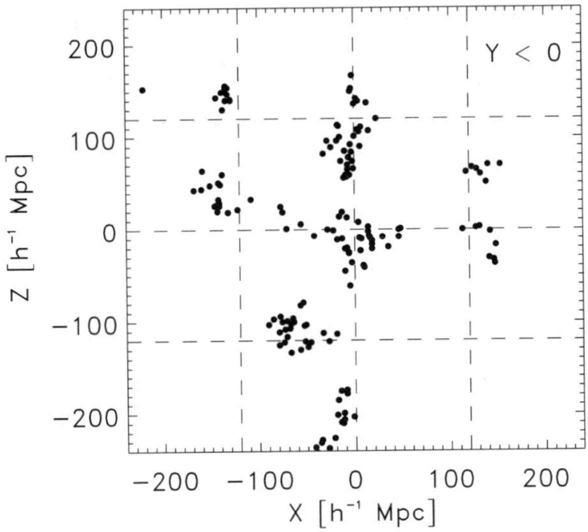

Figure 10.24 *The distribution of 319 clusters in 25 very rich superclusters with at least 8 members in the southern galactic hemisphere (in supergalactic coordinates). The grid with the step size* $120\,h^{-1}$ *Mpc corresponds approximately to distances between high-density regions across voids. (Reproduced, with permission, from Einasto et al. 1997b, Nature, 385, 139–141. Macmillan Publishers Ltd.)*

The standard methods of finding periodicities in astronomical time series are based on period folding. We also can fold multidimensional density distributions, as Fig. 10.25 shows.

To do that we cover the sample space with the element of specific geometry (cubes of size P for the regularity in Fig. 10.24, squares in Fig. 10.25) and calculate for every sample point its phase coordinates $\mathbf{x}_P$. These coordinates are obtained from the data coordinates $\mathbf{x}$ by applying a folding and scaling transform:

$$\mathbf{x}_P = \mathbf{x}/P - \lfloor \mathbf{x}/P \rfloor,$$

where the floor function $\lfloor x \rfloor$ is defined to be the largest integer smaller than or equal to x. If there is a (cubic) periodicity of period P in the distribution of the sample points, it will show up as an inhomogeneous density distribution in the phase cube $C_P = (0 \leq (x_P)_i \leq 1;\ i = 1, 2, 3)$.

If the folding method is applied for searching periodicities in astronomical time series, different variants of the analysis of variance are used as test statistics. A similar statistic that quantifies the notion of "inhomogeneous density" in the multidimensional case is the total variance of density in the phase cube. This statistic (regularity) was proposed by J. Pelt and analyzed and applied to the cluster distribution by Toomet et al. (1999).

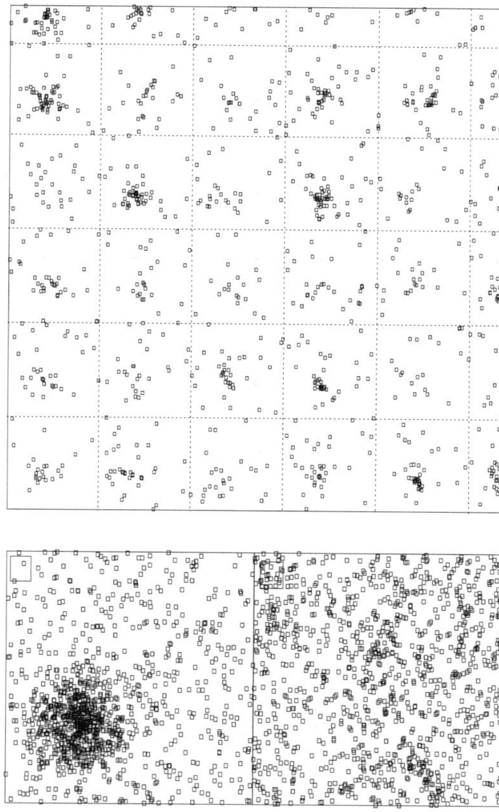

Figure 10.25 *Planar folding. The sample space (upper panel) is divided into smaller trial squares (dotted lines) and all these squares are stacked together (lower panels). If the size of the trial squares is close to the real period, there will be a clear density enhancement in the stacked distribution (lower left panel); otherwise, the point distribution will be almost random (lower right panel).*

For a continuous distribution the regularity κ is defined as

$$\kappa(P) = \frac{1}{\bar{n}_P^2} \int_{C_P} n_P^2(\mathbf{x}_P)\, d^3 x_P,$$

where $n_P(\mathbf{x}_P)$ is the number density in the phase cube (the phase density) and $\bar{n}_P$ is the mean phase density. We can write this integral as

$$\int_{C_P} n_P^2 d^3 x_P = \int_{C_P} n_P dN,$$

where N is the number of points in the sample. Because the phase space has a

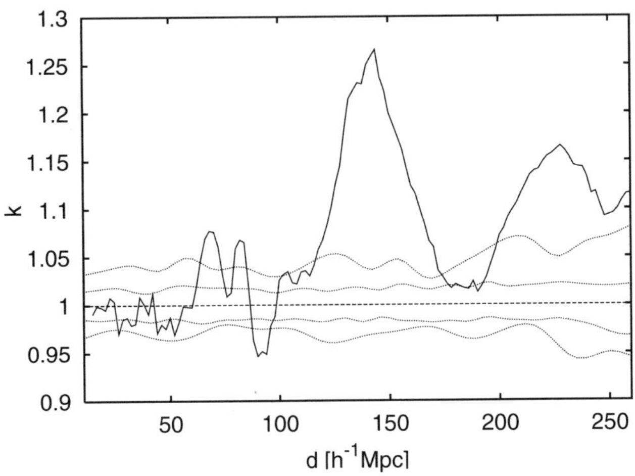

Figure 10.26 *The solid line shows the regularity periodogram for the sample of galaxy clusters in rich superclusters. Dotted lines show the 90 and 99% confidence regions.*

unit volume, $\bar{n}_P = N$, we can write the regularity statistic for a discrete point distribution as

$$\kappa(P) = \frac{1}{N(N-1)} \sum_i^N \sum_j^N K(\mathbf{x}_i, \mathbf{x}_j), \tag{10.10}$$

where $\mathbf{x}_i$ are the phase coordinates (we have dropped the index P here to avoid clutter), and we substitute $N(N-1)$ for N^2 to avoid bias. The kernel function $K(\mathbf{x}_i, \mathbf{x}_j)$ is used to estimate the phase density at sample points:

$$n_P(\mathbf{x}_i) = \sum_j^N K(\mathbf{x}_i, \mathbf{x}_j). \tag{10.11}$$

The kernels we can use here are the usual density kernels (cubic, Gaussian, Epanechikov) described in Chapter 3.

We usually search for small-amplitude signals, when the population distribution is almost Poissonian. Toomet et al. (1999) show that the sample distribution of the statistic (10.10) for a Poisson population can be approximated by the χ^2_ν distribution, where the number of degrees of freedom $\nu = 1/V_\varepsilon$ and V_ε is the phase volume of the density kernel. The mean value of the statistic is $E\{\kappa\} = 1$ for all N, and its sample variance is

$$\operatorname{Var} \kappa = \frac{2\nu}{N^2}.$$

The smallness of this variance allows detection of very weak cubic signals. Toomet et al. (1999) recommend choosing a constant width of the density kernel ε in phase

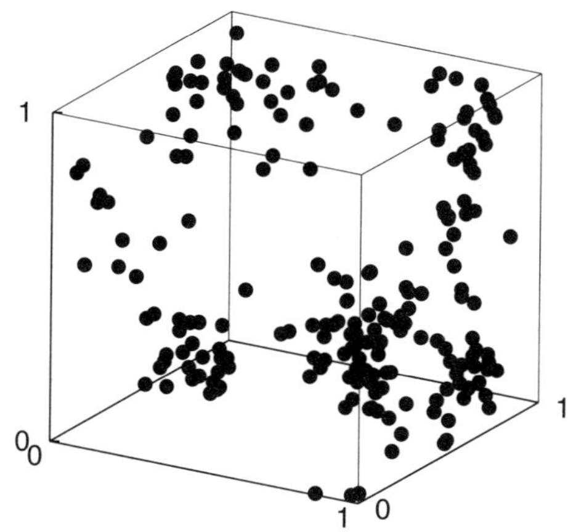

Figure 10.27 *Phase cube for the best period 130* h^{-1}*Mpc for galaxy clusters in rich superclusters in the southern galactic hemisphere. A projection of their spatial distribution is shown in Fig. 10.24.*

space for all trial periods P; this keeps the sample variance of the test statistic constant. Another recommendation is to discard in the density sum (10.11) those points that are close to the point i in data space; this eliminates the effect of local correlations. The weighting caused by the variable mean density of real samples (the selection function) can be taken into account by normalizing the test statistic not by $1/N^2$, as in (10.10), but by the value of the test statistic for a binomial point process of a much larger total number of points, but with the same geometry and selection function as the real sample. Because this test is anisotropic, the search space for the periodogram maxima is four dimensional: periodograms depend on the value of the trial period and on the three Euler angles, which are needed to specify the orientation of a cube in space.

As an application of this method to the real samples, Fig. 10.26 shows the regularity periodogram for the full cluster sample in high-density regions (the southern galactic subcone of this sample is shown in Fig. 10.24). A periodogram is defined as the dependence of a test statistic on the trial period; thus, the regularity periodogram is the function $\kappa(P)$.

Fig. 10.26 shows also two acceptance regions (the inner region for the 90% acceptance level, the outer region for the 99% acceptance level) for a binomial point process with the same number of points as the number of clusters, and with the same sample geometry and selection function. Comparison of these regions with the sample periodogram shows that some of the peaks are significant with a level $\alpha \ll 1\%$. The most significant period around 130–150 h^{-1}Mpc is clearly seen, but there also indications of a harmonic at 60–80 h^{-1}Mpc and of a subharmonic at $\sim 240h^{-1}$Mpc, as it should be in the case of a cubic arrangement. The specific significance levels could be changed by accounting for the multiple comparison problems, but for the present case the phase cube was oriented along the supergalactic Cartesian coordinate axis without previous directional search, so the correction is not too large.

Fig. 10.27 shows the spatial distribution of points in the phase cube for the best period for the half-sample shown in Fig. 10.24.

This phase cube shows that the cubic regularity exists; the density distribution in the phase cube is highly inhomogeneous. The cubic signal is mainly due to a couple of rich point clusters plus several filamentary features; such a configuration can be obtained by folding a lattice populated by points with higher concentration at corners.

We note that the existence of such a regularity is still controversial. The significance of the test would be smaller if we compared the observed sample with a Poisson cluster process, and the period is too close to the size of the cluster samples (about one third of the depth of a subcone). Finally, there is no known theoretical explanation for a cubic arrangement in the large-scale structure.

This all could mean that what we observe is a statistical fluke. The method, however, exists.

APPENDIX A

Coordinate transformations

A.1 Introduction

Positions of objects in the sky are described in various spherical coordinate systems. When studying galaxy catalogs, we frequently need to transform from one system to another, and these formulae can be difficult to find. To make this task easier, we describe here the main coordinate transformations and give the necessary formulae. We also list a few of the more popular map projections that are used to visualize the distribution of objects in the sky.

Observing the stars and other objects in the sky, we can imagine that they lie on a sphere (the celestial sphere). The projection of the Earth's rotational axis intersects the celestial sphere at the *north celestial pole* (NCP) and the *south celestial pole* (SCP). Likewise, projections of the terrestrial meridians and parallels onto the sky form the celestial meridians and parallels.

A.2 The equatorial system

The equatorial system is the basic system used to give the positions of objects in catalogs. The projection of the equator of the Earth onto the celestial sphere is the *celestial equator*. This great circle is the basis of the equatorial system of astronomical coordinates used to assign positions to the stars and other celestial objects.

The *declination* of a star, δ, is the angular distance between the star and the celestial equator. The declination takes values from 0° to 90° for objects in the northern celestial hemisphere and from 0° to -90° for objects in the southern hemisphere. This spherical coordinate plays the role of the terrestrial latitude.

To define the second coordinate, which should be similar to the terrestrial longitude, we need to fix the zero meridian, similar to the role played by the Greenwich meridian. This is done using another great circle on the celestial sphere, called the *ecliptic*. The ecliptic is the great circle described by the Sun's trajectory as a consequence of the orbital motion of the Earth around the Sun in the course of a year. The celestial equator and the ecliptic form an angle of $\epsilon = 23^\circ\,26'\,29'' \simeq 23.5^\circ$. Therefore, these two great circles cross each other at two points. One of this points, the *Vernal equinox*, is the one in which the Sun lies when it crosses the equator from the southern hemisphere to the northern hemisphere, at the spring equinox. This is the direction used for defining the zero point of the second spherical coordinate of the equatorial system. It is called the *right ascension*, α, and for a given star, it is defined as the angle measured counterclockwise along the celestial equator from the Vernal equinox to the celestial meridian passing through the star.

Catalogs of celestial objects, stars, galaxies, etc. list the declination and right ascension. This last coordinate is usually given in hours, minutes, and seconds, keeping in mind that 24 h = $360°$. Due to the precession of the Earth,* the Vernal equinox retrogrades $50.27''$ per year. This fact obviously affects the equatorial coordinates of the objects. Following an International Astronomical Union (IAU) agreement, catalogs list star positions in equatorial coordinates referring to the Vernal equinox at a given instant, or *epoch*. The epochs used in current catalogs are the beginning of the year 1950 and the beginning of the year 2000.

A.3 Galactic coordinates

The shape of our own Galaxy provides us with a very important coordinate system for the study of extragalactic objects, the galactic system. Now the reference plane is the plane of the Milky Way, a central plane crossing the center of the galactic disc. The projection of this plane on the celestial sphere defines the *galactic equator*. A line perpendicular to this plane intersects the celestial sphere at the north galactic pole, NGP (in the northern hemisphere), and at the south galactic pole, SGP (in the southern hemisphere). The equatorial coordinates of the NGP are $\alpha = 12$ h 49 m and $\delta = 27.4°$.

The *galactic latitude*, b, is the angular distance to the galactic equator, measured from $0°$ to $90°$ northward and from $0°$ to $-90°$ southward.

The *galactic longitude*, l, is the angle measured along the galactic equator counterclockwise from the direction of the galactic center to the great circle passing through the object and the galactic poles.

The galactic center ($l = 0°$, $b = 0°$) lies in Sagittarius, with equatorial coordinates $\alpha = 17$ h 42.4 m and $\delta = -28.92°$. Fig. A.1 shows the relation between the equatorial and galactic coordinates for a particular object. We see that the angle between the two great circles, the celestial equator and the galactic equator, is rather large (roughly $63.5°$).

A.3.1 The supergalactic coordinates

The supergalactic coordinates (SGB, SGL) are similar to the galactic coordinates (b, l), but are based on the preferred plane of the Local Supercluster (Supergalaxy). The north supergalactic pole (NSP) has the galactic coordinates $l = 47.37°$ and $b = 6.32°$, and the zero point (origin) for the supergalactic longitude lies in the galactic plane ($b = 0°$ and $l = 137.37°$). Supergalactic Cartesian

* *Precession* is a slow turning of the Earth's rotation axis with a period of about 26,000 years.

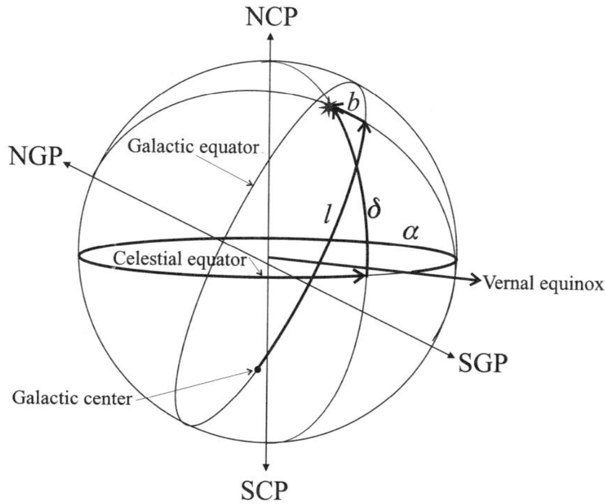

Figure A.1 *Relation between the galactic and the equatorial coordinates.*

coordinates (SGX, SGY, SGZ) are also frequently used, with the supergalactic z-axis directed toward the NSP, the x-axis toward the origin, and the y-axis perpendicular to the first two.

A.4 Coordinate transformations

By means of simple relations from spherical trigonometry, it is straightforward to obtain the equations to transform coordinates from one system to another. We follow here another approach used by the Groningen Image Processing System (see its home page at *Gipsy*). It uses Cartesian 3×3 transformation matrices to convert coordinates. Let us suppose that we want to change the coordinates (α, β) of an object to a different system (α', β'), where α is the longitude and β is the latitude. The unit vector **r** pointing to the direction of (α, β) has coordinates

$$\begin{aligned} r_1 &= \cos\alpha\cos\beta, \\ r_2 &= \sin\alpha\cos\beta, \\ r_3 &= \sin\beta. \end{aligned} \tag{A.1}$$

The coordinates of this vector in the new system will be $\mathbf{s} = \mathbf{T}\,\mathbf{r}$, where $\mathbf{T}$ is the (unitary) coordinate rotation matrix. From **s** we can obtain (α', β') by means of the inverse transformation of (A.1),

$$\begin{aligned} \alpha' &= \arctan(s_2/s_1) \\ \beta' &= \arcsin(s_3) \end{aligned}$$

taking care to get α' in an appropriate sector of the circle.

The matrices $\mathbf{T}$ for useful coordinate transformations are the following (taken from the tables in *skyco.c* in the *Gipsy* source code):

From equatorial 1950.0 to equatorial 2000.0

$$\mathbf{T} = \begin{pmatrix} 0.9999257453 & -0.0111761178 & -0.0048578157 \\ 0.0111761178 & 0.9999375449 & -0.0000271491 \\ 0.0048578157 & -0.0000271444 & 0.9999882004 \end{pmatrix}.$$

From equatorial 2000.0 to galactic

$$\mathbf{T} = \begin{pmatrix} -0.0548808010 & -0.8734368042 & -0.4838349376 \\ 0.4941079543 & -0.4448322550 & 0.7469816560 \\ -0.8676666568 & -0.1980717391 & 0.4559848231 \end{pmatrix}.$$

From equatorial 2000.0 to supergalactic

$$\mathbf{T} = \begin{pmatrix} 0.3751891698 & 0.3408758302 & 0.8619957978 \\ -0.8982988298 & -0.0957026824 & 0.4288358766 \\ 0.2286750954 & -0.9352243929 & 0.2703017493 \end{pmatrix}.$$

From galactic to equatorial 2000.0

$$\mathbf{T} = \begin{pmatrix} -0.0548808010 & 0.4941079543 & -0.8676666568 \\ -0.8734368042 & -0.4448322550 & -0.1980717391 \\ -0.4838349376 & 0.7469816560 & 0.4559848231 \end{pmatrix}.$$

From galactic to supergalactic

$$\mathbf{T} = \begin{pmatrix} -0.7353878609 & 0.6776464374 & 0.0000000002 \\ -0.0745961752 & -0.0809524239 & 0.9939225904 \\ 0.6735281025 & 0.7309186075 & 0.1100812618 \end{pmatrix}.$$

From supergalactic to equatorial 2000.0

$$\mathbf{T} = \begin{pmatrix} 0.3751891698 & -0.8982988298 & 0.2286750954 \\ 0.3408758302 & -0.0957026824 & -0.9352243929 \\ 0.8619957978 & 0.4288358766 & 0.2703017493 \end{pmatrix}.$$

From supergalactic to galactic

$$\mathbf{T} = \begin{pmatrix} -0.7353878609 & -0.0745961752 & 0.6735281025 \\ 0.6776464374 & -0.0809524239 & 0.7309186075 \\ 0.0000000002 & 0.9939225904 & 0.1100812618 \end{pmatrix}.$$

A.5 Sky projections

Several sky projections are used in astronomy to visualize the arrangement of objects in the sky. We describe below a few of the most popular projections. More information on map projections can be found in Bugayevskiy and Snyder (1995).

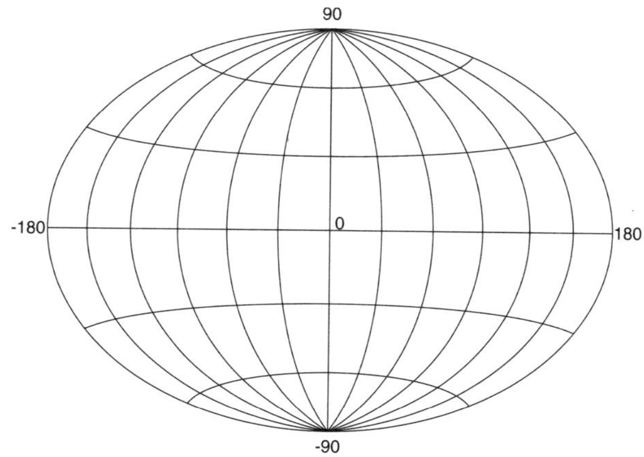

Figure A.2 *The Hammer–Aitoff full-sky projection. The coordinates are given in degrees, and the parallels and meridians are drawn with a* 30° *step.*

The formulae below are taken mostly from the Groningen Image Processing System (*Gipsy*) manual. We designate the sky coordinates as the longitude α and the latitude δ, and the Cartesian map coordinates as x and y.

The first class of projections are those used for full-sky maps. The most popular all-sky projection is the Hammer–Aitoff projection. It is an equal-area elliptical projection (the lines of both constant longitude and latitude are ellipsoidal arcs) with minimal distortions:

$$\begin{aligned} x &= 2R\sqrt{2}\frac{\cos\delta\sin\frac{\alpha}{2}}{\sqrt{1+\cos\delta\cos\frac{\alpha}{2}}}, \\ y &= R\sqrt{2}\frac{\sin\delta}{\sqrt{1+\cos\delta\cos\frac{\alpha}{2}}}. \end{aligned}$$

Here and below R is the scaling factor. Fig. A.2 shows the coordinate grid for the Hammer–Aitoff projection. All equal-area projections distort angles.

A simpler equal-area projection is the Mollweide projection. It is a pseudo-cylindrical projection where the lines of constant latitude (parallels) are represented by straight lines in the map:

$$\begin{aligned} x &= 2R\sqrt{2}\frac{\alpha}{\pi}\cos\psi, \\ y &= R\sqrt{2}\sin\psi, \end{aligned}$$

where the auxiliary angle ψ is obtained by solving the equation

$$2\psi + \sin 2\psi = \pi\sin\delta.$$

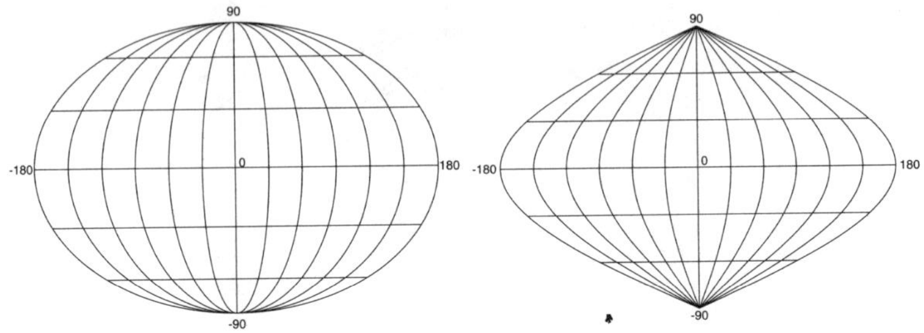

Figure A.3 *The Mollweide (left panel) and Sanson (right panel) full-sky projections. The coordinates are given in degrees, and the parallels and meridians are drawn with a* $30°$ *step.*

The Mollweide projection has more severe distortions than the Hammer–Aitoff projection.

A simpler projection is the Sanson sinusoidal projection, a pseudo-cylindrical equal-area projection where the parallels are equally spaced:

$$\begin{aligned} x &= R\alpha\cos\delta, \\ y &= R\delta. \end{aligned}$$

This projection is more distorted than both previous projections. Fig. A.3 shows both the Mollweide projection (left panel) and the Sanson projection (right panel).

The above formulae have been written for the case when the map pole (the center of the map $x = y = 0$) corresponds to the point $\alpha = \delta = 0$ in the sky. The map pole can be easily shifted to any other point on the celestial sphere.

For partial sky maps azimuthal projections are used, where the points on the sphere are projected to the map plane that is tangential to the sphere at a specific pole, using rays extending from the projection pole.

For small regions of the sky the gnomonic projection is used:

$$\begin{aligned} x &= R\frac{\cos\delta\sin(\alpha-\alpha_0)}{\sin\delta\sin\delta_0+\cos\delta\cos\delta_0\cos(\alpha-\alpha_0)}, \\ y &= R\frac{\sin\delta\cos\delta_0-\cos\delta\sin\delta_0\cos(\alpha-\alpha_0)}{\sin\delta\sin\delta_0+\cos\delta\cos\delta_0\cos(\alpha-\alpha_0)}. \end{aligned}$$

Because azimuthal projections are used for isolated patches of the sky, we explicitly include the coordinates α_0, δ_0 of the map pole in the formulae. This projection distorts both areas and angles.

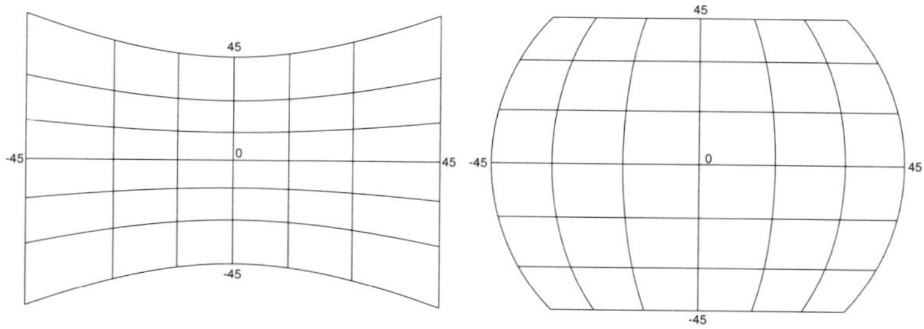

Figure A.4 *The gnomonic (left panel) and orthographic (right panel) partial sky projections. The coordinates are given in degrees, and the parallels and meridians are drawn with a* $15°$ *step.*

In radio astronomy the orthographic projection, an azimuthal projection where the projection pole lies at infinity, is sometimes used. It is given by

$$\begin{aligned} x &= R\cos\delta\sin(\alpha-\alpha_0), \\ y &= R\sin\delta\cos\delta_0 - \cos\delta\sin\delta_0\cos(\alpha-\alpha_0). \end{aligned}$$

This projection distorts both areas and angles. Fig. A.4 shows the gnomonic and orthographic projections of a $90° \times 90°$ patch of the sky.

For the stereographic projection the projection pole is moved to the point on the sphere opposite to the map pole. This gives a conformal map, where all angles are locally preserved (the areas are distorted). The stereographic projection is frequently used to represent whole hemispheres:

$$\begin{aligned} x &= 2R\frac{\cos\delta\sin(\alpha-\alpha_0)}{1+\sin\delta\sin\delta_0+\cos\delta\cos\delta_0\cos(\alpha-\alpha_0)}, \\ y &= 2R\frac{\sin\delta\cos\delta_0-\cos\delta\sin\delta_0\cos(\alpha-\alpha_0)}{1+\sin\delta\sin\delta_0+\cos\delta\cos\delta_0\cos(\alpha-\alpha_0)}. \end{aligned}$$

Fig. A.5 shows the stereographic projection for two map poles, for the usual $\alpha = \delta = 0$ pole in the left panel and for the north pole of the sky ($\delta = 90°$) in the right panel.

Another projection that is frequently used for projections of whole hemispheres is the Lambert azimuthal equal-area projection. The map pole is usually chosen to coincide with the celestial north or south pole, and the formulae for this case are:

$$\begin{aligned} x &= R\sqrt{2-2|\sin\delta|}\cos\alpha, \\ y &= R\sqrt{2-2|\sin\delta|}\sin\alpha. \end{aligned}$$

Fig. A.6 shows the Lambert azimuthal equal-area projection for the northern sky.

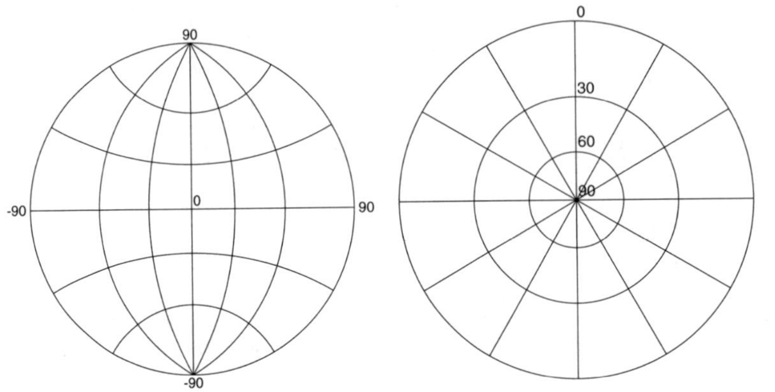

Figure A.5 *The stereographic projection. The left panel illustrates the case where the map pole is at the origin of the spherical coordinates ($\alpha = \delta = 0$). The right panel shows the projection when the map pole coincides with the celestial north pole ($\delta = 90^\circ$). The latitudes of the parallels are given in degrees, and the meridians are drawn with a 30° step.*

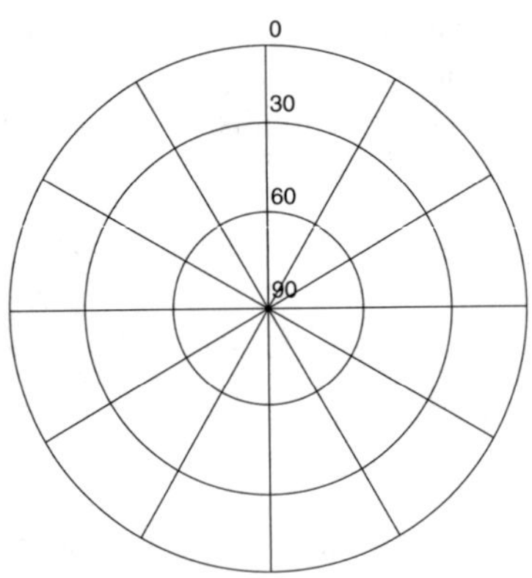

Figure A.6 *The Lambert azimuthal equal-area projection. The map pole coincides with the celestial north pole ($\delta = 90^\circ$). The latitudes of the parallels are given in degrees, and the meridians are drawn with a 30° step.*

APPENDIX B

Some basic concepts in statistics

B.1 Introduction

This appendix is meant to refresh statistical concepts and terminology for astronomers. We list only those concepts that are used in this book or that are closely connected with its subject. For further reading we recommend Barlow (1989) (especially for students) and Marriott (1990) and Kendall and Stuart (1967) for reference. For a review of the Bayesian approach we recommend Loredo (1990).

B.2 General definitions

The conclusions we want to make when using statistics always concern the whole universe (all the objects of a kind in the universe). Such a set of objects is called *the parent population* in statistics.

Example: All galaxies in the universe.

The set of objects in our catalogs is always smaller. They are chosen based on some criterion from the parent population; such collections are called *samples*. The number of objects in a sample is called the *sample size*.

Example: The APM galaxy survey.

For simple *random sampling*, the successive drawings are independent; every possible sample of the same size has equal probability of selection. This is not always strictly possible in astronomy because we make observations only from the Earth or from nearby artificial satellites. Thus, we justify our methods by adopting some extra assumptions, e.g., that of the homogeneity and isotropy of the universe.

Example: The galaxy distribution can be considered a spatial point process with the properties of *homogeneity* or *stationarity* (invariant under translations) and *isotropy* (invariant under rotations).

We also know how to account for the sampling rules by using selection functions, which measure the probability that a galaxy with a given property (e.g., distance) is included in the sample. An additional assumption is the *ergodicity*, which implies that a large enough sample is sufficient to obtain statistically reliable results. The three assumptions — homogeneity, isotropy, and ergodicity — are termed in cosmology the *fair sample hypothesis*.

B.3 Estimation

We are usually interested in finding information about the properties of a whole parent population, but we have only the data in our sample. The procedure that allows us to do that is called *estimation*.

A rule or a method of estimating a parameter of the parent population is called an *estimator*. It is usually given as a function of observables (sample values). A value of an estimator, found for a given sample, is called the *estimate*.

Example: The sample mean $\bar{x}$ can be used to estimate the population mean μ:

$$\widehat{\mu} = \bar{x} = \frac{1}{N}\sum_{i=1}^{N} x_i, \tag{B.1}$$

where N is the size of the sample $\{x_1, \ldots, x_N\}$ and $\widehat{\mu}$ denotes, by convention, the estimator of μ.

Any function of observables alone is called a *statistic*. Estimators are also statistics, but the class of statistics is larger.

Example: The F-statistic is used to compare the variances of two populations, using two samples:

$$F = \frac{V_1}{V_2},$$

where the variances V are given by

$$V = \frac{1}{N-1}\sum_{i=1}^{N}(x_i - \bar{x})^2$$

for either sample.

Because a statistic is a function of sample values, it is itself a random variable, and its probability distribution is called the *sampling distribution*. This distribution can be derived from the probability distribution of the parent population.

Example: For a sample of size N drawn from a population with the Gaussian probability distribution $N(\mu, \sigma^2)$ with mean μ and variance σ^2, the sampling distribution of the sample mean $\bar{x}$ (B.1) is also a Gaussian $N(\mu, \sigma^2/N)$.

The *sampling variance* of an estimator is the variance of its sampling distribution.

Example: For the sample mean in the previous example the sampling variance is σ^2/N.

If the sampling distribution for the sample $(x_1, \ldots, x_n)$, which depends on a parameter θ, is

$$dF(x_1, \ldots, x_n; \theta) = f(x_1, \ldots, x_n; \theta)\, dx_1, \ldots, dx_n,$$

then the function $f(\cdot)$ as a parameter of θ for the fixed sample values is called the *likelihood function* or the likelihood of the sample. It is usually designated as L, and it is the probability density of selection of that particular sample for a given θ. For simplicity, we shall consider below mainly the case of one parameter; in the case of a larger number of parameters k we can always substitute θ with a parameter set $(\theta_1, \ldots, \theta_k)$.

Example: The likelihood function for a random sample from the population with the probability density $f(x; \theta)$ is

$$L(\theta) = \prod_i f(x_i; \theta).$$

Given the sampling distribution $dF(x_1, \ldots, x_n)$, we can find the *expectation value* $E\{t\}$ of any statistic t by

$$E\{t\} = \int t\, dF(x_1, \ldots, x_n).$$

The expectation value is the mean value of a statistic over all possible samples of the same size. Note that a more extended notation in statistics for (B.3) is ET with capital T for the estimator.

B.4 Properties of estimators

The goodness of an estimator is judged on the basis of several desirable properties. An estimator should be *consistent*, meaning that it should converge in probability to the true value of the population parameter when the sample size increases. More exactly, any estimator t_n for a parameter θ, found from a sample of size n, is consistent, if for any positive small ε and η there is some N such that the probability that

$$|t_n - \theta| > \varepsilon$$

is greater than $1 - \eta$ for all $n > N$. Consistency can also be defined for other types of convergence.

Example: The sample mean (B.1) is a consistent estimator for the population mean, if the distribution is Gaussian, as seen from the previous example.

Example: The sample mean for a sample drawn from a population with the Cauchy distribution

$$dF(x) = \frac{1}{\pi} \frac{dx}{\left[1 + (x - \theta)^2\right]} \tag{B.2}$$

has the same distribution (B.2) as the parent population, which does not depend on the sample size. Thus, in this case the sample mean is not a consistent estimator for the parameter θ. Note that the mean of the Cauchy distribution is undefined, while higher moments diverge.

A consistent estimator converges to the parameter value for large sample sizes. An estimator t of θ is called *unbiased* if provides, on average, a correct estimate of the value of the parameter θ for any sample size.

$$E\{t\} = \theta, \tag{B.3}$$

If the above relation (B.3) is not satisfied, the difference $b(t)$, defined by

$$b(t) = E\{t\} - \theta,$$

is called the *bias*. If we can calculate the bias for an estimator, we can correct that estimator to make it unbiased. If there exists an unbiased estimator for a parameter, the parameter is called U-estimable.

Example: Consider a sample of size N from a population with mean μ and variance σ^2. The expectation value of the sample mean $\bar{x}$ is

$$E\{\bar{x}\} = E\left\{\frac{1}{N}\sum_i x_i\right\} = \frac{1}{N}\sum_i E\{x_i\} = \frac{1}{N}\sum_i \mu = \mu.$$

Thus, the sample mean is an unbiased estimator.

Example: The expectation value of the sample variance V is

$$\begin{aligned} E\{V\} &= E\left\{\frac{1}{N}\sum_i (x_i - \bar{x})^2\right\} = E\left\{\frac{1}{N}\sum_i \left(x_i - \frac{1}{N}\sum_i x_i\right)^2\right\} \\ &= E\left\{\frac{N-1}{N^2}\sum_i x_i^2 - \frac{1}{N^2}\sum_i\sum_{j\neq i} x_i x_j\right\} \\ &= \frac{N-1}{N}E\{x^2\} - \frac{N-1}{N}\mu^2 = \frac{N-1}{N}\sigma^2. \end{aligned}$$

Thus, the sample variance V is a biased estimator of the population variance σ^2, and the unbiased estimator is

$$\widehat{V} = \frac{1}{N-1}\sum_i (x_i - \bar{x})^2.$$

For an unbiased estimator, we should not expect that half of the estimates are, on average, above θ and half below, because the central measure used in the definition of unbiasedness in relation (B.3) is the mean and not the median.

An estimator should also be given as a function of *sufficient statistics*. A statistic t is sufficient for a parameter θ if it contains all the relevant information in the sample about θ and no other statistic can add anything to that. This condition is more useful if written for the likelihood function L:

$$L(x_1, \ldots, x_n; \theta) = g(t; \theta)h(x_1, \ldots, x_n), \tag{B.4}$$

where the function $g(t;\theta)$ depends only on t and θ. If the number of population parameters is larger than one (say, k), we have to consider instead of a single statistic t a set of sufficient statistics $(t_1, \ldots, t_m)$. It is easily seen from (B.4) above that the observations themselves always constitute a set of sufficient statistics. Examples of sufficient statistics are given below.

An estimator with a large sample variance is not a good estimator. Therefore, a desirable property for an estimator is to present the smallest possible variance. In practical cases, a compromise has to be reached between unbiasedness and minimum variance. For example, if we select a freely chosen constant as an estimator, its variance is trivially zero, but it might be strongly biased (Meyer 1975). For some parameters it is possible to find the *minimum variance unbiased estimator* (MVUE). The MVUE does not have to exist but it does exist under certain regularity conditions (see Lehmann and Casella 1998). For an unbiased estimator t of the parameter θ the minimum variance $\text{var}_{\text{min}}(t)$ is given by the *Cramér–Rao inequality*

$$\text{var}(t) \geq \text{var}_{\text{min}} = E\left\{\frac{1}{\left(\frac{\partial \ln L}{\partial \theta}\right)^2}\right\} = -E\left\{\frac{1}{\frac{\partial^2 \ln L}{\partial \theta^2}}\right\}. \tag{B.5}$$

If one estimator has a smaller sample variance than another, it is said to be more efficient. If the MVUE exists, the *efficiency* of an estimator t is defined as the ratio of the minimum variance var_{min} to its sample variance $\text{var}(t)$:

$$\text{eff}(t) = \frac{\text{var}_{\text{min}}}{\text{var}(t)}.$$

If the sample variance of an estimator coincides (for large samples) with the MVUE, the estimator is said to be *efficient*.

Example: The sample mean is an efficient estimator for the population mean.

As the MVUE are, in general, difficult to find, another way to measure the performance of an estimator is with its *mean square error* (MSE), which equals the square of the bias plus the variance: $b(t)^2 + \text{var}(t)$.

A general efficient estimator of θ is the *maximum likelihood estimator* (MLE), defined as the value $\widehat{\theta}$ that maximizes the likelihood function:

$$\left(\frac{\partial L}{\partial \theta}\right)_{\theta=\widehat{\theta}} = 0, \qquad \left(\frac{\partial^2 L}{\partial \theta^2}\right)_{\theta=\widehat{\theta}} < 0.$$

The efficiency of the ML estimator allows the use of (B.5) for estimating its variance when the sample is large. Maximum likelihood estimators are invariant under parameter transformations (minimizing the likelihood function with a respect of a function $g(\theta)$ will give the same result as minimizing it with respect of θ), but they can be biased. However, ML estimators are consistent, so they are asymptotically unbiased. In addition, if a sufficient statistic exists, the ML estimator is a function of it. For a review of ML methods we recommend Le Cam (1990).

Example: The likelihood function L for a sample drawn from a univariate (one-dimensional) Gaussian population is

$$L = \frac{1}{(2\pi)^{N/2}\sigma^N} \exp\left[-\frac{1}{2\sigma^2}\sum_i (x_i - \mu)^2\right],$$

where N is the size of the sample, and μ and σ^2 are the population mean and variance, respectively. Instead of maximizing the likelihood function L it is usually easier to maximize the *log-likelihood function*

$$\mathcal{L} = \ln L = -\frac{1}{2\sigma^2}\sum_i (x_i - \mu)^2 - N\left(\ln\sigma + \frac{\ln(2\pi)}{2}\right).$$

Minimizing $-\mathcal{L}$ with respect to μ and σ gives us two equations:

$$\begin{aligned} \sum_i (x_i - \widehat{\mu}) &= 0, \\ \sum_i \frac{(x_i - \widehat{\mu})^2}{\widehat{\sigma}^3} - \sum_i \frac{1}{\widehat{\sigma}} &= 0. \end{aligned}$$

The solution of that system is

$$\begin{aligned} \widehat{\mu} &= \frac{1}{N}\sum_i x_i = \bar{x}, \\ \widehat{\sigma}^2 &= \frac{1}{N}\sum_i (x_i - \bar{x})^2. \end{aligned}$$

The second estimator is only asymptotically unbiased; nevertheless, its MSE is smaller than that of $\widehat{V}$.

If the likelihood function in the previous example is chosen for a sample where different sample members are drawn from parent populations of different means $f(x_i;\theta)$ and variances σ_i (as could occur when fitting a curve to observed points), the maximum likelihood method leads to the requirement to find a minimum of the sum of squares:

$$\sum_i \left[\frac{x_i - f(x_i;\theta)}{\sigma_i}\right]^2.$$

This estimator is called the *least squares estimator* and it is frequently used as an independent postulate, without looking for justification by the postulate of maximum likelihood.

The last desirable property of an estimator is *robustness*. This property means that the estimate does not depend too much on small variations of the underlying probability distribution or the presence of outliers in the data (such as misinterpreted observations).

B.5 Confidence intervals and tests

There are two types of estimators: point estimators, when an estimate is found for a parameter without considering its statistical properties, and *interval estimators*, where an interval is found that is likely to be close to or contain the true parameter value. We usually use interval estimators. To specify the word "likely" above, we have to clarify the notion of probability.

B.5.1 Probability

It may come as a surprise to many astronomers that the notion of *probability* does not have a generally accepted meaning.

We quote below from *A Dictionary of Statistical Terms* (Marriott 1990), originally published by M. Kendall and W. Buckland in 1957 for the International Statistical Institute, and which has remained a standard reference for almost 50 years:

> Probability — a basic concept which may be taken either as undefinable, expressing in some way a "degree of belief," or as the limiting frequency in an infinite random series. Both approaches have their difficulties and the most convenient axiomatization of probability theory is a matter of personal taste. Fortunately both lead to much the same calculus of probabilities.

Most astronomers have been taught probability in the sense of a limiting frequency, but the above quotation should not surprise us — probability is the point where mathematics meets the real world.

Barlow (1989) lists three different notions of probability:

1. *Frequentist probability*, where the probability of a event is determined by the ratio of the number of times M the event occurred in N experiments, $p = M/N$, when $N \to \infty$. This definition requires that the experiment should be repeatable. Thus, probability is not the property of a particular event, but the property of the *ensemble* or parent population and the experiment. This is the "orthodox" definition of probability.

2. *Objective probability*, or *propensity*, which belongs to an object — the probability that a good die, when thrown, will show the numbers 1 to 6 with equal probability, can be considered the property of that die.

3. *Subjective probability* is a "degree of belief," allowing us to assign probability to our everyday predictions or to our theories. The methods using that notion of probability are grouped under the term *Bayesian statistics*.

B.5.2 Bayesian methods

Bayesian statistics starts with prior distributions that are updated in light of new data by means of *Bayes theorem*. The theorem states that the *posterior probability density* of a statistical model (the probability we can assign to our model after an experiment or observations) is:

$$f(\theta|x_1,\ldots,x_n;I) = f(\theta|I)\frac{f(x_1,\ldots,x_n|\theta,I)}{f(x_1,\ldots,x_n|I)}, \tag{B.6}$$

where $f(\theta|I)$ denotes the conditional distribution of the parameter θ, given a value of I (similarly for other distributions), and I denotes the prior information we have about the universe (and about similar data sets, in particular). The first factor on the right-hand side of (B.6) is the *Bayesian prior* (or simply *prior*), which includes our prior knowledge about the values of the parameter θ. The numerator is the *likelihood function* (note that it can also depend on the prior information, in principle) and the denominator is the normalization constant (*global likelihood*). The latter frequently can be difficult to find, because it gives the probability of data conditional on the knowledge of all possible values of the parameters. The Bayesian formula (B.6) shows how an observation (described by the likelihood function) can modify our previous knowledge (the prior) of a theory.

Whereas in the frequentist approach the model (population) parameters are considered as fixed, and we calculate probabilities for an observed sample (we did that already in several places above), the Bayesian approach considers the observed sample as fixed, an ultimate truth, and assigns probabilities to models (sample parameters).

The problems usually associated with the Bayesian approach, the need to define the prior distribution $f(\theta|I)$ and difficulties in calculating the global likelihood, really stress the need to define exactly the statistical model we compare with observations. Using the frequentist approach, we usually define a general statistic and study its properties for different classes of statistical models, which takes at least the same amount of work. If we do not have any prior information, a non-informative prior that does not depend on the model, for example, a *flat prior*, is used. In this case the Bayesian methodology gives the maximum posterior probability for the maximum likelihood solution.

B.5.3 Confidence intervals

Different notions of probability cause different (and conflicting) understandings of interval estimation.

For the frequentist approach, the estimators are random variables, and there is only one value of the population parameter, although unknown. There are only two values for the probability that the estimate t is larger than the parameter θ, 0 or 1. But because the estimator t is a random variable, it may take different values for different samples, and we can find an interval of estimates (t_-, t_+) that will include the value of the parameter θ with a given confidence α. Such an interval is called a *confidence interval*. We can choose many different confidence intervals for a given *confidence level* α. The most popular choices are:

1. The *central interval:* the probabilities above and below the interval are equal.
2. The *symmetric interval:* $t_+ - \mu = \mu - t_-$.
3. The *shortest interval:* for a given α, $t_+ - t_-$ is minimal.

The assertion that the parameter θ lies in a confidence interval of confidence level α will be true, on average, in a proportion α of the samples of the same size.

Example: For a Gaussian sampling distribution, all three types of intervals coincide; the confidence interval $[\widehat{\mu} - \widehat{\sigma}, \widehat{\mu} + \widehat{\sigma}]$ corresponds to the 68% confidence level; the interval $[\widehat{\mu} - 2\widehat{\sigma}, \widehat{\mu} + 2\widehat{\sigma}]$ corresponds to the 95.4% confidence level; and the interval $[\widehat{\mu} - 2.58\widehat{\sigma}, \widehat{\mu} + 2.58\widehat{\sigma}]$ corresponds to the 99% confidence level.

We can also define the *one-tailed limits*, the upper and lower limits. At a given confidence level α and upper limit t_+, the assertion that $\theta < t_+$ will be true, on average, for a proportion α of similar samples.

Example: For a Gaussian sampling distribution, the upper limit for the 68% confidence level is $\widehat{\mu} + 0.47\widehat{\sigma}$; the upper limit for the 95.4% confidence level is $\widehat{\mu} + 1.68\widehat{\sigma}$; and the upper limit for the 99% confidence level is $\widehat{\mu} + 2.33\widehat{\sigma}$. Compare this with the previous example.

We sometimes read about *fiducial intervals* (Wang 2000). In the fiducial approach we consider the likelihood function for a sample as the probability density for the (random) population parameters, given the sample values. In the fiducial approach the assertion that the parameter θ lies in a fiducial interval of the level α is true with the probability α for a given sample. The fiducial approach has been shown to meet with paradoxes, especially in multivariate cases (see, e.g., Kendall and Stuart 1967); the Bayesian approach, which also considers population parameters as random, is free of those paradoxes.

In the Bayesian approach, confidence values for parameter intervals (*Bayesian intervals*) are obtained by integrals of the posterior probability over the intervals. When an interval contains all points with the posterior probability density above a certain level, it is called the *highest posterior density interval.* When the posterior probability density is found, the intervals are called the *credible intervals* or *Bayesian confidence intervals.*

In order to find the posterior probability density, we have to find the normalization factor (global likelihood), which is frequently difficult to calculate. In this case the ratio of integrals of the likelihood over parameter intervals (subvolumes in the multidimensional case) is used to find the *odds* that one parameter interval is more likely than the other, given the observations.

In the multidimensional case, if we are interested only in a subset of model parameters, confidence values can be found by integrating the posterior distribution over the remaining parameters; this is called *marginalization*.

B.5.4 Testing hypotheses

This is a branch of statistics that is closely connected with determining confidence intervals; it deals with comparing statistical hypotheses (Lehmann 1959).

When testing hypotheses, we have to carefully formulate the statistical hypotheses. An important class of hypotheses are *simple hypotheses*, which completely specify the probability distribution of the test statistic.

If we test for a certain effect (e.g., difference of population means), the hypothesis that there is no effect (the population means do not differ), is called the *null hypothesis* and is usually designated as H_0. Any other hypothesis that is different from the null hypothesis is called the *alternative hypothesis* and is usually designated as H_1.

If we have defined a hypothesis (say, H_0), we can either accept or reject it on the basis of the value of the test statistic. If we reject a true hypothesis, we make a *type I error*. If we accept a false hypothesis, we make a *type II error*. We can do that only if we have formulated the alternative hypothesis.

Example: We want to automatically select clusters of galaxies from a huge Sloan digital sky survey galaxy database (Nichol et al. 2000). We calculate for each galaxy a test statistic p and find from the whole sample the probability distribution of the test statistic under the null hypothesis that a galaxy with a given value of p does not belong to a cluster (i.e., is a field galaxy). Type I error considers a galaxy as belonging to a cluster, when it is, in fact, a field galaxy.

The decision to accept or reject a hypothesis is based on the division of the sample space into two mutually exclusive regions, the *region of acceptance* and the *region of rejection* (or the *critical region*).

The *level of significance* of a test determines the probability of making a type I error. Given the boundary t_1 that separates the two regions, the significance level is defined as the probability $\alpha = Pr(t \geq t_1)$, if the critical region lies at $t > t_1$, or as $\alpha = Pr(t \leq t_1)$, if the critical region lies at $t < t_1$ (see Fig. B.1). In the first case t_1 is called the upper α significance point, and in the second case it is called the lower α significance point. The significance levels are given in percentages, and the usual choice is 5, 1, and 0.1%.

The *power of the test* is defined as the probability of *not making* a type II error, $1 - \beta$ in Fig. B.1. A statistical test for a given level of significance should be formulated by choosing the critical region that maximizes the power of the test. A good test is where the probability distributions of the test statistic under the hypotheses H_0 and H_1 are as different as possible (see Fig. B.1). This allows us to choose a rejection level t_1 with a small significance and a high power.

A Dictionary of Statistical Terms (Marriott 1990) also defines a *type III error*; we make this error when we choose a false test for our problem.

Example: If we have to make *multiple comparisons*, as in the previous example, the total number of type I errors we make (falsely including a field galaxy in the

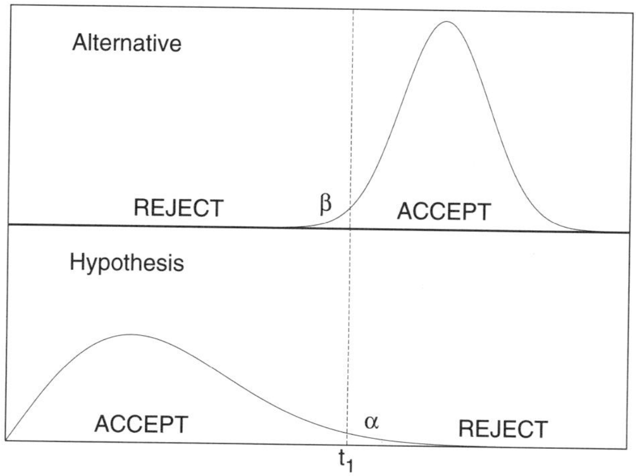

Figure B.1 *Critical regions and type I and type II errors. The rejection level t_1 divides the sample space into critical regions. The probability of making a type I error is given by the area α under the probability density curve in the rejection region of the hypothesis in the lower panel; it is the significance of the test. The power of the test is determined by the probability of making a type II error (area β under the probability density curve in the rejection region of the alternative test in the upper panel); the power of the test is $1 - \beta$.*

population of cluster galaxies) may become large, constituting a large percentage of the true cluster population. A recent method that allows to control the ratio of the number of type I errors (erroneously rejected field galaxies) to the total number of rejections (the cluster galaxies) is called the "false discovery rate" (FDR) (Nichol et al. 2000; Miller et al. 2001).

The generalized χ^2 test (Meyer 1975) measures the *goodness-of-fit* of the observed distribution of the sample to a given proposed distribution function of the population. If the probability distribution function depends on unknown parameters the problem is a composite hypothesis test. As above, we consider a single unknown parameter θ.

The N observation data points, $\{x_i\}_{i=1}^N$, are binned into n intervals, with f_j the number of observations lying in the jth interval. Therefore $\sum_{j=1}^n f_j = N$. Once the parameter has been estimated, for example, by the maximum likelihood method, it is possible to calculate the probability of observing a point x in the interval j for the proposed distribution

$$p_j(\widehat{\theta}) = \int_j f(x;\widehat{\theta})dx.$$

$Np_j(\widehat{\theta})$ is, therefore, the expected frequency of points lying in the interval j. The statistic

$$X^2 = \sum_{j=1}^{n} \frac{[f_j - Np_j(\widehat{\theta})]^2}{Np_j(\widehat{\theta})} \tag{B.7}$$

is, for large values of n, approximately distributed according to a χ^2 distribution function with $n - 2$ degrees of freedom (the degrees of freedom are, in general, $n - k - 1$, where k is the number of estimated parameters). The hypothesis that the distribution of the population is the proposed one is rejected with a level of significance α if $X^2 > \chi^2_{\alpha,n-k-1}$.

A special kind of maximum likelihood method for parameter estimation is the χ^2 minimization method (Meyer 1975; Press et al. 1992). This is very popular in cosmology, particularly in the analysis of the CMB data. The starting point here is a set of N points with the form (x_i, y_i, σ_i). σ_i is the rms deviation of the measurement $y_i(x_i)$ that is assumed to be normally distributed with mean value $f(x_i;\theta)$ and variance σ_i^2. The χ^2 statistic is defined as

$$\chi^2 = \sum_{i}^{N} \frac{[y_i - f(x_i;\theta)]^2}{\sigma_i^2}. \tag{B.8}$$

Assuming that the measurements y_i are independently distributed, the statistic (B.8) follows a χ^2 distribution function with the number of degrees of freedom $n = N$ ($n = N - 1$ if we found the estimate $\hat{\theta}$ from the data and $n = N - k$ for k parameters).

References

Abell G.O. (1958) The distribution of rich clusters of galaxies. *Astrophys. J. Suppl. Ser.* **3**, 211–288.

Abell G.O., Corwin H.G., Jr. and Olowin R.P. (1989) A catalog of rich clusters of galaxies. *Astrophys. J. Suppl. Ser.* **70**, 1–138.

Abramowicz M.A. and Stegun I.A. (1965) *Handbook of Mathematical Functions*. Dover Publications, New York.

Adler R.J. (1981) *The Geometry of Random Fields*. John Wiley & Sons, New York.

Aikio J. and Mähönen P. (1998) A simple void-searching algorithm. *Astrophys. J.* **497**, 534–540.

Alimi J.M., Valls-Gabaud D. and Blanchard A. (1988) A cross-correlation analysis of luminosity segregation in the clustering of galaxies. *Astron. Astrophys.* **206**, L11–L14.

Amendola L. and Palladino E. (1999) The scale of homogeneity in the Las Campanas redshift survey. *Astrophys. J. Lett.* **514**, L1–L4.

Argoul F., Arnéodo A., Elezgaray J. and Grasseau G. (1989) Wavelet transform of fractal aggregates. *Phys. Lett. A* **135**, 327–336.

Arnéodo A., Grasseau G. and Holschneider M. (1988) Wavelet transform of multifractals. *Phys. Rev. Lett.* **61**, 2281–2284.

Avnir D., Biham O., Lidar D. and Malcai O. (1998) Is the geometry of nature fractal? *Science* **279**, 39–40.

Babu G.J. and Feigelson E.D. (1996a) Spatial point processes in astronomy. *J. Statis. Planning Inference* **50**, 311–326.

Babu G.J. and Feigelson E.D. (1996b) *Astrostatistics*. Chapman & Hall, London.

Baddeley A. (1999) Spatial sampling and censoring. In *Stochastic Geometry. Likelihood and Computation* (eds. O.E. Barndorff-Nielsen, W.S. Kendall and M.N.M. van Lieshout), Chapman & Hall, Boca Raton, pp. 37–78.

Baddeley A.J., Kerscher M., Schladitz K. and Scott B. (2000) Estimating the J function without edge correction. *Stat. Neerlandica* **54**, 1–14.

Baddeley A.J., Moyeed R.A., Howard C.V. and Boyde A.S. (1993) Analysis of a three-dimensional point pattern with replication. *Appl. Statist.* **42**, 641–668.

Bahcall N.A. and Soneira R.M. (1983) The spatial correlation function of rich clusters of galaxies. *Astrophys. J.* **270**, 20–38.

Bahcall N.A. and West M.J. (1992) The cluster correlation function: Consistent results from an automated survey. *Astrophys. J.* **392**, 419–423.

Baleisis A., Lahav O., Loan A.J. and Wall J.V. (1998) Searching for large-scale structure in deep radio surveys. *Mon. Not. R. Astr. Soc.* **297**, 545–558.

Balian R. and Schaeffer R. (1989) Scale-invariant matter distribution in the universe. I. Counts in cells. *Astron. Astrophys.* **220**, 1–29.

Bardeen J.M., Bond J.R., Kaiser N. and Szalay A.S. (1986) The statistics of peaks of Gaussian random fields. *Astrophys. J.* **304**, 15–61.

Barlett M.S. (1964) The spectral analysis of two-dimensional point processes. *Biometrika* **51**, 299–311.

Barlow R. (1989) *Statistics. A Guide to the Use of Statistical Methods in Physical Sciences*. The Manchester Physics Series, John Wiley & Sons, Chichester.

Barnsley M. (1988) *Fractals Everywhere*. Academic Press, London.

Barrow J.D. (1992) Some statistical problems in cosmology with discussion by S.P. Bhavsar and F.L. Bookstein. In *Statistical Challenges in Modern Astronomy I* (eds. G.J. Babu and E.D. Feigelson), Springer-Verlag, New York, pp. 21–55.

Barrow J.D., Sonoda D.H. and Bhavsar S.P. (1984) A bootstrap resampling analysis of galaxy clustering. *Mon. Not. R. Astr. Soc.* **210**, 19P–23P.

Barrow J.D., Sonoda D.H. and Bhavsar S.P. (1985) Minimal spanning trees, filaments and galaxy clustering. *Mon. Not. R. Astr. Soc.* **216**, 17–35.

Bartelmann M. and Schneider P. (2001) Weak gravitational lensing. *Phys. Rep.* **340**, 291–472.

Baugh C.M. (1996) The real space correlation function measured from the APM galaxy survey. *Mon. Not. R. Astr. Soc.* **280**, 267–275.

Baugh C.M. and Efstathiou G. (1993) The three-dimensional power spectrum measured from the APM galaxy survey — I. Use of the angular correlation function. *Mon. Not. R. Astr. Soc.* **265**, 145–156.

Baugh C.M. and Efstathiou G. (1994) The three-dimensional power spectrum measured from the APM galaxy survey — II. Use of the two-dimensional power spectrum. *Mon. Not. R. Astr. Soc.* **267**, 323–332.

Beisbart C. and Kerscher M. (2000) Luminosity and morphology dependent clustering of galaxies. *Astrophys. J.* **545**, 6–25.

Benítez N. and Sanz J.L. (1999) Measuring Ω/b with weak lensing. *Astrophys. J. Lett.* **525**, L1–L4.

Bennett C.L., Banday A., Gorski K.M., Hinshaw G., Jackson P., Keegstra P., Kogut A., Smoot G.F., Wilkinson D.T. and Wright E.L. (1996) 4-year Cobe DMR cosmic microwave background observations: Maps and basic results. *Astrophys. J. Lett.* **464**, L1–L4.

Benson A.J., Baugh C.M., Cole S., Frenk C.S. and Lacey C.G. (2000) The dependence of velocity and clustering statistics on galaxy properties. *Mon. Not. R. Astr. Soc.* **316**, 107–119.

Bernardeau F. and van de Weygaert R. (1996) A new method for accurate estimation of velocity field statistics. *Mon. Not. R. Astr. Soc.* **279**, 693–711.

Bertschinger E. (1987) Path integral methods for primordial density perturbations — Sampling of constrained Gaussian random fields. *Astrophys. J. Lett.* **323**, L103–L106.

Bertschinger E. (1992) Large-scale structures and motions: Linear theory and statistics. In *New Insights into the Universe, Lecture Notes in Physics 408* (eds. V.J. Martínez, M. Portilla and D. Sáez), Springer-Verlag, Berlin, pp. 65–126.

Bertschinger E. (1998) Simulations of structure formation in the universe. *Ann. Rev. Astron. Astrophys.* **36**, 599–654.

Bertschinger E. and Dekel A. (1989) Recovering the full velocity and density fields from large-scale redshift–distance samples. *Astrophys. J. Lett.* **336**, L5–L8.

Bhavsar S.P. and Ling E.N. (1988) Are the filaments real? *Astrophys. J. Lett.* **331**, L63–L68.

Binggeli B., Tarenghi M. and Sandage A. (1990) The abundance and morphological segregation of dwarf galaxies in the field. *Astron. Astrophys.* **228**, 42–60.

Binney J. and Tremaine S. (1987) *Galactic Dynamics*. Princeton University Press, Princeton.

Bistolas V. and Hoffmann Y. (1998) Non-linear constrained realizations of the large scale structure. *Astrophys. J.* **492**, 439–491.

Blumenfeld R. and Mandelbrot B. (1997) Lévy dusts, Mittag–Leffler statistics, mass fractal lacunarity, and perceived dimension. *Phys. Rev. E* **56**, 112–118.

Bond J.R., Cole S., Efstathiou G. and Kaiser N. (1991) Excursion set mass functions for hierarchical Gaussian fluctuations. *Astrophys. J.* **379**, 440–460.

Bond J.R. and Efstathiou G. (1984) Cosmic background radiation anisotropies in universes dominated by nonbaryonic dark matter. *Astrophys. J. Lett.* **285**, L45–L48.

Bond J.R. and Myers S.T. (1996a) The peak–patch picture of cosmic catalogs. I. Algorithms. *Astrophys. J. Suppl.* **103**, 1–39.

Bond J.R. and Myers S.T. (1996b) The peak–patch picture of cosmic catalogs. II. Validation. *Astrophys. J. Suppl.* **103**, 41–62.

Bonometto S.A., Iovino A., Guzzo L., Giovanelli R. and Haynes M. (1993) Correlation functions from the Perseus-Pisces redshift survey. *Astrophys. J.* **419**, 451–458.

Bonometto S.A. and Sharp N.A. (1980) On the derivation of higher order correlation functions. *Astron. Astrophys.* **92**, 222–224.

Borgani S. (1993) The multifractal behaviour of hierarchical density distributions. *Mon. Not. R. Astr. Soc.* **260**, 537–549.

Borgani S. (1995) Scaling in the universe. *Phys. Rep.* **193**, 1–152.

Borgani S. and Guzzo L. (2001) X-ray clusters of galaxies as tracers of structure in the Universe. *Nature* **409**, 39–45.

Borgani S., Martínez V.J., Pérez M.A. and Valdarnini R. (1994) Is there any scaling in the cluster distribution? *Astrophys. J.* **435**, 37–48.

Bouchet F.R., Colombi S., Hivon E. and Juszkiewicz R. (1995) Perturbative Lagrangian approach to gravitational instability. *Astr. Astrophys.* **296**, 575–608.

Böhringer H., Schuecker P., Guzzo L., Collins C.A., Voges W., Schindler S., Neumann D.M., Cruddace R.G., De Grandi S., Chincarini G., Edge A.C., MacGillivray H.T. and Shaver P. (2001) The ROSAT-ESO flux limited X-ray (REFLEX) galaxy cluster survey. I. The construction of the cluster sample. *Astron. Astrophys.* **369**, 826–850.

Braun J. and Sambridge M. (1995) A numerical method for solving partial differential equations on highly irregular evolving grids. *Nature* **376**, 655–660.

Bridle S.L., Hobson M.P., Lasenby A.N. and Saunders R. (1998) A maximum-entropy method for reconstructing the projected mass distribution of gravitational lenses. *Mon. Not. R. Astr. Soc.* **299**, 895–904.

Broadhurst T.J., Ellis R.S., Koo D.C. and Szalay A.S. (1990) Large-scale distribution of galaxies at the galactic poles. *Nature* **343**, 726–728.

Buchert T. (1992) Lagrangian theory of gravitational instability of Friedman-Lemaitre cosmologies and the 'Zeldovich approximation.' *Mon. Not. R. Astr. Soc.* **254**, 729–737.

Buchert T. and Martínez V.J. (1993) Two-point correlation function in pancake models and the fair sample hypothesis. *Astrophys. J.* **411**, 485–500.

Bugayevskiy L.M. and Snyder J.P. (1995) *Map Projections: A Reference Manual*. Taylor & Francis, London.

Bunde A. and Havlin S. (eds.) (1994) *Fractals in Science*. Springer-Verlag, Berlin.

Bunn E.F. and White M. (1997) The four-year COBE normalization and large-scale structure. *Astrophys. J.* **480**, 6–21.

Burrus C.S., Gopinath R.A. and Guo H. (1998) *Introduction to Wavelets and Wavelet Transforms. A Primer*. Prentice-Hall, Upper Saddle River.

Burstein D. and Heiles C. (1982) Reddenings derived from H I and galaxy counts — Accuracy and maps. *Astronom. J.* **87**, 1165–1189.

Buryak O. and Doroshkevich A. (1996) Correlation function as a measure of the structure. *Astron. Astrophys.* **306**, 1–8.

Canavezes A., Springel V., Oliver S.J., Rowan-Robinson M., Keeble O., White S.D.M., Saunders W., Efstathiou G., Frenk C.S., McMahon R.G., Maddox S., Sutherland W. and Tadros H. (1998) The topology of the IRAS Point Source Catalogue Redshift Survey. *Mon. Not. R. Astr. Soc.* **297**, 777–793.

Cappi A., Benoist C., da Costa L.N. and Maurogordato S. (1998) Is the Universe a fractal? Results from the Southern Sky Redshift Survey 2. *Astron. Astrophys.* **335**, 779–788.

Carroll B.W. and Ostlie D.A. (1996) *Modern Astrophysics*. Addison-Wesley, Reading.

Catelan P., Lucchin F., Matarrese S. and Porciani C. (1998) The bias field of dark matter halos. *Mon. Not. R. Astr. Soc.* **297**, 692–712.

Cen R. (1998) Gaussian peaks and clusters of galaxies. *Astrophys. J.* **509**, 494–516.

Christensen L.L. (1996) *Compound Redshift Catalogues and their Application to Redshift Distortions of the Two-Point Correlation Function.* Master's thesis, University of Copenhagen, Copenhagen.

Chui C.K. (1992) *An Introduction to Wavelets*. Academic Press, San Diego.

Coleman P.H. and Pietronero L. (1992) The fractal structure of the Universe. *Phys. Rep.* **213**, 311–391.

Coles P. (1992) Analysis of patterns in galaxy clustering with discussions by F. L. Bookstein and N.K. Bose. In *Statistical Challenges in Modern Astronomy I* (eds. G.J. Babu and E.D. Feigelson), Springer-Verlag, New York, pp. 57–81.

Coles P. (1993) Galaxy formation with local bias. *Mon. Not. R. Astr. Soc.* **262**, 1065–1075.

Coles P. and Barrow J.D. (1987) Non-Gaussian statistics and the microwave background radiation. *Mon. Not. R. Astr. Soc.* **228**, 407–426.

Coles P. and Chiang L.Y. (2000) Non-linearity and non-Gaussianity through phase information. *http://arXiv/org/abs/astro-ph/0010521*.

Coles P. and Jones B. (1991) A lognormal model for the cosmological mass distribution. *Mon. Not. R. Astr. Soc.* **248**, 1–13.

Coles P. and Lucchin F. (1995) *Cosmology: The Formation and Evolution of Cosmic Structure*. John Wiley & Sons, New York.

Coles P., Pearson R.C., Borgani S., Plionis M. and Moscardini L. (1998) The cluster distribution as a test of dark matter models. IV: Topology and geometry. *Mon. Not. R. Astr. Soc.* **294**, 245–258.

Colombi S., Pogosyan D. and Souradeep T. (2000) Tree structure of the percolating Universe. *Phys. Rev. Let.* **26**, 5515–5518.

Couchman H.M.P., Thomas P.A. and Pearce F.R. (1995) Hydra: An adaptive-mesh implementation of P^3M-SPH. *Astrophys. J.* **452**, 797–813.

Cox D.R. and Isham V. (1980) *Point Processes*. Chapman & Hall, London.

Cramér H. and Leadbetter M.R. (1967) *Stationary and Related Stochastic Processes. Sample Function Properties and Their Applications*. John Wiley & Sons, New York.

Cressie N. (1991) *Statistics for Spatial Data*. John Wiley & Sons, New York.

Croft R.A.C., Dalton G.B., Efstathiou G., Sutherland W.J. and Maddox S.J. (1997) The richness dependence of galaxy cluster correlations: Results from a redshift survey of rich APM clusters. *Mon. Not. R. Astr. Soc.* **291**, 305–313.

Croft R.A.C. and Efstathiou G. (1994) The correlation function of rich clusters of galaxies in CDM-like models. *Mon. Not. R. Astr. Soc.* **267**, 390–400.

Croft R.A.C. and Gaztañaga E. (1997) Reconstruction of cosmological density and velocity fields in the Lagrangian Zel'dovich approximation. *Mon. Not. R. Astr. Soc.* **285**, 793–805.

da Costa L.N., Geller M.J., Pellegrini P.S., Latham D.W., Fairall A.P., Marzke R.O., Willmer C.N.A., Huchra J.P., Calderon J.H., Ramella M. and Kurtz M.J. (1994) A complete southern sky redshift survey. *Astrophys. J. Lett.* **424**, L1–L4.

da Costa L.N., Pellegrini P.S., Davis M., Meiksin A., Sargent W.L.W. and Tonry J.L. (1991) Southern Sky Redshift Survey — The catalog. *Astrophys. J. Suppl. Ser.* **75**, 935–964.

da Costa L.N., Willmer C.N.A., Pellegrini P.S., Chaves O.L., Rité C., Maia M.A.G., Geller M.J., Latham D.W., Kurtz M.J., Huchra J.P., Ramella M., Fairall A.P., Smith C. and Lípari S. (1998) The Southern Sky Redshift Survey. *Astronom. J.* **116**, 1–7.

Daley D.J. and Vere-Jones D. (1988) *An Introduction to the Theory of Point Processes*. Springer-Verlag, New York.

Dalton G.B., Maddox S.J., Sutherland W.J. and Efstathiou G. (1997) The APM Galaxy Survey — V. Catalogues of galaxy clusters. *Mon. Not. R. Astr. Soc.* **289**, 263–284.

Davis M. (1997) Is the universe homogeneous on large scales? In *Critical Dialogues in Cosmology* (ed. N. Turok), World Scientific, Singapore, pp. 13–23.

Davis M. and Geller M.J. (1976) Galaxy correlations as a function of morphological type. *Astrophys. J.* **208**, 13–19.

Davis M. and Huchra J.P. (1982) A survey of galaxy redshifts. III. The density field and the induced gravity field. *Astrophys. J.* **254**, 437–450.

Davis M., Meiksin A., Strauss M.A., da Costa L.N. and Yahil A. (1988) On the universality of the two-point galaxy correlation function. *Astrophys. J. Lett.* **333**, L9–L12.

Davis M. and Peebles P.J.E. (1983) A survey of galaxy redshifts — V. The two-point position and velocity correlations. *Astrophys. J.* **267**, 465–482.

de Bernardis P., Ade P.A.R., Bock J.J., Bond J.R., Borrill J., Boscaleri A., Coble K., Contaldi C.R., Crill B.P., Troia G.D., Farese P., Ganga K., Giacometti M., Hivon E., Hristov V.V., Iacoangeli A., Jaffe A.H., Jones W.C., Lange A.E., Martinis L., Masi S., Mason P., Mauskopf P.D., Melchiorri A., Montroy T., Netterfield C.B., Pascale E., Piacentini F., Pogosyan D., Polenta G., Pongetti F., Prunet S., Romeo G., Ruhl J.E. and Scaramuzzi F. (2001) Multiple peaks in the angular power spectrum of the cosmic microwave background: Significance and consequences for cosmology. *http://arXiv/org/abs/astro-ph/0105296*.

de Lapparent V., Geller M.J. and Huchra J.P. (1986) A slice of the universe. *Astrophys. J. Lett.* **302**, L1–L5.

de Lapparent V., Geller M.J. and Huchra J.P. (1989) The luminosity function for the CfA redshift survey slices. *Astrophys. J.* **343**, 1–17.

de Vaucouleurs G. (1970) The case for a hierarchical cosmology. *Science* **167**, 1203–1213.

Dekel A. (1997) Cosmological implications of large-scale flows. In *Galaxy Scaling Relations: Origins, Evolution and Applications* (eds. L. da Costa and A. Renzini), Springer-Verlag, Berlin, pp. 245–285.

Dekel A. and Bertschinger E. (1990) Potential, velocity and density fields from sparse and noisy redshift–distance samples: Method. *Astrophys. J.* **364**, 349–369.

Dekel A., Blumenthal G.R., Primack J.R. and Stanhill D. (1992) Large-scale periodicity and Gaussian fluctuations. *Mon. Not. R. Astr. Soc.* **257**, 715–730.

Dekel A., Eldar A., Kolatt T., Yahil A., Willick J.A., Faber S.M., Courteau S. and Burstein D. (1999) POTENT reconstruction from Mark III velocities. *Astrophys. J.* **522**, 1–38.

Dekel A. and Lahav O. (1999) Stochastic nonlinear galaxy biasing. *Astrophys. J.* **520**, 24–34.

Dicke R.H., Peebles P.J.E., Roll P.G. and Wilkinson, D.T. (1965) Cosmic black-body radiation. *Astrophys. J.* **142**, 414–419.

Diggle P. (1983) *Statistical Analysis of Point Processes*. Chapman & Hall, London.

Diggle P. (1985) A kernel method for smoothing point process data. *Appl. Statist.* **34**, 138–147.

Djorgovski S. and Davis M. (1987) Fundamental properties of elliptical galaxies. *Astrophys. J.* **313**, 59–68.

Dodelson S. and Gaztañaga E. (2000) Inverting the angular correlation function. *Mon. Not. R. Astr. Soc.* **312**, 774–780.

Doguwa S. and Upton G.J.G. (1989) Edge-corrected estimators for the reduced second moment measure of point processes. *Biom. J.* **31**, 563–576.

Domínguez-Tenreiro R., Gómez-Flechoso M.A. and Martínez V.J. (1994) Scaling analysis of the distribution of galaxies in the CfA catalog. *Astrophys. J.* **424**, 42–58.

Domínguez-Tenreiro R. and Martínez V.J. (1989) Multidimensional analysis of the large-scale segregation of luminosity. *Astrophys. J. Lett.* **339**, L9–L11.

Domínguez-Tenreiro R., Roy L.J. and Martínez V.J. (1992) On the multifractal character of the Lorenz attractor. *Prog. Theor. Phys.* **87**, 1107–1118.

Doroshkevich A.G. (1970) Spatial structure of perturbations and the origin of rotation of galaxies in the fluctuation theory. *Astrofizika* **6**, 581–600.

Doroshkevich A.G., Müller V., Retzlaff J. and Turchaninov V. (1999) Superlarge-scale structure in N-body simulations. *Mon. Not. R. Astr. Soc.* **306**, 575–591.

Durrer R., Eckmann J.P., Sylos Labini F., Montuori M. and Pietronero L. (1997) Angular projections of fractal sets. *Europhys. Lett.* **40**, 491–496.

Efron B. and Tibshirani R. (1993) *An Introduction to the Bootstrap*. Chapman & Hall, London.

Efstathiou G., Davis M., Frenk C.S. and White S.D.M. (1985) Numerical techniques for large cosmological N-body simulations. *Astrophys. J. Suppl. Ser.* **57**, 241–260.

Efstathiou G., Ellis R.S. and Peterson B.A. (1988) Analysis of a complete galaxy redshift survey. II. — The field-galaxy luminosity function. *Mon. Not. R. Astr. Soc.* **232**, 431–461.

Efstathiou G., Fall S.M. and Hogan C. (1979) Self-similar gravitational clustering. *Mon. Not. R. Astr. Soc.* **189**, 203–220.

Efstathiou G. and Moody S.J. (2001) Maximum likelihood estimates of the two- and three-dimensional power spectra of the APM galaxy survey. *Mon. Not. R. Astr. Soc.* **325**, 1603–1615.

Ehlers J., Geren P. and Sachs R.K. (1968) Isotropic solutions of the Einstein–Liouville equations. *J. Math. Phys.* **9**, 1344–1349.

Einasto J., Einasto M., Frisch P., Gottloeber S., Mueller V., Saar V., Starobinsky A.A., Tago E., Tucker D. and Andernach H. (1997a) The supercluster–void network — II. An oscillating cluster correlation function. *Mon. Not. R. Astr. Soc.* **289**, 801–812.

Einasto J., Einasto M., Gottloeber S., Mueller V., Saar V., Starobinsky A.A., Tago E., Tucker D., Andernach H. and Frisch P. (1997b) A 120 Mpc periodicity in the three-dimensional distribution of galaxy superclusters. *Nature* **385**, 139–141.

Einasto J., Einasto M. and Gramann M. (1989) Structure and formation of superclusters. IX – Self-similarity of voids. *Mon. Not. R. Astr. Soc.* **238**, 155–177.

Einasto J. and Gramann M. (1993) Transition scale to a homogeneous universe. *Astrophys. J.* **407**, 443–447.

Einasto J., Joeveer M. and Saar E. (1980) Structure of superclusters and supercluster formation. *Mon. Not. R. Astr. Soc.* **193**, 353–375.

Einasto J., Klypin A.A., Saar E. and Shandarin S.F. (1984) Structure of superclusters and supercluster formation — III. Quantitative study of the Local Supercluster. *Mon. Not. R. Astr. Soc.* **206**, 529–558.

Einasto M., Tago E., Jaaniste J., Einasto J. and Andernach H. (1997) The supercluster-void network I. The supercluster catalogue and large-scale distribution. *Astr. Astrophys. Suppl. Ser.* **123**, 119–133.

Eisenstein D.J. and Hu W. (1999) Power spectra for cold dark matter and its variants. *Astrophys. J.* **511**, 5–15.

El-Ad H. and Piran T. (1997) Voids in the large-scale structure. *Astrophys. J.* **491**, 421–435.

El-Ad H., Piran T. and da Costa L.N. (1997) A catalogue of the voids in the IRAS 1.2-Jy survey. *Mon. Not. R. Astr. Soc.* **287**, 790–798.

Escalera E. and Mazure A. (1992) Wavelet analysis of subclustering — An illustration, Abell 754. *Astrophys. J.* **388**, 23–32.

Faber S.M. and Jackson R.E. (1976) Velocity dispersions and mass-to-light ratios for elliptical galaxies. *Astrophys. J.* **204**, 668–683.

Fairall A. (1998) *Large-Scale Structures in the Universe*. John Wiley & Sons in association with Praxis Publishing, Chichester.

Falco E.E., Kurtz M.J., Geller M.J., Huchra J.P., Peters J., Berlind P., Mink D.J., Tokarz S.P. and Elwell B. (1999) The Updated Zwicky Catalog (UZC). *PASP* **111**, 438–452.

Falconer K.J. (1985) *The Geometry of Fractal Sets*. Cambridge University Press, Cambridge.

Falconer K.J. (1990) *Fractal Geometry. Mathematical Foundations and Applications*. John Wiley & Sons, Chichester.

Farge M. (1992) Wavelet transforms and their application to turbulence. *Annu. Rev. Fluid Mech.* **24**, 395–457.

Feder J. (1988) *Fractals*. Plenum Press, New York.

Feldman H.A., Frieman J.A., Fry J.N. and Scoccimarro R. (2001) Constraints on galaxy bias, matter density, and primordial non-Gaussianity from the PSCz galaxy redshift survey. *Phys. Rev. Let.* **86**, 1434–1437.

Feldman H.A., Kaiser N. and Peacock J.A. (1994) Power-spectrum analysis of three-dimensional redshift surveys. *Astrophys. J.* **426**, 23–37.

Felten J.E. (1977) Study of the luminosity function for field galaxies. *Astronom. J.* **82**, 861–878.

Fiksel T. (1988) Edge-corrected density estimators for point processes. *Statistics* **19**, 67–76.

Fisher K.B., Davis M., Strauss M.A., Yahil A. and Huchra J.P. (1993) The power spectrum of IRAS galaxies. *Astrophys. J.* **402**, 42–57.

Fisher K.B., Huchra J.P., Strauss M.A., Davis M., Yahil A. and Schlegel D. (1995) The IRAS 1.2 Jy survey: Redshift data. *Astrophys. J. Suppl. Ser.* **100**, 69–103.

Fisher K.B., Lahav O., Hoffman Y., Lynden-Bell D. and Zaroubi S. (1994) Wiener reconstruction of density, velocity, and potential fields from all-sky galaxy redshift surveys. *Mon. Not. R. Astr. Soc.* **272**, 885–910.

Fisher K.B., Strauss M.A., Davis M., Yahil A. and Huchra J.P. (1992) The density evolution of IRAS galaxies. *Astrophys. J.* **389**, 188–195.

Fixsen D.J., Cheng E.S., Gales J.M., Mather J.C., Shafer R.A. and Wright E.L. (1996) The cosmic microwave background spectrum from the Full COBE FIRAS data set. *Astrophys. J.* **473**, 576–587.

Folkes S., Ronen S., Price I., Lahav O., Colless M., Maddox S., Deeley K., Glazebrook K., Bland-Hawthorn J., Cannon R., Cole S., Collins C., Couch W., Driver S.P., Dalton G., Efstathiou G., Ellis R.S., Frenk C.S., Kaiser N., Lewis I., Lumsden S., Peacock J., Peterson B.A., Sutherland W. and Taylor K. (1999) The 2dF Galaxy Redshift Survey: spectral types and luminosity functions. *Mon. Not. R. Astr. Soc.* **308**, 459–472.

Freysz E., Pouligny B., Argoul F. and Arnéodo A. (1990) Optical wavelet transform of fractal aggregates. *Phys. Rev. Lett.* **64**, 745–748.

Fry J.N. (1984) The galaxy correlation function hierarchy in perturbation theory. *Astrophys. J.* **279**, 499–510.

Fry J.N. (1985) Cosmological density fluctuations and large-scale structure. From N-point correlation functions to the probability distribution. *Astrophys. J.* **289**, 10–17.

Gamow G. (1948) The origin of elements and the separation of galaxies. *Phys. Rev.* **74**, 505–506.

Gaztañaga E. (1992) N-point correlation functions in the CfA and SSRS redshift distribution of galaxies. *Astrophys. J. Lett.* **398**, L17–L20.

Gaztañaga E. (1994) High-order galaxy correlation functions in the APM galaxy survey. *Mon. Not. R. Astr. Soc.* **268**, 913–924.

Geller M.J. and Huchra J.P. (1989) Mapping the universe. *Science* **246**, 897–903.

Ghez J.M. and Vaienti S. (1989) On the wavelet analysis of multifractal sets. *J. Stat. Phys.* **57**, 415–420.

Giavalisco M., Steidel C.S., Adelberger K.L., Dickinson M.E., Pettini M. and Kellogg M. (1998) The angular clustering of Lyman-break galaxies at redshift $z \sim 3$. *Astrophys. J.* **503**, 543–552.

Giovanelli R. and Haynes M.P. (1991) Redshift surveys of galaxies. *Ann. Rev. Astron. Astrophys.* **29**, 499–541.

Giovanelli R., Haynes M.P. and Chincarini G.L. (1986) Morphological segregation in the Pisces-Perseus supercluster. *Astrophys. J.* **300**, 77–92.

Gorski K. (1988) On the pattern of perturbations of the Hubble flow. *Astrophys. J. Lett.* **332**, L7–L11.

Gott J. R. I., Dickinson M. and Melott A.L. (1986) The sponge-like topology of large-scale structure in the universe. *Astrophys. J.* **306**, 341–357.

Gramann M., Cen R. and Gott J.R. (1994) Recovering the real density field of galaxies from redshift space. *Astrophys. J.* **425**, 382–391.

Groth E.J. and Peebles P.J.E. (1977) Statistical analysis of catalogs of extragalactic objects — VII. Two- and three-point correlation functions for the high-resolution Shane-Wirtanen catalog of galaxies. *Astrophys. J.* **217**, 385–405.

Gunn J.E. (1995) The Sloan Digital Sky Survey. In *American Astronomical Society Meeting*, vol. 186, p. 4405.

Guzzo L. (1997) Is the universe homogeneous? (On large scales). *New Astronomy* **2**, 517–532.

Guzzo L., Bartlett J.G., Cappi A., Maurogordato S., Zucca E., Zamorani G., Balkowski C., Blanchard A., Cayatte V., Chincarini G., Collins C.A., Maccagni D., MacGillivray H., Merighi R., Mignoli M., Proust D., Ramella M., Scaramella R., Stirpe G.M. and Vettolani G. (2000) The ESO Slice Project (ESP) galaxy redshift survey. VII. The redshift and real-space correlation functions. *Astron. Astrophys.* **355**, 1–16.

Guzzo L., Iovino A., Chincarini G., Giovanelli R. and Haynes M.P. (1991) Scale-invariant clustering in the large-scale distribution of galaxies. *Astrophys. J. Lett.* **382**, L5–L9.

Guzzo L., Strauss M.A., Fisher K.B., Giovanelli R. and Haynes M.P. (1997) Redshift-space distortions and the real-space clustering of different galaxy types. *Astrophys. J.* **489**, 37–48.

Hadwiger H. (1957) *Vorlesungen über Inhalt, Oberfläche und Isoperimetrie*. Springer-Verlag, Berlin.

Hall P. (1988) *Introduction to the Theory of Coverage Processes*. John Wiley & Sons, Chichester.

Halsey T.C., Jensen M.H., Kadanoff L.P., Procaccia I. and Shraiman B.I. (1986) Fractal measures and their singularities: The characterization of strange sets. *Phys. Rev. A* **33**, 1141–1151.

Hamilton A.J.S. (1988) Evidence for biasing in the CfA survey. *Astrophys. J. Lett.* **331**, L59–L62.

Hamilton A.J.S. (1992) Measuring Omega and the real correlation function from the redshift correlation function. *Astrophys. J. Lett.* **385**, L5–L8.

Hamilton A.J.S. (1993a) Ω from the anisotropy of the redshift correlation function in the IRAS 2 Jansky survey. *Astrophys. J. Lett.* **406**, L47–L50.

Hamilton A.J.S. (1993b) Toward better ways to measure the galaxy correlation function. *Astrophys. J.* **417**, 19–35.

Hamilton A.J.S. (1998) Linear redshift distortions: A review. In *The Evolving Universe*, vol. 231 of *Astrophysics and Space Science Library*. Kluwer Academic Publishers, Dordrecht, pp. 185–275.

Hamilton A.J.S. (2000) Uncorrelated modes of the nonlinear power spectrum. *Mon. Not. R. Astr. Soc.* **312**, 257–284.

Hamilton A.J.S. and Culhane M. (1996) Spherical redshift distortions. *Mon. Not. R. Astr. Soc.* **278**, 73–86.

Hamilton A.J.S., Kumar P., Lu E. and Matthews A. (1991) Reconstructing the primordial spectrum of fluctuations of the universe from the observed nonlinear clustering of galaxies. *Astrophys. J. Lett.* **374**, L1–L4.

Hamilton A.J.S. and Tegmark M. (2000) Decorrelating the power spectrum of galaxies. *Mon. Not. R. Astr. Soc.* **312**, 285–294.

Hamilton A.J.S. and Tegmark M. (2000) The real space power spectrum of the PSCz survey from 0.01 to 300 $h\,\mathrm{Mpc}^{-1}$. *http://arXiv/org/abs/astro-ph/0008392*.

Hamilton A.J.S., Tegmark M. and Padmanabhan N. (2000) Linear redshift distortions and power in the PSCz survey. *Mon. Not. R. Astr. Soc.* **317**, L23–L27.

Hanisch K.H. (1981) On classes of random sets and point processes. *Serdica* **7**, 160–167.

Harrison E. (2000) *Cosmology. The Science of the Universe*. Cambridge University Press, Cambridge.

Hatton S. (1999) Approaching a homogeneous galaxy distribution: Results from the Stromlo–APM redshift survey. *Mon. Not. R. Astr. Soc.* **310**, 1128–1136.

Hauser M.G. and Peebles P.J.E. (1973) Statistical analysis of catalogs of extragalactic objects. II. The Abell catalog of rich clusters. *Astrophys. J.* **185**, 757–785.

Heavens A.F. and Taylor A.N. (1995) A spherical harmonic analysis of redshift space. *Mon. Not. R. Astr. Soc.* **275**, 483–497.

Heck A. and Perdang J.M. (eds.) (1991) *Applying Fractals in Astronomy*, vol. 3 of *Lecture Notes in Physics*. Springer-Verlag, Berlin.

Heinrich L., Körner R., Mehlhorn N. and Muche L. (1998) Numerical and analytical computation of some second-order characteristics of spatial Poisson-Voronoi tesselations. *Statistics* **31**, 235–262.

Hentschel H.G.E. and Procaccia I. (1983) The infinite number of generalized dimensions of fractals and strange attractors. *Physica D* **8**, 435–444.

Hermit S., Santiago B.X., Lahav O., Strauss M.A., Davis M., Dressler A. and Huchra J.P. (1996) The two-point correlation function and morphological segregation in the Optical Redshift Survey. *Mon. Not. R. Astr. Soc.* **283**, 709–720.

Hernquist L. (1987) Performance characteristics of tree codes. *Astrophys. J. Suppl. Ser.* **64**, 715–734.

Hockney R.W. and Eastwood J.W. (1988) *Computer Simulation Using Particles*. Adam Hilger, Bristol.

Hoffman Y. and Ribak E. (1991) Constrained realizations of Gaussian random fields: A simple algorithm. *Astrophys. J. Lett.* **380**, L5–L8.

Holschneider M. (1988) On the wavelet transformation of fractal objects. *J. Stat. Phys.* **50**, 963–993.

Hubble E. (1934) The distribution of extra-galactic nebulae. *Astrophys. J.* **79**, 8–76.

Hubble E. and Humason M.L. (1931) The velocity-distance relation among extra-galactic nebulae. *Astrophys. J.* **74**, 43–80.

Hubble E.P. (1936) *The Realm of Nebulae*. Yale University Press, New Haven.

Huchra J., Davis M., Latham D. and Tonry J. (1983) A survey of galaxy redshifts — IV. The data. *Astrophys. J. Suppl. Ser.* **52**, 89–119.

Huchra J.P., Geller M.J., de Lapparent V. and Corwin H. G. J. (1990) The CfA redshift survey — Data for the NGP + 30 zone. *Astrophys. J. Suppl. Ser.* **72**, 433–470.

Icke V. (1984) Voids and filaments. *Mon. Not. R. Astr. Soc.* **206**, 1P–3P.

Icke V. and van de Weygaert R. (1987) Fragmenting the universe. *Astron. Astrophys.* **184**, 16–32.

Jenkins A., Frenk C.S., Pearce F.R., Thomas P.A., Colberg J.M., White S.D.M., Couchman H.M.P., Peacock J.A., Efstathiou G. and Nelson A.H. (1998) Evolution of structure in cold dark matter universes. *Astrophys. J.* **499**, 20–40.

Jensen L.G. and Szalay A.S. (1986) N-point correlations for biased galaxy formation. *Astrophys. J. Lett.* **305**, L5–L9.

Jensen M.H., Paladin G. and Vulpiani A. (1991) Multiscaling in multifractals. *Phys. Rev. Lett.* **67**, 208–211.

Jing Y.P. and Börner G. (1998) The three-point correlation function of galaxies determined from the Las Campanas redshift survey. *Astrophys. J.* **503**, 37–47.

Jones B.J.T., Coles P. and Martínez V.J. (1992) Heterotopic clustering. *Mon. Not. R. Astr. Soc.* **259**, 146–154.

Joyce M., Montuori M., Sylos Labini F. and Pietronero L. (1999) Comment on the paper "The ESO Slice Project [ESP] galaxy redshift survey: V. Evidence for a D=3 sample dimensionality." *Astr. Astrophys.* **344**, 387–392.

Kaiser N. (1984) On the spatial correlations of Abell clusters. *Astrophys. J. Lett.* **284**, L9–L12.

Kaiser N. (1986) A sparse-sampling strategy for the estimation of large-scale clustering from redshift surveys. *Mon. Not, R. Astr. Soc.* **219**, 785–790.

Kaiser N. (1987) Clustering in real space and in redshift space. *Mon. Not, R. Astr. Soc.* **227**, 1–21.

Kaiser N. and Peacock J.A. (1991) Power-spectrum analysis of one-dimensional redshift surveys. *Astrophys. J.* **379**, 482–506.

Kaiser N. and Squires G. (1993) Mapping the dark matter with weak gravitational lensing. *Astrophys. J.* **404**, 441–450.

Kauffmann G., Colberg J.M., Diaferio A. and White S.D.M. (1999) Clustering of galaxies in a hierachical universe: I. Methods and results at $z = 0$. *Mon. Not. R. Astr. Soc.* **303**, 188–206.

Kendall M.G. and Stuart A. (1963) *The Advanced Theory of Statistics. Distribution Theory*, vol. I. Charles Griffin & Co. Ltd., London.

Kendall M.G. and Stuart A. (1967) *The Advanced Theory of Statistics. Inference and Relationship*, vol. II. Charles Griffin & Co. Ltd., London.

Kepner J., Fan X., Bahcall N., Gunn J., Lupton R. and Xu G. (1999) An automated cluster finder: The adaptive matched filter. *Astrophys. J.* **517**, 78–91.

Kerscher M. (1999) The geometry of second-order statistics — Biases in common estimators. *Astron. Astrophys.* **343**, 333–347.

Kerscher M. (2000) Statistical analysis of large-scale structure in the Universe. In *Statistical Physics and Spatial Statistics: The Art of Analyzing and Modelling Spatial Structures and Pattern Formation* (eds. R.K. Mecke and D. Stoyan), *Lecture Notes in Physics*, vol. 554, Springer-Verlag, Berlin, pp. 36–71.

Kerscher M. (2001) Constructing, characterizing, and simulating Gaussian and higher-order point distributions. *Phys. Rev. E* **64**, 056109.

Kerscher M. and Martínez V.J. (1998) Galaxy clustering and spatial point processes. *Bull. Int. Statist. Inst.* **57-2**, 363–366.

Kerscher M., Pons-Bordería M.J., Schmalzing J., Trasarti-Battistoni R., Buchert T., Martínez V.J. and Valdarnini R. (1999) A Global Descriptor of Spatial Pattern Interaction in the Galaxy Distribution. *Astrophys. J.* **513**, 543–548.

Kerscher M., Schmalzing J., Buchert T. and Wagner H. (1998) Fluctuations in the IRAS 1.2 Jy catalogue. *Astron. Astrophys.* **333**, 1–12.

Kerscher M., Schmalzing J., Retzlaff J., Borgani S., Buchert T., Gottlober S., Muller V., Plionis M. and Wagner H. (1997) Minkowski functionals of Abell/ACO clusters. *Mon. Not. R. Astr. Soc.* **284**, 73–84.

Kerscher M., Szapudi I. and Szalay A.S. (2000) A comparison of estimators for the two-point correlation function. *Astrophys. J. Lett.* **535**, L13–L16.

Khintchine A.Y. (1955) *Mathematical Methods of Queueing Theory* (in Russian). Papers of the Mathematical Institute of V.A. Steklov, Moscow.

King C.R. and Ellis R.S. (1985) The evolution of spiral galaxies and uncertainties in interpreting galaxy counts. *Astrophys. J.* **288**, 456–464.

Kirshner R.P. (1996) Galaxy redshift surveys. In *Dark Matter in the Universe* (eds. S. Bonometto, J.R. Primack and A. Provenzale), IOS Press, Amsterdam, pp. 36–48.

Kirshner R.P., Oemler A. J., Schechter P.L. and Shectman S.A. (1981) A million cubic megaparsec void in Bootes. *Astrophys. J. Lett.* **248**, L57–L60.

Kirshner R.P., Oemler A. J., Schechter P.L. and Shectman S.A. (1987) A survey of the Bootes void. *Astrophys. J.* **314**, 493–506.

Klypin A. and Shandarin S.F. (1993) Percolation technique for galaxy clustering. *Astrophys. J.* **413**, 48–58.

Kneib J.P., Ellis R.S., Smail I., Couch W.J. and Sharples R.M. (1996) Hubble space telescope observations of the lensing cluster Abell 2218. *Astrophys. J.* **471**, 643–656.

Kolatt T., Dekel A., Ganon G. and Willick J.A. (1996) Simulating our cosmological neighborhood: Mock catalogs for velocity analysis. *Astrophys. J.* **458**, 419–457.

Kruskal J.B. (1956) On the shortest spanning subtree of a graph and the traveling salesman problem. *Proc. Amer. Math. Soc.* **7**, 48–50.

Krzewina L.G. and Saslaw W.C. (1996) Minimal spanning tree statistics for the analysis of large-scale structure. *Mon. Not. R. Astr. Soc.* **278**, 869–876.

Sylos Labini F. and Montuori M. (1998) Scale invariant properties of the APM–Stromlo survey. *Astr. Astrophys.* **331**, 809–814.

Lahav O., Lilje P.B., Primack J.R. and Rees M.J. (1991) Dynamical effects of the cosmological constant. *Mon. Not. R. Astr. Soc.* **251**, 128–136.

Landy S.D. and Szalay A.S. (1993) Bias and variance of angular correlation functions. *Astrophys. J.* **412**, 64–71.

Layzer D. (1956) A new model for the distribution of galaxies in space. *Astronom. J.* **61**, 383–385.

Le Cam L. (1990) Maximum likelihood: an introduction. *Int. Statist. Rev.* **58**, 153–171.

Lehmann E.L. (1959) *Testing Statistical Hypotheses*. John Wiley & Sons, New York.

Lehmann E.L. and Casella G. (1998) *Theory of Point Estimation*. Springer-Verlag, New York.

Lemson G. and Sanders R.H. (1991) On the use of the conditional density as a description of galaxy clustering. *Mon. Not. R. Astr. Soc.* **252**, 319–328.

Lin H., Kirshner R.P., Shectman S.A., Landy S.D., Oemler A., Tucker D.L. and Schechter P.L. (1996) The luminosity function of galaxies in the Las Campanas redshift survey. *Astrophys. J.* **464**, 60–78.

Loredo T.J. (1990) From Laplace to supernova SN 1987A: Bayesian inference in astrophysics. In *Maximum Entropy and Bayesian Methods* (ed. P.F. Fougère), Kluwer Academic Publishers, Dordrecht, pp. 81–142.

Loveday J. (1996) The APM bright galaxy catalogue. *Mon. Not. R. Astr. Soc.* **278**, 1025–1048.

Loveday J., Maddox S.J., Efstathiou G. and Peterson B.A. (1995) The Stromlo–APM redshift survey. 2: Variation of galaxy clustering with morphology and luminosity. *Astrophys. J.* **442**, 457–468.

Loveday J., Peterson B.A., Maddox S.J. and Efstathiou G. (1996) The Stromlo–APM redshift survey. IV. The redshift catalog. *Astrophys. J. Suppl. Ser.* **107**, 201–214.

Lucy L.B. (1974) An iterative technique for the rectification of observed distributions. *Astronom. J.* **79**, 745–754.

Luo S. and Vishniac E. (1995) Three-dimensional shape statistics: Methodology. *Astrophys. J. Suppl. Ser.* **96**, 429–460.

Luo S., Vishniac E.T. and Martel H. (1996) Three-dimensional shape statistics: Applications. *Astrophys. J.* **468**, 62–74.

Lynden-Bell D., Faber S.M., Burstein D., Davies R.L., Dressler A., Terlevich R.J. and Wegner G. (1988) Spectroscopy and photometry of elliptical galaxies. V. Galaxy streaming toward the new supergalactic center. *Astrophys. J.* **326**, 19–49.

Maddox S.J., Efstathiou G. and Sutherland W.J. (1996) The APM Galaxy Survey — III. An analysis of systematic errors in the angular correlation function and cosmological implications. *Mon. Not. R. Astr. Soc.* **283**, 1227–1263.

Maddox S.J., Efstathiou G., Sutherland W.J. and Loveday J. (1990a) The APM galaxy survey — I. APM measurements and star-galaxy separation. *Mon. Not. R. Astr. Soc.* **243**, 692–712.

Maddox S.J., Efstathiou G., Sutherland W.J. and Loveday J. (1990b) Galaxy correlations on large scales. *Mon. Not. R. Astr. Soc.* **242**, 43P–47P.

Mandelbrot B.B. (1975) Sur un modèle décomposable d'Univers hiérarchisé: déduction des corrélations galactiques sur la sphère céleste. *C.R. Acad. Sci. (Paris) A* **280**, 1551–1554.

Mandelbrot B.B. (1982) *The Fractal Geometry of Nature.* Revised edition of *Fractals* (1977). W.H. Freeman, San Francisco.

Mann R.G., Heavens A.F. and Peacock J.A. (1993) The richness dependence of cluster correlations. *Mon. Not. R. Astr. Soc.* **263**, 798–816.

Margon B. (1999) The Sloan Digital Sky Survey. *Phil. Trans. R. Soc. London A* **357**, 93–103.

Marriott F.H.C. (1990) *A Dictionary of Statistical Terms.* 5th ed., Longman Scientific & Technical, Harlow.

Martel H. (1991) Linear perturbation theory and spherical overdensities in $\Lambda \neq 0$ Friedmann models. *Astrophys. J.* **377**, 7–13.

Martínez V.J. (1991) Fractal aspects of galaxy clustering. In *Applying Fractals in Astronomy*, vol. 3 of *Lecture Notes in Physics* (eds. A. Heck and J.M. Perdang J.M.), Springer-Verlag, Berlin, pp. 135–159.

Martínez V.J. (1996) Measures of galaxy clustering. In *Dark Matter in the Universe* (eds. S. Bonometto, J.R. Primack and A. Provenzale), IOS Press, Amsterdam, pp. 255–267.

Martínez V.J. (1997) Recent advances in large-scale structure statistics. In *Statistical Challenges in Modern Astronomy II* (eds. G.J. Babu and E.D. Feigelson), Springer-Verlag, New York, pp. 153–171.

Martínez V.J. (1999) Is the universe fractal? *Science* **284**, 445–446.

Martínez V.J. (Non)-fractality on large scales. In *IAU Symp. 201: New Cosmological Data and the Value of the Fundamental Parameters, ASP Conference Series.* (eds. A. Lasenby and A. Wilkinson), vol. 201, in press.

Martínez V.J. and Coles P. (1994) Correlations and scaling in the QDOT redshift survey. *Astrophys. J.* **437**, 550–555.

Martínez V.J. and Jones B.J.T. (1990) Why the universe is not a fractal. *Mon. Not. R. Astr. Soc.* **242**, 517–521.

Martínez V.J., Jones B.J.T., Domínguez-Tenreiro R. and van de Weygaert R. (1990) Clustering paradigms and multifractal measures. *Astrophys. J.* **357**, 50–61.

Martínez V.J., López-Martí B. and Pons-Bordería M.J. (2001) Does the galaxy correlation length increase with the sample depth? *Astrophys. J. Lett.* **554**, L5–L8.

Martínez V.J., Paredes S., Borgani S. and Coles P. (1995) Multiscaling properties of large-scale structure in the universe. *Science* **269**, 1245–1247.

Martínez V.J., Paredes S. and Saar E. (1993) Wavelet analysis of the multifractal character of the galaxy distribution. *Mon. Not. R. Astr. Soc.* **260**, 365–375.

Martínez V.J., Pons-Bordería M., Moyeed R.A. and Graham M.J. (1998) Searching for the scale of homogeneity. *Mon. Not. R. Astr. Soc.* **298**, 1212–1222.

Martínez V.J., Portilla M., Jones B.J.T. and Paredes S. (1993) The galaxy clustering correlation length. *Astron. Astrophys.* **280**, 5–19.

Matarrese S., Verde L. and Heavens A.F. (1997) Large-scale bias in the Universe: Bispectrum method. *Mon. Not. R. Astr. Soc.* **290**, 651–662.

Matérn B. (1960) Spatial Variation. *Medd. fran Statens Skogsforskningsinst.* **49**, 1–144.

Matsubara T., Szalay A.S. and Landy S.D. (2000) Cosmological parameters from the eigenmode analysis of the Las Campanas redshift survey. *Astrophys. J. Lett.* **535**, L1–L4.

Maurogordato S. and Lachièze-Rey M. (1987) Void probabilities in the galaxy distribution — Scaling and luminosity segregation. *Astrophys. J.* **320**, 13–25.

McCauley J.L. (1998) The galaxy distributions: Homogeneous, fractal, or neither? *Fractals* **6**, 109–119.

McCracken H.J., Metcalfe N., Shanks T., Campos A., Gardner J.P. and Fong R. (2000) Galaxy number counts — IV. Surveying the Herschel deep field in the near-infrared. *Mon. Not. R. Astr. Soc.* **311**, 707–718.

Meakin P. (1987) Diffusion-limited aggregation on multifractal lattices: A model for fluid-fluid displacement in porous media. *Phys. Rev. A* **36**, 2833–2837.

Mecke K.R., Buchert T. and Wagner H. (1994) Robust morphological measures for large-scale structure in the Universe. *Astr. Astrophys.* **288**, 697–704.

Mecke K.R. and Wagner H. (1991) Euler characteristic and related measures for random geometric sets. *J. Stat. Phys.* **64**, 843–850.

Melott A.L. (1990) The topology of the large-scale structure in the universe. *Phys. Rep.* **193**, 1–39.

Melott A.L. and Dominik K.G. (1993) Tests of smoothing methods for topological study of galaxy redshift surveys. *Astrophys. J. Suppl. Ser.* **86**, 1–4.

Meyer S.L. (1975) *Data Analysis for Scientists and Engineers*. John Wiley & Sons, New York.

Miller C.J. and Batuski D.J. (2001) The power spectrum of rich clusters on near-gigaparsec scales. *Astrophys. J.* **551**, 635–642.

Miller C.J., Genovese C., Nichol R.C., Wasserman L., Connolly A., Reichart D., Hopkins A., Schneider J. and Moore A. (2001) Controlling the false discovery rate in astrophysical data analysis. *http://arXiv/org/abs/astro-ph/0107034*.

Miller C.J., Nichol R.C. and Batuski D.J. (2001) Possible detection of baryonic fluctuations in the large-scale structure power spectrum. *Astrophys. J.* **555**, 68–73.

Milne E.A. (1933) World-structure and the expansion of the universe. *Zeitschr. für Astrophys.* **6**, 1–35.

Mo H.J., Jing Y.P. and Börner G. (1992) On the error estimates of correlation functions. *Astrophys. J.* **392**, 452–457.

Mo H.J., Peacock J.A. and Xia X.Y. (1993) The cross-correlation of IRAS galaxies with Abell clusters and radio galaxies. *Mon. Not. R. Astr. Soc.* **260**, 121–131.

Mo H.J. and White S.D.M. (1996) An analytic model for the spatial clustering of dark matter haloes. *Mon. Not. R. Astr. Soc.* **282**, 347–361.

Mo H.J., Ying Y.P. and White S.D.M. (1997) High-order correlations of peaks and halos: A step toward understanding galaxy biasing. *Mon. Not. R. Astr. Soc.* **284**, 189–201.

Møller J., Syversveen A.R. and Waagepetersen R.P. (1998) Log Gaussian Cox processes. *Scand. J. Statist.* **25**, 451–482.

Mollerach S., Martínez V.J., Diego J.M., Martínez-González E., Sanz J.L. and Paredes S. (1999) The roughness of the last scattering surface. *Astrophys. J.* **525**, 17–24.

Monaco P. (1999) Dynamics in the cosmological mass function (or, why does the Press & Schechter work?). In *Observational Cosmology: The Development of Galaxy Systems* (eds. G. Giuricin, M. Mezzetti and P. Salucci), vol. 176, Astronomical Society of the Pacific, San Francisco, pp. 186–196.

Monaco P. and Efstathiou G. (1999) Reconstruction of cosmological initial conditions from galaxy redshift catalogues. *Mon. Not. R. Astr. Soc.* **308**, 763–779.

Monaco P., Efstathiou G., Maddox S.J., Branchini E., Frenk C.S., McMahon R.G., Oliver S.J., Rowan-Robinson M., Saunders W., Sutherland W.J., Tadros H. and White S.D.M. (2000) The 1-point PDF of the initial conditions of our local universe from the IRAS PSC redshift catalogue. *Mon. Not. R. Astr. Soc.* **318**, 681–692.

Monaghan J.J. (1992) Smoothed particle hydrodynamics. *Ann. Rev. Astron. Astrophys.* **30**, 543–574.

Monin A.S. and Yaglom A.M. (1975) *Statistical Fluid Mechanics*. MIT Press, Cambridge.

Moore B., Frenk C.S., Weinberg D.H., Saunders W., Lawrence A., Ellis R.S., Kaiser N., Efstathiou G. and Rowan-Robinson M. (1992) The topology of the QDOT IRAS redshift survey. *Mon. Not. R. Astr. Soc.* **256**, 477–499.

Müller V., Arbabi-Bidgoli S., Einasto J. and Tucker D. (2000) Voids in the Las Campanas redshift survey versus cold dark matter models. *Mon. Not. R. Astr. Soc.* **318**, 280–288.

Narayan R. and Bartelmann M. (1999) Lectures on gravitational lensing. In *Formation of Structure in the Universe*, Cambridge University Press, pp. 360–421.

Narayanan V.K. and Croft R.A.C. (1999) Recovering the primordial fluctuations: A comparison of methods. *Astrophys. J.* **515**, 471–486.

Neyman J. and Scott E.L. (1952) A theory of the spatial distribution of galaxies. *Astrophys. J.* **116**, 144–163.

Neyman J. and Scott E.L. (1955) On the inapplicability of the theory of fluctuations to galaxies. *Astronom. J.* **60**, 33.

Neyman J. and Scott E.L. (1958) Statistical approach to problems of cosmology. *J. Roy. Statist. Soc. B* **20**, 1–43.

Nichol R., Miller C.J., Connolly A., Chong S.S., Genovese C., Moore A., Reichart D., Schneider J., Wasserman L., Annis J., Brinkman J., Bohringer H., Castander F., Kim R., McKay T., Postman M., Sheldon E., Szapudi I., Romer K. and Voges W. (2000) SDSS–RASS: Next generation of cluster-finding algorithms. In *http://arXiv/org/abs/astro-ph/0011557*.

Nilson P. (1973) Uppsala general catalogue of galaxies. *Nova Acta Regiae Soc. Sci. Upsaliensis Ser. V* **1**, 1–447.

Nusser A. and Dekel A. (1990) Filamentary structure from Gaussian fluctuations using the adhesion approximation. *Astrophys. J.* **362**, 14–24.

Nusser A., Dekel A., Bertschinger E. and Blumenthal G. (1991) Cosmological velocity–density relation in the quasi-linear regime. *Astrophys. J.* **379**, 6–18.

Nusser A. and Lahav O. (2000) The Lyman-α forest in a truncated hierarchical structure formation. *Mon. Not. R. Astr. Soc.* **313**, L39–L42.

Ohser J. and Stoyan D. (1981) On the second-order and orientation analysis of planar stationary point processes. *Biom. J.* **23**, 523–533.

Okabe A., Boots B., Sugihara K. and Chiu S.N. (1999) *Spatial Tessellations. Concepts and Applications of Voronoi Diagrams*. John Wiley & Sons, Chichester.

Padmanabhan N., Tegmark M. and Hamilton A.J.S. (2000) The power spectrum of the CfA/SSRS UZS galaxy redshift survey. *Astrophys. J.* **550**, 52–64.

Padmanabhan T. (1993) *Structure Formation in the Universe*. Cambridge University Press, Cambridge.

Pan J. and Coles P. (2000) Large-scale cosmic homogeneity from a multifractal analysis of the PSCz catalogue. *Mon. Not. R. Astr. Soc.* **318**, L51–L54.

Paredes S., Jones B.J.T. and Martínez V.J. (1995) The clustering of galaxy clusters — Synthetic distributions and the correlation function amplitude. *Mon. Not. R. Astr. Soc.* **276**, 1116–1130.

Park C., Vogeley M.S., Geller M.J. and Huchra J.P. (1994) Power spectrum, correlation function, and tests for luminosity bias in the CfA redshift survey. *Astrophys. J.* **431**, 569–585.

Paul T. (1985) Ph.D. thesis, Univ. Aix-Marseille II, Luminy, Marceille, France.

Peacock J.A. (1999) *Cosmological Physics*. Cambridge University Press, Cambridge.

Peacock J.A. (2000) Clustering of mass and galaxies. In *http://arXiv/org/abs/astro-ph/0002013*.

Peacock J.A., Cole S., Norberg P., Baugh C.M., Bland-Hawthorn J., Bridges T., Cannon R.D., Colless M., Collins C., Couch W., Dalton G., Deeley K., Propris R.D., Driver S.P., Efstathiou G., Ellis R.S., Frenk C.S., Glazebrook K., Jackson C., Lahav O., Lewis I., Lumsden S., Maddox S., Percival W.J., Peterson B.A., Price I., Sutherland W. and Taylor K. (2001) A measurement of the cosmological mass density from clustering in the 2dF galaxy redshift survey. *Nature* **410**, 169–173.

Peacock J.A. and Dodds S.J. (1996) Non-linear evolution of cosmological power spectra. *Mon. Not. R. Astr. Soc.* **280**, L19–L26.

Peacock J.A. and Heavens A.F. (1990) Alternatives to the Press–Schechter cosmological mass function. *Mon. Not. R. Astr. Soc.* **243**, 133–143.

Peacock J.A. and Smith R.E. (2000) Halo occupation numbers and galaxy bias. *Mon. Not. R. Astr. Soc.* **318**, 1144–1156.

Peebles P.J.E. (1973) Statistical analysis of catalogs of extragalactic objects. I. Theory. *Astrophys. J.* **185**, 413–440.

Peebles P.J.E. (1978) Stability of a hierarchical clustering pattern in the distribution of galaxies. *Astron. Astrophys.* **68**, 345–352.

Peebles P.J.E. (1980) *The Large-Scale Structure of the Universe*. Princeton University Press, Princeton.

Peebles P.J.E. (1989) Tracing galaxy orbits back in time. *Astrophys. J. Lett.* **344**, L53–L56.

Peebles P.J.E. (1993) *Principles of Physical Cosmology*. Princeton University Press, Princeton.

Peebles P.J.E. (1998) The standard cosmological model. In *Les Rencontres de Physique de la Vallée d'Aoste, Results and Perspectives in Particle Physics* (ed. M. Greco), Poligrafica Laziale s.r.l., Frascati, pp. 39–56.

Peebles P.J.E. (2001) The galaxy and mass N-Point correlation functions: a blast from the past. In *Historical Development of Modern Cosmology, ASP Conference Series* (eds. V.J. Martínez, V. Trimble and M.J. Pons-Bordería), vol. 252, Astronomical Society of the Pacific, San Francisco, pp. 201–218.

Peebles P.J.E. and Hauser M.G. (1974) Statistical analysis of catalogs of extragalactic objects. III. The Shane-Wirtanen and Zwicky catalogs. *Astrophys. J. Suppl. Ser.* **28**, 19–36.

Peitgen H.O. and Saupe D. (eds.) (1988) *The Science of Fractal Images*. Springer-Verlag, New York.

Penzias A.A. and Wilson, R.W. (1965) A measurement of excess antenna temperature at 4080 Mc/s. *Astrophys. J.* **142**, 419–420.

Percival W.J., Baugh C.M., Bland-Hawthorn J., Bridges T., Cannon R., Cole S., Colless M., Collins C., Couch W., Dalton G., Propris R.D., Driver S.P., Efstathiou G., Ellis R.S., Frenk C.S., Glazebrook K., Jackson C., Lahav O., Lewis I., Lumsden S., Maddox S., Moody S., Norberg P., Peacock J.A., Peterson B.A., Sutherland W. and Taylor K. (2001) The 2dF galaxy redshift survey: The power spectrum and the matter content of the universe. *http://arXiv/org/abs/astro-ph/0105252*.

Perlmutter S., Aldering G., Goldhaber G., Knop R.A., Nugent P., Castro P.G., Deustua S., Fabbro S., Goobar A., Groom D.E., Hook I.M., Kim A.G., Kim M.Y., Lee J.C., Nunes N.J., Pain R., Pennypacker C.R., Quimby R., Lidman C., Ellis R.S., Irwin M., McMahon R.G., Ruiz-Lapuente P., Walton N., Schaefer B., Boyle B.J., Filippenko A.V., Matheson T., Fruchter A.S., Panagia N., Newberg H.J.M. and Couch W.J. (1999) Measurements of Ω and Λ from 42 high-redshift supernovae. *Astrophys. J.* **516**, 565–586.

Phillipps S. and Shanks T. (1987) On the variation of galaxy correlations with luminosity. *Mon. Not. R. Astr. Soc.* **229**, 621–626.

Pierpaoli E., Scott D. and White M. (2001) Power spectrum normalization from the local abundance of rich clusters of galaxies. *Mon. Not. R. Astr. Soc.* **325**, 77–88.

Pietronero L. (1987) The fractal structure of the Universe: correlations of galaxies and clusters and the average mass density. *Physica A* **144**, 257–284.

Pisani A. (1996) A non-parametric and scale-independent method for cluster analysis — II. The multivariate case. *Mon. Not. R. Astr. Soc.* **278**, 697–726.

Politzer H.D. and Wise M.B. (1984) Relations between spatial correlations of rich clusters of galaxies. *Astrophys. J. Lett.* **285**, L1–L3.

Pons-Bordería M.J., Martínez V.J., Stoyan D., Stoyan H. and Saar E. (1999) Comparing estimators of the galaxy correlation function. *Astrophys. J.* **523**, 480–491.

Postman M., Huchra J.P. and Geller M.J. (1992) The distribution of nearby rich clusters of galaxies. *Astrophys. J.* **384**, 404–422.

Press W.H. and Davis M. (1982) How to identify and weigh virialized clusters of galaxies in a complete redshift catalog. *Astrophys. J.* **259**, 449–473.

Press W.H. and Schechter P. (1974) Formation of galaxies and clusters of galaxies by self-similar gravitational condensation. *Astrophys. J.* **187**, 425–438.

Press W.H., Teukolsky S.A., Vetterling W.T. and Flannery B.P. (1992) *Numerical Recipes, Second Edition*. Cambridge University Press, Cambridge.

Prim R.C. (1957) Shortest connection networks and some generalizations. *Bell System Tech. J.* **288**, 1389–1401.

Provenzale A., Guzzo L. and Murante G. (1994) Clustering properties from finite galaxy samples. *Mon. Not. R. Astr. Soc.* **266**, 555–566.

Reiprich T.H. and Böhringer H. (1999) The empirical X-ray luminosity-gravitational mass relation for clusters of galaxies. *Astronom. Nachr.* **320**, 296.

Rényi A. (1970) *Probability Theory*. North-Holland, Amsterdam.

Riess A.G., Press W.H. and Kirshner R.P. (1996) A Precise Distance Indicator: Type Ia Supernova Multicolor Light-Curve Shapes. *Astrophys. J.* **473**, 88–109.

Ripley B.D. (1976) The second-order analysis of stationary point processes. *J. Appl. Prob.* **13**, 255–266.

Ripley B.D. (1981) *Spatial Statistics*. John Wiley & Sons, New York.

Ripley B.D. (1987) *Stochastic Simulation*. John Wiley & Sons, New York.

Ripley B.D. (1988) *Statistical Inference for Spatial Processes*. Cambridge University Press, Cambridge.

Rivolo A.R. (1986) The two-point galaxy correlation function of the Local Supercluster. *Astrophys. J.* **301**, 70–76.

Roukema B.F. (2001) On the comoving distance as an arc-length in four dimensions. *Mon. Not. R. Astr. Soc.* **325**, 138–142.

Rowan-Robinson M. (1985) *The Cosmological Distance Ladder: Distance and Time in the Universe*. W.H. Freeman, New York.

Rowan-Robinson M. (1996) IRAS Galaxy Redshift Surveys. In *Mapping, Measuring, and Modelling the Universe, ASP Conference Series* (eds. P. Coles, V.J. Martínez and M.J. Pons-Bordería), vol. 94, Astronomical Society of the Pacific, San Francisco, pp. 171–176.

Rowan-Robinson M., Lawrence A., Saunders W., Crawford J., Ellis R., Frenk C.S., Parry I., Xiaoyang X., Allington-Smith J., Efstathiou G. and Kaiser N. (1990) A sparse-sampled redshift survey of IRAS galaxies — I. The convergence of the IRAS dipole and the origin of our motion with respect to the microwave background. *Mon. Not. R. Astr. Soc.* **247**, 1–18.

Rowan-Robinson M., Saunders W., Lawrence A. and Leech K. (1991) The QMW IRAS galaxy catalogue — A highly complete and reliable IRAS 60-micron galaxy catalogue. *Mon. Not. R. Astr. Soc.* **253**, 485–495.

Rybicki G.B. and Press W.H. (1992) Interpolation, realization, and reconstruction of noisy, irregularly sampled data. *Astrophys. J.* **398**, 169–176.

Saar E. (1973) Evolution of density contrast in underdense regions. *Publ. Tartu Astrophys. Obs.* **41**, 30–40.

Sahni V. and Coles P. (1995) Approximation methods for non-linear gravitational clustering. *Phys. Reports* **262**, 1–136.

Sahni V., Sathyaprakash B.S. and Shandarin S.F. (1997) Probing large scale structure using percolation and genus curves. *Astrophys. J. Lett.* **476**, L1–L6.

Sahni V., Sathyaprakash B.S. and Shandarin S.F. (1998) Shapefinders: a new shape diagnostic for large-scale structure. *Astrophys. J. Lett.* **495**, L5–L8.

Salzer J.J., Hanson M.M. and Gavazzi G. (1990) The relative spatial distributions of high- and low-luminosity galaxies toward Coma. *Astrophys. J.* **353**, 39–50.

Santaló L. (1976) *Integral Geometry and Geometric Probability*. Addison-Wesley, Reading.

Sanz J.L., Herranz D. and Martínez-Gónzalez E. (2001) Optimal detection of sources on a homogeneous and isotropic background. *Astrophys. J.* **552**, 484–492.

Saslaw W.C. (2000) *The Distribution of the Galaxies. Gravitational Clustering in Cosmology*. Cambridge University Press, Cambridge.

Saunders W. and Ballinger B.E. (2000) Interpolation of discretely-sampled density fields. In *http://arXiv/org/abs/astro-ph/0005606*.

Saunders W., Frenk C., Rowan-Robinson M., Lawrence A. and Efstathiou G. (1991) The density field of the local universe. *Nature* **349**, 32–38.

Saunders W., Rowan-Robinson M., Lawrence A., Efstathiou G., Kaiser N., Ellis R.S. and Frenk C.S. (1990) The 60-μm and far-infrared luminosity functions of the IRAS galaxies. *Mon. Not. R. Astr. Soc.* **242**, 318–337.

Saunders W., Sutherland W.J., Maddox S.J., Keeble O., Oliver S.J., Rowan-Robinson M., McMahon R.G., Efstathiou G.P., Tadros H., White S.D.M., Frenk C.S., Carramiñana A. and Hawkins M.R.S. (2000) The PSCz catalogue. *Mon. Not. R. Astr. Soc.* **317**, 55–64.

Scaramella R., Guzzo L., Zamorani G., Zucca E., Balkowski C., Blanchard A., Cappi A., Cayatte V., Chincarini G., Collins C., Fiorani A., Maccagni D., MacGillivray H., Maurogordato S., Merighi R., Mignoli M., Proust D., Ramella M., Stirpe G.M. and Vettolani G. (1998) The ESO Slice Project [ESP] galaxy redshift survey: V. Evidence for a D=3 sample dimensionality. *Astr. Astrophys.* **334**, 404–408.

Schaap W.E. and van de Weygaert R. (2000) Continuous fields and discrete samples: reconstruction through Delaunay tessellations. *Astron. Astrophys.* **363**, L29–L32.

Schaap W.E. and van de Weygaert R. (2001) Delaunay recovery of cosmic density and velocity probes. *http://arXiv/org/abs/astro-ph/0109261*.

Schechter P. (1976) An analytic expression for the luminosity function for galaxies. *Astrophys. J.* **203**, 297–306.

Scherrer R.J. and Bertschinger E. (1991) Statistics of primordial density perturbations from discrete seed masses. *Astrophys. J.* **381**, 349–360.

Schlegel D.J., Finkbeiner D.P. and Davis M. (1998) Maps of dust infrared emission for use in estimation of reddening and cosmic microwave background radiation foregrounds. *Astrophys. J.* **500**, 525–553.

Schmalzing J. and Górski K.M. (1998) Minkowski functionals used in the morphological analysis of cosmic microwave background anisotropy maps. *Mon. Not. R. Astr. Soc.* **297**, 355–365.

Schmalzing J., Kerscher M. and Buchert T. (1996) Minkowski functionals in cosmology. In *Dark Matter in the Universe* (eds. S. Bonometto, J.R. Primack and A. Provenzale), IOS Press, Amsterdam, pp. 281–291.

Schmoldt I.M., Saar V., Saha P., Branchini E., Efstathiou G.P., Frenk C.S., Keeble O., Maddox S., McMahon R., Oliver S., Rowan-Robinson M., Saunders W., Sutherland W.J., Tadros H. and White S.D.M. (1999) On density and velocity fields and β from the IRAS PSCz survey. *Astronom. J.* **118**, 1146–1160.

Schneider P., Ehlers J. and Falco E.E. (1992) *Gravitational Lenses.* Springer-Verlag, New York.

Schneider P., van Waerbeke L., Jain B. and Kruse G. (1998) A new measure for cosmic shear. *Mon. Not. R. Astr. Soc.* **296**, 873–892.

Schramm D.N. and Turner M.S. (1998) Big-bang nucleosynthesis enters the precision era. *Rev. Mod. Phys.* **70**, 303–318.

Scoccimarro R. (2000) The bispectrum: From theory to observations. *Astrophys. J.* **544**, 597–615.

Scoccimarro R., Feldman H.A., Fry J.N. and Frieman J.A. (2001) The bispectrum of IRAS redshift catalogs. *Astrophys. J.* **546**, 652–664.

Scoccimarro R. and Frieman J.A. (1999) Hyperextended cosmological perturbation theory: Predicting non-linear clustering amplitudes. *Astrophys. J.* **520**, 35–44.

Scoccimarro R. and Sheth R.K. (2001) PTHalos: A fast method for generating mock galaxy distributions. *http://arXiv/org/abs/astro-ph/0106120.*

Seldner M., Siebers B., Groth E.J. and Peebles P.J.E. (1977) New reduction of the Lick catalog of galaxies. *Astronom. J.* **82**, 249–256.

Seljak U. (1998) Cosmography and power spectrum estimation: A unified approach. *Astrophys. J.* **503**, 492–501.

Seljak U. (2000) Analytic model for galaxy and dark matter clustering. *Mon. Not. R. Astr. Soc.* **318**, 203–213.

Seljak U. (2001) Redshift space bias and β from the halo model. *Mon. Not. R. Astr. Soc.* **325**, 1359–1364.

Seljak U. and Zaldarriaga M. (1996) A line-of-sight integration approach to cosmic microwave background anisotropies. *Astrophys. J.* **469**, 437–444.

Shandarin S.F. (2001) Testing non-Gaussianity in CMB maps by morphological statistics. *http://arXiv/org/abs/astro-ph/0107319.*

Shandarin S.F. (1983) Percolation theory and the cell-lattice structure of the universe. *Sov. Astr. Lett.* **9**, 100–102.

Shane C.D. and Wirtanen C.A. (1967) The Lick catalog of galaxies. *Publ. Lick Obs.* **22**, 1.

Shanks T. (1990) Galaxy count models and the extragalactic background light. In *IAU Symp. 139: The Galactic and Extragalactic Background Radiation* (eds. S. Bowyer and C. Leinert), vol. 139, pp. 269–281.

Sheth R.K. (1995) Constrained realizations and minimum variance reconstruction of non-Gaussian random fields. *Mon. Not. R. Astr. Soc.* **277**, 933–944.

Sheth R.K. and Lemson G. (1999) The forest of merger history trees associated with the formation of dark matter haloes. *Mon. Not. R. Astr. Soc.* **305**, 946–956.

Sheth R.K., Mo H.J. and Tormen G. (2001) Ellipsoidal collapse and an improved model for the number and spatial distribution of dark matter haloes. *Mon. Not. R. Astr. Soc.* **323**, 1–12.

Sheth R.K. and Tormen G. (1996) Large scale bias and the peak background split. *Mon. Not. R. Astr. Soc.* **308**, 119–126.

Sibson R. (1981) A brief description of natural neighbor interpolations. In *Interpreting Multivariate Data* (ed. V. Barnett), John Wiley & Sons, Chichester, pp. 153–171.

Sigad Y., Eldar A., Dekel A., Strauss M.A. and Yahil A. (1998) IRAS versus POTENT density fields on large scales: Biasing and Omega. *Astrophys. J.* **495**, 516–532.

Silberman L., Dekel A., Eldar A. and Zehavi I. (2001) Cosmological density and power spectrum from peculiar velocities: Nonlinear corrections and principal component analysis. *Astrophys. J.* **557**, 102–116.

Silverman B.W. (1986) *Density Estimation for Statistics and Data Analysis*. Chapman & Hall, London.

Slezak E., Bijaoui A. and Mars G. (1990) Identification of structures from galaxy counts — Use of the wavelet transform. *Astron. Astrophys.* **227**, 301–316.

Slezak E., de Lapparent V. and Bijaoui A. (1993) Objective detection of voids and high-density structures in the first CfA redshift survey slice. *Astrophys. J.* **409**, 517–529.

Snethlage M. (1999) Is bootstrap really helpful in point process statistics? *Metrika* **49**, 245–255.

Snethlage M., Martínez V.J., Stoyan D. and Saar E. Point field models for the galaxy point pattern. *Astron. Astrophys.*, submitted.

Soneira R.M. and Peebles P.J.E. (1978) A computer model universe — Simulation of the nature of the galaxy distribution in the Lick catalog. *Astronom. J.* **83**, 845–849.

Starck J.L., Murtagh F. and Bijaoui A. (1998) *Image Processing and Data Analysis. The Multiscale Approach*. Cambridge University Press, Cambridge.

Stein M.L. (1993) Asymptocially optimal estimation of the reduced second moment measure of point processes. *Biometrika* **80**, 443–449.

Stein M.L. (1997) Discussion by Michael L. Stein. In *Statistical Challenges in Modern Astronomy II* (eds. G.J. Babu and E.D. Feigelson), Springer-Verlag, New York, pp. 166–171.

Stoyan D. (1984a) On correlations of marked point processes. *Math. Nachr.* **116**, 197–207.

Stoyan D. (1984b) Correlations of the marks of marked point processes — statistical inference and simple models. *J. Inf. Process. Cybern.* **20**, 285–294.

Stoyan D. (1994) Caution with "fractal" point patterns! *Statistics* **25**, 267–270.

Stoyan D. (2000) Basic ideas of spatial statistics. In *Statistical Physics and Spatial Statistics: The Art of Analyzing and Modelling Spatial Structures and Pattern Formation, Lecture Notes in Physics* (eds. R.K. Mecke and D. Stoyan), vol. 554, Springer-Verlag, Berlin, pp. 3–21.

Stoyan D., Kendall W.S. and Mecke J. (1995) *Stochastic Geometry and Its Applications*. John Wiley & Sons, Chichester.

Stoyan D. and Stoyan H. (1994) *Fractals, Random Shapes and Point Fields*. John Wiley & Sons, Chichester.

Stoyan D. and Stoyan H. (2000) Improving ratio estimators of second-order point process characteristics. *Scand. J. Statist.* **27**, 641–656.

Strauss M.A., Davis M., Yahil A. and Huchra J.P. (1990) A redshift survey of IRAS galaxies — I. Sample selection. *Astrophys. J.* **361**, 49–62.

Strauss M.A., Huchra J.P., Davis M., Yahil A., Fisher K.B. and Tonry J. (1992) A redshift survey of IRAS galaxies — VII. The infrared and redshift data for the 1.936 Jansky sample. *Astrophys. J. Suppl. Ser.* **83**, 29–63.

Strauss M.A. and Willick J.A. (1995) The density and peculiar velocity fields of nearby galaxies. *Phys. Reports* **261**, 271–431.

Sutherland W. and Efstathiou G. (1991) Correlation functions of rich clusters of galaxies. *Mon. Not. R. Astr. Soc.* **248**, 159–167.

Sylos Labini F. and Montuori M. (1998) Scale invariant properties of the APM-Stromlo survey. *Astr. Astrophys.* **331**, 809–814.

Sylos Labini F., Montuori M. and Pietronero L. (1998) Scale-invariance of galaxy clustering. *Phys. Rep.* **293**, 61–226.

Szalay A.S., Matsubara T. and Landy S.D. (1998) Redshift space distortions of the correlation function in wide angle galaxy surveys. *Astrophys. J. Lett.* **498**, L1–L4.

Szapudi I. (1998) A new method for calculating counts in cells. *Astrophys. J.* **497**, 16–20.

Szapudi I. (2000) Cosmic statistics of statistics: N-point correlations. *http://arXiv/org/abs/astro-ph/0008224*.

Szapudi I., Branchini E., Frenk C.S., Maddox S. and Saunders W. (2000) The luminosity dependence of clustering and higher order correlations in the PSCz survey. *Mon. Not. R. Astr. Soc.* **318**, L45–L50.

Szapudi I. and Colombi S. (1996) Cosmic error and statistics of large-scale structure. *Astrophys. J.* **470**, 131–148.

Szapudi I., Colombi S. and Bernardeau F. (1999) Cosmic statistics of statistics. *Mon. Not. R. Astr. Soc.* **310**, 428–444.

Szapudi I. and Szalay A.S. (1993) Higher order statistics of the galaxy distribution using generating functions. *Astrophys. J.* **408**, 43–56.

Szapudi I. and Szalay A.S. (1998) A new class of estimators for the N-point correlations. *Astrophys. J. Lett.* **494**, L41–L44.

Tadros H., Ballinger W.E., Taylor A.N., Heavens A.F., Efstathiou G., Saunders W., Frenk C.S., Keeble O., McMahon R., Maddox S.J., Oliver S., Rowan-Robinson M., Sutherland W.J. and White S.D.M. (1999) Spherical harmonic analysis of the PSCz galaxy catalogue: Redshift distortions and the real-space power spectrum. *Mon. Not. R. Astr. Soc.* **305**, 527–546.

Takayasu H. (1989) *Fractals in the Physical Sciences*. Manchester University Press, Manchester.

Taylor A. and Valentine H. (1999) The inverse redshift-space operator: Reconstructing cosmological density and velocity fields. *Mon. Not. R. Astr. Soc.* **306**, 491–503.

Tegmark M. (1995) A method for extracting maximum resolution power spectra from galaxy surveys. *Astrophys. J.* **455**, 429–438.

Tegmark M. (1997) How to measure CMB power spectra without losing information. *Phys. Review D* **55**, 5895–5907.

Tegmark M., Hamilton A.J.S., Strauss M.A., Vogeley M.S. and Szalay A.S. (1998) Measuring the galaxy power spectrum with future redshift surveys. *Astrophys. J.* **499**, 555–576.

Tegmark M., Taylor A.N. and Heavens A.F. (1997) Karhunen–Loèwe eigenvalue problems in cosmology: How should we tackle large data sets? *Astrophys. J.* **480**, 22–35.

Tegmark M., Zaldarriaga M. and Hamilton A.J.S. (2001) Towards a refined cosmic concordance model: Joint 11-parameter constraints from CMB and large-scale structure. *Phys. Rev. D* **63**, 043007.

Toomet O., Andernach H., Einasto J., Einasto M., Kasak E., Starobinsky A.A. and Tago E. (1999) The supercluster–void network V. Alternative evidence for its regularity. *http://arXiv/org/abs/astro-ph/9907238.*

Tucker D.L., Oemler A., Kirshner R.P., Lin H., Shectman S.A., Landy S.D., Schechter P.L., Muller V., Gottlober S. and Einasto J. (1997) The Las Campanas redshift survey galaxy-galaxy autocorrelation function. *Mon. Not. R. Astr. Soc.* **285**, L5–L9.

Tully R.B. and Fisher J.R. (1977) A new method of determining distances to galaxies. *Astron. Astrophys.* **54**, 661–673.

Turner E.L. and Gott J.R. (1976) Groups of galaxies. I. A catalog. *Astrophys. J. Suppl. Ser.* **32**, 409–427.

Upton G. and Fingleton B. (1985) *Spatial Data Analysis by Example*, vol. 1. John Wiley & Sons, Chichester.

Valentine H., Saunders W. and Taylor A. (2000) Reconstructing PSCz with a generalized PIZA. *Mon. Not. R. Astr. Soc.* **319**, L13–L17.

van de Weygaert R. (1991) *Voids and the Geometry of Large Scale Structure*. Ph.D. thesis, University of Leiden, Leiden.

van de Weygaert R. (1994) Fragmenting the Universe. 3: The constructions and statistics of 3-D Voronoi tessellations. *Astron. Astrophys.* **283**, 361–406.

van de Weygaert R. and Bertschinger E. (1996) Peak and gravity constraints in Gaussian primordial density fields: An application of the Hoffman–Ribak method. *Mon. Not. R. Astr. Soc.* **281**, 84–118.

van de Weygaert R. and Hoffman Y. (2000) The structure of the local universe and the coldness of the cosmic flow. In *Cosmic Flows Workshop, ASP Conference Series* (eds. S. Courteau and J. Willick), vol. 201, Astronomical Society of the Pacific, San Francisco, pp. 169–176.

van de Weygaert R. and Icke V. (1989) Fragmenting the universe — II. Voronoi vertices as Abell clusters. *Astron. Astrophys.* **213**, 1–9.

van de Weygaert R., Jones B.J.T. and Martínez V.J. (1992) The minimal spanning tree as an estimator for generalized dimensions. *Phys. Lett. A* **169**, 145–150.

van Haarlem M. (1996) Projection effects in the Abell catalogue. In *Mapping, Measuring, and Modelling the Universe, ASP Conference Series* (eds. P. Coles, V.J. Martínez and M.J. Pons-Bordería), vol. 94, Astronomical Society of the Pacific, San Francisco, pp. 191–196.

van Lieshout M.N.M. and Baddeley A. (1996) A non-parametric measure of spatial interaction in point patterns. *Stat. Neerlandica* **50**, 344–361.

van Waerbeke L., Mellier Y., Radovich M., Bertin E., Dantel-Fort M., McCracken H.J., Fèvre O.L., Foucaud S., Cuillandre J.C., Erben T., Jain B., Schneider P., Bernardeau F. and Fort B. (2001) Cosmic shear statistics and cosmology. *Astr. Astrophys.* **374**, 757–769.

Verde L., Heavens A.F., Matarrese S. and Moscardini L. (1998) Large-scale bias in the Universe II: redshift space bispectrum. *Mon. Not. R. Astr. Soc.* **300**, 747–757.

Verde L., Wang L., Heavens A.F. and Kamionkowski M. (2000) Large-scale structure, the cosmic microwave background, and primordial non-Gaussianity. *Mon. Not. R. Astr. Soc.* **313**, 141–145.

Viana P.T.P. and Liddle A.R. (1999) Galaxy clusters at $0.3 < z < 0.4$ and the value of Ω_0. *Mon. Not. R. Astr. Soc.* **303**, 535–546.

Vogeley M.S., Park C., Geller M.J., Huchra J.P. and Gott J.R.I. (1994) Topological analysis of the CfA redshift survey. *Astrophys. J.* **420**, 525–544.

Vogeley M.S. and Szalay A.S. (1996) Eigenmode analysis of galaxy redshift surveys. I. Theory and methods. *Astrophys. J.* **465**, 34–53.

Voronoi G. (1908) Nouvelles applications des parametres continus a la theorie des fromes quadratiques. *J. Reine Angew. Math.* **134**, 198–287.

Walker J.S. (1998) *A Primer on Wavelets and Their Scientific Applications*. Chapman & Hall, Boca Raton.

Wand M.P. and Jones M.C. (1995) *Kernel Smoothing*. Chapman & Hall, London.

Wang Y.H. (2000) Fiducial intervals: what are they? *Amer. Statistician* **54**, 105–111.

Wälder U. and Stoyan D. (1996) On variograms and point process statistics. *Biom. J.* **38**, 895–905.

Webb S. (1999) *Measuring the Universe. The Cosmological Distance Ladder*. Springer-Verlag in association with Praxis Publishing, Chichester.

Weinberg D.H., Gott J.R.I. and Melott A.L. (1987) The topology of large-scale structure — I. Topology and the random phase hypothesis. *Astrophys. J.* **321**, 2–27.

White M. (2001) The redshift space power spectrum in the halo model. *Mon. Not. R. Astr. Soc.* **321**, 1–3.

White S.D.M. (1979) The hierarchy of correlation functions and its relation to other measures of galaxy clustering. *Mon. Not. R. Astr. Soc.* **186**, 145–154.

Willick J.A., Courteau S., Faber S.M., Burstein D., Dekel A. and Strauss M.A. (1997) Homogeneous velocity–distance data for peculiar velocity analysis. III. The Mark III catalog of galaxy peculiar velocities. *Astrophys. J. Suppl.* **109**, 333–366.

Willick J.A., Strauss M., Dekel A. and Kolatt T. (1997) Maximum likelihood comparisons of Tully–Fisher and redshift data: Constraints on Ω and biasing. *Astrophys. J.* **486**, 629–664.

Witten T.A. and Sander L.M. (1981) Diffusion-limited aggregation, a kinetic critical phenomenon. *Phys. Rev. Lett.* **47**, 1400–1403.

Wu K.K.S., Lahav O. and Rees M.J. (1999) The large-scale smoothness of the Universe. *Nature* **397**, 225–235.

Yahil A., Strauss M.A., Davis M. and Huchra J.P. (1991) A redshift survey of *IRAS* galaxies. II. Methods for determining self-consistent velocity and density fields. *Astrophys. J.* **372**, 380–393.

Yoshida N., Colberg J.M., White S.D.M., Evrard A.E., MacFarland T.J., Couchman H.M.P., Jenkins A., Frenk C.S., Pearce F.R., Efstathiou G., Peacock J.A. and Thomas P.A. (2001) Simulation of deep pencil-beam redshift surveys. *Mon. Not. R. Astr. Soc.* **325**, 803–816.

Zaroubi S., Hoffman Y. and Dekel A. (1999) Wiener reconstruction of large-scale structure from peculiar velocities. *Astrophys. J.* **520**, 413–425.

Zaroubi S., Hoffman Y., Fisher K.B. and Lahav O. (1995) Wiener reconstruction of the large scale structure. *Astrophys. J.* **449**, 446–459.

Zeldovich Y.B. (1970) Gravitational instability: An approximate theory for large density perturbations. *Astr. Astrophys.* **5**, 84–89.

Ziman J.M. (1979) *Models of Disorder: The Theoretical Physics of Homogeneously Disordered Systems*. Cambridge University Press, Cambridge.

Zwicky F., Herzog E. and Wild P. (1961–1968) *Catalogue of Galaxies and of Clusters of Galaxies*. California Institute of Technology (CIT), Pasadena.

Web site references

2dFGRS: *http://www.mso.anu.edu.au/2dFGRS*. The home page of the ongoing 2dF redshift survey. A partial release of the data has been already made.

2QZ: *http://www.2dfquasar.org/*. The home page of the 2dF quasar redshift survey, with the first results and data release.

ADS: *http://adswww.harvard.edu/*. A database for everything astronomical, including almost all published astronomical articles.

APM: *http://ast.cam.ac.uk/~apmcat/*. The APM sky catalogs.

ArXiv: *http://xxx.lanl.gov*. A good site to search for recent preprints and articles.

BIPS: *http://astrosun.tn.cornell.edu/staff/loredo/bayes/*. Collection of Web resources on application of Bayesian inference in physical sciences.

CMBFast: *http://www.sns.ias.edu/~matiasz/CMBFAST/cmbfast.html*. Source code for the CMBFAST algorithm.

Cosmics: *http://arcturus/mit/edu/cosmics/*. A program package for generating initial conditions for cosmological modeling, including constrained realizations.

Dust: *http://astron.berkeley.edu/davis/dust/index.html*. The Web page for the galactic dust distribution data.

FFTlog: *http://casa.colorado.edu/~ajsh/FFTLog*. Source code for the FFTLog algorithm.

Gipsy: *http://www.astro.rug.nl/~gipsy/*. The Groningen Image Processing System home page.

Hu 2001: *http://background.uchicago.edu/*. Wayne Hu's Web page, a treasure trove of information on the physics of cosmic microwave background.

Hydra: *http://coho.mcmaster.ca/hydra/hydra_consort.html*. Web page of a numerical cosmology project. Simulation data and free simulation packages are available.

LCRS: *http://manaslu.astro.utoronto.ca/~lin/lcrs.html*. The Las Campanas redshift survey home page, with publicly available data.

MarkIII: *http://redshift.stanford.edu/MarkIII/*. The Mark III galaxy peculiar velocity catalog can be found here.

PSCz: *http://www-astro.physics.ox.ac.uk/~wjs/pscz.html*. The IRAS PSCz Redshift Survey home page, with the catalog.

ROSAT: *http://wave.xray.mpe.mpg.de/rosat*. The data from the ROSAT X-ray satellite.

SDSS: *http://www.sdss.org/*. The Sloan Digital Sky Survey home page. A partial release of the data has been already made.

Spanky: *http://spanky.triumf.ca/*. Collection of fractal software.

Szapudi: *http://www.cita.utoronto.ca/~szapudi/istvan.html*. Istvan Szapudi's Web page, with the FORCE package for counts-in-cells error analysis.

TDC: *http://tdc-www.harvard.edu/*. The Smithsonian Astrophysical Observatory Telescope Data Center, with the updated Zwicky catalog of galaxies and the CfA redshift survey data.

Virgo: *http://star-www.dur.ac.uk/~frazerp/virgo/virgo.html*. The home page for the Virgo numerical cosmology project.

Virtual: *http://www.srl.caltech.edu/nvo/*. The home page for the National Virtual Observatory project.

Index